Hartmut Zabel
Medical Physics

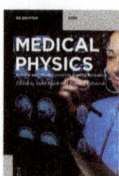

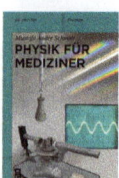

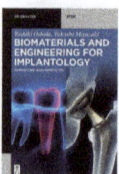

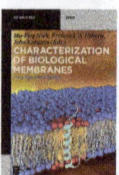

Hartmut Zabel

Medical Physics

Volume 1: Physical Aspects of the Human Body

2nd Edition

DE GRUYTER

Author
Prof. Dr. Dr. h. c. Hartmut Zabel
Ruhr-Universität Bochum
Fak. für Physik und Astronomie
44780 Bochum
hartmut.zabel@ruhr-uni-bochum.de

ISBN 978-3-11-075691-3
e-ISBN (PDF) 978-3-11-075695-1
e-ISBN (EPUB) 978-3-11-075698-2

Library of Congress Control Number: 2022946976

Bibliographic information published by the Deutsche Nationalbibliothek
The Deutsche Nationalbibliothek lists this publication in the Deutsche Nationalbibliografie;
detailed bibliographic data are available on the internet at http://dnb.dnb.de.

www.degruyter.com

Preface

The second edition of the textbook on *Medical Physics* is now published in three instead of two volumes. This is because the new edition is not only corrected and updated but also considerably expanded. The enhancement is primarily due to many new features designed to increase the usefulness of this textbook as a learning companion to regular courses. Each chapter concludes with a detailed summary, questions, exercises, and a self-assessment of the knowledge gained. All chapters contain info boxes for in-depth information on specific topics and math boxes for deriving central mathematical concepts. References to current literature and further reading recommendations lead to reviews in the subject area. Answers to questions and solutions to tasks can be found in the Appendix.

The textbook is aimed at bachelor's or master's students in the first year of medical physics. It is recommended that students first acquire some basic knowledge of modern physics before proceeding with reading this *Medical Physics* textbook.

The range of topics in *Medical Physics* is extensive. For this reason, most textbooks in this field have been written by several authors. Covering all topics on your own is both a challenge and an opportunity to design the various chapters as coherently as possible and to relate them to one another. I would not have been able to meet this challenge without giving respective lectures.

Compared to the first edition, the chapters have been rearranged to allow a clear distinction between the physical aspects of the human body in the first volume, the physical aspects of diagnostics in the second volume, and the physical aspects of therapy in the third volume.

The first volume begins with classical mechanics, including forces, moments, and energy, concepts applied to the human body. The overarching importance of the action potential and signal transmission in the body's functioning is presented in Chapters 5 and 6, followed by a discussion of those organs, systems, and senses that have a clear connection to physics, such as the cardiovascular and the respiratory systems. The last two chapters deal with the primary sensory organs of the body, the visual sense, and the auditory sense.

The second volume deals exclusively with imaging methods, distinguishing between those without ionizing radiation (ultrasound, endoscopy, magnetic resonance imaging) and those with ionizing radiation (x-rays, SPECT, PET). As an introduction to these radiological chapters, essential facts about x-ray and nuclear physics are presented, the interactions of radiation with matter are explained, and measures for radiation protection are discussed.

The third volume is devoted to the physical fundamentals of radiotherapeutic procedures using x-rays, protons, neutrons, and γ-rays. These chapters are introduced by first comparing the radiation response of benign cells and malignant tumor cells. The last two chapters deal with laser processes and highlight the physics of

https://doi.org/10.1515/9783110756951-202

nanoparticles in diagnostics and therapy. An additional chapter on medical statistics rounds off the third volume.

All these additions, guidance, and exercises will hopefully make studying this medical physics textbook a valuable and informative companion to your regular coursework. Questions can be directed to the editor or author and will be answered promptly. Corrections are very welcome and will be posted on the book's website.

Bochum, December 14, 2022

Acknowledgements

My first thanks go to all those who pointed out errors and made suggestions for improvements to the first edition. Constructive criticism is always very helpful in improving the text, correcting mistakes, and clearing up misunderstandings. I am very grateful to Dr. Alexey Saphoznik, who took the time to read the entire first volume and suggested many modifications. My special thanks go to Professor Birge Kollmeier (University of Oldenburg), who made valuable recommendations for improving the content, particularly concerning sound and sound perception. I would also like to acknowledge my ophthalmologist Dr. Elbracht-Hülseweh, from whom I learned a lot during many years of treatment. Thanks also go to my colleagues who helped and guided me during the preparation of the first edition.

My special thanks go to the editorial staff at De Gruyter Verlag and, in particular, to Kristin Berber-Nerlinger, who encouraged me in the first place to prepare a second edition and gave me valuable advice on the implementation of this project. Nadja Schedensack helped in all stages, and Kathleen Prüfer did an excellent editorial job. I am very grateful to the entire publishing team of De Gruyter.

Last but not least, I would like to thank my family and, in particular, Rosemarie, who accompanied this project with much patience, understanding, and encouragement.

https://doi.org/10.1515/9783110756951-203

Contents

Appendix

1 Brief overview of body parts and functions

Table of important physical properties of the human body	
Distinctive systems in the body	11
Average cell size	20 μm
Range of cell sizes	4 μm to 1 mm
Total number of cells (rough estimate)	3×10^{13}
Number of genes (estimate)	25 000
Number of proteins (estimate)	2×10^4
Cell types	210
Outer skin area	~2 m^2
Senses of the body	More than 8

1.1 Introduction

It is well known that body parts and functions are controlled by chemical and biochemical processes. What is perhaps less obvious is that our bodies also follow strict physical principles. However, this becomes clear when we examine different parts of our body. Every movement requires forces, moments, and mechanical stability. The body's energy balance follows the principles of thermodynamics. The sensory system, including hearing and vision, is based on acoustic and optical principles. The respiratory system, circulatory system, and kidneys all obey the laws of diffusion, hydrostatics, and hydrodynamics. The propagation of signals along nerve fibers can be described by electrical circuits. Chapters 2–12 are devoted to the description of some organs, sensors, and systems and emphasize their connection to fundamental physical principles. Attempting such a description inevitably leads to simplification. However, the aim of this textbook is not to underpin the complexity of biological systems but rather to draw attention to the connections with physical principles. Therefore, the chapters in this volume are not a substitute for textbooks on physiology.

In this first chapter we give a very brief overview of the main organs and systems of the human body, which will be deepened in later chapters. Most of these organs are not specific to humans but can be found in all mammals and even in most vertebrates.

https://doi.org/10.1515/9783110756951-001

1.2 Overview

1.2.1 Cell

Cells with an average size of 20 μm are the fundamental building blocks of living matter [1–3]. The human body consists of about 3×10^{13} *eukaryotic cells* [2], i.e., cells that contain a nucleus. They are stacked together like bricks of a house. The cross section of a typical cell is schematically shown in Fig. 1.1. All cells have a wall to distinguish between inside and outside. The chemical and electrical gradients across the wall are essential for cell functioning and intercell communication. Inside the cells, there is another smaller nucleus that holds the genetic material in the form of a double-stranded helical chain, known as deoxyribonucleic acid (DNA). Each DNA molecule contains a sequence of paired nucleobases, which read like letters of a book that constitutes the genetic code. Genes are like chapters in this book. About 1000 pairs constitute one chapter. Estimates are that 80% of about 25 000 genes in the DNA encode about 2×10^4 different functional proteins in the human body that do their daily job [4]. Proteins are made up of 21 different amino acids arranged in chains and folded up to complex three-dimensional structures. They build ion channels and molecular motors; form receptors, enzymes, and hormones; take care of oxygen transport; strengthen tissues and bones in the body; regulate water and ion concentrations; and are responsible for many more tasks. Proteins vary greatly in size; some hold more than 100 000 atoms [5]. Only about 5000 proteins have been described at an atomic resolution so far [6]. Although almost all cells contain complete and identical genetic information, they specialize in different tasks. Muscle cells develop a surprising tensile force, liver cells specialize in performing important tasks for food metabolism, and nerve cells transmit electrical signals as fast as 100 m/s. We can distinguish between about 210 different cell types with specific tasks [7]. Only stem cells are not specialized. The specialization of all other cells for particular tasks is only possible through a high degree of self-organization and communication between the cells.

Comparing cells with bricks is, after all, a gross oversimplification. When engineers build a complex system, they start with simple building blocks that fit together to form something more complex. In contrast, in organisms, the complexity does not begin at the cell level but already at the molecular level of DNA, ribosomes, lysosomes, and many more. This incredible complexity organizes life. First, cells organize to form tissues: epithelial tissue, connective tissue, muscular tissue, and nervous tissue. In turn, tissues assemble to form organs with characteristic and distinct shapes and functions like the liver and the kidneys. Several organs work together in a system, such as the digestive system that involves 10 different organs. The body contains 11 distinct systems: 1. cardiovascular/circulatory system; 2. respiratory system; 3. digestive system/excretory system; 4. endocrine system; 5. lymphatic system/immune system; 6. sensory system; 7. locomotor system; 8. nervous

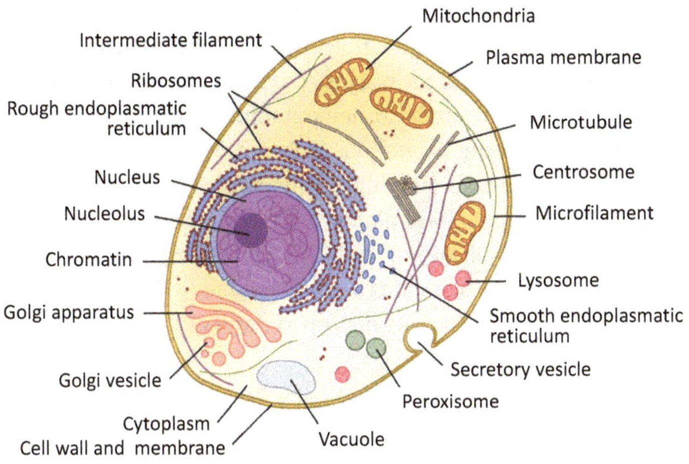

Intermediate filament

Ribosomes

Rough endoplasmatic
reticulum

Nucleus

Nucleolus

Chromatin

Golgi apparatus

Golgi vesicle

Cytoplasm

Cell wall and membrane

Mitochondria

Plasma membrane

Microtubule

Centrosome

Microfilament

Lysosome

Smooth endoplasmatic
reticulum

Secretory vesicle

Peroxisome

Vacuole

Fig. 1.1: Schematics of a human cell. The cell membrane separates the cytoplasm from the extracellular space. The cytoplasm contains, among many other parts, the genetic information inside the nucleus and the mitochondria with its genetic information, acting as a powerhouse for the cell functions. The cell membrane is a double lipid layer perforated by many ion channels to maintain an electrical potential across the cell membrane or change it by depolarization upon a stimulus (adapted from OpenStax Anatomy and Physiology, 2016, © creative commons).

system; 9. renal system/urinary system; 10. integumentary system; and 11. reproductive system. All 11 systems interact and are responsible for the human body's life. These interactions take place under the promise of constancy in a variable environment. For instance, core body temperature, arterial blood pressure, and partial oxygen/carbon dioxide pressures of blood are kept constant by an active negative feedback system, like a thermostat. This control mechanism is known as homeostasis, according to Bernard,[1] who first stated this principle [8, 9]. Body temperature homeostasis is one example that is further discussed in Chapter 4, and further homeostatic systems are described in the following chapters.

Finally, cells do not live forever. Cells have a life cycle, comprising phases of growth, rest, cell division, and cell death. Understanding the cell cycle's controlling mechanisms is essential for coping with cancer and treating tumors through radiation therapy. These topics are considered in Chapters 8–10 of Vol. 2.

1.2.2 Circulation

For maintaining all body functions, blood *circulation* is essential. As the name implies, blood circulation is a closed circuit, where the flow is mechanically powered by

---

1 Claude Bernard (1813–1878), French physiologist and pharmacologist.

the heart acting as a pump. A dense mesh of blood vessels penetrates any body part. Blood circulation is a transport system for oxygen, nutrients, and heat to their destinations, including muscles, bones, and all other organs. Oxygen is taken up from the surrounding air during inhalation and binds to hemoglobin by diffusing through membranes in the lung. In return, during expiration, carbon dioxide, as a residue of combustion, is exhaled. Figure 1.2 shows a simplified schematic of blood circulation. It has an oxygen-rich and an oxygen-poor part, along with a high-pressure part and a low-pressure part. The color coding (red: oxygen-rich, blue: oxygen-poor) originates from our visual perception. Oxygen-rich blood has a much brighter red color than oxygen-poor blood. Circulation is discussed in more depth in Chapter 8.

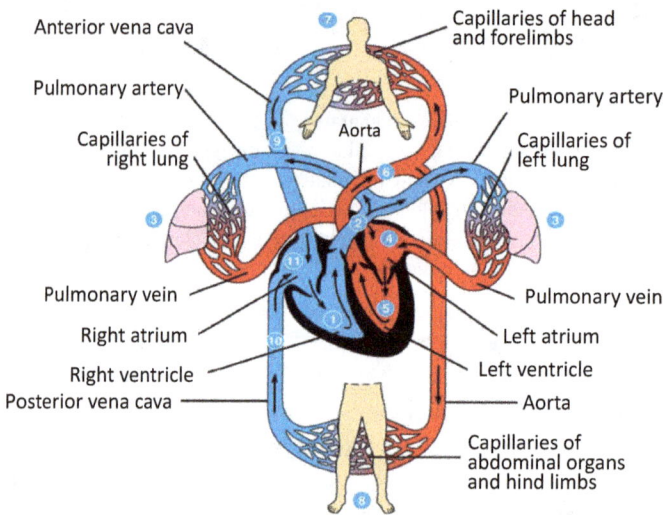

Fig. 1.2: Schematic of the human circulatory system. The right ventricle of the heart (1) takes in oxygen-poor blood from the extremities (9, 10) and pumps it into the lungs (2, 3) for oxygen enrichment. After returning from the lung into the left atrium and ventricle of the heart (4, 5), the blood is ejected into the periphery (6–8) to supply all body tissues with oxygen. Blood pressures in the left circulation (red) are higher than in the right circulation (blue) at similar locations. The left ventricle muscle is also stronger than the right ventricle muscle (reproduced from http://www.online-sciences.com/, © creative commons).

1.2.3 Heart

The *heart* has an electrical and a mechanical component. The electrical component takes care of self-stimulating the sinus-atrial node, and the distribution of the excitation over the entire heart muscle (myocardium) for sequential contraction. The contracting part of the myocardium increases the blood pressure and performs volume work, i.e., it ejects the cardiac volume periodically into the arteries, thereby maintaining a pulsatile blood flow. Figure 1.3 shows a cross section of the heart. Heart, vessels, and blood form one unit, known as the *cardiovascular system*. Their electrical and mechanical components are topics of Chapters 7 and 8, respectively.

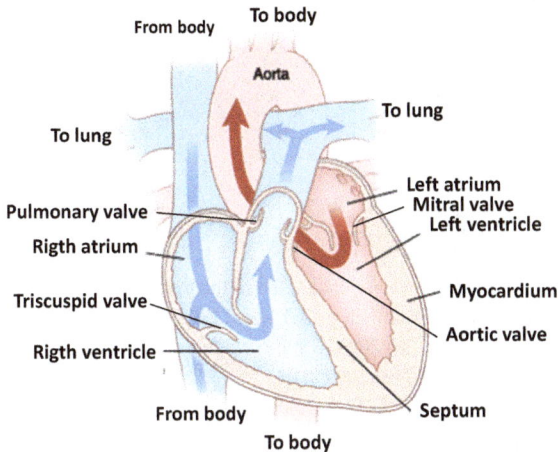

Fig. 1.3: Main parts of the heart, consisting of four chambers (left atrium, right atrium, left ventricle, and right ventricle) and four valves (tricuspid, pulmonary, mitral, and aortic).

1.2.4 Kidneys

The *kidneys* are part of the urinary system that contains, in addition, ureters, bladder, and urethra. The kidneys' main task is filtration of the blood to remove residual products of metabolism. In particular, the kidneys eliminate toxic substances resulting from protein breakdowns, such as urea and urea acid. Everything that is filtered out and not reabsorbed goes into the urinary tract. The central location of the kidneys in the body is shown in Fig. 1.4. Another important task of the kidneys is maintaining the sodium and potassium ion concentration, the acid–base (pH value) regulation, the water balance, and the osmotic pressure of the blood. The kidneys' third task is the production of hormones that are involved in the regulation of blood pressure and blood flow. Kidneys and their various physical properties of filtration and clearance are discussed in Chapter 10, including artificial filtering via dialysis.

1.2.5 Respiratory system

The *respiratory system*, schematically shown in Fig. 1.5, consists of two parts. The upper part takes care of inhalation through the nose, oral cavity, and throat. The lower part transports the air through the trachea and the bronchial tree down into the lungs.

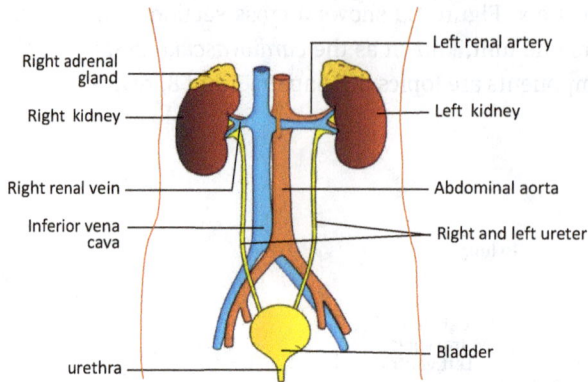

Fig. 1.4: Location of the kidneys in the body. The kidneys are connected to the blood circulatory system receiving about 20% of the cardiac output. Per day they filter about 180 l of blood. After filtering, most of the fluid is resorbed, only 1–2 l/day is collected in the bladder and disposed through the urethra (adapted from Wikimedia, © creative commons).

On exhalation, air flows through both parts in reverse order. Inspiration and expiration require a combined action of the chest cage and the diaphragm to exert the thorax's volume work. The main task of breathing is the exchange of gases in the lungs, i.e., the uptake of oxygen by the blood and the release of carbon dioxide back into the air. Breathing works by generating a difference between atmospheric gas pressure and gas pressure in the thorax (intrapulmonary pressure difference). The flow rate and the flow resistance are governed by the laws of aerodynamics, as discussed further in Chapter 9.

Fig. 1.5: The respiratory system consists of an upper part for inspiration and expiration through the mouth and nose and a lower part for the exchange of oxygen and carbon dioxide in the capillaries of the lungs. Thorax and diaphragm generate overpressure for expiration and under-pressure for inspiration.

1.2.6 Digestive system

The *digestive system* comprises all organs involved in taking up food for breaking it down into nutrients (carbohydrates, proteins, fats, and vitamins). The nutrients are then transported from the intestines via blood circulation to cells throughout the body for combustion with oxygen and energy production. The digestive system combines most of the body's organs (Fig. 1.6): oral cavity, including teeth and tongue, esophagus, stomach, liver, spleen, gallbladder, pancreas, small intestines, colon, rectum, and anus. A vital organ is the liver. The liver is responsible for the production of vitally important proteins such as the synthesis of albumin and clotting factors, for storage of carbohydrates and vitamins, for the synthesis of bile, for the inactivation of medications, for detoxification of the blood, and for the production of biochemicals that break down nutrition components. The liver is unique in its multitude of vital tasks. In case of malfunction, the liver can presently only be replaced by a donor but not substituted. The energy production of nutrients and the body's energy household are further discussed in Chapter 4, while the liver itself is not a topic of this text because it has mainly biochemical tasks.

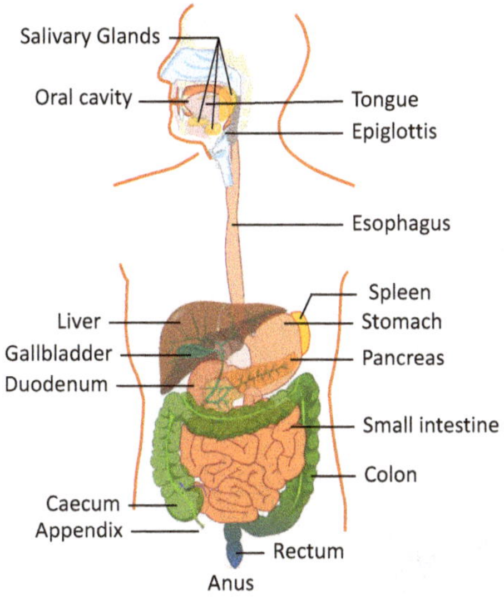

Fig. 1.6: The digestive system is responsible for the body's energy supply. This task encompasses the largest number of organs: mouth, esophagus, stomach, liver, spleen, gallbladder, intestines, pancreas, colon, and rectum (adapted from OpenStax Anatomy and Physiology, 2016, © creative commons).

1.2.7 Sensory organs

The body is equipped with various detectors (*sensory organs*) that continuously provide information on the external environment and the inner organs' status for processing in the central nervous system (CNS). The most important sensory organs are the eyes for visual information, the ear for auditory perception, and the vestibular organ for sensing changes in position and acceleration. Furthermore, we have a nose for smelling, a tongue for tasting, and a large skin area for feeling temperature, pressure, humidity, air flow, water, and any kind of injury. Fingers can determine the surface finish (smoothness and roughness) and the shape of objects. Also, we have sensors for the pressure in the bladder and colon, sensors for hunger and thirst, and sensors for sexual appetite. Some sensory organs are shown in Fig. 1.7. These sensors (also called receptors) require nerves (neurons) for signal transmission and the brain for signal processing, including decision-making. However, we do not have sensors for electric and magnetic fields like some fish and birds have or sensors for altitude. The altitude is measured indirectly through lack of oxygen and depression in the middle ear. We discuss the physical principles of signal transmission via neurons in Chapter 6, and three sensors in more detail: the visual perception in Chapter 11, the auditory perception, and the balance in Chapter 12.

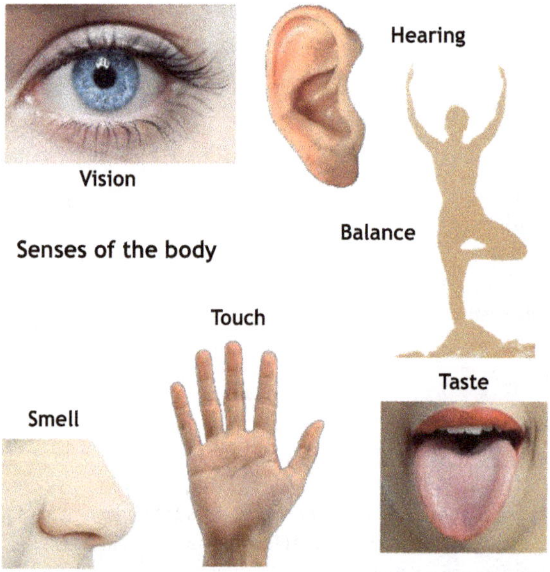

Hearing

Vision

Senses of the body

Balance

Touch

Taste

Smell

Fig. 1.7: Some senses are shown which provide information on the external environment. The human body contains many more sensors, the biggest one is the skin.

1.2.8 Nervous system

Brain, spinal cord, and nerve fibers belong to the *nervous system* with three main functions. It collects information from the environment by receiving signals through specialized receptors. The information is transmitted along nerve fibers to the brain and spinal cord for integration and processing. After processing and decision-making, signals are sent from the brain through the spinal cord to muscles, inner organs, and glands for motion. The control of inner organs, such as the heart, lung, and digestive system, runs semiautomatically. Therefore, we distinguish between a *somatic* nervous system for conscious sensation and deliberate movement and a *vegetative* nervous system for the autonomous regulation of the inner organ functions. Both nervous systems encompass sensory (*afferent*) and motor (*efferent*) connections. Afferent connections transmit signals from the periphery to the center (brain or spinal cord). Efferent connections conduct signals from the center to the periphery. The center is called the CNS, and the peripheral nervous system includes all somatic and vegetative nerves. An overview is shown in Fig. 1.8.

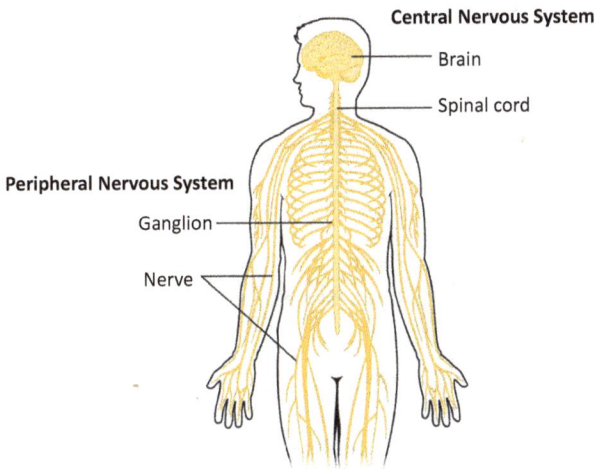

Fig. 1.8: The nervous system reaches all parts of the body. It automatically controls the inner organs and the deliberate reaction to external stimuli. The nervous system also executes decisions made in the brain's cognitive center, like standing up, walking, and throwing a ball (reproduced from OpenStax Anatomy and Physiology, 2016, © creative commons).

Another distinction is required for the vegetative nervous system: the sympathetic and the parasympathetic nervous system. Both nerve fibers intertwine and antagonistically control the function of most internal organs, such as heart rate and blood flow. Sympathetic nerves stimulate performance enhancement and rapid reactions in dangerous circumstances, while parasympathetic nerves decrease activity. Both systems are necessary for optimal control of organ functions. In Chapters 5 and 6, the basic principles of resting potentials, action potentials, and signal transmission in neurons are discussed.

1.2.9 Locomotor system

The body's locomotor system consists of all movable and supporting bones, joints, skeletal muscles, tendons, cartilage, and connective tissues that enable the body to move. In contrast to insects and crustaceans, where a hard shell protects the internal organs, the skeletal system offers less protection but more mobility and flexibility. Only the brain and the spinal cord are well protected in the skull and spinal canal. The skeletal muscles (Fig. 1.9) are triggered by nerve fibers and respond by contraction to the transmission of electrical signals (action potentials) from the brain to the muscles. The controlled sequence of action potentials leads to a consciously initiated series of movements. Chapters 2 and 3 discuss the mechanical, elastomechanical, and kinematic aspects of the locomotor system.

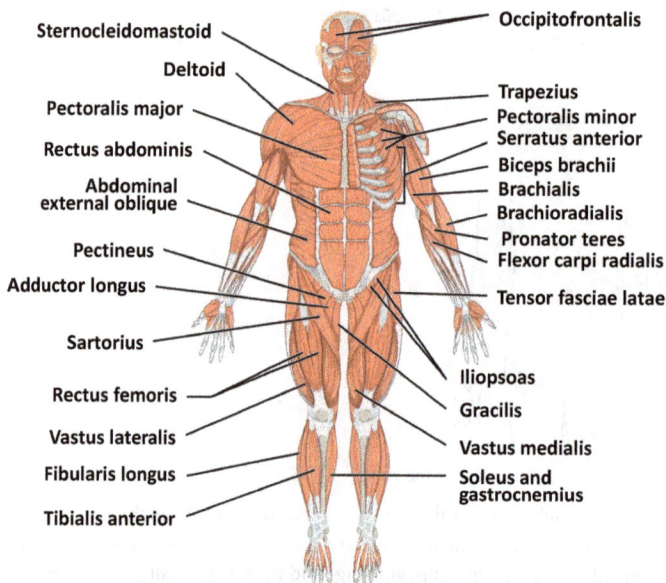

Fig. 1.9: The locomotor system consists of all movable and supporting bones, joints, skeletal muscles, tendons, cartilage, and connective tissues (reproduced from Textbook OpenStax Anatomy and Physiology, 2016, © creative commons).

1.2.10 Skin

The *skin* is the largest organ of the body, spanning about 2 m^2 for adults. The skin is part of the *integumentary system* that also comprises hair and nails. Figure 1.10 shows a cross section of the skin. It is not only a sensor for temperature, pressure, and humidity, as already mentioned. It is also an essential part of the body's temperature regulation through radiation, and in extreme cases, through sweating for additional heat loss and shivering for heat production. Furthermore, the skin protects the inner organs from mechanical impact, short period temperature variations, microorganisms, chemicals, and radiation. Melanin in the skin protects against UV radiation, which has a hazardous ionization effect. Moreover, the skin is an exchange system: excretion of waste products on the one hand and production of various compounds and essential vitamins on the other hand. The skin consists of two main layers, the highly vascular dermis covered by the avascular epidermis. There is a gradient in the skin from nourished deeper layers to unnourished and dying cells on the surface of the skin, called the epidermis. The nerve fibers and sensory receptors are located in the dermis. Hair follicles together with arrector pili muscles attached to them are also embedded in the dermis. Beneath the dermis, the hypodermis contains fat that helps to insulate the rest of the body from temperature fluctuations. Finally, it should be noted that the skin is not only an outer protective layer of the body but also covers all internal organs and cavities such as the oral cavity or intestines. The moist skin on the inside is called the mucous membrane or mucosa, which is not further discussed in this text.

1.2.11 Reproductive system

Sexual differences are present all over the body but are not essential, aside from the reproductive system. The *reproductive organs* of men and women are fundamentally different and are grouped in inner and outer parts. The reproductive organs of males comprise the penis, testis, epididymis, spermatic duct, and prostate. The reproductive organs of females contain two ovaries, two fallopian tubes, uterus, vagina, and female breast (mamma) for newborns' nutrition. It is interesting to note that prostate and mamma are the organs most often affected by cancerous tissues. Cancer and various cancer treatment methods are topics of Chapters 12–15.

This brief and partial overview of body parts and functions will be deepened in Chapters 2–12 for organs that exhibit clear physical aspects, such as the eye, ear, heart, circulation, and respiratory system. The intention is not a physiological description for which excellent textbooks exist; some are recommended under "Further reading." The intention is a description that clarifies the physical principles that determine these organs' functionality. Whatever aspect of the body we touch upon, we will immediately recognize its incredible complexity. The complexity starts already at the level of the building blocks, the cells, and continues on any level of the body up

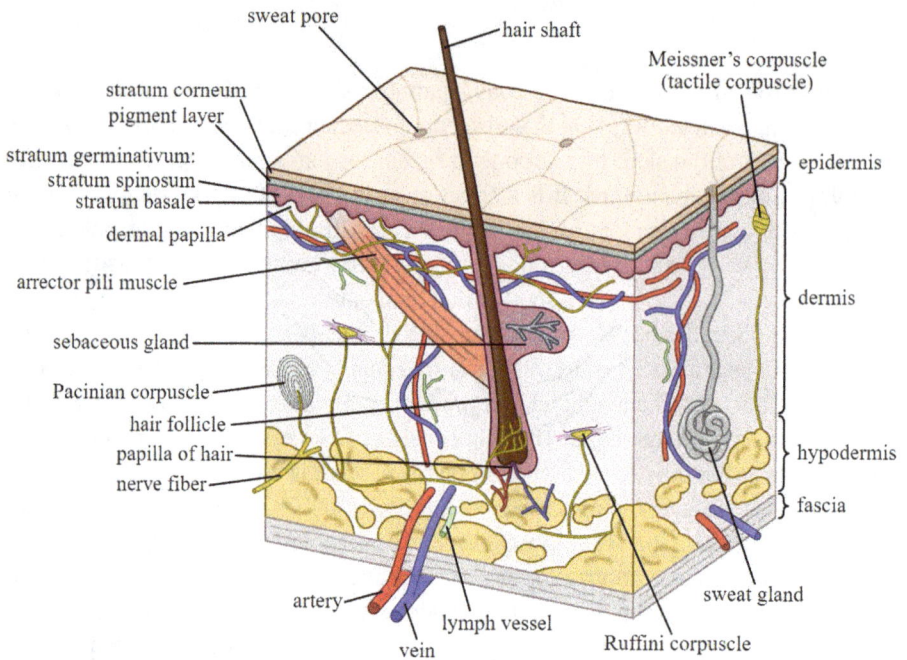

Fig. 1.10: Layers of the skin. Only the dermis and hypodermis layers are supplied with oxygen through the blood vessels. The epidermis is a "dead" layer of cells, eventually stripped off and replaced by dermis cells (reproduced from Wikimedia, Human skin, https://en.wikipedia.org/wiki/Human_skin© creative commons).

the brain that provides consciousness. Although we like to believe in our uniqueness, the blueprint for body parts and functions is more or less the same in all mammals and even all vertebrates. Therefore, animal studies are an essential part of medical and physiological investigations. This research has provided us with tremendous insight into all physiological aspects of our body and will continue to do so.

Medical physics has focused in the past mainly on instrumentation for diagnostic imaging modalities and therapeutic radiology. New fields emerge and will become increasingly important in the future, such as studies of interfaces between the nervous system and electronics, biocompatible materials, intelligent prostheses, and more specific local therapies utilizing functionalized nanoparticles. Some of these aspects are discussed in Vol. 2 and Vol. 3.

1.3 Summary

S1.1   Cells are complex building blocks of the body. Eukaryotic cells contain in the nucleus the complete genetic code.

S1.2 Mitochondria in cells combust nutrients with oxygen for energy production.

S1.3 Cells organize themselves into extended tissues, tissues assemble into organs, and several organs work together in a system.

S1.4 In the body, we can distinguish 11 systems.

S1.5 The heart is a muscle stimulated via an electrical signal to function as a mechanical pump and to maintain blood flow in the circulatory system.

S1.6 The circulatory system provides oxygen and nutrients to cells throughout the body to generate energy and return carbon dioxide as a metabolic end product for exhaling through the lungs.

S1.7 The kidneys filter the blood, control the blood pressure and viscosity, and take care of the electrolyte and water balance.

S1.8 The respiratory system is responsible for the uptake of oxygen into the blood and the expiration of carbon dioxide.

S1.9 The digestive system is responsible for the energy supply to the body by digesting of food and breaking it down into its main nutrients, while waste is filtered out and excreted via urine and feces.

S1.10 The sensory system consists of specialized receptors that inform the body about the body's internal status and the environment. Nerve fibers transmit signals from the receptors to the CNS (center) for data processing.

S1.11 The nervous system connects the sensors to the center and the center to organs and muscles.

S1.12 The locomotive system, consisting of muscles, tendons, and bones, allows us to move around and do mechanical work.

S1.13 The skin is the largest organ of the body that acts as a sensor for various physical parameters and protection against mechanical impact, temperature fluctuations, chemicals, microorganisms, and radiation.

S1.14 The reproductive system allows us to have a partnership and maintain the human race.

Questions

?

Some of the following questions go beyond the preceding text's information. Please consult the books and webpages listed under "Further reading" if you do not know the answer. After trying yourself, you will find the correct answer in the appendix.

Q1.1 What is the function of cell membranes?

Q1.2 What are cell membranes made of?

Q1.3 How many cells are in the body?

Q1.4 How do cells organize themselves?

Q1.5 What is the difference between organs and systems?

Q1.6 A negative feedback system controls the body temperature. Which other negative or positive feedback systems are operational in the body?

Q1.7 What keeps the circulatory system going?

Q1.8 Do lung and heart work synchronously?

Q1.9 The heart pumps about 6 l/min blood through the left ventricle. How much blood does the heart pump through the right ventricle?
Q1.10 Why is oxygen-rich blood more reddish than oxygen-poor blood?
Q1.11 How many muscles are in the heart?
Q1.12 What are the three main tasks of the kidneys?
Q1.13 When does blood have bluish color and more reddish color?
Q1.14 Why is blood red at all?
Q1.15 What is the purpose of the lung?
Q1.16 Does the expiration of carbon dioxide hinder oxygen uptake during inspiration?
Q1.17 Why do we have two different nervous systems, afferent and efferent?
Q1.18 Which organs belong to the digestive system?
Q1.19 Which body system removes solid waste and which removes liquid waste?
Q1.20 How many sensors does the body contain? List at least six.
Q1.21 Where does the information processing of the sensors take place?
Q1.22 How does the sensors' receptor information get to the processing location?
Q1.23 If muscles can only contract, how can we move our extremities back and forth?
Q1.24 What are the two main pigments in the skin and how do they determine the color of the skin?
Q1.25 The skin is the largest organ of the body. What functions does it have?
Q1.26 Ejaculation is a sympathetic or parasympathetic reflex coordinated by the lumbar portion of the spinal cord?
Q1.27 Where are the female reproductive system's ovaries located, inside or outside of the pelvic cavity?

⚡ Attained competence checker + 0 −

I know the main components of cells.

I can name the 11 different systems in the body.

I can distinguish between organs and systems.

I realize the difference between somatic and vegetative nervous systems.

I understand that the combustion of nutrients in mitochondria requires oxygen.

Blood circulation is a transport system for several items, which I can name.

I know that breathing requires the combined action of the chest and diaphragm.

I know that the kidneys filter waste products from the blood.

I am aware of the fact that muscles only move in response to electrical signals.

It is clear that the heart acts as a pump maintaining a pressure difference between the oxygenated and deoxygeneted parts of the blood circulation.

I recognize that the skin as a sensory organ has many different additional functions.

Suggestion for home experiments

HE1.1   Use a transmission microscope to observe cells, cell walls, and nucleuses. A magnification of 100–200 should be sufficient. You may find large and transparent cells by peeling off the thin upper skin of a white onion or garlic. The picture shows an example taken from an onion skin.

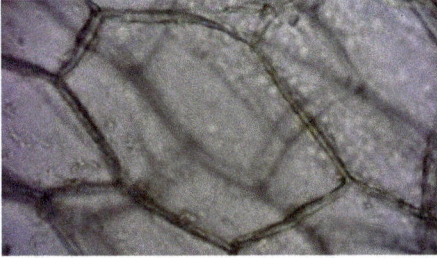

Exercises

E1.1   **Cell types:** Describe the difference between eukarytic cells and prokaryotic cells. Give examples.

E1.2   **Number of cells:** Knowing the average diameter of cells, make a rough estimate of the number of cells in the human body.

E1.3   **Density:** Assuming an average person weight of 700 N, determine its density and compare it with the density of water.

E1.4   **Body measurements:** Which body measurements can you take at home without visiting a medical doctor? Distinguish between periodically reoccurring properties on a timescale of a day and time constant properties. Do not list sports activities. Take notes of your measurement results.

E1.5   **Homeostasis:** Which properties can you identify as being part of the homeostasis of your body?

E1.6   **Feedback systems:** Describe briefly a positive and a negative feedback system. What distinguishes them? Give exampes.

References

[1]   Marth JD. A unified vision of the building blocks of life. Nat Cell Biol. 2008; 10: 1015–1016.
[2]   http://book.bionumbers.org/how-big-is-a-human-cell/
[3]   Cadart C, Venkova L, Recho P, Lagomarsino MC, Piel M. The physics of cell-size regulation across timescales. Nat Phys. 2019; 15: 993–1004.
[4]   Omenn GS, Lane L, Lundberg EK, Beavis RC, Overall CM, Deutsch EW. Metrics for the Human Proteome Project: Progress on identifying and characterizing the human proteome, including post-translational modifications. J Proteome Res. 2016; 15: 3951–3960.
[5]   http://book.bionumbers.org/how-big-is-the-average-protein/
[6]   Goodsell DS. The protein data bank: Exploring biomolecular structure. Nat Educ. 2010; 3: 39.

[7]   O'Connor CM, Adams JU. Essentials of cell biology. Cambridge, MA: NPG Education, 2010, available at: https://www.nature.com/scitable/ebooks/cntNm-14749010/
[8]   Bernard C. An introduction to the study of experimental medicine. New York: Dover Publications, Inc., Dover Books on Biology; 1957.
[9]   Libretti S, Puckett Y. Physiology, homeostasis. Treasure Island, Florida: StatPearls Publishing; 2020 [https://www.ncbi.nlm.nih.gov/books/NBK559138/].

Further reading

Textbook on anatomy
Faller A, Schünke M, Schünke G. The human body. An introduction to structure and function. Stuttgart, New York: Thieme Verlag; 2004.
Marieb EN, Hoehn KN. Human anatomy and physiology. 9th edition. London, New York: Pearson; 2013.
Seeley R, Vanputte C, Russo A. Seeley's anatomy and physiology. New York: McGraw Hill Book Co.; 2016.
Martini FH, Nath J, Bartholomew EF. Essentials of anatomy and physiology. 7th edition. London, New York: Pearson; 2017.
Tortora GJ, Derrickson B. Principles of anatomy and physiology. 14th edition. Hoboken (New Jersey): John Wiley & Sons; 2015.

Textbooks on physiology
Boron WF, Boulpaep EL. Medical physiology. 2nd edition. Philadelphia, London, New York, St. Louis, Sydney, Toronto: Saunders. Elsevier; 2012.
Guyton AC, Hall JE. Textbook of medical physiology. 11th edition. Philadelphia, Pennsylvania, USA: Elsevier Inc., Elsevier Saunders; 2006.
Pape H-C, Kurtz A, Silbernagel S, eds. Physiologie. 7th edition. Stuttgart, New York: Thieme Verlag; 2014.

Textbooks on biology
Campbell. Biology. 11th edition. New York: Pearson; 2011.
Alberts BE, Johnson A, Lewis J, Morgan D, Raff M, Roberts K, Walter P. Molecular biology of the cell. 6th edition. New York: Garland Science; 2014.

Literature of general interest

Watson JD. The double helix: A personal account of the discovery of the structure of DNA. 3rd edition. New York, London, Toronto, Sydney, Singapour: Simon & Schuster; 1968.
Thomas L. The lives of a cell: Notes of biology watcher. 11th edition, Toronto, New York, London: Bantam Books, Inc.; 1980.

Useful websites

https://openstax.org/details/books/anatomy-and-physiology (confirmed November 23, 2022)
https://openstax.org/details/books/biology (confirmed 23.11.2022)
3D Anatomy of the human body: https://human.biodigital.com/m/anatomy/(confirmed 13.11.2022)

2 Body mechanics and muscles

Microscopy image of sarcomeres

Physical properties of muscles

Bone density	1.2 (g/cm^3)
Muscle density	1.04 (g/cm^3)
Other soft (fatty) tissues density	0.92 (g/cm^3)
Force exerted by one myosin-actin bridge	2–5 pN
Force per sacromere	~2.5 nN
Maximum muscle tension	~40 N/cm^2
Sarcomere diameter	10–100 μm
Sarcomere length	2.5–3 μm
Actin filament diameter	5 nm
Myosin filament diameter	16 nm
Number of myosin heads in one myosin filament	500
Types of muscles in the body	4
Number of skeletal muscles in the body	650
Number of muscle hierarchy levels	4
Number of muscle fibers in a motor unit	20–2000

2.1 Basic mechanical terms

When it comes to body mechanics, muscles and bones play a dominant role. As with all biological systems, we have to distinguish between microscopic and macroscopic scales. On the macroscopic scale are the forces and moments acting on different parts of the body. On a microscopic scale, we want to understand how muscle contraction creates forces. In this chapter, we first introduce basic mechanical terms, such as forces, torques, levers, and stability conditions. These forces and torques tell us how strong muscles must be to move our body and do mechanical work. Obviously, muscles, tendons, and bones work together to perform the activity we expect them to do. In the second part, we describe the hierarchical structure of muscles and try to understand the mechanism that leads to muscle tension. The

https://doi.org/10.1515/9783110756951-002

analysis is continued in Chapter 3, where we examine the bone structure, and in Chapter 6, where we learn that action potentials tell muscles to contract. This order of discussion is somewhat reversed when compared to what actually happens in the body. First, we make a decision in our brain. Then we send an action potential to the muscle, and finally, the muscle contracts to lift a bone. At least until the end of Chapter 6, we have gathered the main pieces of the information to get some insight into the complexity of body mechanics.

2.1.1 Body density

From a mechanical point of view, the human body is extended, irregular in shape, anisotropic in its mechanical properties, and inhomogeneous concerning its mass distribution. Nonetheless, we can ascribe to the human body volume, weight, density, center of mass, the momentum of inertia about different axes, torques, and various levers.

The average mass density of the body and the local density are important physical parameters. In x-ray and proton radiation therapy, knowledge of the local mass and electron densities is required to simulate the x-ray attenuation and the stopping power in the tissue, respectively. On average, the density of the human body has three main contributions: bones, muscles, and fatty tissues, including all membranes and collagens. Thus, by determining the average density of a body, we can estimate, for instance, the proportion of fatty tissues.

To determine the average mass *density* ρ_{body}, we need to know the total mass m_{body} and the body volume V_{body}:

$$\rho_{body} = \frac{m_{body}}{V_{body}}. \tag{2.1}$$

Here, the body mass is the sum of the following:

$$m_{body} = \sum_i m_i = m_{muscle} + m_{bone} + m_{tissue}. \tag{2.2}$$

It is trivial to determine the body mass, but it is less trivial to determine an irregular body volume. One method would be to submerge a person wholly and briefly into a tub brimful of water and measure the water overflow. Another slightly more elegant method follows Archimedes' procedure, which directly measures the density without determining the volume. First, the person's weight $W_1 = m_1g$ is measured under normal atmospheric conditions and then again after immersion into water, yielding $W_2 = m_2g$, where g is the gravitational acceleration of the Earth. From these two measurements, the average density of the body follows according to (see Exercise E2.1)

$$\rho_{\text{body}} = \rho_{\text{water}} \frac{W_1}{W_1 - W_2}. \tag{2.3}$$

The densities of the body's main components, relative proportions, and weights are listed in Tab. 2.1. This procedure is referred to as hydrostatic density testing.

The average mass density of the body, composed of bones, muscles, and the rest, varies between 1.03 and 1.08 g/cm^3. Thus, the body is almost floating in water, and only little effort is required to keep it aloft. Infobox I explains how the fatty fraction of the body can be determined from the same hydrostatic body density measurement. Thus, a single hydrostatic measurement of the body's weight in water yields not only its average density but also the fraction of fatty tissue.

Tab. 2.1: Average values of densities, relative mass proportion, mass, and volume for a person with a total mass of 75 kg and a volume of 0.07 m^3.

	Density (g/cm^3)	Relative mass proportion (%)	Mass (kg)	Volume (m^3)
Bones	1.2	12	9	0.0075
Muscles	1.04	29	22	0.0209
Other soft (fatty) tissues	0.92	59	44	0.048

Sexual differences are not considered (from [1]).

Infobox I: Hydrostatic body fat testing

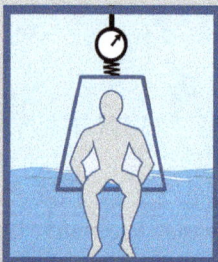

If we want to determine the fatty fraction of the body f_{fat}, we just need to take the density ratio $\rho_{\text{body}}/\rho_{\text{water}}$ according to eq. (2.3), and to make a reasonable assumption about the average density of the remaining body parts: $\rho_{\text{rest}} = \rho_{\text{bones}} + \rho_{\text{muscles}}$. For instance, we may assume $\rho_{\text{rest}}/\rho_{\text{water}} = 1.1$. Then the ratios $\rho_{\text{fat}}/\rho_{\text{water}} = 0.9$ and $\rho_{\text{rest}}/\rho_{\text{fat}} = 1.22$ are known. If we measure $W_2 = 0.05W_1$, then $\rho_{\text{body}}/\rho_{\text{water}} = 1.052$ and $\rho_{\text{rest}}/\rho_{\text{body}} = (\rho_{\text{rest}}/\rho_{\text{water}})(\rho_{\text{body}}/\rho_{\text{water}})^{-1} = 1.1/1.052 = 1.046$.

The total body volume is $V_{body} = m_{body}/\rho_{body}$. The body volume can be split into fatty fraction f and the rest:

$$\frac{m_{body}}{\rho_{body}} = \frac{m_{fat}}{\rho_{fat}} + \frac{m_{rest}}{\rho_{rest}} = \frac{fm_{body}}{\rho_{fat}} + \frac{(1-f)m_{body}}{\rho_{rest}}, \tag{2.4}$$

yielding

$$\frac{1}{\rho_{body}} = \frac{1}{\rho_{fat}} + \frac{1-f}{\rho_{rest}}. \tag{2.5}$$

Solving for the fraction f:

$$f = \frac{1 - \rho_{rest}/\rho_{body}}{1 - \rho_{rest}/\rho_{fat}}. \tag{2.6}$$

Inserting numbers from above:

$$f = \frac{1 - 1.046}{1 - 1.22} = 0.21.$$

Therefore, in this example, the proportion of fatty tissue in the body is 21%. The method described here may seem a bit rough, but it is more accurate than determining the percentage of fat with the help of a body scale. Some scales measure the body's electrical resistance, from which the fat percentage is derived based on a number of assumptions.

2.1.2 Symmetry planes

We use a Cartesian coordinate system adapted to the body's symmetry planes. In the upright position, the bilaterally symmetric human body can be divided into three orthogonal planes, shown in Fig. 2.1:
– xz-plane or *sagittal plane*,
– yz-plane or *coronal plane*,
– xy-plane or *transverse plane*.

The medial line is defined by crossing the coronal plane and the sagittal plane. Furthermore, we distinguish between the following directions:
– front (ventral or anterior) and back (dorsal or posterior),
– up (cranial) and down (caudal),
– close (proximal) or distant (distal), and
– between left and right, both lateral directions.

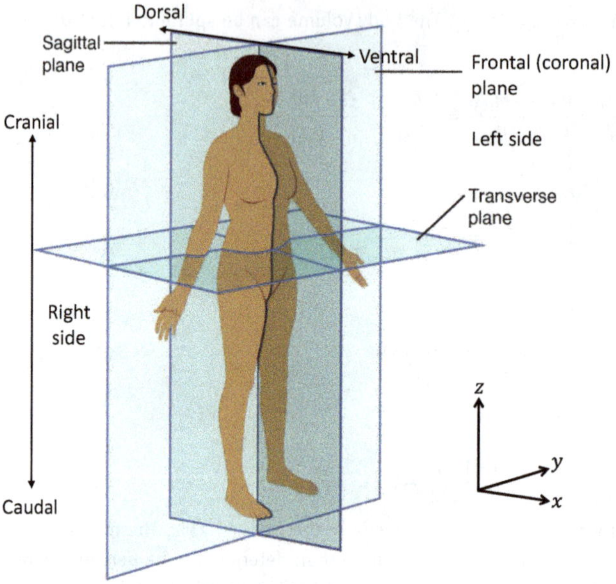

Fig. 2.1: Three principal symmetry planes of the human body: Coronal plane, sagittal plane, and transverse plane (reproduced from openstax.org/details/books/anatomy-and-physiology © creative commons).

2.1.3 Center of gravity

The *center of mass* or the *center of gravity (cg)* of the human body is not only difficult to determine; its location varies by changing posture.

Considering the body's symmetry planes, we expect to find the cg along the medial line when standing up or laying down. We determine the cg by first taking the weight of the body W_{body}, then placing the person on a flat board, supported on one side by a fulcrum and on the other side by a scale.

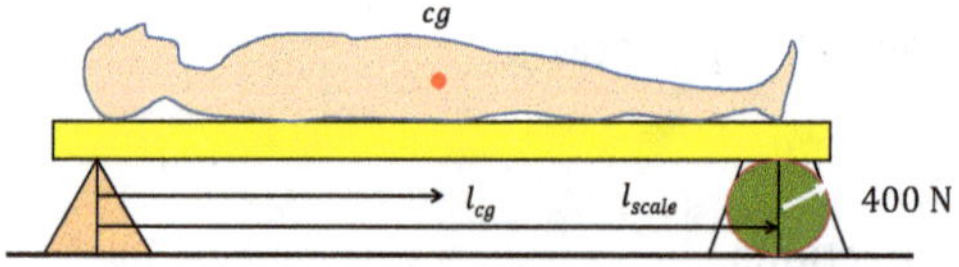

Fig. 2.2: Arrangement for determining the center of gravity of a person.

Using the lever rule, we obtain

$$l_{cg} W_{body} = l_{scale} W_{scale},\qquad(2.7)$$

where l_{cg} is the lever arm that spans from the fulcrum to the cg, and l_{scale} is the lever arm extending from the fulcrum to the scale (s. Fig. 2.2). Reading the weight displayed on the scale and correcting for the board's weight, we will find the cg, which – as suspected – is in the sagittal plane at the position of the pelvic space (see Figs. 2.2 and 2.3).

> The cg of the human body is in the sagittal plane at the position of the pelvic space. **!**

It is useful to distinguish between a cg of the upper body half (cranial cg) and the lower body half (caudal cg). Those points are indicated in Fig. 2.3. When a person bends over, the cg will shift. The new position of the cg can be found by connecting the cranial and the caudal cg via a straight line. The body's cg then lies halfway in between and outside the body, as shown in Fig. 2.3.

Knowing the location of the cg and controlling it is extremely important for sport activities, ballet dancing, ice skating, and various other body activities. For instance, in high jumping, the cg passes below the bar while the high jumper crosses over the bar, as sketched in Fig. 2.4.

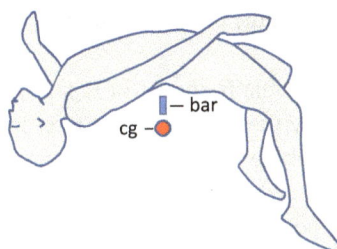

Fig. 2.3: Center of gravity of the body in upright position and after bending over (adapted from Ref. [1]).

Fig. 2.4: The center of mass of a high jumper crosses below the bar (adapted from Ref. [1]).

2.2 Body mechanics

2.2.1 Mechanical models

The body's mechanics is a very complex affair, requiring, in general, the coordinated action of many muscles. Nevertheless, we may model the body mechanics by a highly simplifying approach, where bones are represented by sticks counterbalanced by strings representing muscles. In all static cases, the mechanical equilibrium is governed by the vector sums:

$$\sum_n \vec{F}_n = 0, \tag{2.8}$$

$$\sum_n \vec{T}_n = 0. \tag{2.9}$$

Here $\vec{F}_n$ represents forces, such as the body weight $\vec{W} = m\vec{g}$, and $\vec{T}_n$ are the acting torques like the lever torque $\vec{T} = \vec{l}_{\text{scale}} \times \vec{W}_{\text{scale}}$ used in eq. (2.4).[1] The SI unit of force is Newton,[2,3] (N), and the SI unit of torque is Newton meter (Nm). A simple case is an upright posture. The force of the body weight $\vec{F}_1$ is counterbalanced by the opposing reaction force $\vec{F}_2$ exerted by the floor, the person is standing on, such that the total force is $\vec{F}_1 + \vec{F}_2 = 0$.

Now we take the body's weight as 1.0 W. Then the upper part of the body has about 70% of the total weight or 0.7 W. This weight is balanced by two legs, each one supporting 0.35 W. If we lift one leg, the other one has to support the upper weight of 0.7 W plus one leg's weight (0.15 W), which adds up to 0.85 W or a factor of 2.4 higher than standing on both feet. However, balancing on one leg requires the contraction of the hip's adductor muscles, which amplifies the force on the hip, specifically on the femur, by a factor of $2.5 \times W$, i.e., seven times more than during normal posture!

Another example is shown in Fig. 2.5. Bending over without dipping requires a muscle action by the erector spinae muscle connecting the middle of the spine with the lumbar vertebrae to balance the upper body's weight. This situation can be modeled with *sticks and strings*, as indicated on the right panel of Fig. 2.5. The muscle is fixed between points D and C. The upper body's weight is composed of W_1 ($\approx$ ½ W) for the chest and W_2 ($\approx$ 1/4 W) for head and arms. Both weights hang on the lever stretching from points A to B with the length l_{AB}. In equilibrium, the torques $\vec{T}_1$ and $\vec{T}_2$ exerted by weights W_1 and W_2 must be counterbalanced by the muscle torque $\vec{T}_m$ acting on point D:

$$\vec{T}_m - \vec{T}_1 - \vec{T}_2 = 0. \tag{2.10}$$

$\vec{T}_m$ acts clockwise, therefore, has a positive sign; $\vec{T}_1$ and $\vec{T}_2$ act counterclockwise, therefore, have negative signs. Assuming that point E is halfway between points A and B, and point D is 2/3 this distance and using an upper-body bending angle of

---

1 Here and in the following, we designate a general force by the vector $\vec{F}$ and the body weight by a scalar W or vector $\vec{W}$, depending on the circumstances.

2 $1\,\text{N} = 1\,\text{kg}\,\text{m/s}^2$.

3 Isaac Newton (1642–1726), English mathematician, astronomer, and theologian.

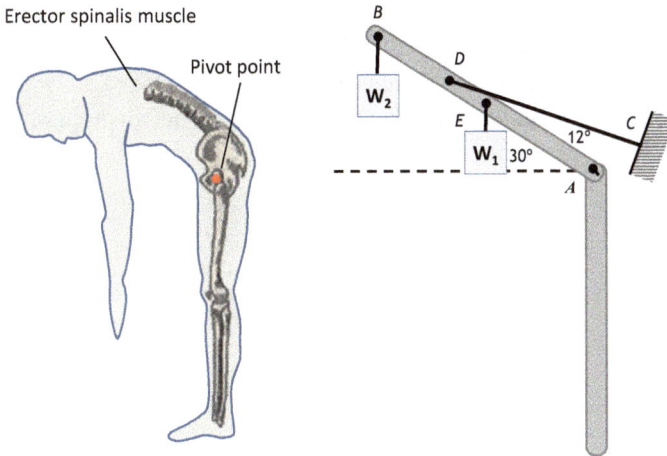

Fig. 2.5: Stick and string model for the action of muscles to counterbalance a particular posture of the body. *A* is a fixed fulcrum (adapted from Ref. [2]).

30° counterbalanced by the erector spinae muscle acting at an angle of 12°, we find for the magnitudes of the torques:

$$\left|\vec{T}_1\right| = \frac{1}{2}W \times \frac{1}{2}l_{AB} \times \sin(90° - 30°);$$

$$\left|\vec{T}_2\right| = \frac{1}{4}W \times l_{AB} \times \sin(60°);$$

$$|\vec{T}_m| = F_m \times \frac{2}{3}l_{AB} \times \sin(12°).$$

We are interested in the magnitude of $|\vec{F}_m|$, which is then

$$F_m = \frac{3}{4}W\frac{\sin 60°}{\sin 12°} = 3.12 \; W.$$

This is a surprising result: leaning over by about 30° without dropping requires a muscle tension of roughly three times the bodyweight!

The mechanical equivalent of other postures can be discussed similarly, which hints on various muscle strengths. In general, the muscle strength can be estimated to be about 20–30 N/cm^2. Concerning posture, it should be reminded that the body does not act in terms of a static balance. Instead, the body balance is always dynamic, requiring the activity of many muscles incorporated into a control and feedback system that also involves the vestibular organ (Chapter 12).

2.2.2 Levers

Many levers in the body help us to be in mechanical equilibrium at rest or do work, such as climbing stairs, lifting weights, or eating apples. However, before discussing some of these activities, we first introduce the levers' mechanical advantage concept.

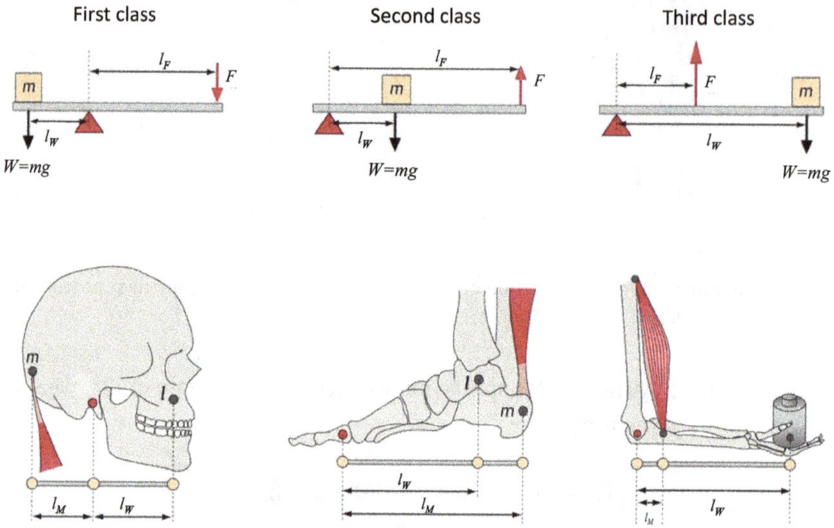

Fig. 2.6: Top panels: Levers of first, second, and third class. Bottom panels: Examples of different corresponding levers in the body. l_W= load arm, l_F = force arm, l_M = muscle arm. Red triangles and dots indicate the fulcrum of the particular lever (adapted from [3] with permission of Thieme Verlag).

We distinguish between three classes of levers sketched in Fig. 2.6. They are different by the position of the load $W = mg$ with respect to the fulcrum and the position of the counteracting lifting force F. Case (a) is termed a *two-armed lever* or a *first-class lever*. Here the load and the lift arm are on opposite sides of the fulcrum. The lever is in equilibrium when the torques are equal:

$$l_W \times W = l_F \times F,$$ (2.11)

such that the force F required to lift the mass m is

$$F = \frac{l_W}{l_F} W \ll W.$$ (2.12)

Hence, the lift force F is much smaller than the weight W, as long as the load arm l_W is shorter than the lift arm l_F. Clearly, this situation is of mechanical advantage.[4] An example of the first-class lever in the body is our jaw. Since here $l_W \cong l_F$, the mechanical advantage is, however, not big.

Second-class levers have their load and lift arms on the same side of the fulcrum. Therefore, they are termed *single-armed levers*. Pushcarts are good examples for second-class levers. The equilibrium condition and the mechanical advantage are the same as for the two-armed lever. The foot is a second-class lever, if the toes act as the fulcrum. Indeed, the foot and leg are particularly strong. We can easily lift our complete body by standing on our toes.

Third-class levers are also single-armed. But in contrast to second-class levers, load and lift arms are exchanged so that $l_L > l_F$. Therefore,

$$F = \frac{l_W}{l_F} W \gg W, \tag{2.13}$$

which, from a mechanical point of view, is not of much help. We find more often third-class levers in the body than first or second classes. The most prominent example is our arm combined with the forearm. As sketched in Fig. 2.6 (right panel), the forearm acting as the load arm l_W is much longer than the muscle arm l_M. The muscle arm is only the short distance between the anchoring of the biceps at the ulna and the elbow joint.

The *mechanical advantage* is defined as the ratio: F/W. For first- and second-class levers, the ratio is larger than 1; for the third-class levers, it is smaller than 1.

Now we come back to the discussion of weight lifting by the forearm. The arm has two muscles attached to the forearm by tendons, one in front of the elbow joint (biceps brachii) and one in the back (triceps brachii). The biceps and triceps muscles are called antagonistic muscles for reasons that become clear later on. The terms *biceps* and *triceps* refer to the number of tendons that attach the muscle at their tapered end to the bones. For holding some weight in hand (Fig. 2.7(a)), the biceps contracts, exerting a force F_M, while the triceps is relaxed. The contraction of the biceps results in a torque $T_M = l_M F_M$, counterbalancing the torque by the weight held in hand: $T_W = l_W F_W$. In equilibrium $T_M = T_W$. Now we turn over the hand and push down on a bar or a table. Then the triceps contracts and the biceps is relaxed. Furthermore, the lever is now of first class but still of low mechanical advantage since the muscle arm of the triceps to the elbow joint l_M is now even shorter than for the biceps. Pushing down, therefore, is even less effective than lifting up.

---

4 In this and the following equations we may neglect the symbols for the vector properties and cross-product as long as forces and lever arms are oriented perpendicular to each other.

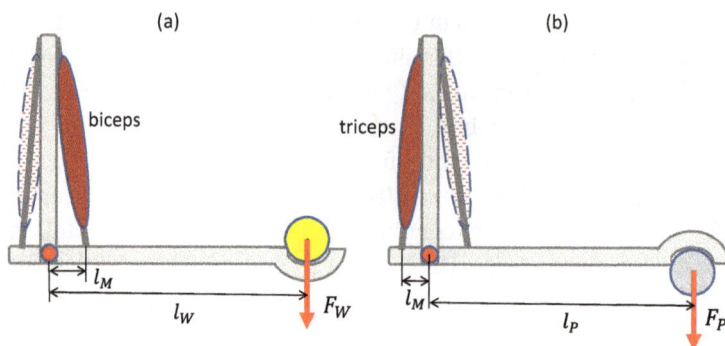

Fig. 2.7: Mechanical model for lifting some weight with the hand (a), and for pushing down a bar by turning over the hand (b). Full colored muscle indicates tension, and dashed lines indicate a relaxed state of the muscle. F_W is the force acting by weight, F_P is the force exerted by pushing down. l_W and l_P are the corresponding lever arms, and l_M is the lever arm of the muscles.

We already noted that levers in the body are primarily of third class. Although the mechanical advantage is minor for third-class levers, they provide us with much higher mobility and speed than first- and second-class levers would do. Thus, our body is not designed for lifting heavy weights but for increased agility. These examples also show that muscles exert force and torque only by contracting their lengths. Therefore, a pair of two muscles must move the lever arm up or push down.

> **!** Most levers in the body are of third kind, providing us agility but not much mechanical advantage.

2.2.3 Mechanical stability

A rectangular cuboid is stable when the cg projection onto the supporting surface lies between the left (Lf) or right fulcrum (Rf). Then the resulting torque is oriented opposite to the tilt angle α and turns the cuboid back to an upright position, as illustrated in Fig. 2.8(a). If the cg projection lies outside these limits, a torque occurs $(T = lF \sin(\alpha))$, which points in the same direction as the tilt angle and turns the cuboid to the side (Fig. 2.8(b)). Similar considerations also hold for the body (Fig. 2.8(c)). As long as the projection of the body's cg lies within the area of the feet, the posture is stable. Otherwise, it becomes unstable. Using a cane or a crutch considerably increases the area of stability (Fig. 2.8(d)).

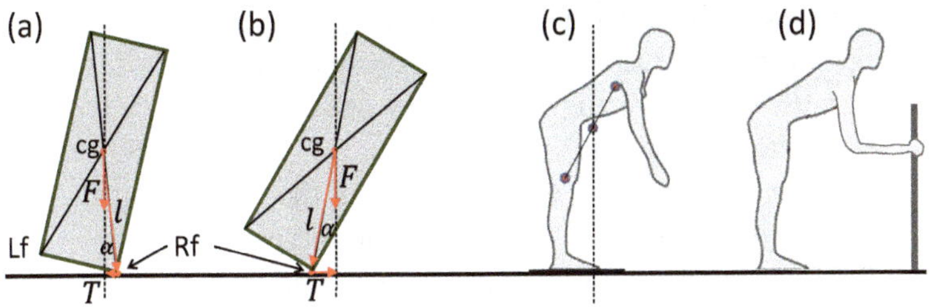

Fig. 2.8: The stability of a body on a flat floor depends on the center of gravity's projection with respect to the pivot point: (a) A torque turns the cuboid back again; (b) the torque points in the direction of the tilt, tipping over the cuboid; (c) projection of cg is within the stable foot area; and (d) area of stability is increased by using a cane. Lf, left fulcrum; Rf, right fulcrum.

2.2.4 Femur

In the context of torques and levers, the *femur* warrants special attention. The femur is the longest, heaviest, and strongest bone in the body. The upper body weight lasts on both femurs when standing still, and additional forces act when walking, running, or jumping. The femur experiences extreme forces due to the strength of the hip's muscles and the thigh acting on the femur to move the leg.

The pair of forces F_1 and F_2 from the upper body weight, the muscle tension, and the supporting legs, all together exert a torque (Fig. 2.9):

$$T = 2|\vec{F}|l\sin\alpha. \tag{2.14}$$

This torque is estimated to be about 2–3 Nm for a body standing at rest, but varies between 5 and 6 Nm during walking and even higher during jumping.

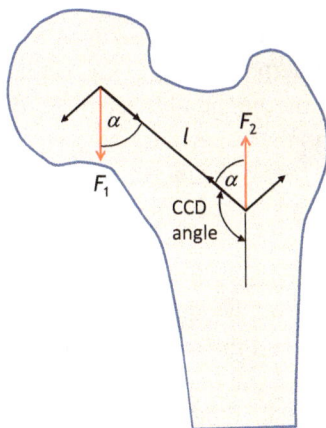

Fig. 2.9: Torque acting on the femur at the upper end of the long bone. CCD is the caput–collum–diaphyseal angle. The CCD angle decreases with age.

The *caput–collum–diaphyseal angle* (CCD) is the angle between the neck and the shaft of the femur in the hip: $CCD = 180° - \alpha$. Because of the constant load on the femur, the CCD angle changes with age. At a young age, the CCD angle is about 140°, decreasing over time to about 115°. A decreasing CCD angle implies an increasing angle α and thus an increasing torque acting on the femur. Considering that the bones' strength decreases with higher age, the increasing torque at the femur is a big problem and often results in cracks or even fractures. Fortunately, a hip fracture is reparable by inserting an artificial hip joint. Bone fracture is reviewed in Chapter 3, and hip replacement is treated in Chapter 13.

2.2.5 Degrees of freedom

An extended body has six *degrees of freedom* for the motion: three translational and three rotational. In the case of the human body, we do not need to discuss the motion of the whole body but the motion of parts of the body that lends the body flexibility, agility, and fine motor skills better than a robot. For instance, the head can be bent forward, backward, sideways, and can be rotated. This makes a total of four degrees of freedom. A special joint, called condyloid joint (Fig. 2.10(5)), connecting the atlas with the occipital bone (back side of the skull), provides this greater range of motion than the rest of the vertebrae.

Six different types of joints can be identified in the body. Aside from the condyloid joint, the other joints are ball-and-socket joint, hinge joint, pivot joint, saddle joint, and plane joint. These joints are sketched in Fig. 2.10 and examples are shown where they can be located in the body. They have the following properties:

1. Pivot joints have one axis for left–right rotation. An example is the radioulnar joint.
2. Hinge joints, similar to pivot joints, have one axis of rotation, providing back and forward movement. An example is the elbow joint and the knee.
3. Saddle points comprise two main axes of rotation perpendicular to each other and allow four directions of movement. Examples are the thumb's saddle joint, i.e., the wrist between the first metacarpal bone and the trapezium.
4. Ball-and-socket joints have three axes of rotation perpendicular to each other and therefore provide six directions of movement. Typical examples are the hip and the shoulder joints.
5. Condyloid joints are distinct from ball-and-socket joints by their elliptical shape. They have two perpendicular axes of rotation, allowing four main movements. Examples are the joints between the forearm and the wrist and the joint between the atlas and the occipital condyles, as already mentioned.
6. Plane joints have no axis of rotation but a glide plane, allowing translational motion. Examples are the small joints of the vertebrae.

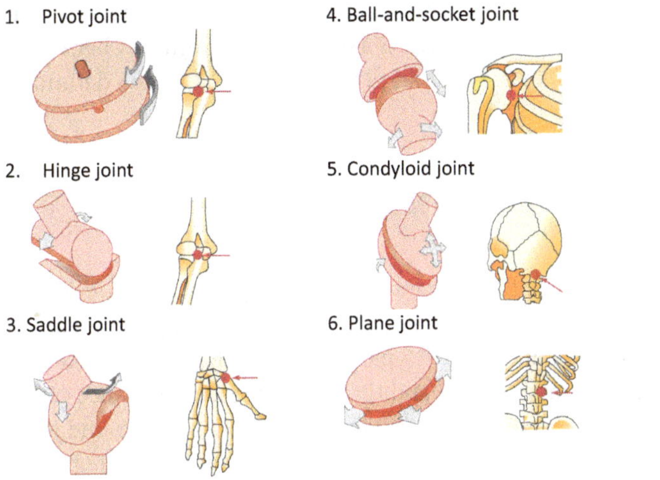

1. Pivot joint

4. Ball-and-socket joint

2. Hinge joint

5. Condyloid joint

3. Saddle joint

6. Plane joint

Fig. 2.10: Different types of joints in the skeleton (adapted from Ref. [3] with permission of Thieme Verlag).

Joints would not work unless muscles and tendons tighten them. Tendons attach to bones and muscles move bones through contraction and relaxation. In Fig. 2.7, for example, the forearm moves upward by contraction of the biceps and simultaneous relaxation of the antagonistic triceps. Pressing down the forearm requires the contraction of the triceps and the relaxation of the biceps. Muscles, tendons, and joints work together to provide torque to balance the load and allow various body movements. Figure 2.11 shows examples of the mobility of the vertebral column

0°
30–40°

0°
30–35°

90–100°

30°

0°

30°

Lateral flexion   Rotation   Flexion and extension

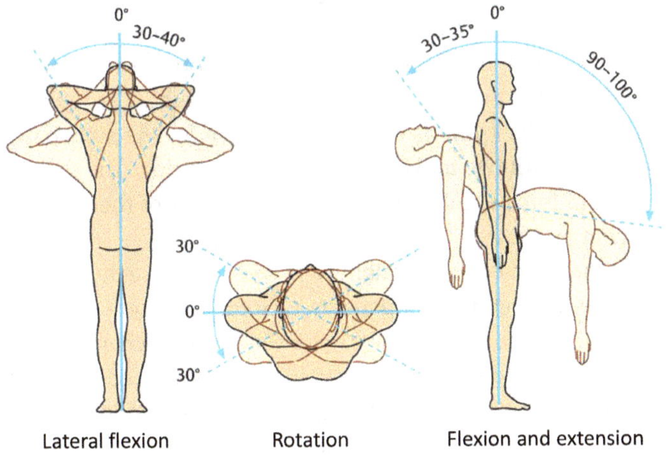

Fig. 2.11: Mobility of the vertebral column. The maximum extension for each movement from zero position is given in degrees (adapted from Ref. [2] with permission of Thieme Verlag).

(spine) via three main directions: left: lateral bending or lateral flexion in the coronal plane; middle: rotation around the vertical axis; right: forward and backward bending (flexion and extension) in the sagittal plane (see Fig. 2.1 for the definition of the planes).

2.2.6 Biomechanics of walking

Normal walking (gait) is a complex three-dimensional activity, requiring the coordination of several joints, bones, and muscles [4, 5]. The cyclic gait pattern can be subdivided into 2 major phases and up to 16 partial phases. In Fig. 2.12, the two major phases are shown: the stance phase and the swing phase. The stance phase starts as soon as the heel of one foot (here the right foot) touches the ground (heel strike, HS) and lasts until the same foot is retaken off the ground (toe off, TO). At this moment, the swing phase starts, supported by the contralateral leg. The stance phase takes about 60% of a cycle time, and the swing phase about 40%. There is an inevitable overlap between the left and right leg phases.

The stance phase can be further subdivided into four additional phases: initial HS, contralateral toe off (CTO), mid-stance (MS), and contralateral heal off (CHO). During the latter three phases, the body is supported by just one leg (single support phase). Correspondingly during this phase, the force and the torque on the femur and knee reach maximum values. During the swing phase, we can distinguish three additional phases: toe off, mid-swing, and HS, which closes the cycle.

The force pattern during one gait cycle is shown in Fig. 2.13 for the foot striking the ground (red line) and the force at the joint between femur and tibia (tibiofemoral force) (blue line). The ground reaction force shows two characteristic maxima: the first maximum at CTO is due to upward acceleration of the body toward the MS; the second maximum is from deceleration as the body drops down from MS to CHO. The maximum ground force is about 120% of the body weight. The red dots indicate the cg and the dashed black line is a guide to the eye for the sinusoidal movement of the cg during gait with two maxima at MS and mid-swing. The tibiofemoral force mimics the ground force at a higher force level because of the simultaneous muscle tension reaching up to $2 \times W$.

In the past, tests were conducted using a straight walk path with a centered force plate that can measure three orthogonal components of force and torque on the ground [4]. Evaluating the joint acting forces is then achieved by biomechanical simulation using three-dimensional models of the lower limb [5]. More recently, the joints' forces and torques have been measured with strain gauges inserted in artificial hips and knees, telemetrically connected to the outside [6]. The force on the hip can be as high as six times the body weight. This force is a combination of total static body weight, accelerated body weight when hitting the ground, and muscle tension. The gait pattern is repeated about 5000–8000 times per day for an average person.

Walking seems easy to a healthy person. But we should be aware that walking requires the coordinated action of 6 joints, 28 bones, 33 muscles, and more than 100 tendons [2]. When any part fails due to an accident or disease, we become aware of the gait's complexity.

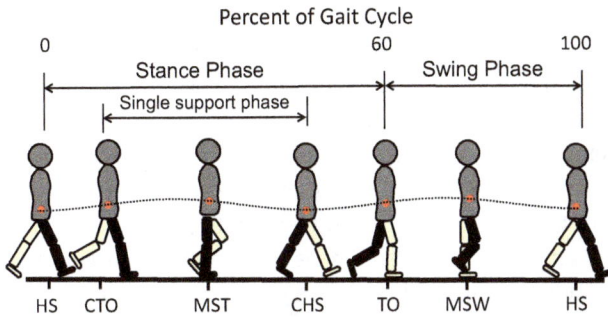

Fig. 2.12: Phases of a normal gait consists of a stance phase and a swing phase. HS, heels strike; CTO, contralateral toe off; MST, mid-stance; CHS, contralateral heel strike; TO, toe off; MSW, mid-swing. The red dots indicate the body center of gravity and the black dashed line is a guide to the eye.

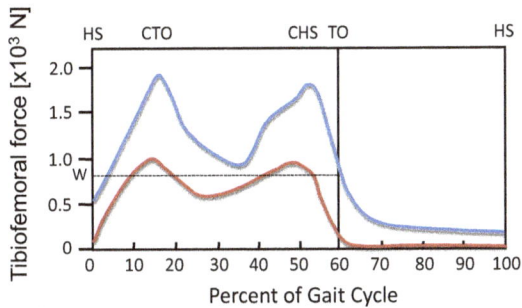

Fig. 2.13: Red curve: Vertical ground reaction force of the right leg during normal gait. Blue curve: Forces acting on the tibiofemoral joint during a gait cycle (according to [5]). The dashed line indicates the body weight (*W*); the other acronyms have the same meaning as in Fig. 2.12.

2.3 Skeletal muscles

2.3.1 Structure of skeletal muscles

When we talk about muscles, we often naturally visualize skeletal muscles. However, in the body, we can identify four types of muscles: (a) muscles that are responsible for the functioning of large parts of internal organs such as the intestines; (b) skeletal muscles that control the movement of bones in the body; (c) the heart muscle (myocardium), which regulates the contraction of the heart; and (d) circular muscles (sphincter) for controlling among others, mouth, urethra, anus, and for accommodating the lens and iris. Here we focus on the skeletal muscles. The myocardium is discussed in Chapters 7 and 8, and the circular muscles are touched upon again in Chapters 10 (kidneys) and 11 (vision). These differently shaped muscles also differ in their appearance, as shown in Fig. 2.14. Muscles of the internal organs are smooth muscles. Skeletal muscles have a characteristic cross-stripe appearance and are known as *striated muscles*. The cardiac muscle is also cross-striped, but differently organized compared to skeletal muscles. In the following, we limit ourselves to the discussion of skeletal muscles.

About 650 different skeletal muscles make up roughly 30–40% of a person's body weight [7]. Skeletal muscles consist of various shaped muscle bellies (macroscopic shape of a contracted muscle) that narrow down and merge into tendons on either end. The tendons, in turn, are attached to different bones. The origin of muscles may be single, double, triple, or quadruple headed but always ends in a single tendon for moving a particular bone. Accordingly, these muscles have additional names such as biceps and triceps as already mentioned.

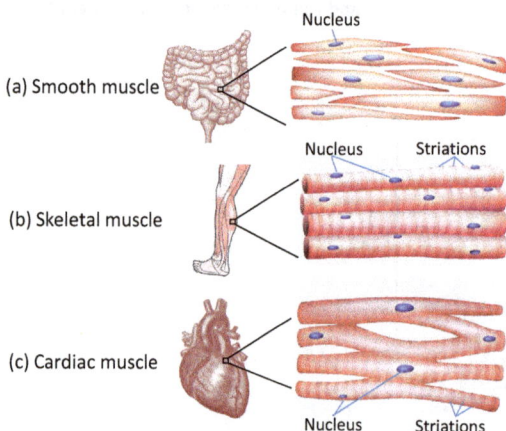

Fig. 2.14: Three types of muscles: (a) smooth muscle; (b) skeletal muscle; and (c) cardiac muscle (adapted from OpenStax Anatomy and Physiology, ©Creative Commons).

The movement of the skeletal bones is always achieved through contraction of the muscle fibers. Unlike a spring, muscles do not relax when released. An antagonistic pair of muscles is needed to return to the resting position. We have already seen that the forearm's biceps and triceps muscles precisely work according to this principle (see Fig. 2.7). Both contract: the muscle biceps brachii moves the forearm up, and the muscle triceps brachii moves it down.

! The movement of the long bones requires the action of pairwise antagonistic muscles.

2.3.2 Hierarchical organization

The skeletal muscles have a hierarchically organized structure on four levels, illustrated in Fig. 2.15 [8]:

- The first structural level consists of several bundles, also known as the *fascicles* (Latin: *fasciculus*). These can be distinguished with the naked eye and give the muscle its striped appearance. Each bundle is separated from other bundles by a layer of connective tissue (mostly collagen) called the *perimysium*. All bundles together are wrapped in an *epimysium*.
- About 150 muscle fibers within a bundle (fascicle) mark the second level. The fibers are separated by a connective tissue called the *endomysium*. Each muscle fiber is an elongated multinucleated cell that contains hundreds of *myofibrils*.
- Connective tissue, called *sarcoplasm*, separates myofibrils on the third level. The myofibrils are wrapped in a *sarcolemma* that also contains a *sarcoplasmic reticulum*. The sarcoplasmic reticulum, in turn, is important as a reservoir for Ca^{2+} ions necessary for fiber contraction. The connective tissue is needed to transmit the force developed in the sarcomeres to the body.
- On the fourth and final level we recognize in each *myofibril* hundreds of *myofilaments* organized in *sarcomeres*. Sarcomeres are themselves highly ordered and organized. This unusual fibrous cell structure extends for several millimeters to centimeters from one end of the tendon to the other with a diameter of 10–100 µm. Blood capillaries and somatic motor neurons round off the muscle content as a discrete organ. Oxygen is needed for the metabolism of the muscles; and somatic motor neurons (Chapter 6), which attach to muscle fibers, tell the sarcomeres when to contract.

In summary, the four levels of the muscle hierarchy from macroscale to nanoscale are as follows: 1. bundles; 2. fibers; 3. myofibrils; and 4. myofilaments.

> Muscles are hierarchically organized on four levels: 1. bundles; 2. fibers; 3. myofibrils; and 4. myofilaments. !

The complexity of the muscle structure, its function, and its feedback control mechanism is remarkable. For a comprehensive discussion of all details, the reader is advised to consult one of the standard anatomy or physiology texts listed at the end of this chapter. Here we focus on the biomechanical aspects of muscles.

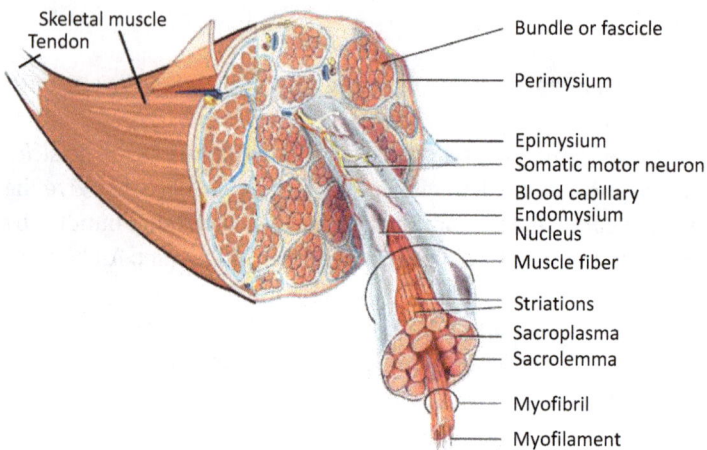

Fig. 2.15: An anatomical cross section through a skeletal muscle showing the hierarchical order of bundles, fibers, myofibrils, and myofilaments (adapted from OpenStax Anatomy and Physiology, ©Creative Commons).

2.3.3 Muscle contraction

Now we focus our attention on a single myofibril, the machine room of muscle action. One myofibril is sketched in Fig. 2.16(a). Myofibrils consist of an ordered hexagonal array of myofilaments (a). The hexagonal order indicated in panel (d) is so perfect that sharp Bragg reflections can be seen by x-ray diffraction (see Infobox II) [9, 10]. Furthermore, myofibrils are periodically structured by membranes perpendicular to the long axis called z-*disks* (b). The z-disks form a transverse structure, giving the myofibril a bamboo-like, striated appearance. The distance from one z-disk to the next is called a sarcomere. The sarcomeres line up parallel to the long axis of muscle cells and between z-disks. Panel (e) illustrates the self-organized sarcomere structure in skeletal muscles. The distance between z-disks is about 2.5 μm at rest. But when activated, the sarcomere can stretch and contract considerably. Because of the serial arrangement, the extensions and contractions add up to macroscopic length changes. One sarcomere comprises thin *actin filaments* (diameter 5 nm) and thick *myosin filaments* (diameter 16 nm). Together they form a contractible *molecular motor*.

The actin filaments are attached to successive z-disks, while the myosin filament can move back and forth within the surrounding actin filaments. One myosin filament consists of about 400–500 myosin molecules. Each molecule features two heads and one tail. The tails of different myosin molecules are twisted together, forming a rod-like structure in the filament, while the globular (rounded) heads stick out of the rod. The heads are subdivided into two parts: top and neck. The top binds to the actin filament and hydrolyzes ATP to ADP;[5] the necks act as a lever during muscle contraction.

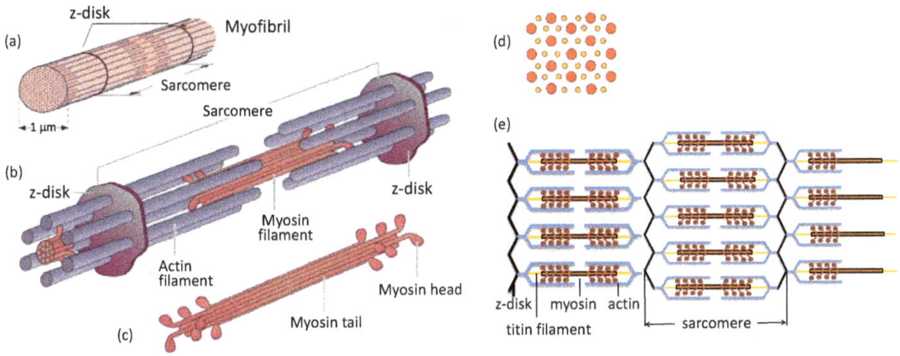

Fig. 2.16: (a) Schematics of one myofibril consisting of many sarcomeres. (b) Each sarcomere consists of fixed actin filaments and movable myosin filaments. (c) The heads of myosin filaments bond to the actin filaments; their movement pulls together the entire sarcomere. (d) Hexagonal order of myosin (red) and actin filaments (yellow) in an edge-on view. (e) Schematics of the sarcomere structure (reproduced in part from Ref. [2] with permission of Thieme Verlag).

When activated by motor neurons, the *myosin heads* make a swinging and sliding movement along the surrounding actin fibers. All 500 myosin heads in one sarcomere must move orchestrated and in unison in one direction. All sarcomeres in one myofibril must act together, as well as all myofibrils in a bundle. The result is a contraction in the overall sarcomere length while maintaining the myosin filament extension. When we focus on just one myosin–actin pair, the sliding filament cycle can be broken down into five steps (Fig. 2.17), known as the *Lymn–Taylor cycle*[6,7] [11]:

Step 1: Hydrolysis of ATP to ADP plus inorganic phosphate (Pi), which allows binding of myosin to actin, forming a cross-bridge. The hydrolysis of ATP to ADP is described in Section 5.7.

Step 2: Pi splits off, initiating conformational changes of the myosin head, rotation of the lever arm, and a first power stroke, which results in a partial movement of myosin against actin by about 4 nm.

Step 3: The remaining ADP also splits off, which provides a second power stroke and an additional movement by about 7 nm in the same direction toward the z-disk.

---

5 ATP, adenosine triphosphate; ADP, adenosine diphosphate. More details are provided in Section 5.7.
6 Edwin W. Taylor, American biochemist and biophysicist.
7 R. W. Lymn, American biochemist and biophysicist.

Step 4:   Binding of a freshly recruited ATP molecule to the active site of the myo-
          sin head, thereby detaching myosin from actin.

Step 5:   Identical to the first one, but the myosin head is shifted by one position or
          11 nm toward the z-disk.

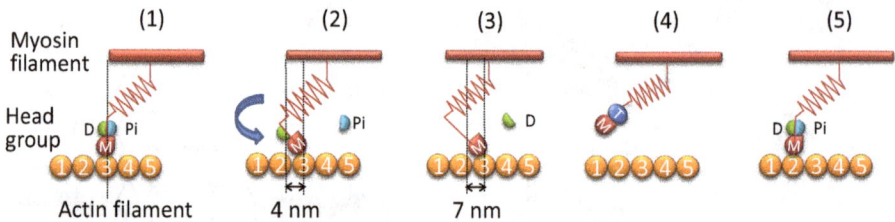

Fig. 2.17: Sliding filament cycle of sarcomeres. (1) Hydrolysis of ATP to ADP + Pi and weak binding to actin; (2) release of Pi and rotation of head group, first power stroke; (3) release of ADP and second power stroke; (4) bond break with the help of ATP; (5) hydrolysis of ATP to ADP + Pi and binding to next actin site. Short notation: M, myosin; T, ATP; D, ADP; springs indicate elastic properties of myosin head group (adapted from [15]).

On a macroscopic scale, sarcomere contraction leads to a pull on the tendons on both ends of a muscle. The total contraction is approximately 50% of the original sarcomere length. The total contraction is, however, reached in small steps. If all 500 myosin heads of one filament bind at once, they together can shorten a sarcomere by only about 1%. For a 50% contraction, a rapid, incremental binding and releasing motion is required, at least 50 times in rapid succession. If each cycle moves the sarcomere by 11 nm, after 50 steps the sarcomere is shortened by 0.5 μm. Each step requires mechanical energy of about 5 $pN \times 11$ nm $= 5.5 \times 10^{-20}$ J. ATP molecules deliver an energy of 30.5 kJ/mol or 5×10^{-20} J/ATP. Therefore, the energy consumed and the energy delivered perfectly match.

The sliding filament model of muscle contraction was first put forward by Huxley[8] and Hanson[9] [12]. Historical notes on the discovery of the sliding actin–myosin mechanism are posted in [13].

ⓘ Infobox II: X-ray diffraction from myofibrils

The highly ordered arrangement of the myofibrils in the axial direction invites detailed analysis using x-ray diffraction [14]. X-rays with a wavelength λ of about 0.1 nm are used for crystal structure analysis. The scattering of x-rays from periodic atomic structures leads to constructive and destructive interference. Condition for constructive interference of waves 1 and 2 is expressed in the Bragg equation: $2d_{hkl} \sin \theta = \lambda$. Here, d is spacing of lattice planes and hkl are indices which characterize

---

8 A. Hugh Huxley (1924–2013), British molecular biologist and biophysicist.
9 Jean Hanson (1919–1973), American zoologist.

their spatial symmetry. The myosin fibrils, when viewed edge on, show hexagonal symmetry, surrounded by actin filaments forming a honeycomb lattice. A typical d-spacing may be estimated to be about 50 nm. Bragg reflections occur then at scattering angles $\theta < 1°$. The x-ray picture to the right is taken at small angles and shows a regular interference pattern, from which the spacing of the actin filaments can be calculated. The x-ray picture is reproduced from [10] with permission.

Sarcomeres are contractile molecular motors with fixed myosin filament lengths. The sliding motion of myosin along actin filaments causes muscle contraction.

2.3.4 Muscle tension

The myosin heads are the active molecular motors that develop force when they contract. According to estimates made by optical tweezer experiments, a myosin head generates a force of approx. 1–5 pico-Newtons (pN) [16]. About 500 myosin heads are in one myosin filament and about 10^6 sarcomeres per bundle. Hence, an estimated 10^9 active actin/myosin cross-bridges per bundle jointly develop a maximum force of roughly 1 mN. The total force a muscle can develop scales with the number of bundles per cross section. For a typical bundle density, the tension (force per area) is approximately 10^4 N/m^2. This tension is not reached at the beginning or the end of the contraction but at a sarcomere contraction of about 50%, as we will see next.

Figure 2.18 shows the development of tension during fiber contraction. In the elongated state (c), there is little overlap between myosin heads and actin fibers, and therefore, the tension in this state is rather low. Highest tension is developed at point (b) when there is an optimal overlap between myosin heads and actin fibers with the highest density of cross-bridges. Further contraction as in state (a) is contraproductive since actin filaments start to overlap, losing potential cross-bridges. Under optimal tension, sarcomeres have a length of approximately 2.2–2.5 μm. Longer or shorter sarcomeres result in lower tension.

There are two types of muscle tension that we have to distinguish: isometric and isotonic. Both are highlighted in Fig. 2.19. Isometric conditions are realized when the muscle is contracted without changing its length from inactive at rest to active under tension. If we were to measure the force with a newton meter, we would notice an increasing force with no change in length. For example, imagine trying to lift a heavy rock without lifting it off the ground.

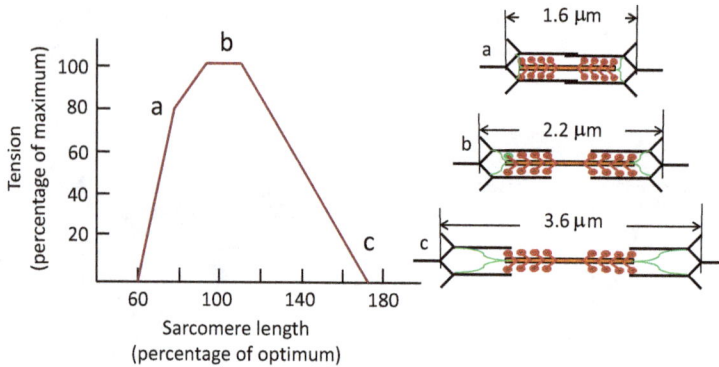

Fig. 2.18: Development of tension as a function of sarcomere length. The labels a, b, and c in the plot correspond to the sketches of the sarcomere lengths.

In contrast, during isotonic muscle contraction, the muscle shortens its length without changing its tension. This situation occurs, for instance, when lifting an object of constant weight.

Fig. 2.19: Isometric and isotonic muscle tension. Isometric action implies an increase of tension without a change of muscle length; isotonic action refers to a muscle contraction without change of tension.

Isometric contraction is important for maintaining posture, holding joints together, and holding objects in a fixed position. Isotonic contraction is vital for body movement and performing work like shuffling or bicycling. From a physical point of view, only isotonic contraction shows positive work. However, in a physiological sense, also isometric contraction consumes energy, although body movement is not involved.

We distinguish between isometric and isotonic muscle contraction. Isometric contraction is for posture, isotonic contraction for movement. !

2.3.5 Muscle activation

Skeletal muscles are activated by somatic motor neurons; i.e., those neurons that control intended movements.[10] A single axon branches out when it reaches a muscle and connects with up to 150 muscle fibers, i.e., all fibers within a bundle. Each fiber is connected to only one axon, but one axon connects to many fibers within a bundle and to fibers in different bundles. One axon junction, together with all fibers it can elicit, forms one *motor unit*. Motor neurons and muscle fibers are outlined in Fig. 2.20. Motor units are to be distinguished from action units; the latter is identical to a sarcomere. When an action potential stimulates a motor unit, all fibers connected to that particular junction contract synchronously.[11] We refer to the electrically stimulated contraction of muscles as an *electromechanical coupling*. The fibers that contract according to one motor unit are not grouped together. Instead, they are distributed over several fibers in a bundle and to fibers in different bundles.

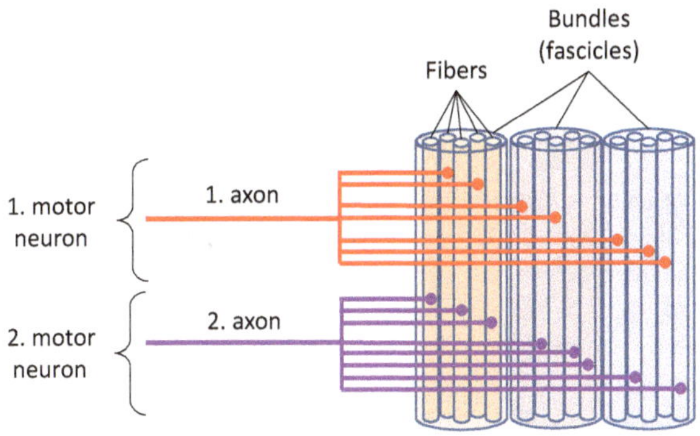

Fig. 2.20: Fibers in bundles are excited by axons. Two axons with their branches to different fibers are shown. Axons, together with their branches, form a motor unit. Each muscle fiber is innervated by only one muscle fiber. However, any motor neuron can innervate many muscle fibers.

The number of motor units per muscle depends on the required accuracy of movement. Fine motoric entails a low number of fibers per motor unit. For example, motor

---

10 For better comprehension of the content of this section, it is recommended to study this part again after reviewing Chapter 6 on signal transmission in nerves.
11 Action potentials are discussed in Chapters 5 and 6. In brief, action potential refers to the depolarization of a nerve cell for a short time period. The depolarized state can then propagate along the nerve fiber with high speed to its destination at the axon junction.

units that control eye movement connect to only 10–20 muscle fibers. In contrast, skeletal muscles responsible for large movements like the arm or leg have up to 2000 muscle fibers connected to a single motor unit.

As an example we consider the contraction of just one bundle in a muscle. The fibers in the bundle are innervated by several motor units (three in Fig. 2.21). No contraction occurs as long as the motor neuron's stimulus strength (action potential) is below the threshold level. At the threshold level, some motor units stimulate fibers to contract. As more motor units stimulate fibers to contract, tension increases up to a maximum, beyond which saturation is reached. At the same time, the diameter of the muscle belly increases.

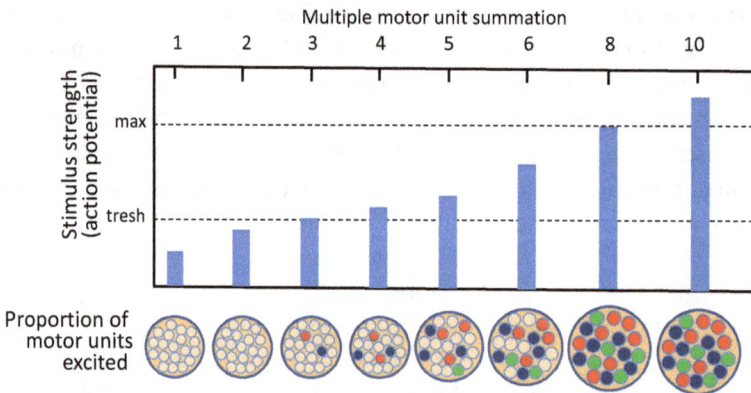

Fig. 2.21: Stimulus strength of the motor unity as a function of the number of motor units activated. The lower panel shows several muscle fibers (small circles) in a bundle (large circle). Three representative motor units innervate the fibers. When their action potential is higher than the threshold, the fibers change color to red, green or blue depending on which motor unit they connect. With increasing stimulus strength and frequency, the number of contracted fibers increases up to saturation. Simultaneously, the cross section of the fibers increases.

As we have seen, muscle fibers contract according to their stimulus. The stimulus has some strength (amplitude) and a time structure. Let us assume that the stimulus is beyond the threshold. Then the resulting contraction depends on the time sequence of the action potentials. A *twitch contraction* is a brief contraction of all the fibers within a motor unit in response to an action potential arriving at a somatic motor unit. Action potentials can arrive as a single pulse or in short pulse sequences. We first discuss a single action potential. An action potential lasts 5 ms (see Chapter 5), the reacting twitch may last between 20 and 200 ms. First, there is an inevitable delay between triggering an action potential and the beginning of the contraction, the so-called latency period (green line in Fig. 2.22). During the latency period, the action potential spreads across the sarcolemma, and Ca^{2+} ions are released from the sarcoplasmic reticulum. Next follows the contraction period, which lasts about 10–100 ms (blue line). During this time, Ca^{2+} ions trigger a chemical reaction that allows myosin heads to bind to actin filaments, otherwise hampered by a protein called troponin. The third period is the relaxation period, which lasts another 10–100 ms (purple line). During this time, Ca^{2+} ions are actively pumped back into the sarcoplasmic reticulum, myosin heads detach from the actin fibers, and the tension in the fibers is released. As already mentioned, ATP is required to break myosin–actin bonds. Fast muscle movements such as the movement of the eye have contraction and relaxation periods of only 10 ms; skeletal muscles react more slowly.

When two stimuli occur in very short time intervals, shorter than the *refractory period* of an action potential, the muscle reacts to the first but not the second. During a refractory period of about 5 ms, the muscle fiber is immune to another stimulus. Refractory periods can be up to 5 ms for skeletal muscles and up to 300 ms for the heart muscle (see Chapter 7). However, suppose a second stimulus arrives after the refractory period of the first and before relaxation. In that case, the second contraction adds to the first, and together, the contraction is stronger than any single one. This phenomenon is known as *wave summation* (see Fig. 2.22). When the stimulation rate is 20–30 Hz, the fiber can only partially relax, and the resulting undulating contraction is known as *unfused* or *incomplete tetanus*. If the stimulation frequency is increased to 80–100 Hz, the fiber will not relax, resulting in a *fused* or *complete tetanus*. Individual twitches can no longer be distinguished, and the tensile strength reaches a maximum of about 5–10 times stronger than the peak tension after a single twitch. Only skeletal muscles have the ability to tetanic contraction to increase tension. In contrast, the heart muscle cannot produce a tetanus because of the long refractory period during which the heart relaxes completely before a new contraction can occur (more details on the heart can be found in Chapter 7).

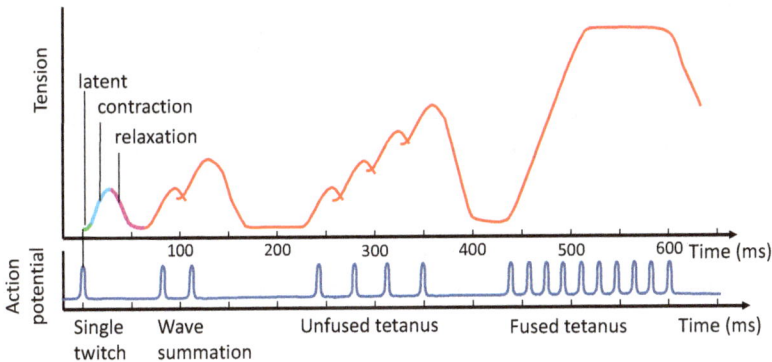

Fig. 2.22: Tension development in response to a single twitch, a double twitch, multiple but low-frequency twitches resulting in unfused tetanus, and high-frequency twitches causing a fused tetanus.

2.3.6 Coordination and feedback

Muscle and muscle contraction are only part of the story. The other part is the coordination. Any movement requires an efferent nerve signal from the central nervous system (CNS) to the muscle to contract and an inverted signal from the muscle that reports back the progress of the contraction. The back-reporting is achieved by *spindles* attached to the muscle fibers. The spindle information is sent by afferent nerve fibers to the CNS for adjustment. To illustrate this, let us consider balancing on one foot. This is definitely not an equilibrium position. It requires the other foot and arms to balance. In addition, the eyes and the vestibular organ (Chapters 11 and 12) provide important information about the position, which eventually needs to be adjusted in order to avoid dropping. The brain has to process this information and sends signals back to the muscles in order to maintain the upright position. Now we close our eyes and try to balance blindly on one foot. It is still possible, but much harder. If we could turn off the vestibular organ, balancing would not be possible at all. Brain injuries, strokes, or other degenerations such as Alzheimer's and Parkinson's also dramatically impair our ability to coordinate. Coordination efficiency may also be severely reduced by drug abuse.

2.4 Summary

S2.1 Determination of body density is important for gaining information on the proportion of fat.

S2.2 The cg of an upright standing person lies in the pelvic space region; the cg can be shifted up or down or even to an outside position upon changing the body's posture.

S2.3 Forces on bones are often much greater than the gravitational force due to muscle tension.

S2.4 Most of the levers in the body is of the third class. They do not offer a mechanical advantage, but they lend increased mobility and agility to the body.

S2.5 Muscles cause movement only via contraction of their length. The up-and-down movement of bones, therefore, requires a pair of antagonistic muscles.

S2.6 For movement of the body, six different types of joints are available that provide the body with a maximum degree of freedom.

S2.7 Muscles have a hierarchical structure of four levels: bundles, fibers, myofibrils, and myofilaments.

S2.8 Sarcomeres consist of an actin helix and myofilaments. Together they form the basic unit of a molecular motor, called the action unit.

S2.9 Contraction of sarcomeres leads to the development of tension.

S2.10 Each myosin–actin pair delivers only one pico-Newton of force. The cooperation of all pairs produces the force required.

S2.11   Isometric action means an increase in tension without a change in muscle length.

S2.12   Isotonic action refers to muscle contraction without a change in tension.

S2.13   Somatic efferent motor neurons initiate the contraction of muscle fibers.

S2.14   Afferent neurons starting at muscle spindles give information on the progress of the movement.

S2.15   A motor unit consists of a neuromuscular junction and all fibers that it can elicit.

S2.16   Following an action potential, a twitch contraction occurs in all fibers belonging to one motor unit.

S2.17   Two action potentials at time increments shorter than the refractory time will not result in additional tension.

S2.18   Two action potentials at time increments larger than the refractory time will cause a wave summation.

S2.19   Stimulation at high frequency may cause unfused or fused tetanus.

Questions

Q2.1   The density of the body is an important physical parameter. Why? Which activities depend on the density?

Q2.2   Why do bones have a higher density than muscles and fatty tissue?

Q2.3   The coronal plane divides the body into which parts?

Q2.4   The transverse plane divides the body into which parts?

Q2.5   Discuss the movement of the cg of a long jumper. What trajectory does it follow?

Q2.6   Why do most levers in the body have a mechanical advantage lower than 1?

Q2.7   Identify levers in the body that have a mechanical advantage higher than 1.

Q2.8   It was determined that the maximum force that the biceps can exert is about 2600 N. Why is it then not possible to lift a weight of 2600 N with one hand?

Q2.9   What is the reason for permanent plastic deformation of the femur that changes the CCD angle and increases the torque?

Q2.10   Give examples for ball-and-socket joints in the body.

Q2.11   How can the joint of the knee be characterized?

Q2.12   How many phases can you identify for normal walking?

Q2.13   Name the antagonistic muscles which are responsible for bending the knee.

Q2.14   How many hierarchy levels does a muscle have? Name them.

Q2.15   On which hierarchical level does muscle contraction take place?

Q2.16   How is muscle contraction achieved?

Q2.17   Does the force that a muscle can produce depend on its length or its cross section?

Q2.18   What is the difference between isometric and isotonic tension?

Q2.19   How are muscles activated to contract?

Q2.20   Which ion exchange is important for the contraction of muscles? What is required for its release?

Q2.21   What do you understand a twitch to be?

Q2.22   What is fused or unfused tetanus?

Q2.23   Is the formation of tetanus possible for the myocardium?

Attained competence checker + 0 −

| I know how to determine the density of a human body. |
| I can locate the body's center of gravity. |
| I can distinguish between three classes of levers. |
| I can identify movements of the body with classes of levers. |
| I appreciate that muscles only contract. |
| I realize that for up and down movement, a pair of antagonistic muscles is required. |
| I can distinguish between different types of muscles in the body. |
| The hierarchy of skeletal muscle structure is known to me, and I can identify the four different levels. |
| I know that actin filaments and myosin filaments together form a molecular motor called a sarcomere. |
| I know when the highest tension of a sarcomere is achieved. |
| I can distinguish between isometric and isotonic muscle contraction. |
| I know that muscle strength can be increased by a fused tetanus. |

Suggestions for home experiments

H2.1 Try to flex your body as shown in Fig. 2.11 and measure or estimate your maximum flexion angle for specific movements.

H2.2 Try to lift a heavy object (rock) without lifting it off the ground. Watch your muscle belly increase in diameter.

H2.3 Lean against a wall with your arm straight. How long can you hold your arm? What kind of muscle tension is required to hold your posture?

H2.4 Find out how muscles work in insects.

ℹ️ **Exercises**

E2.1 **Body density:** Show that eq. (2.1) holds.

E2.2 **Proportion of fat:** How can you determine the proportion of fat in the body by simple methods?

E2.3 **Force on the lumbar vertebra:** Determine the force on the lumbar vertebra for the position indicated in Fig. 2.5. Assume that the persons' weight is 700 N holding a weight of 100 N in hand. This extra weight projects through point D. Compare the force on the lumbar vertebra with and without extra weight.

E2.4 **Torque at forearm:** Referring to Fig. 2.7(a), what is the torque if the forearm is not at the right angle but bent down to an angle of 120°?

E2.5 **Standing on the tiptoe:** What is the muscle force required for lifting up yourself on the tiptoe? What type of lever is this? Assume an angle of the foot against the floor of 15°, a foot length 0.25 m, and a body mass of 70 kg. Which tendon and muscle is responsible for lifting you up?

E2.6 **Lifting speed:** Daily experience tells us that we can lift a single apple faster than a sack of apples. Thus weight and speed of contraction are inversely related. This relation becomes clear on a microscopic level. Explain this observation on a microscopic level.

E2.7 **Myosin–actin pairs:** Determine the number of myosin–actin pairs to hold up a weight of 10 N in the hand.

References

[1]   Levangie PK, Norkin CC. Joint structure and function. 5th edition. Philadelphia: Davis; 2011.

[2]   Faller A, Schuenke M. The human body. An introduction to structure and function. Stuttgart: Georg Thieme Verlag KG; 2004.

[3]   Zabel H. Kurzlehrbuch Physik. Stuttgart: Georg Thieme Verlag KG: 2016.

[4]   Mann RA, Hagy J. Biomechanics of walking. Am J Sports Med. 1980; 8: 345–350.

[5]   Shelburne KB, Torry MR, Pandy MG. Muscle, ligament, and joint-contact forces at the knee during walking. Med Sci Sports Exerc. 2005; 37: 1948–1956.

[6]   Damm P, Graichen F, Rohlmann A, Bender A, Bergmann G. Total hip joint prosthesis for in vivo measurement of forces and moments. Med Eng Phys. 2010; 32: 95–100.

[7]   Janssen I, Heymsfield SB, Wang ZM, Ross R. Skeletal muscle mass and distribution in 468 men and women aged 18–88 yr. J Appl Physiol. 2000; 89: 81–88.

[8]   Seeley R, Vanputte C, Russo A. Seeley's anatomy and physiology. Boston: McGraw Hill Book Co; 2016.

[9]   Geeves MA, Holmes KC. The molecular mechanism of muscle contraction. Advances in protein chemistry. 2005; 71: 161–193.

[10] Huxley HE, Brown W. The low-angle x-ray diagram of vertebrate striated muscle and its behavior during contraction and rigor. J Mol Biol. 1967; 30:383–434.

[11] Lymn R, Taylor E. Mechanism of adenosine triphosphate hydrolysis by actomyosin Biochemistry. 1971; 10: 4617–4624

[12] Huxley HE. The mechanism of muscular contraction. Science 1969; 164:1356–1365.

[13] Hitchcock-degregori SE, Irving TC. Hugh E. Huxley: The compleat biophysicist. Biophys J. 2014; 107: 1493–1501.

[14] Reconditi M. Recent improvements in small angle x-ray diffraction for the study of muscle physiology. Rep Prog Phys. 2006; 69: 2709–2759.

[15] Caruel M, Lev Truskinovsky L. Physics of muscle contraction. Rep Prog Phys, IOP Publishing. 2018; 81: 036602.

[16] Mehta AD, Rief M, Spudich JA, Smith DA, Simmons RM. Single-molecule biomechanics with optical methods. Science. 1999; 283: 1689.

Further reading

Herman IP. Physics of the human body. Berlin, Heidelberg: Springer; 2008.

Guyton AC, Hall JE. Textbook of medical physiology. 11th edition. Philadelphia, Pennsylvania, USA: Elsevier Inc., Elsevier Saunders; 2006.

Tortora GJ, Derrickson B. Principles of anatomy and physiology. 14th edition. Hoboken (New Jersey): John Wiley & Sons; 2015.

Martini FH, Nath J, Bartholomew EF. Essentials of anatomy and physiology. 7th edition. New York: Pearson; 2017.

Marieb EN, Hoehn KN. Human anatomy and physiology. 9th edition. New York: Pearson; 2013.

Seeley R, Vanputte C, Russo A. Seeley's anatomy and physiology. Boston: McGraw Hill Book Co.; 2016.

Pape H-C, Kurtz A, Silbernagel S, eds. Physiologie. 7th edition. Stuttgart, New York: Thieme Verlag; 2014.

Scanlon VC, Sanders T. Essentials of anatomy and physiology. 5th edition. Philadelphia: Davies Company FA; 2007. https://openstax.org/details/books/anatomy-and-physiology

3 Elastomechanics: beams, bones, and fractures

Physical properties of bones	
Number of bones at adulthood	206
Density of cortical bone	1.9 g/cm^3
Density of trabecular bone	0.43 g/cm^3
Hierarchy levels of bones	7
Types of cells	4
Porosity of cortical bone	0.05–0.1
Porosity of trabecular bone	0.75–0.95
Young's modulus for bone compression	15–34 GPa
Yield stress for bone compression	165–210 MPa

3.1 Introduction

Soft contractile muscles move hard bones for mobility. This fundamental principle applies to the human body and all vertebrates. In a way, muscles and bones work together like the ropes and boom arms of a crane. Both perform mechanical work with the help of levers. However, hard bones have many more roles. They give the body rigidity and stability and protect important organs such as the central nervous system. Muscle and bone are two examples of biomaterials with very different mechanical properties. In this chapter, we aim to answer some of the most obvious questions: why are bones hard? Can bones be bent? When do bones break? How and why do bones age? Bones have a rich hierarchical structure from the nanoscale to the macroscale. Describing bones at all levels is a complex, multidisciplinary task with contributions from mechanical engineering, materials science, crystallography, and biochemistry. We start by reviewing some basic elastic, plastic, and viscoelastic properties of materials before analyzing the different levels of bone structure.

3.2 Elastic deformation

3.2.1 Strain and stress

Any solid body can be deformed by a pair of forces $\vec{F}_1$ and $\vec{F}_2$ acting on opposite surfaces and in opposite directions. In such a situation, the body cannot move since the resulting force cancels: $\vec{F}_1 + \vec{F}_2 = 0$. For the following it is sufficient to consider the magnitude of the force component $F = |\vec{F}_1| = |\vec{F}_2|$ acting perpendicular to a surface area A. The ratio of the force F per surface area A yields a pressure $p = F/A$, also called stress or tension:

https://doi.org/10.1515/9783110756951-003

$$\sigma = F/A. \tag{3.1}$$

If the stress is exerted in only one direction, whereas all other body surfaces are free, the stress is called *uniaxial*. Depending on the direction of forces, the uniaxial stress can be either tensile or compressive (see Fig. 3.1(a) and (b)). If all body surfaces experience the same pressure (stress), the acting pressure is called *hydrostatic* (Fig. 3.1(c)).[1] Hydrostatic pressure plays a decisive role in many body parts, such as the cardiovascular system, respiration, kidneys, and eyeballs.

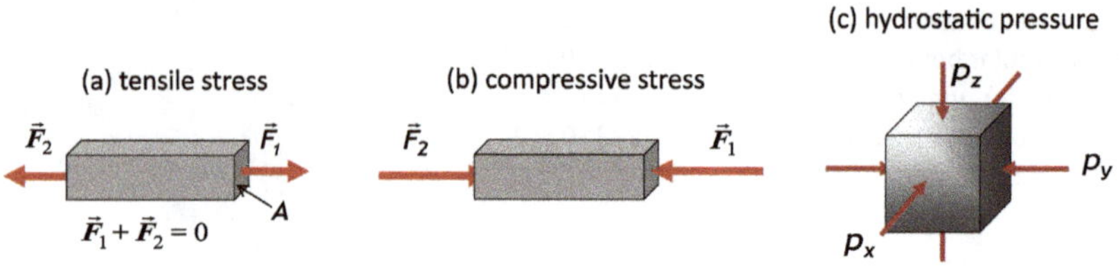

Fig. 3.1: (a) and (b) Tensile stress and compressive stress are due to pairs of forces acting on opposite surfaces of area *A* in one spatial direction; and (c) hydrostatic pressure is due to pairs of forces acting on surfaces in all three spatial directions.

If an extended solid body is under uniaxial external pressure (*stress*), it will react by changing its length *l*. The resulting deformation is called *strain*. Strain ε is measured in terms of a relative length change:

$$\varepsilon = \Delta l/l, \tag{3.2}$$

where *l* is the original length of the body. ε is usually expressed in percent.

3.2.2 Hooke's law

Tensile stress results in elongation, *compressive stress* causes contraction. For small deformations, the relation between stress σ and strain ε is linear, which is known as *Hooke's*[2] *law* (Fig. 3.2):

$$\sigma = Y\,\varepsilon. \tag{3.3}$$

---

1 The term "hydrostatic" refers to the fact that water can transmit pressure homogeneously to all sides of a body when exerted by a piston from only one side. However, hydrostatics is not limited to water as a fluid.
2 Robert Hooke (1835.1703), English scientist and architect.

The proportionality constant Y is the *elastic modulus*, also known as *Young's*[3] *modulus*.[4] The elastic modulus Y characterizes a material as elastically soft or hard, easy to deform like rubber, or hard to deform like steel. The stress $\sigma = F/A$ has the unit $[\sigma] = N/m^2 = $ Pascal:[5]

$$1 \text{ N/m}^2 = 1 \text{ Pascal (Pa)} = 1 \text{ kg}/\text{ms}^2,$$

$$1 \text{ bar} = 10^5 \text{ Pa} = 100 \text{ kPa} = 1000 \text{ hPa}.$$

The strain ε is dimensionless. Therefore, the unit of σ and the unit of Y are identical. Typical values of Y range from 0.5 GPa for rubber to 200 GPa for steel. Some elastic moduli are listed in Tab. 3.1.

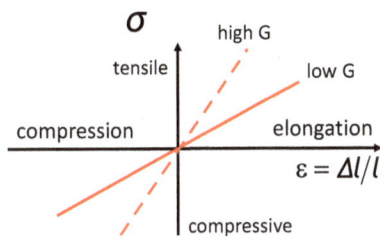

Fig. 3.2: Solid red line: Hooke's law for materials with low elastic modulus like wood or bones; dashed red line: for materials with a large elastic modulus like steel.

Hooke's law is explicitly limited to the *elastic deformation* of materials. This limitation implies three fundamental properties: first, the relationship is linear without offset, i.e., for $\sigma = 0$ follows $\varepsilon = 0$. Second, the deformation is completely reversible, i.e., any release of stress results in a fully reversible strain without memory effect. Third, the strain relaxation is instantaneous, i.e., there is no time lapse between a stress release and strain relaxation. Elastic deformation is clearly distinguished from plastic deformation, the latter features irreversible deformations, as discussed further.

Tab. 3.1: Elastic moduli of some selected materials.

Material	Y (GPa)
Aluminum	70
Steel	200
Wood	13
Bones	15
Rubber	0.5

3 Thomas Young (1773–1829), English physicist, physiologist, and language analyst.
4 The linear relationship expressed in Hooke's law is a simplification. Actually, the stress and the strain have tensor properties, as the reactive strain often has components not parallel to the stress. Here we neglect the tensor property, because it is not important for the discussion of the elastic properties of bones.
5 Blaise Pascal (1623–1662), French mathematician, physicist, philosopher, and theologian.

> ⚠ The elastic deformation of solids implies that the strain is linearly proportional to the applied stress. After stress relief, the strain is fully restored.

In addition to the length change parallel to the applied stress, solid bodies react by changing their thickness t in two perpendicular directions. For instance, tensile stress causes an elongation $\Delta l/l$ in the stress direction and a contraction $\Delta t/t$ in the two directions perpendicular to the stress line. The ratio:

$$\mu = \frac{\Delta t/t}{\Delta l/l} \tag{3.4}$$

is called the *Poisson*[6] *ratio*. The total volume change upon uniaxial stress is then

$$\frac{\Delta V}{V} = \frac{\sigma}{Y}(1-2\mu). \tag{3.5}$$

The factor of 2 is due to the two perpendicular directions to the applied stress. The volume change is zero for $\mu = 0.5$. Typical μ-values range from 0.2 to 0.4. As the Poisson contraction does not play a role for the human body, we will neglect this effect further on.

3.2.3 Shear deformation

Other forms of elastic deformation occur by applying tangential forces at opposite sides of a body that change the angles of the body but not its volume. The tangential force is the force component projected into the surface area A. Pairs of forces act such that the body does not gain angular momentum. We differentiate between shear deformation, torsional deformation, and bending deformation. These three types are sketched in Fig. 3.3. In simple terms, the elastic shear deformation is expressed by a linear equation, which has the same form as Hooke's law for uniaxial stress:

$$\tau = G \cdot \alpha, \tag{3.6}$$

where $\tau = F/A$ is the shear stress and α is the shear angle. The proportionality constant G is the *shear modulus*. Shear is particularly important for bone fracture. When *bending* a beam, as illustrated in Fig. 3.3(c), there is a strain gradient from top to bottom, from tension to compression, separated by a neutral axis (plane) where the strain is zero. The strain as a function of distance z above and below the neutral line is

$$\varepsilon_z = \frac{z}{R}, \tag{3.7}$$

---

6 Siméon Denis Poisson (1781–1840), French mathematician, physicist, and statistician.

where R is the bending radius of the beam, and the stress is accordingly:

$$\sigma_z = Y\varepsilon_z = Y\frac{z}{R}. \tag{3.8}$$

Through bending, the *elastic stiffness* k of a material can be tested. The stiffness is defined as the ratio of force F applied to the end of the beam, and the deflection δ:

$$k = F/\delta. \tag{3.9}$$

A high k-value indicates a stiff material. Note the difference between stiffness k and elastic modulus Y: Y is an intrinsic material property, while k depends on the geometry of the beam and the boundary condition imposed.

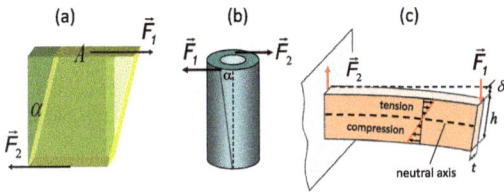

Fig. 3.3: Elastic deformations by tangential forces: (a) shear; (b) torsion; and (c) bending. Note that pairs of shear forces are applied parallel to a surface in all three cases. By bending of a beam, the counterforce F_2 is exerted by a fixture.

The three elastic constants E, G, and μ are not independent but connected via the equation that holds for isotropic bodies:

$$G = \frac{Y}{1+\mu}. \tag{3.10}$$

Since $\mu \geq 0$, it follows that $G < Y$. Therefore we conclude that shear deformation is more likely to occur than compression, eventually resulting in fracture, as discussed further on.

3.3 Elastic properties of beams

Bones have a hard shell and an open pore structure in the interior, making them lightweight without losing elastic stiffness. Long bones resemble hollow cylindrical-shaped beams. We can appreciate the mechanical advantage of hollow beams by considering the torque required for bending a beam:

$$\vec{T} = \vec{L} \times \vec{F}. \tag{3.11}$$

Here L is the length and F is the applied tangential force at the end of a clamped beam on one side and free on the other side, as indicated in Fig. 3.3(c). The torque can also be expressed in scalar form and in terms of a *moment of resistance M* times the bending stress σ_z:

$$T = M\sigma_z. \tag{3.12}$$

The moment of resistance about the neutral axis is defined by the integral [1]:

$$M = \frac{1}{h} \int_{-h/2}^{+h/2} z^2 dA, \tag{3.13}$$

where z is the distance from the neutral line, $dA = tdz$, t is the beam thickness, h is the beam height, and A is the cross-sectional area. M depends on the geometry of the beam, and for a rectangular beam (Fig. 3.4(a)), we obtain by integration:

$$M = \frac{1}{12} th^2. \tag{3.14}$$

This equation tells us that edgewise beams have higher bending resistance than sideways beams because M increases quadratically with h but only linearly with t. This property is used in civil engineering to construct, for instance, houses and bridges, and it can easily be confirmed by bending a plastic ruler edgewise and sideways. If we consider a cylindrical beam (Fig. 3.4(b)) of the same cross-sectional area instead and with a radius $R = h$, the moment of resistance is even higher than for the rectangular beam:

$$M = \frac{\pi}{2} R^3. \tag{3.15}$$

Finally, we consider a hollow cylinder of an outer radius R and an inner radius r (Fig. 3.4(c)). For not too thin shells, the approximate moment of resistance is

$$M = \frac{\pi}{2} \frac{R^4 - r^4}{R} \approx 2\pi R^2 \Delta R. \tag{3.16}$$

This equation shows that a hollow cylinder has a bending moment of resistance similar to a bulk cylinder, with the decisive advantage that the hollow cylinder is much lighter. Long bones can be viewed as hollow cylinders using exactly this mechanical advantage. Also, the pore structure in the trabecular part of bones makes them lightweight without compromising elastic stiffness. Moreover, long bones such as the femur have to withstand compressive stress at both ends. Hollow cylinders are just as suitable for this task as solid cylinders.

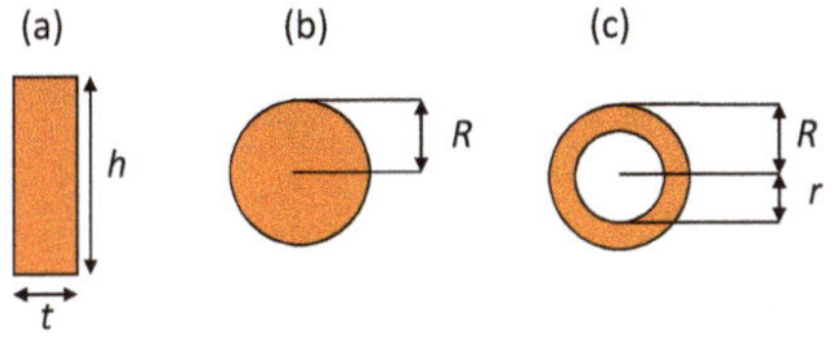

(a)   (b)   (c)

Fig. 3.4: Moment of resistance is calculated for rectangular beams (a), full cylindrical beams (b), and hollow cylindrical beams (c).

> Long bones can be viewed as hollow cylinders, being lightweight without compromising bending resistance. **!**

3.4 Plastic deformation and fracture

As long as the stress does not exceed a critical value, the strain is completely reversible, i.e., the deformation relaxes after releasing the stress. This is the validity range of Hooke's law from $\sigma = 0$ up to the yield point at $\sigma = \sigma_Y$. The yield stress σ_Y at the yield point defines the *strength* of the material.

Elastic strain changes the distance between atoms by a tiny amount but does not change the atoms' arrangement in materials. However, beyond a critical stress value known as yield stress, ductile materials start to deform by plastic flow, i.e., atoms move to new locations, and dislocations occur (Fig. 3.5). Dislocations enable entire crystal planes to move against one another. Then elastic strain is no longer linearly related to the stress, and the deformation is no more reversible: after removing the stress, a residual strain remains (dashed line in Fig. 3.5). If we want to remove the residual plastic deformation, opposite stress is required to bring the material back to its original shape. The possibility to plastically deform materials such as metal sheets is immensely important for the fabrication of all kinds of goods such as pots and pans, automobiles, trains, and aircraft.

Now we follow the applied stress along the red line in Fig. 3.5 into the light blue area. Here, progressive deformation leads to work-hardening of metals, i.e., dislocations become entangled and hinder more plastic flow. Work-hardening is very important for metals to be used as construction materials or tools since defect-free metals are usually too soft for practical applications.

Note that the yield stress σ_Y is lower than the maximum stress σ_{max} that develops after work hardening. A further rise in stress beyond the work hardening range of ductile materials eventually leads to fracture. Crack initiation, crack propagation, and fracture result from material thinning via elongation or material fatigue from repeated load.

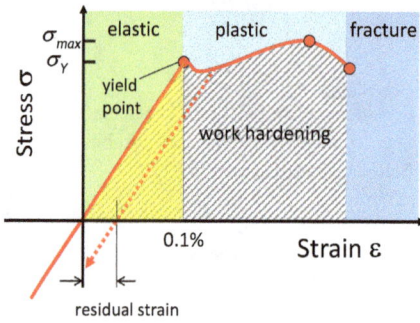

Fig. 3.5: Stress–strain relationship is shown beyond the elastic regime into the plastic deformation, ending by fracture. The yield point is the maximum stress without plastic deformation. The dashed red line indicates the irreversibility of strain after the release of stress beyond the yield point. The yellow-shaded area defines the resilience, and the white-shaded area the toughness of materials.

Ductility is the ability of a solid material to deform under tensile or compressive stress. An essential prerequisite of ductile deformation is a charge distribution, which does not change when atomic planes in the material start to flow under the action of stress. This is indeed the case for metals, but not for materials with ionic or covalent bonds. The latter materials turn out to be *brittle*. In brittle materials, stress exceeding the yield point results in immediate fracture. For many materials, the yield stress is reached at a strain of 0.1%. We experience the difference between ductile and brittle materials by dropping cups of different materials on a hard floor: the metal cup may eventually deform but not break, the porcelain cup will not deform but shatter into pieces. Bones are not ductile materials, fortunately, because any impact would result in a permanent bump, and during our lifetime, we would collect quite a number of bumps. Bones tend to be brittle. But elastic and plastic properties of bones are much more complex because they are composite and inhomogeneous materials, as we will see in Sections 3.5 and 3.6.

A material's *resilience* is defined as the energy density (energy per volume, E/V) stored in materials under load up to the yield point. Accordingly, the resilience is given by the integral:

$$\left(\frac{E}{V}\right)_{\text{resil}} = \int_{0}^{\varepsilon_Y} \sigma d\varepsilon, \qquad (3.17)$$

and indicated by the yellow-shaded area in Fig. 3.5. In contrast, the *toughness* of a material is defined as the energy density stored in materials upon plastic deformation up to the point of fracture and is given by the integral:

$$\left(\frac{E}{V}\right)_{\text{tough}} = \int_{\varepsilon_{\text{Y}}}^{\varepsilon_{\text{frac}}} \sigma d\varepsilon. \tag{3.18}$$

The toughness integral corresponds to the light gray-shaded area in Fig. 3.5. The bigger the area, the tougher the material. Ductile materials are tough, since they can absorb large amounts of strain energy. Brittle materials show resilience but no toughness. They break already at the yield point. On the other hand, brittle materials may have a higher *strength* because they resist high stress (yield point) before breaking. In contrast, ductile materials tend to have lower strength, while the strength increases after work-hardening.

> Ductile materials are plastically deformable beyond the yield point. Brittle materials crack at the yield point. They have resilience but no toughness. !

3.5 Viscoelastic materials

Biomaterials such as bones often show viscoelastic properties that differ from elastic and plastic behavior. As we have seen for the Hooke's regime, elastic materials respond to tension with immediate strain. Conversely, the stress relief causes an immediate stress relaxation along the same path in the stress–strain diagram. Plastically deformed materials return along paths different from the initial strain curve. The result is a hysteretic stress–strain curve as shown in Fig. 3.6(a). The area enclosed by the hysteresis corresponds to the stored deformation energy. Viscoelastic materials combine properties of elastic and plastic materials. When applying stress, the ascending load curve (red in Fig. 3.6(b)) differs from the descending unloading curve (blue in Fig. 3.6(b)). The unloading curve usually returns to the starting point, but generally with a certain time lag. At the end, we describe a hysteretic closed loop. The loop area again corresponds to the energy density that is required to run through the loop. Consequently, viscoelastic materials warm up by cycling.

Moreover, the viscoelastic response depends on the frequency of stress pulses, the amplitude, and the total number of cycles. At low rates, the reaction is quasi-static; at high rates, the strain cannot follow the applied stress because the diffusion processes evoked have their intrinsic relaxation time constants. Therefore, the initial slope of the ascending curve, which provides the Young's modulus, is not a constant but depends on the stress rate.

A typical elastic response of a viscoelastic material is shown in Fig. 3.7. Stress and strain are plotted as a function of time. The panel (a) shows the stress applied with a constant amplitude σ_0 for some time between t_0 and t_1. The viscoelastic

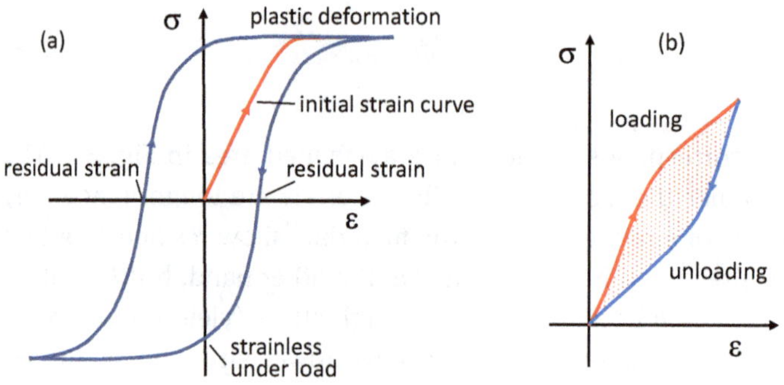

Fig. 3.6: Stress–strain hysteresis for (a) ductile materials and (b) viscoelastic materials.

material reacts to the stress (panel b) by an instantaneous strain with amplitude ε_0, followed by a slow exponential strain increase up to the maximum strain ε_{max}.

After releasing the stress at t_1, the strain reacts in reverse order: an immediate strain relaxation by the same amplitude ε_0, followed by a slow exponential strain relaxation.

The time dependence of the strain indicates that the viscoelastic material is composed of at least two components: one, which reacts on stress like an elastic solid, and another one which is controlled by diffusion, viscous flow, or creep that has its intrinsic time constant τ, determining the exponential strain relaxation. Given enough time ($|t_1 - t_0| \gg \tau$), the system will reach the equilibrium state at a strain $\varepsilon = \varepsilon_{max}$. However, for shorter times, equilibrium is not reached. After repeated stress applications for shorter time periods, as indicated in panel (c), the response is always incomplete. But with more cycles, the system will eventually reach saturation (panel d). This graph also shows that with increasing stress rate, the strength apparently increases: by fast impact, the material appears stiffer than it is.

Viscoelastic properties can be modeled by mechanical equivalents: springs for the elastic part and a dissipative element (dashpot) for the damping part. These elements may be arranged in series (Maxwell[7] model, Fig. 3.8(a)), in parallel (Kelvin[8]–Voigt[9] model, Fig. 3.8(b)), or in combinations (standard linear model, Fig. 3.8(c)). The viscoelastic behavior displayed in Fig. 3.7 is best represented by the Kelvin–Voigt model with a spring and a dashpot in parallel. The first-order differential equation is an adequate description of the stress response:

$$\sigma = Y\varepsilon + \eta\dot{\varepsilon}. \tag{3.19}$$

---

7 James Clerk Maxwell (1831–1879), Scottish mathematician and theoretical physicist.
8 William Thomson Kelvin (1824–1907), British physicist and thermodynamicist.
9 Woldemar Voigt (1850–1919), German physicist.

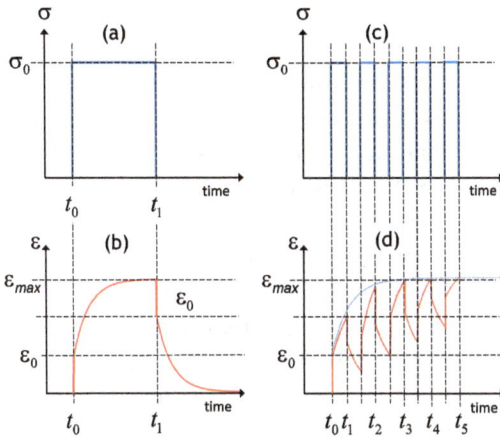

Fig. 3.7: Stress–strain behavior as function of time of a viscoelastic material. (a) Application of stress for a time period t larger than the intrinsic relaxation time τ. (b) Strain response with two components: immediate and exponential relaxation. (c) Repeated stress application for time periods shorter than the intrinsic relaxation time τ. (d) The strain response is always incomplete. The system eventually reaches a maximum strain state.

The first term is the well-known elastic Hooke response (eq. (3.3)); the second term describes the viscous behavior with a viscosity coefficient η. The solution for the time-dependent strain has the form:

$$\varepsilon(t) = \varepsilon_0 + \Delta\varepsilon(1 - e^{-t/\tau}).$$ (3.20)

Here ε_0 is the time-independent strain state and $\Delta\varepsilon$ is the amplitude of the viscoelastic response.

Concerning bones, all three mechanical models turn out to be rather poor descriptions of the bones' complex mechanical properties. Nevertheless, they are useful for discussing partial aspects of bone's viscoelastic behavior [2].

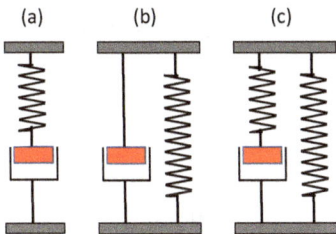

Fig. 3.8: Three mechanical models of viscoelasticity consisting of a spring and a dissipative element indicated by a dashpot in red: (a) Maxwell model; (b) Kelvin–Voigt model; (c) standard linear model.

3.6 Structure of bones

In the body, we have some 270 bones when born. After fusing, some 206 distinct bones are left in adulthood [3]. Those 206 bones are of very different sizes and shapes. We distinguish long and hollow bones for the extremities; short bones for hands and feet; irregularly shaped bones for the spine and knee; flat bones for the skull, shoulder blade (scapula), ribs, and breastbone (sternum). All bones share a hard shell (compact or cortical shell) and a spongy bone interior (trabecular bone). The hardness of bones makes them good fossils that last for millions of years. Figure 3.9 shows an example of a long bone (femur) featuring the characteristic bone structure of a hard cortical shell and a spongy trabecular interior. But the femur has, in addition, a medullary cavity, which is characteristic only for long bones. Bones have many tasks. Bones protect vital organs like the heart and the lung. They lend stability and mobility to the body via joints between bones and attachment points for tendons of muscles and ligaments. And they act as production and storage centers. They produce blood cells in the bone marrow and store minerals and lipids released on demand.

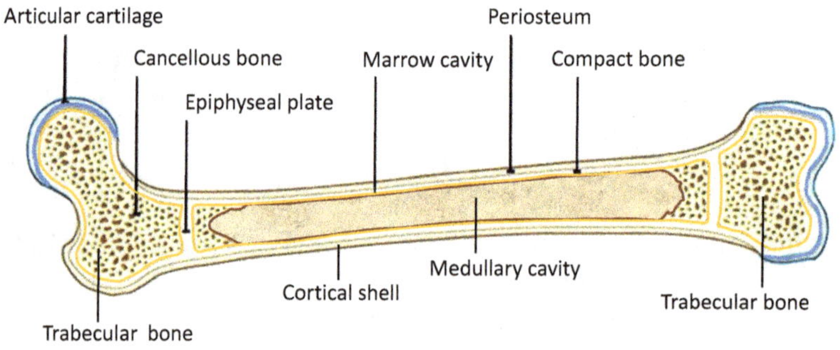

Fig. 3.9: Structure of a long bone (femur).

> **!** Bones have a heterogeneous structure: a hard cortical shell envelops a spongy trabecular interior.

3.6.1 Hierarchical structure of bones

Bones are organized in a hierarchical order similar to the muscle tissue discussed in Chapter 2. Up to seven levels are distinguishable from the molecular level up to the macroscopic scale [4–6].

On the molecular level, we find collagen and minerals. Collagen is an elongated fiber-like protein that serves as connective tissue in most parts of the body, delivering strength and protection. Collagen fibers are shown schematically in Fig. 3.10.

The fundamental structural unit is a right-handed triple helix consisting of three coiled and intertwined polymer chains. The chains are held together by hydrogen bonds. The collagen triple helix is 300 nm long and only 1.5 nm wide.

Fig. 3.10: Right-handed triple helix structure of collagen. The individual strands are differently colored (reproduced from http://www.rcsb.org/pdb/home/home.do).

Thousands of collagen fibers are packed laterally and one on top the other to form cylindrical fibrils. Fibrils have a 25–500 nm diameter, depending on the collagen type and the number of fibers (Fig. 3.13). The ends of adjacent collagen fibers are shifted by a distance of 67 nm, producing a stairlike appearance visible in electron micrographs. In bones, the gaps between the fibers are filled by nanocrystalline minerals of calcium hydroxylapatite ($Ca_5(PO_4)_3OH$)) (short notation: HA). The crystal structure is shown in Fig. 3.11. HA is also known as bone mineral. Bones contain HA up to 50% by volume and 70% by weight [7]. Note that HA only contains atoms that are abundant in the body. In particular, the phosphate group PO_4^{3-} is also present in ATP, in phospholipid membranes, and in the genetic materials DNA and RNA.

Collagen fibers and HA minerals together form organic–inorganic composite biomaterials that constitute the building blocks of fibrils in the cortical bone structure. The collagen fibers provide bending resistance (eq. (3.16)) to bones, whereas the minerals give the bones hardness[10] as well as a high elastic modulus (eq. (3.3)) for the compressional load. The organic collagen material lends bones its flexibility, while the inorganic HA material is responsible for its resilience (eq. (3.17)). In other words, collagen prevents HA from brittle cracking, while HA prevents collagen from yielding. The right mix is essential. Figure 3.12 shows strikingly what happens when one or the other component is not optimized. Without HA minerals, bones bend; without collagen, bones shatter.

Together the mineralized collagen fibers show viscoelastic properties, which are considered in more detail later. Nanocrystals of HA are also present in dental enamel.

---

10 Hardness is a measure of resistance to localized plastic deformation induced by, for instance, mechanical indentation with a sharp pin.

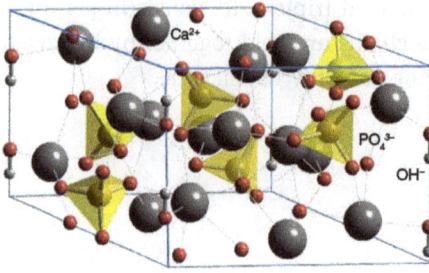

Fig. 3.11: Crystal structure of naturally occurring calcium hydroxylapatite ($Ca_5(PO_4)_3OH$), short HA. In bones nonstoichiometric and calcium-deficient forms are found, which form plate-like crystallites (reproduced from http://www.chemtube3d.com/images/craigimages/CraigMichael/i624fg51.png © creative commons).

Furthermore, HA can be injected into the skin for correcting fold depressions. HA as a biocompatible material is also crucial for the design of nanoparticles (see Vol. 3).

> **!** Bones are composite materials combining soft collagen fibers and hard nanocrystals.

Fig. 3.12: A long bone (a) is treated such that either the HA minerals are dissolved, leaving a flexible bone in panel (b), or the collagen fiber density is reduced, resulting in a cracked and shattered material after impact in panel (c).

Figure 3.13 gives an overview of the hierarchical architecture of bones. Collagen fibers intertwine and form a triple helix structure. As already mentioned, the helical structure leaves gaps that are filled with HA nanocrystals. Many fibers combine to collagen fibrils. In turn, the collagen fibrils are the building blocks for the next level in the hierarchical structure that constitutes cylindrical lamellas forming *osteons* in the compact part of bones. Osteons have a diameter of about 100 µm. The osteons are cemented together and form highly regular structures in the cortical part of bones. Each cylindrical osteon has in its center a Haversian canal that contains blood vessels and nerve fibers. The longitudinal Haversian[11] canals are interconnected by transverse Volkmann[12] canals, which link different osteons. The blood vessels carry

---

11 Clopton Havers (1657–1702), English physician and anatomist.
12 Alfred Wilhelm Volkmann (1801–1877), German physiologist and anatomist.

away and distribute newly generated blood cells from the bone marrow and minerals from the osteons.

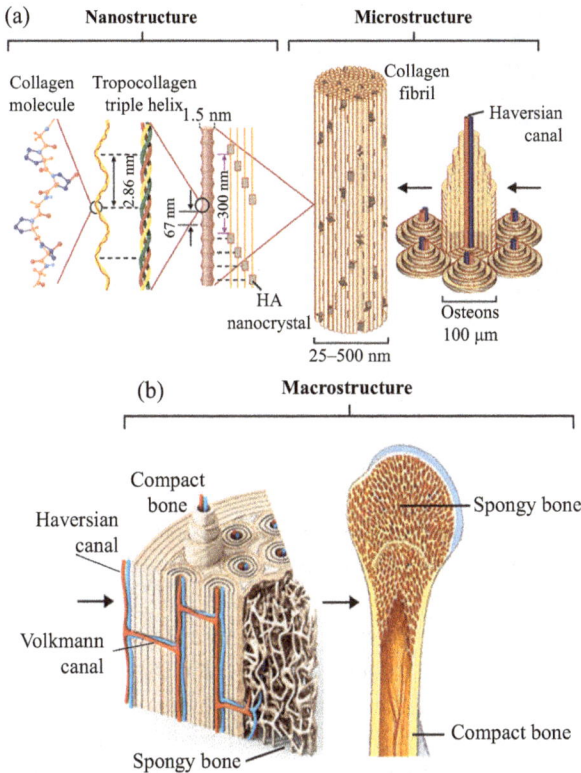

Fig. 3.13: Hierarchical structure of compact bones from nanostructural level (panel a) to the macroscopic level (panel b) (reproduced from [5, 6] with permission of Macmillan Publishers Limited).

3.6.2 Cellular structure of bones

In bones, four types of cells can be distinguished: osteoblast, osteoclast, osteocytes, and osteogenic cells. Osteogenic cells are stem cells that develop osteoblast cells. Osteoblast cells are responsible for the construction of bone material. They segregate calcium phosphate and calcium carbonate, which crystallize in a watery environment along the collagen fibrils and form HA nanocrystals. Upon secretion, the osteoblast cells become trapped in pores of the matrix called lacunae. Here they transform themselves into osteocyte cells that no longer can divide (red ellipses in Fig. 3.14). The tissue hardens, constituting the typical bone structure. A transverse top view of this structure is shown in Fig. 3.14. The osteocyte cells are interconnected by channels called canaliculi (blue lines). They transport minerals across the bone and sense any damage due to fracture.

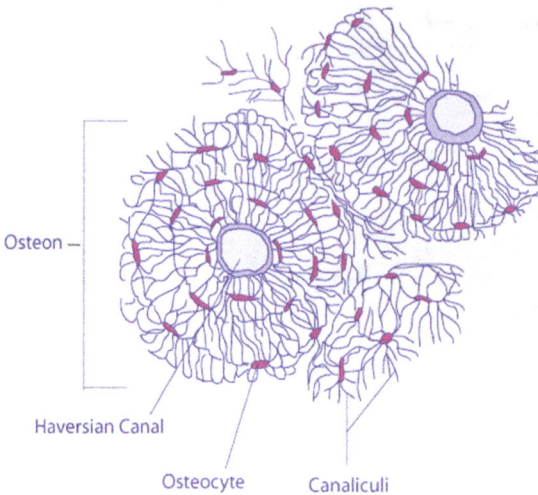

Fig. 3.14: Cross section of a long bone, showing the bone structure including osteocytes (from https://en.wikipedia.org/wiki/File:Transverse_Section_Of_Bone.png © creative commons).

The antagonist of osteoblast cells are osteoclast cells. The latter ones eliminate old bone structures to be replaced by new ones from osteoblast cells. Osteoclast cells also deliver Ca-ions to the body when needed, for instance, in muscles.

There is a dynamic equilibrium between the formation and annihilation of bone cells. This dynamical equilibrium is also responsible for morphological changes of bone structures, such as changes of the femur's CCD angle upon sustained load. With increasing age, the osteoclast cells outbalance the osteoblast cells. Then the bone structure becomes weaker, and the loss of minerals results in osteoporosis [8]. Remodeling in response to long-term mechanical load is a unique feature of living bones and distinctively different from engineering materials. The observation of bone's adaptation to changes in mechanical conditions was made first by Wolff,[13] who called it the "law of transformation of bones." Although osteoclasts were discovered as multinucleated cells as early as 1873 by Koelliker,[14] the full complexity of bone development at the cellular level has only unfolded in recent years. A historical note can be found in [9].

---

13 Julius Wolff (1836–1902), German surgeon and orthopedist.
14 Rudolf Albert von Koelliker (1817–1905), Swiss-German physiologist.

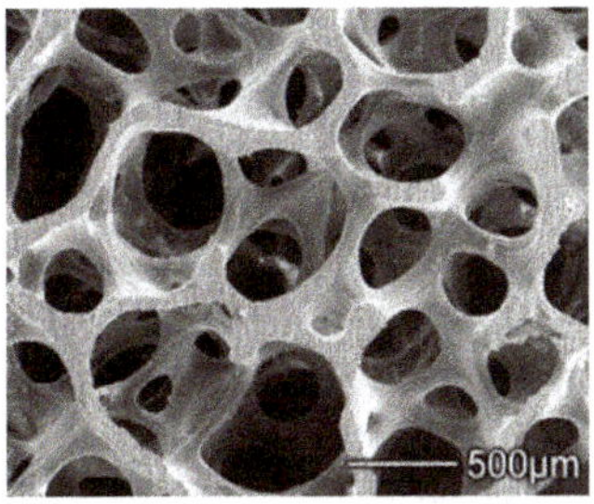

Fig. 3.15: Scanning electron microscopy image of human trabecular bone. In vivo, the trabecular meshwork is covered by a cortical shell, and the pores are filled with marrow (reproduced from [9] with permission of Elsevier Publisher, Inc.).

The trabecular or spongy part of bones has the appearance of an open-pore latticework. Figure 3.15 shows a scanning electron micrograph of the trabecula meshwork [10]. This part contains mineralized collagen fibrils for reinforcement, just like in the cortical fibrils; however, they do not enclose Haversian canals. In contrast to cortical tissue, the fibrils are not parallel to each other and not closely packed. But they do contain three of the four bone cells: osteocytes, osteoblasts, and osteoclasts, as well as lacunae and canaliculi. The main difference to the cortical part is the open meshwork structure with a much lower density.

Bones contain four types of cells: three for building bone structure (osteogenic, osteoblast, and osteocyte), and one for eliminating the old bone material (osteoclast). !

3.7 Elastic and plastic properties of bones

From an elastomechanical point of view, bones are composite biomaterials consisting of mineralized collagen fibrils. The mineral-reinforced fibrils constitute the elementary building blocks for a large variety of bone structures. They may be parallelly aligned and closely packed as in the bone's cortical part or randomly arranged to form a trabecular meshwork as shown in Fig. 3.15. Therefore, bones are heterogeneous materials with different densities and different elastic properties. The main elements "collagen, fibrils, and osteons" are schematically displayed in Fig. 3.16. The bone's elastic properties and fracture formation have been studied on the macroscopic and microscopic scales. We will review the main results for both scales. Any description of the bone's elastomechanical properties has to consider differences in the elastic response of collagen and minerals and differences in the three-dimensional packing of fibrils, cortical versus trabecular. A description on a macroscopic level is much more complex as all strains from different components and their spatial distribution have to be considered, although they are not visible on the macroscale.

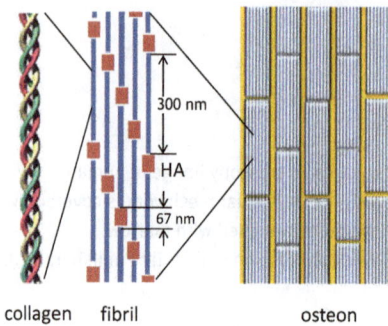

collagen   fibril   osteon

Fig. 3.16: Schematic overview of the bones' main building blocks. We find triple helix collagen fibers on the molecular level, which line up in a stacked fashion on the fibrillary level, leaving gaps that are filled by mineral nanocrystals. Fibrils, in turn, bundled together in lamellar structures, form osteons.

3.7.1 Macroscopic level

For a general characterization of the bones' elastomechanical properties, macroscopic strain–stress tests are justified because of their similarity to in vivo behavior. However, they do not provide any insight into the mechanism of strain resistance or reasons for failure. This in mind, we can test long bone's elastic properties. Long bones have an intrinsic shape anisotropy. The tensile and compressive load can be determined when applying a force parallel to the long axis. The bending load is probed by applying forces perpendicular to the long axis. The yield stress and the Young's modulus follows from such static measurements. The results are listed in Tab. 3.2 from Ref. [11]. We notice that, on average, the strain resistance is lower for bending load than for compressional load. Therefore, we expect that fractures are more likely to occur due to bending than due to compressional or tensile load. The torsional load is particularly likely to cause fracture with the lowest yield stress and elastic modulus. Different types of macroscopic fractures associated with the loads discussed are highlighted in Fig. 3.17.

Tab. 3.2: Yield stress and Young's modulus of bones for torsional load, bending, compression, and tension.

	Torsion	Bending	Compression	Tension	Shear
Yield stress (MPa)	65–71	103–238	167–213	107–170	50–60
Young's modulus (GPa)	3.1–3.7	9.8–15.7	14.7–34.3	11.4–29.2	

Range of values from various test measurements are reproduced from Ref. [11].

The complex structure of bones, which combines hard minerals and collagen fibrils with high tensile strength, leads to a viscoelastic reaction on impact. *Viscoelasticity* is characterized by an elastic modulus Y that depends on the rate of tension [12]. The faster the stress increases, the higher is the elastic modulus (slope) and the yield stress, as shown in Fig. 3.18.

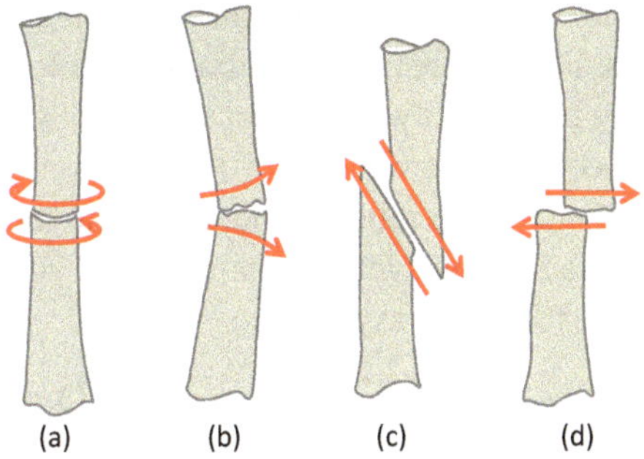

Fig. 3.17: Fracture of long bones due to different types of loads: (a) torsional load; (b) bending load; (c) compressional load; and (d) shear load.

The lowest curve in Fig. 3.18 shows the stress–strain curve for a rate of $0.001 \, \text{s}^{-1}$, equivalent to a static situation. It consists of a linear regime up to a stress of about 130 MPa and a strain of 0.9%, followed by a post-yield regime, where the stress levels off, but the strain still increases. This behavior may indicate that a decoupling of two bone components occurs, for instance, between fibrils and the extrafibrillar matrix. With increasing strain rate, the strength increases, the yield point goes up, but the resilience decreases. In the event of a rapid impact such as an accident, the bone's apparent elastic strength increases dramatically and protects against fracture.

Bones also show *fatigue*, as indicated by the dashed line in Fig. 3.18 [13]. Fatigue is a general material property, meaning that the strength decreases with the number of load cycles applied. Fatigue depends on many different parameters such as the magnitude of load, type of load, the rate applied, temperature, and humidity. In contrast to most other materials, fatigue in bones occurs already after a few cycles, not after thousands or millions of cycles as normally observed. This indicates that microfractures arise at an early stage that dramatically reduces the strength of bones. If no time is provided for self-repair, these microfractures accumulate, resulting finally in failure [14].

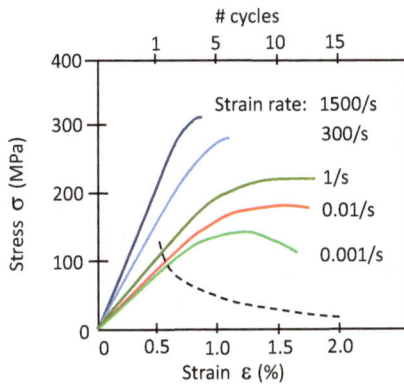

Fig. 3.18: Viscoelastic modulus of bones depends on the rate of applied stress. The lower dashed line shows the yield stress as a function of number of load cycling indicated on the top scale (adapted from [11]).

Elastic properties have also been tested for cortical and trabecular bones independently by taking samples from both parts. First, there is quite a dramatic difference in structure (compare Figs. 3.9 and 3.13) and density. The cortical bone has a much higher density ($1.9 \, \text{g/cm}^3$) than the trabecular bone ($0.43 \, \text{g/cm}^3$) [12]. This correlates well with differences in porosity P defined as the ratio of void volume to the total volume. It is more convenient to determine the bone volume than the void volume. Therefore, we redefine the porosity by the following expression:

$$P = 1 - BV/TV, \tag{3.21}$$

where BV is the bone volume and TV is the total volume. For cortical bone, the porosity $P = 0.05$–0.1, whereas for trabecular bone it is 0.75–0.95. Accordingly, the elastic properties are quite distinct. Cortical bones are strong but not tough. In contrast, trabecular bones are less strong but much tougher. The difference can easily be recognized in Fig. 3.19, which compares the stress–strain relationships of both bone structures. In both cases, the *bone mineral density* (BMD) is important. If the BMD is higher than average, bones become brittle. If the BMD is lower than average, bones lose strength and become soft, as shown in Fig. 3.12. The correlation between bone strength and mineralization of fibrils has again recently be confirmed [14].

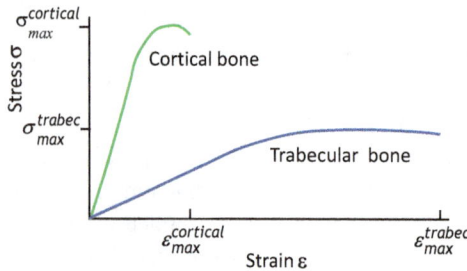

Fig. 3.19: Stress–strain relationship for cortical bone and trabecular bone. Note the different slopes and different toughness of both bone structures (adapted from Ref. [11]).

> ❗ The elastic properties of bones depend on the mineral density and is different for cortical and trabecular parts. The strength of bones increases with increasing strain rate but decreases with the number of load cycles.

Although the meshwork of trabecular fibers appears random, it is actually not. A good example is the complex stress distribution of the femur shown in Fig. 3.20. Using samples from different parts of the trabecular meshwork, it could be shown that the elastic properties are anisotropic. Most of the fibrils align along the lines of greatest stress, while the remaining fibrils are used for cross-linking [10, 11].

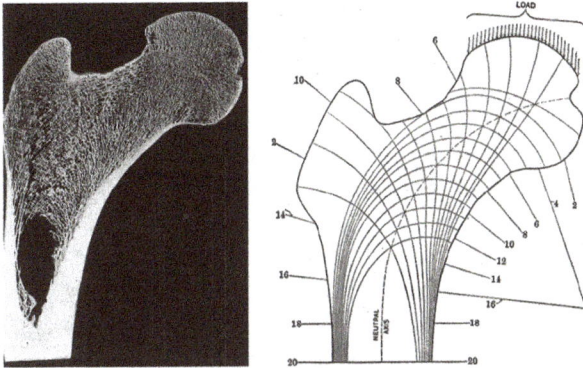

Fig. 3.20: Left panel: Cross section of the upper femur. Right panel: Diagram of the lines of stress according to the femur's mathematical stress analysis (reproduced from [15]).

3.7.2 Microscopic level

It has been suggested that bone toughness is due to a molecular slip mechanism, explaining the observed differences in biomechanical and elastomechanical properties. Slipping can break weak bonds and stretch the composite without destroying it [16]. Alternatively, bone minerals have been made responsible for the toughness of the bones. The mineral crystallites are considered too small to fracture and thus contribute to the overall strength of the bones. More recently, the mechanical behavior of

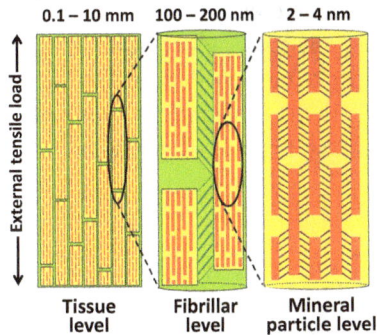

Fig. 3.21: The hierarchical structure of bones imposes a hierarchical deformation transfer from tissue level to mineral particle level. Yellow cylinders denote the mineralized collagen fibrils in longitudinal sections of bone tissue such as in osteons. Red tablets denote the mineral apatite crystallites embedded within the collagenous matrix of the fibrils. Green background denotes interfibrillar matrix and slanted lines indicate cross-links between fibrils as well as between mineral plates. The strain decreases from tissue level to mineral particle level in a ratio of approximately 12:5:2 (adapted from [16] with permission of National Academy of Sciences, USA, Copyright (2006)).

bones at the molecular level has been elucidated using various imaging modalities, x-ray scattering, and nanoscale stress–strain test methods [17–19].

Strain in response to stress on bone tissue varies dramatically from the microscale to the nanoscale. This has been revealed by measuring strain via x-ray scattering independently in the filaments and in the apatite nanocrystallites [17]. The measurements yield a surprising result: for the same macroscopic strain, tissues, fibrils, and mineral particles are exposed to successively lower levels of local strain in a ratio of 12:5:2. Therefore only about 42% of the strain at the tissue level are transmitted to the fibrils, and only 12% of the original strain arrives at mineral particles. This implies that fibrils and minerals are much less strained in comparison to the actually applied strain. The authors explain this apparent discrepancy by the hierarchy of the bone structure, displayed schematically in Fig. 3.21. Much of the strain is taken up by shear forces of cross-linked fibrils in the interfibrillar matrix. On the next lower level, the minerals in the fibrils are again cross-linked, dissipating the tensile strain into shear strain. The brittle apatite minerals remain shielded from overload by this shear transfer mechanism. Nevertheless, the strain at the mineral level is still excessively high by a factor of 2–3 with respect to the yield strain. Here size is indeed important; crack formation is prevented by lack of nucleation points within these nanocrystals. Therefore, the mineral particles can withstand stress and strain beyond the yield point.

> **!** On a microscopic scale, compressional strain on bones is transformed into shear strain by cross-links in the interfibrillar matrix.

Using scanning electron microscopy and atomic force microscopy, researchers have studied the crack formation in bones [18]. They conclude that opposite sides of scissions in mineralized collagen fibrils are held together by some "glue" (see Fig. 3.22). The glue's nature is not clear at present, but it appears to be formed by an unmineralized organic but nonfibrillar material. It is most likely the same interfibrillar matrix responsible for transmitting shear stress between the fibrils. Independent of the

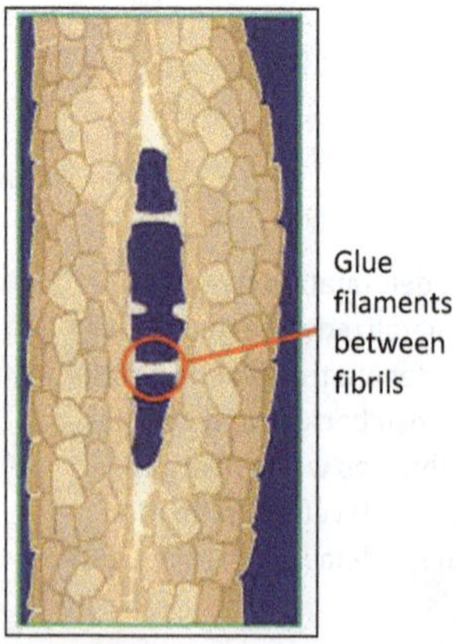

Glue
filaments
between
fibrils

Fig. 3.22: When fibrils are pulled apart, they are bonded by "glue" filaments between the fibrils (adapted from [18] with permission of Macmillan Publishers Limited).

glue's true molecular nature, it is evident that the glue contributes to bone toughness before it finally ruptures. The glue increases the energy required to stretch and ultimately break the tissue. Conversely, the bonds can reform when removing the load, providing an energy dissipation mechanism during repeated load cycles. The timescale for bond reformation is not clear yet, but this mechanism could eventually explain the viscoelastic properties and bones' fatigue on a molecular level.

> Bones are two-component (soft and hard), heterogeneous (cortical and trabecular), anisotropic (long and narrow), viscoelastic, and self-renewable biomaterials. **!**

Concluding, in this chapter we have emphasized that bones are hierarchically organized materials. While this statement is true, it is not specific to bones. Muscles and tissues are also organized hierarchically. Moreover, we found that bones combine two components, soft and hard materials. Roughly speaking, this also applies to the whole body, which consists of soft tissue and hard bones. Both combine structure and function and create connections between the two. For example, muscles exhibit myofibrillar structures and sarcomere-contractile function. Bones form composites of collagen and minerals and have a supportive and protective function. The connection between muscles and bones are tendons, which lend the body movement. Complex hierarchical structures, multifunctionality, and connections between the organs are the hallmarks of the body's blueprint.

3.8 Summary

S3.1   We distinguish between elastic deformation and plastic deformation.

S3.2   Hooke's law describes the linear elastic response between stress and strain up to the yield point, where Young's modulus is the proportionality constant.

S3.3   Stiffness of materials is determined by deflection upon load. The stiffness constant depends on material's properties and boundary conditions.

S3.4   Materials can be characterized as ductile or brittle.

S3.5   Straining a material beyond the yield point causes plastic deformation or fracture.

S3.6   Toughness is the amount of stored mechanical energy during plastic deformation.

S3.7   Brittle materials have a high yield stress but low toughness. Ductile materials have lower yield stress but higher toughness.

S3.8   Viscoelastic materials have a stress–strain curve that depends on the strain rate.

S3.9   Long bones resemble hollow cylinders in their middle part; they display a moment of resistance similar to solid cylinders while being much lighter.

S3.10   Bones are complex composite biomaterials.

S3.11 On a macroscopic level we distinguish between cortical bone and trabecular bone. Cortical bone forms the hard shell of bones, and trabecular bone is an open meshwork of fibers that is found inside.

S3.12 Cortical bone has high density and low porosity. Trabecular bone has low density and high porosity.

S3.13 Bone tissue is organized on several hierarchical levels.

S3.14 The most important ingredients of bones on the nanoscale are minerals (apatite) and collagen, which stack together to form fibrils.

S3.15 Fibrils build lamellar structures, which make up the osteons.

S3.16 Osteons contain channels for blood vessels and nerve fibers.

S3.17 Two antagonistic cell types are present in osteons, osteoblast cells, and osteoclast.

S3.18 Osteoblast cells generate bone material, and osteoclast cells eliminate old bone structures.

S3.19 Creation by osteoblast cells and annihilation by osteoclast cells are in a dynamical equilibrium.

S3.20 Bones have (visco)elastic properties up to a yield point, but no plastic deformation upon sudden impact.

S3.21 Large compressional strain on bones is transformed into shear strain by cross-links in the interfibrillar matrix; glue filaments between fibrils help to absorb strain energy and to suppress fracture.

Questions

Q3.1 What do you understand by Hooke's law?
Q3.2 How can a hydrostatic pressure be exerted on a body?
Q3.3 How can elastic behavior be distinguished from plastic behavior?
Q3.4 What is the difference between ductile and brittle materials?
Q3.5 Which materials are usually ductile, and which ones are usually brittle?
Q3.6 How can you recognize whether a material is ductile or brittle?
Q3.7 What does toughness mean? When is a material tough, when not?
Q3.8 What type of fractures occur mostly in bones?
Q3.9 Explain the term "viscoelasticity."
Q3.10 How can you distinguish between elastic and viscoelastic materials?
Q3.11 Why does the stress–strain curve of viscoelastic materials depend on the strain rate?
Q3.12 What is the basic structural unit of bones?
Q3.13 Describe the hierarchical bone structure.
Q3.14 Which antagonistic cells are present in bones? What are their tasks?
Q3.15 What is the microscopic reason for osteoporosis?
Q3.16 What makes bones lightweight but strong?
Q3.17 What is missing in the spongy part of the bone, which is present in the cortical part of bones?
Q3.18 Why do bones show a viscoelastic behavior?
Q3.19 Why is the stress–strain curve different for cortical bone structure and for trabecular bone structure?
Q3.20 What is the microscopic mechanism of plastic deformation in bones?

Attained competence checker	+	0	–	
I know that Hooks' law relates linearly stress to strain.				
I know what happens to materials when the stress exceeds the yield point.				
I know that plastic deformation is not reversible.				
I appreciate why hollow cylinders have a mechanical advantage.				
I know how the elastic modulus changes with the strain rate in viscoelastic materials.				
I know that all bones are composed of two types with different fiber arrangement and porosity on the inside and outside				
I know that the two types of bones feature different structural and elastic properties.				
I recognize that bones display a hierarchical structure from the molecular level to the macroscopic level, and I can name at least four levels.				
I realize that bones feature two different types of cells.				
I can name different bone cells and I know about their antagonistic function.				
I can name the most important components of bones on the nanoscale.				
I know which component lends resilience to the bone and which provides hardness.				
I know what the main reasons for osteoporosis are.				

Suggestions for home experiments

1. Compression of toilet paper rolls
Compare the compression of rolls in upright and prone position. Which orientation is more resilient to compression?

2. Bending chicken bones
You can test the two-component structure of bones yourself. Take a chicken bone and dip it in vinegar essence for a day or two. This dissolves the HA crystals and the bone becomes soft and easy to bend. You can even determine the stiffness of the bone before and after the vinegar treatment by clamping the bone on one side and measuring the deflection with a suitable weight hanging on the other side.

3. Trabecular meshwork
Take a chicken bone or any other long bone. Use a fretsaw and make a cut about 1 cm from the joint. Take a thin slice of the bone and examine the trabecular meshwork under a microscope.

i **Exercises**

E3.1 **Moment of resistance:** Show the correctness of eq. (3.13) by carrying out the integration in eq. (3.12).

E3.2 **Approximation:** Show that the approximation in eq. (3.15) holds.

E3.3 **Contraction:** Let us assume that a 0.7 m long stick can represent the leg with a cross-sectional area of 3 cm^2. By how much does the leg shrink when a 70 kg person is supported by just one leg?

E3.4 **Beam bending:** A 0.5 m long bone is clamped on one end, and the other free end is exposed to a bending force of 100 N. What are the relative deflection δ/L and the absolute deflection δ if the stiffness of the bone is $k=2 \times 10^3$ N/m?

E3.5 **Critical tensile stress:** A bone may fracture by bending. Let us assume that the bone of diameter $d = 4$ cm is clamped on one side and free to bent on the other side. What is the bending radius just before a crack occurs on the top by a critical tensile stress of 200 MPa?

E3.6 **Pressure in the body:** For which parts of the body is pressure an important physical parameter, and why?

E3.7 **Torque on the femur:** Determine the torque of an 80 kg person on the femur in upright posture at a young age (CCD $= 140°$) and an advanced age (CCD $= 110°$). By how much percent does the torque increase with age?

E3.8 **Crystal components:** Consider the crystal structure of HA shown in Fig. 3.11. Which atoms, molecules, or radicals can you identify as being part of other molecules in the body?

References

[1] Flügge S, eds. Elasticity and plasticity, Handbook of physics. Berlin, Heidelberg: Springer Verlag; 1958.

[2] Herman IP. Physics of the human body. Berlin, Heidelberg: Springer; 2008.

[3] Nordin M, Frankel VH. Basic biomechanics of the musculoskeletal system. 4th edition. Philadelphia, New York, London: Wolters Kluwer & Lippincott Williams & Wilkins; 2012.

[4] Dunlop JWC, Fratzl P. Biological composites. Annu Rev Mater Res. 2010; 40: 1–24.

[5] Grandfield K. Bone, implants, and their interfaces. Phys Today. 2015; 68:40–45.

[6] Wegst UGK, Bai H, Saiz E, Tomsia AP, Ritchie RO. Bioinspired natural materials. Nat Mater. 2014; 14: 23.

[7] Chen Q, Thouas G. Biomaterials: A basic introduction. Boca Raton, London, New York: CRC Press, Taylor & Francis Group; 2015.

[8] Burr DB. Changes in bone matrix properties with aging. Bone. 2019; 120: 85–93.

[9] Martin TJ. Historically significant events in the discovery of RANK/RANKL/OPG. World J Orthop. 2013; 4: 186–197.

[10] Fantner GE, Rabinovych O, Schitter G, Thurner P, Kindt JH, Finch MM, Weaver JC, Golde LS, Morse DE, Lipman EA, Rangelow IW, Hansma PK. Hierarchical interconnections in the nano-composite material bone: Fibrillar cross-links resist fracture on several length scales. Compos Sci Technol. 2006; 66: 1205–1211.

[11] Caeiro JR, González P, Guede D. Biomechanics and bone (& II): Trials in different hierarchical levels of bone and alternative tools for the determination of bone. Rev Osteoporos Metab Miner. 2013; 5: 99–108.

[12]   Endo K, Yamada S, Todoh M, Takahata M, Iwasaki N, Tadano S. Structural strength of cancellous specimens from bovine femur under cyclic compression. Peer J. 2016; 1562: 1–15.
[13]   Hansen U, Zioupos P, Simpson R, Currey JD, Hynd D. The effect of strain rate on the mechanical properties of human cortical bone. J Biomech Eng. 2008; 130: 011011, 1–8.
[14]   Casari D, Kochetkova T, Michler J, Zysset P, Schwiedrzik J. Microtensile failure mechanisms in lamellar bone: Influence of fibrillar orientation, specimen size and hydration. Acta Biomater. 2021; 131: 391–402.
[15]   Gray H. Anatomy of the human body. Philadelphia: Lea & Febiger; 1918. Bartleby.com, 2000. http://www.bartleby.com/107/
[16]   Launey ME, Buehler MJ, Ritchie RO. On the mechanistic origins of toughness in bone. Annu Rev Mater Res. 2010; 40:25–53.
[17]   Gupta HS, Seto J, Wagermaier W, Zaslansky P, Boesecke P, Fratzl P. Cooperative deformation of mineral and collagen in bone at the nanoscale. Proc Natl Acad Sci. 2006; 103: 17741–17746.
[18]   Fantner GE, Hassenkam T, Kindt JH, Weaver JD, Birkedal H, Pechenik L, Cutroni JA, Cidade GA, Stucky GD, Morse DE, Hansma PD. Sacrificial bonds and hidden length dissipate energy as mineralized fibrils separate during bone fracture. Nat Mater. 2005; 4: 612.
[19]   Fratzl P, Weinkamer R. Nature's hierarchical materials, Prog Mater Sci. 2007; 52: 1263–1334.

Further reading

Vincent J. Structural biomaterials, Chapter One: Basic elasticity and viscoelasticity. Princeton, Oxford, Bejing: Princeton University Press; 2012.
Qu H, Fu H, Han Z, Sun Y. Biomaterials for bone tissue engineering scaffolds: A review. RSC Adv. 2019; 9: 26252–26262.
Chen Q, Thouas G. Biomaterials: A basic introduction. Boca Raton, London, New York: CRC Press, Taylor & Francis Group; 2015.
Herman IP. Physics of the human body. Berlin, Heidelberg: Springer; 2008.
Nordin M, Frankel VH. Basic biomechanics of the musculoskeletal system. 5th edition. Philadelphia, New York, London: Wolters Kluwer; 2021.

Useful website

Bone biology: http://hubpages.com/education/Osteoblasts-Osteoclasts-Calcium-and-Bone-Remodeling
Anatomy of the human body: https://www.bartleby.com/107/

4 Energy household of the body

Physical properties of the - energy household

Caloric oxygen equivalent (COE)	20 kJ/l [O_2]
Respiratory exchange ratio (RER)	0.6–1.0
Energy density of food items	15–40 kJ/g
Energy requirement for ATP synthesis from ADP	+48 kJ/mol
Energy gain through ATP–ADP conversion	−30.5 kJ/mol
Basal metabolic rate (BMR) of an average person at rest	8 MJ/day
Oxygen consumption	2–3 l/min
Core temperature	37 ± 0.5 °C
Thermal conductivity of tissue	0.1–0.2 W/K m

4.1 The body as a thermodynamic machine

The human body can be considered as a heat engine that obeys the laws of thermo-dynamics. The first law of thermodynamics states that the stored internal energy dU of a body can be changed either by exchanging heat δQ with a thermal reservoir or by performing work δW:[1]

$$dU = \delta Q - \delta W. \qquad (4.1)$$

The internal energy of a body rises by heat flow from the reservoir to the body ($+\delta Q$) and decreases by heat flow from the body to the environment ($-\delta Q$). In case of the human body, $+\delta Q$ is mainly provided by chemical energy, i.e., through me-tabolizing food, like in a combustion engine, while there are several ways for heat loss ($-\delta Q$), discussed later on.

Similarly, the internal energy of the body changes by adding or subtracting me-chanical work. Lifting up a person by an elevator increases the person's internal energy ($-\delta W$), whereas the work performed by the person, such as chopping wood, consumes internal energy ($+\delta W$).

The *efficiency* of mechanical work is defined by

$$\epsilon = \frac{\partial W}{\partial Q}, \qquad (4.2)$$

---

1 In thermodynamics, the differentials dQ and dW are not defined physical quantities, since their value depends on the path and not only on the initial and the final states. Instead, thermodynamics operates with inexact differentials δQ and δW. However, the total change of the internal energy can be expressed by the exact differential dU. The symbol ∂ expresses partial differentiation with respect to one parameter, while the variables may depend on more than one parameter.

https://doi.org/10.1515/9783110756951-004

i.e., the mechanical work performed per unit of thermal (chemical) energy consumed. As in any thermodynamic machine, this ratio is smaller than 1.

In equilibrium, $\delta Q = \delta W$ and $dU = 0$. However the equilibrium is always a dynamic one, governed by temporal changes of δQ and δW. Thus, it is reasonable to consider the rate of changes:

$$\frac{dU}{dt} = \frac{\partial Q}{\partial t} - \frac{\partial W}{\partial t}. \tag{4.3}$$

Applied to the thermodynamics of the body, the rate of the internal energy $\partial U/\partial t$ is known as the *metabolic rate*. $\partial Q/\partial t$ is the inflow and outflow of heat, and $\partial W/\partial t$ is the mechanical power.[2] For further information on the basics of thermodynamics, please consult any standard physics textbook and books listed under "Further reading."

4.2 Metabolism

The body gets energy by burning food in a process called metabolism. More specifically, metabolism is the combustion of organic compounds with oxygen in living cells to produce energy [1]. Organic compounds are first produced in the photosynthesis process using solar energy, during which CO_2 and H_2O are converted into carbohydrates plus oxygen. Carbohydrates store electrical energy by raising electrons in chemical bonds to higher energy states. These high-energy carbohydrates serve as fuel. Metabolism reverses this process by fragmenting organic food and using the stored energy.

In contrast to the rapid burning of wood or gasoline with oxygen, the metabolic oxidation of carbohydrates takes place in discrete small and controlled steps at body temperature with the help of enzymes. Enzymes are proteins that lower the potential energy barrier to reaction with oxygen. The end products are adenosine triphosphate (ATP) molecules as energy storage (see later) and simple waste products: carbon dioxide (CO_2) and water (H_2O). Subsequently, the waste products are recycled in plants. Figure 4.1 shows the entire cycle from food production in plants to metabolic combustion and back to photosynthesis of carbohydrates. Other high-energy foods like meat and fat also ultimately originate from solar energy, and the metabolism of these products in the body is similar to burning carbohydrates.

Metabolism consists of two parts: *catabolism* and *anabolism* [2]. Large organic molecules are degraded to smaller molecules during the catabolic reaction by releasing energy. During the anabolic process, the released energy is used in cells to

---

2 In the remainder of this chapter, we will use a simplified nomenclature for expressing differentials.

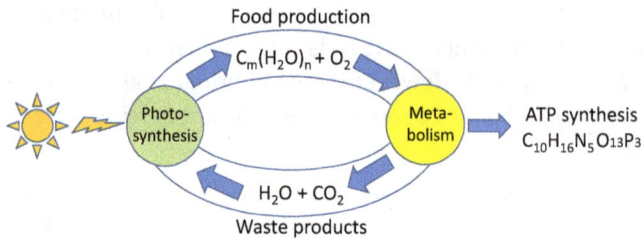

Fig. 4.1: Cycle of food production via photosynthesis from water and carbon dioxide and food consumption by metabolism with the help of oxygen. The metabolic process generates ATP used as an energy source for mechanical, chemical, and transport activities.

synthesize molecules necessary to maintain the cells' functions and perform cellular work, such as powering ion pumps, synthesizing proteins, and contracting myosin–actin filaments in muscles. About 60% of the energy released during catabolic reactions is spent on converting low-energy ADP (adenosine diphosphate) into high-energy ATP (adenosine triphosphate).[3] The remaining 40% dissipates as heat to keep our body temperature. ATP synthesis occurs in the mitochondria (see Fig. 1.1 and Infobox I on mitochondria) of all cells, and the entire process is referred to as *cellular aerobic respiration*. In fact, the oxygen taken up during inspiration and carried by the red blood cells to their destination (see Chapter 9) is eventually consumed within the cell's mitochondria to burn proteins, carbohydrates, and fats. This complex biochemical process is referred to as the *citric acid cycle* or *Krebs* cycle, according to Krebs,[4] who discovered it. Similarly, food is broken down in the intestinal tract, and the nutrients are reabsorbed through blood vessels in the intestine to be delivered to the mitochondria in cells for cellular respiration. The production of ATP supports aerobic mechanical tasks (muscle contraction), ion transport (Na^+–K^+ pump), and chemical reactions (glutamine). Muscle cells host many more mitochondria (up to several thousand) than other cells for efficient mechanical work.

> **!** Metabolism is the combustion of edible food with oxygen. The metabolic end products are carbon dioxide, water, and energy.
>   We distinguish between catabolism and anabolism. Catabolism breaks down molecules into smaller units. Anabolism synthesizes proteins and other macromolecules for cellular tasks.

---

3 ATP contains three high-energy phosphorus groups (Fig. 5.8(a)). In contrast, low-energy ADP comprises only two phosphorus groups.
4 Hans Adolf Krebs (1900–1981), German-British biochemist and Nobel Prize winner in medicine 1953.

Infobox I: Mitochondria

Mitochondria are organelles that are found in large numbers in most cells of eukaryotic organisms, especially in our body cells such as muscles, heart, and liver. Since biochemical processes of respiration and energy generation take place in mitochondria, they are referred to as the "powerhouse of the cells." The size of the mitochondria varies between 0.7 and 3 μm. They consist of a double lipid membrane: an outer membrane is separated from the inner mitochondrial membrane by a membrane gap. The inner membrane is strongly folded inward and forms layers called cristae. The purpose of folding is to increase the membrane surface area to booster the efficiency of ATP production. The ATP generated in the matrix of the crista walls can then diffuse into the cytoplasm of the cell. Mitochondria as the site of oxidative phosphorylation (synthesis of ATP from ADP) in eukaryotes were discovered by Eugene Kennedy (1919–2011) and Albert Lehninger (1917–1986), both American biochemists. It is believed that mitochondria, which carry their own genetic material, may originate from bacteria that have somehow been built into the cytoplasm of eukaryotic cells.

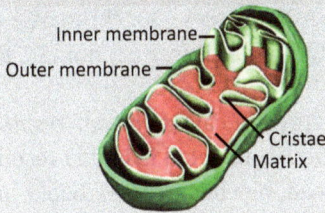

4.3 Caloric oxygen equivalent

As already mentioned, metabolism refers to the combustion of edible organic compounds with oxygen (O_2) into its constituents CO_2 and H_2O. The equivalent oxygen consumption is, therefore, a measure of the metabolic energy gain. A pretty good approximation shows that 1 l of O_2 produces about 20 kJ of energy. This ratio is known as the *caloric oxygen equivalent* (COE):[5]

$$\text{COE} = 1\,\text{l}[O_2] \triangleq 20\,\frac{\text{kJ}}{\text{l}\,[O_2]}. \tag{4.4}$$

We want to confirm the CEO with examples of four main food items, which are:
1. Carbohydrates (sugar, glucose): $C_m(H_2O)_n$
2. Fat (i.e., palmitic acid): $CH_3(CH_2)_{14}COOH$
3. Proteins (i.e., alanine): CH_3-HCNH_2-COOH
4. Alcohol: $C_2H_5(OH)$

We first consider the COE of 1 mol of glucose ($C_6H_{12}O_6$). The physical burning of glucose with oxygen yields the reaction:

---

5 The symbol $\triangleq$ stands for "corresponds to."

$$C_6H_{12}O_6 + 6O_2 \rightarrow 6CO_2 + 6H_2O + 2.78\,MJ/mol_{Glu}.$$

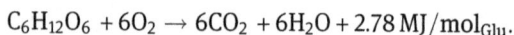

This reaction shows that 1 mol of $C_6H_{12}O_6$ reacts with 6 mol of O_2, resulting in 6 mol of CO_2 and 6 mol of H_2O. The reaction is exothermic and delivers 2.78 MJ/mol_{Glu} of energy.

In contrast, the catabolic burning of glucose occurs in form of the reaction:

$$C_6H_{12}O_6 + 6O_2 + 38\,ADP + 38\,P_i \rightarrow 6CO_2 + 6H_2O + 38\,ATP + 0.956\,MJ/mol_{Glu}.$$

In the catabolic reaction, ATP is synthesized from ADP with the phosphate group PO_4^{3-} (symbolized as P_i). The energy difference is used for the ATP synthesis: 2.78 MJ/mol_{Glu} $-$ 0.956 MJ/mol_{Glu} = 1.824 MJ/mol_{Glu}, or an energy of 1.824 MJ/38 mol_{ATP} = 48 kJ/mol_{ATP}. The remaining energy of 0.956 MJ mol_{Glu} is available as heat.

ATP is an energy storage molecule like a rechargeable battery used for cellular work when needed. The back-conversion of ATP into ADP releases -30.5 kJ/mol_{ATP} of stored energy. The reaction is known as the hydrolysis of ATP (see Section 5.7 for more details).

The breakdown of sugar and conversion to energy are called *cellular respiration*. Studying the various steps of the cellular respiration in detail is highly recommended but goes beyond the scope of this text. Instead, biochemistry textbooks are recommended at the end of this chapter for further studies.

Next, we determine the mole volumes of the reaction partners:

$$22.4\,l \text{ of } C_6H_{12}O_6, \ 134.4\,l \text{ of } O_2, \ 134.4\,l \text{ of } CO_2, \text{ and } 134.4\,l \text{ of } H_2O.$$

The molar mass of the reactants before and after the reaction is

$$180\,g + 192\,g \rightarrow 264\,g + 108\,g.$$

Using the oxygen mole volume, we find the COE of glucose as follows:

$$COE(C_6H_{12}O_6) = \frac{2.78 \ MJ}{134.4\,l[O_2]} = 20.7 \ \frac{kJ}{l[O_2]}, \tag{4.5}$$

thus, confirming the statement made in the beginning. The energy density normalized by the glucose mass is

$$\frac{\Delta E}{m_{mol}(C_6H_{12}O_6)} = \frac{2.78\,MJ}{180\,g_{Glu}} = 15.4 \ \frac{kJ}{g_{Glu}}.$$

All values are listed in Tab. 4.1. Carbohydrates, fats, and alcohols are almost completely oxidized to CO_2 and H_2O in the body, and therefore the physical and physiological energy density are virtually identical. When proteins are catabolized, uric acid is produced in addition to CO_2 and H_2O, such that the physiological value for the energy density is lower than the physical one; the latter is given in a round bracket in Tab. 4.1.

We also determine the efficiency of the catabolic energy generation. Per mole glucose, 1.82 MJ useful chemical energy is stored in ATP, compared to a total of 2.78 MJ energy produced. This yields an efficiency of 65.6%. The efficiency of energy production may also be judged by considering the conversion factor of ATP molecules produced per oxygen molecules inhaled. For glucose this ratio is $38/6 = 6.3$.

The COE is the amount of energy produced by the consumption of 1 l of oxygen at normal atmospheric conditions. **!**

4.4 Respiratory exchange ratio

The *respiratory exchange ratio* (RER) is another characteristic quantity of the metabolic process. It is defined as the ratio of the number of CO_2 moles exhaled to the number of O_2 moles inhaled:

$$RER = \frac{m \; [CO_2]}{n \; [O_2]},$$ (4.6)

Here m and n are the mole numbers. For the glucose reaction, we find an RER:

$$RER_{glucose} = 6CO_2/6O_2 = 1,$$

i.e., each O_2 molecule produces one CO_2 molecule. For all other food items, the RER is lower than 1, around 0.6–0.8. The RER value is used to estimate which type of food is mainly combusted during aerobic exercise. When analyzing the exhaled gas, one can judge whether more meat or more carbon hydrates are used as energy sources. Also, the RER changes with time after a meal. First, carbon hydrates are metabolized and the RER is close to 1. Later, proteins and fat are consumed, and the RER drops to lower values. In Exercise E4.1, we analyze the COE, ATP, RER, and energy density for palmitic acid as another example for metabolic energy conversion.

The RER is the mole ratio of exhaled carbon dioxide to inhaled oxygen. **!**

Tab. 4.1: Energy density, caloric oxygen equivalent (COE), and respiratory exchange rate (RER) for the main food items.

Food item	Energy density (kJ/g)	COE (kJ/l O_2)	RER [CO_2]/[O_2]
Carbon hydrates	15.4	20.7	1
Proteins	17.6 (22)	19.3	0.8
Ethanol	29.8	20.4	0.66
Fat	27–40	19.4	0.7–0.8

4.5 Basal metabolic rate

Having calculated the energy content of food, we consider next the body's energy requirement and the rate of energy consumption. The energy requirement of the body scales with the mass of the body rather than with its surface. The *basal metabolic rate* (BMR) is then defined as follows:

> *Energy consumption per kilogram body weight per hour required for maintaining all functions of the inner organs without performing any physical work.*

For providing numbers, we need first to define a reference frame. The reference frame is a healthy body at rest after actively digesting food (3–4 h after eating) and in a neutral environment that does not require additional body energy for maintaining the body temperature. The use of energy in this state is sufficient only for the basic functioning of vital organs, including the heart, lungs, nervous system, kidneys, liver, intestine, muscles, brain, and skin. An alternative and related measurement with less restrictive conditions is the *resting metabolic rate* (RMR).

> **!** The BMR and RNR refer to the energy consumption of the body at rest under well-defined conditions.

BMR and RMR can be determined by gas analysis of the respiratory system as the COE is proportional to the energy production by combusting carbohydrates, fats, and proteins. The following empirical formula, known as the *Mifflin St Jeor equation* [3], provides the BMR for males and females as a function of mass m in kg, height h in cm, and age a in years. First, the total energy required per day, P, is

$$P = (10m + 6.25h - 5a + s) \times 4.184 \frac{kJ}{day}. \tag{4.7}$$

Here $s = +5$ for males and -160 for females. From this follows the BMR:

$$BMR = \frac{P}{24 \times m}.$$

The average person's BMR is about 7–8 MJ/day or 80 W. This power is broken down into the consumption by the different organs as listed in Tab. 4.2 [4].

Although the brain has only 2% of the total body mass, it consumes 19–20% of the oxygen uptake and therefore 19–20% of the total energy. This consumption is independent of whether we sleep or whether we are awake. The difference of the oxygen consumption is minute, but nevertheless detectable by magnetic resonance imaging (see Vol. 2/Chapter 3). The reason for similar oxygen consumption independent of being awake or sleeping is the mere shift of the brain activity to different cerebral regions. Also, during intense brain activity for solving, for instance, a mathematical problem, the oxygen consumption in this particular "math" region is enhanced, whereas it is lowered in other regions such that the average remains nearly constant.

Tab. 4.2: Energy consumption broken down into the consumption by different organs of the body.

Energy requirement in % of the total body consumption	
Liver	27%
Brain	19%
Skeletal muscles	18%
Kidneys	10%
Heart	7%
Other organs	19%

Tab. 4.3: Body energy consumption for some representative activities (adapted from [4]).

Activity	Oxygen consumption (ml/min · kg)	Equivalent power (W)
Rest	3.5	80
Slow walking	10	230
Bicycling at 16 km/h	20	460–500
Jogging	30–40	500–600
Squash	30	700
Bicycle racing 40 km/h	70	1600

The oxygen uptake and energy consumption rapidly increase with the body's physical activity. Some representative examples are listed in Tab. 4.3. We notice that the body energy consumption is much higher than the actual physical energy produced. This mismatch is due to our body's limited efficiency of mechanical work, discussed further.

4.6 Oxygen uptake

The higher the power for a particular activity, the higher is the demand for oxygen. This is clear from daily experience and documented in Tab. 4.3 for some activities. The maximum oxygen volume $\dot{V}_{max}(O_2)$ per time unit (oxygen consumption) that a healthy person can take up depends certainly on fitness but mainly on age. In general, the oxygen consumption $\dot{V}(O_2)$ is defined as cardiac output (CO)[6] times the blood oxygen concentration difference of arterial and venous blood:

---

6 The acronym CO should not be mixed with the chemical formula CO for carbon monoxide. From the context it should be evident which one is meant.

$$\frac{dV(O_2)}{dt} = \dot{V}(O_2) = \dot{V}_{CO}(O_2) \times (C_a - C_v). \tag{4.8}$$

This equation is also known as Fick's[7] principle (Fick's first equation). The cardiac output $\dot{V}_{CO}(O_2)$ is the product of *stroke volume* (SV) and heart frequency f_{heart}:

$$\dot{V}_{CO}(O_2) = SV \times f_{heart}. \tag{4.9}$$

The stroke volume is the blood volume pumped by the heart during one contraction, which is about 70 ml. The oxygen concentration difference ($\Delta C_{av} = C_a - C_v$) is also known as the *arteriovenous oxygen difference* ΔC_{av} [5]. In the form above, the oxygen consumption $\dot{V}(O_2)$ is rather difficult to determine, as ΔC_{av} and SV are unknown[8].

For increasing the oxygen consumption during aerobic physical exercise, the body adapts by increasing the heart frequency f_{heart}, the breathing frequency f_{breath}, and by increasing the tidal volume TV, i.e., the inspirational air volume during breathing. The stroke volume SV remains constant (see Section 8.2.2). $\dot{V}(O_2)$ and $\dot{V}_{max}(O_2)$ can be determined using a combination of treadmill or ergometer, a spirometer (Section 9.4), and a heart rate monitor. With the treadmill/ergometer, a test person performs physical exercise (walking and bicycling), the spirometer measures the tidal lung volume TV and the breathing frequency f_{breath}, and the heart rate monitor measures f_{heart}. However, this is not sufficient. We also need to know the individual SV and ΔC_{av}. The latter can be determined by two cannulae: one cannula inserted into the pulmonary vein to measure the oxygenated blood concentration; the other intravenous cannula for determining deoxygenated blood concentration. However, this invasive procedure is rather inconvenient. Therefore, ΔC_{av} is usually measured by oxygen sensors in the spirometer. For determining the SV, several methods are available, including Doppler sonography and imaging methods such as MRI. However, the simplest way is the application of the Fick principle quoted earlier. Then the tidal volume is measured via a spirometer and $\dot{V}(O_2)$ is scaled from the known oxygen partial volume during inhaling of air:

$$\dot{V}(O_2) = TV \times f_{breath} \times y_{O_2}, \tag{4.10}$$

where $y_{O_2} = 0.2$ is the mole fraction of oxygen (Section 9.3.1). Therefore, we have for the stroke volume:

$$SV = \frac{\dot{V}_{CO}(O_2)}{f_{heart}} = \frac{\dot{V}(O_2)}{\Delta C_{ac} \times f_{heart}}. \tag{4.11}$$

In this exercise, the maximal oxygen consumption $\dot{V}_{max}(O_2)$ is reached at the maximum heartbeat ($\sim$ 160–180 min^{-1} for athletes) and saturated SV. In Fig. 4.2, $\dot{V}_{max}(O_2)$

---

7 Adolf Fick (1829–1901), German mathematician and physiologist.
8 For further details, we refer to Chapter 8 (eq. 8.3) and Section 9.4.

is plotted as a function of age for males and females [6]. In the left panel, we recognize that $\dot{V}_{max}(O_2)$ first increases up to adulthood and then decreases monotonously. The initial increase is attributed to the gain of body mass during growth. After normalizing by the body mass, we notice a steady decline of oxygen consumption with age in the right panel. This loss may show fluctuations and nonlinearities depending on many factors, including health status and fitness. However, the average loss is on the order of 10% per decade. This decline appears to be universal and predictable for males and females. Fitness training can improve the decline by only a small margin on the order of 2% [7]. Average middle-aged nonathlete men have a maximal oxygen consumption of about 45 ml/kg min, and nonathlete women about 35 ml/kg min. Top athletes can reach 80 and 75 ml/kg min, respectively [8, 9]. As the uptake of oxygen directly relates to the cellular respiration and production rate of ATP, the maximal oxygen consumption is considered the best indicator of cardiorespiratory fitness [9].

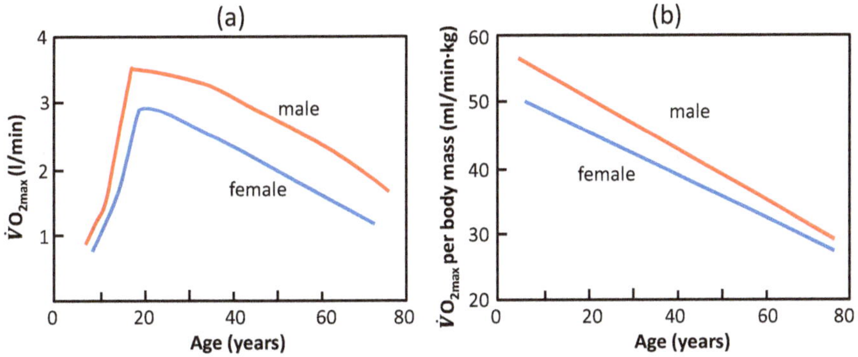

Fig. 4.2: (a) Maximum oxygen volume per time unit as a function of age for male and female (adapted from [5, 6]); (b) maximum oxygen volume per body mass and time unit as a function of age for male and female (adapted from [10]).

Highest performance with highest power can only be delivered for a short period of time. If the muscle power requires more oxygen supply than can be delivered by the blood circulation, then the body taps into an alternative *anaerobic energy reservoir*. During anaerobic exercise, extra energy is taken from splitting creatine phosphate into creatine and phosphate and converting glucose to lactate. This delivers quickly high power, but only for a short period of time. The temporal dependence of power production is shown in Fig. 4.3. The plot indicates that the anaerobic phase may be as short as 1 min before depleting this extra energy reservoir. For more extended periods of high-power physical exercise like bicycle tours or jogging, the oxygen consumption must balance the oxygen uptake, which is the *aerobic* phase of exercise.

The maximum oxygen consumption of a healthy body depends on exercise but mainly on age.

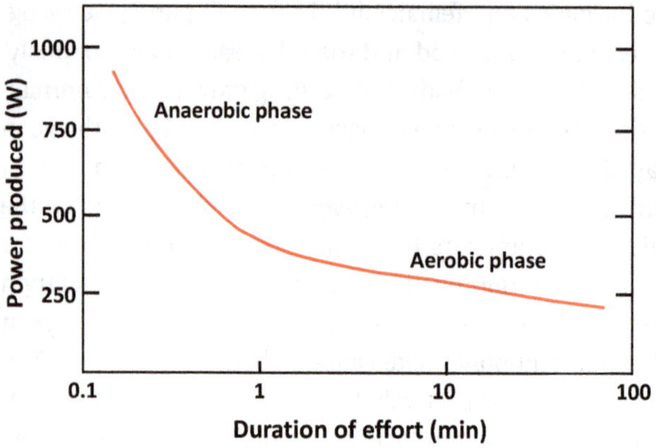

Fig. 4.3: Anaerobic and aerobic power production as a function of duration of effort (adapted from [11]).

4.7 Efficiency of physical work

Next we want to determine the efficiency of human physical work. We take as an example bicycling at 16 km/h, which consumes chemical energy of 460 W according to Tab. 4.3. The mechanical power, when measured with an ergometer, will be on the order of 100 W. Thus the efficiency ε (eq. 4.2) is

$$\frac{dW}{dQ} = \frac{100\,\text{W}}{460\,\text{W}} = 0.22$$

on the order of 20–21%. Bicycling is the most efficient mechanical work that we can perform by our body. A comparison of efficiencies during different tasks is listed in Tab. 4.4.

Tab. 4.4: Efficiency of various human activities.

Task	Efficiency (%)
Bicycling	~20
Swimming at surface	<2
Swimming under water	~4
Shuffling snow	~3

4.8 Metabolic heat production

We can estimate the body's total stored thermal energy, knowing that the equilibrium body temperature is 37 °C or 310 K, and assuming that the body consists of

80% water or water equivalent substances. Water has the highest specific heat of all substances: $c_{water} = 4.2$ kJ/kg K. For the remaining 20% body, we assume a specific heat that is 50% less. Together we estimate the specific heat for a body mass of 70 kg:

$$C_{body} = c_{water} m_{water} + c_{rest} m_{rest} = 265 \; \frac{kJ}{K}, \qquad (4.12)$$

and a total stored thermal energy:

$$\Delta Q_{tot} = C_{body} \Delta T = \frac{265 \, kJ}{K} \Delta T. \qquad (4.13)$$

For a body temperature of 310 K, i.e., $\Delta T = 37$ K, yielding a total stored thermal energy of 9.8 MJ. The metabolic heat production takes care that the stored energy remains constant in time, i.e., any heat loss must be compensated by an equivalent *metabolic heat production* rate (MHR):

$$\dot{Q}_{tot} = \dot{Q}_{MHR}. \qquad (4.14)$$

The MHR is the sum of BMR and heat produced by physical activity:

$$\dot{Q}_{MHR} = \dot{Q}_{BMR} + \dot{Q}_{work}. \qquad (4.15)$$

We have already confirmed that we consume at rest an energy of about 7 MJ/day or roughly 300 kJ/h. The BMR is therefore:

$$P_{BMR} \equiv \dot{Q}_{BMR} = 300 \; \frac{kJ}{h}, \qquad (4.16)$$

and using (4.15), the MHR can be written as follows:

$$\dot{Q}_{MHR} = P_{BMR} + \dot{Q}_{work} = \kappa P_{BMR}. \qquad (4.17)$$

The proportionality constant κ is called the *metabolic activity factor*. κ may vary between 1 (rest) and 20 (heavy work). If the activity factor κ is larger than needed for the work performed, the body tends to overheat. The rate of temperature change dT/dt can be determined from

$$\frac{dT}{dt} = \frac{\dot{Q}_{MHR}}{C_{body}} = \kappa \frac{P_{BMR}}{C_{body}} = \kappa \cdot 300 \left[\frac{kJ}{h}\right] \cdot \frac{1}{265} \left[\frac{K}{kJ}\right] = 1.15 \cdot \kappa \left[\frac{K}{h}\right]. \qquad (4.18)$$

The last equation shows us that the body's temperature increase per hour is about 1.1–1.2 K at rest and much more during activity. Heat overproduction is essential for any temperature control. Without heat loss from the body to the environment, the body would overheat.

4.9 Heat losses of the body

The body features four mechanisms for heat loss to the environment (Fig. 4.4):
1. Heat conduction
2. Radiation
3. Convection or wind chill
4. Evaporation or sweating and shivering

The first three types are conventional strategies of heat loss; the last one is unconventional and only occurs in living matter. Now we want to take a closer look at these four mechanisms.

Conduction   Radiation   Convection   Evaporation

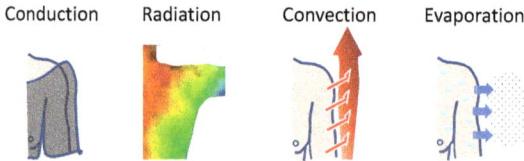

Fig. 4.4: Pictograms of four types of heat loss through thermal conduction between the body and cloth, thermal radiation at body temperature, convection through air flow, and evaporation of sweat droplets on the skin.

> **!** The body has four strategies to regulate its body temperature: heat conduction, radiation, convection, and evaporation.

4.9.1 Heat conduction

Heat conduction requires a temperature gradient and a medium that transports heat from the heat source (higher temperature) to the heat sink (lower temperature). The medium can be gaseous, liquid, or solid. But vacuum is an insulator. Heat conduction can be described by heat flow, where the unit transported is energy. In general, the heat flow density follows the temperature gradient $\vec{\nabla}T$:

$$\vec{j} = -\lambda\vec{\nabla}T, \tag{4.19}$$

where the negative sign indicates the flow direction from higher to lower temperatures.

The unit of $\vec{j}$ is: $[j]$=W/m^2 s. λ is the material-specific thermal conductivity with the unit $[\lambda]$ = W/K m. $\vec{j}$ describes the amount of heat dQ that flows per unit time dt through an area with cross section A. Therefore, $\vec{j}$ can be expressed as follows:

$$\vec{j} = -\frac{1}{A}\frac{dQ}{dt}\vec{n}. \tag{4.20}$$

Here $\vec{n}$ is the unit normal vector to the plane A. For a one-dimensional heat flow through an area A and over a distance Δx between two heat reservoirs at temperatures $T_2 > T_1$ ($\Delta T = T_2 - T_1$), as sketched in Fig. 4.5, eq. (4.20) can be rewritten as follows:

$$jA = \left(\frac{dQ}{dt}\right)_Q = \dot{Q}_Q = \lambda A \frac{\Delta T}{\Delta x} = \frac{\Delta T}{R_Q}. \tag{4.21}$$

The last equation is equivalent to Ohm's law of electrical conductance. Therefore, we call

$$R_Q = \frac{1}{\lambda}\frac{\Delta x}{A}, \tag{4.22}$$

the heat resistance. As for the electrical case, the resistance increases with the length Δx and is inversely proportional to the cross section A. The material-specific thermal conductivity λ is listed in Tab. 4.5 for a few representative materials. A complete list can be found in table works [11]. Note the very high thermal conductivity of metals due to the high mobility of free electrons as heat carriers. In insulating solids, lattice vibrations, i.e., quantized phonons, are responsible for thermal transport. In contrast, heat in air is transported by momentum and energy exchange between the slow and heavy air molecules. In contrast, in water, the heat conductivity is much higher due to the higher density of the water molecules.

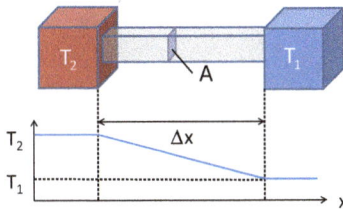

Fig. 4.5: Thermal conduction through a bar of cross-sectional area A and length Δx stretching between two heat reservoirs at constant temperatures T_2 and T_1.

In analogy to electric circuits, thermal resistances can be arranged in a series or in parallel. The total thermal resistance is the sum of all individual ones if they are lined up in series:

$$R_{Q,\,tot} = R_{Q,1} + R_{Q,2} + \cdots. \tag{4.23}$$

If the individual resistances are arranged in parallel, the reciprocal thermal resistances add up:

$$\frac{1}{R_{Q,\text{tot}}} = \frac{1}{R_{Q,1}} + \frac{1}{R_{Q,2}} + \cdots. \qquad (4.24)$$

This has consequences for our clothing strategy. In summer, we may wear only one T-shirt. As winter approaches, we put on extra shirts and coats. These cloth items are arranged in a series and increase the thermal resistance, i.e., reduce heat loss. Air between several fabrics also provides good thermal insulation. However, if the cloth gets wet due to rain or sweating, the water's high thermal conductivity acts as a short circuit through which heat escapes. This becomes especially dangerous if you drop into cold water. Not only is swimming ability impaired with cloth on, hypothermia is also a severe risk. Divers protect themselves with special neoprene overalls having low thermal conductivity. A similar problem occurs when an icy metal pole is touched with bare hands below the freezing temperature. The very high thermal conductivity of metals quickly dissipates heat from the fingers, and at the same time, the moisture from the hand freezes at the interface with the metal, which leads to the fingers sticking to the metal pole. This is dangerous in either case: hold on or pull off. The only promising strategy is to warm up the hand with a heat blower as quickly as possible. An example of a parallel arrangement of thermal resistances are windows in buildings. Each window contributes to the total heat loss in winter, and the more windows, the lower is the heat resistance. Table 4.5 lists the thermal conductivity of some well-known materials.

Tab. 4.5: Thermal conductivity of some typical materials.

Material	Thermal conductivity λ (W/K m)
Air	0.01–0.1
Clothes (cotton)	0.1
Water	0.6
Tissue	0.1–0.2
Metal	200–400

4.9.2 Heat radiation

Heat loss by thermal radiation is omnipresent. We can view our body as a blackbody radiator that absorbs and emits electromagnetic radiation, but neither reflects nor transmits radiation. According to the Stefan–Boltzmann law,[9,10] the heat emitted per second by a blackbody with the total surface A at a temperature T in the solid angle 4π is given by [12]

---

9 Josef Stefan (1835–1893), Austrian physicist.
10 Ludwig Boltzmann (1844–1906), Austrian physicist and thermodynamicist.

$$\dot{Q}_R = \varepsilon\sigma AT^4. \tag{4.25}$$

Here σ (=5.66 × 10^{-8} W/m^2 K^4) is the Stefan–Boltzmann constant, T is the absolute temperature, and ε is the emissivity (a dimensionless number) that characterizes the blackbody surface with respect to its roughness versus smoothness. For the skin, ε is usually taken as 1. The heat radiation increases with the body temperature to the power of 4!

Usually, one blackbody of area A at a temperature T_1 is not alone. There is another one nearby with a temperature T_2. Body 1 emits radiation according to its temperature T_1 and absorbs radiation from body 2 at temperature T_2. Conversely, body 2 emits radiation according to its temperature T_2 and absorbs radiation from body 1. The effective heat flow from body 1 to 2 is then

$$\dot{Q}_R = \varepsilon\sigma A\left(T_2^4 - T_1^4\right) \cong 4\varepsilon\sigma AT^3\Delta T. \tag{4.26}$$

The approximation is made for the case that the temperature difference ΔT between bodies 1 and 2 is only a fraction of their temperatures on the absolute temperature scale. This equation is known as Newton's law of heat loss, which states that the radiation from a blackbody at temperature T_1 in an environment at temperature T_2 is proportional to the third power of temperature T times the temperature difference ΔT. It can be shown that a naked person with a body temperature of 37 °C = 310 K in an environment of 20 °C = 293 K is hypothermic in a short time only through radiation heat loss (see Exercise 4.4). Clothing is therefore an essential protection, at least in the northern hemisphere. In babies, the surface-to-volume ratio is unfavorable because a baby has a large surface for heat radiation but only a small volume for heat storage. The result can be rapid hypothermia if not protected by towels, a warm environment, or newborn incubators.

Heat radiation of the body is also used for detecting inflammations and tumors close to the surface. Those have higher temperatures than the surrounding tissue and can easily be detected using an infrared camera. In general, thermography has proven very useful in noninvasive medical diagnostic and physiology [13].

4.9.3 Convection

Heat conduction assumes that the mediating material such as air or water is at rest. However, if a laminar or turbulent gas or liquid flow touches a hot surface, more heat can be dissipated per unit of time. In an open system, air will always flow from hotter to colder areas, which is the usual chimney effect. In hot countries, the direction of the airflow can be controlled to cool buildings. When applied to the skin, the airflow that dissipates body heat is called wind chill. While wind chill feels comfortable in a summer heat wave, it can be quite dangerous in a winter blizzard and may lead to hypothermia.

As a first approximation, the rate of energy loss due to convection depends linearly on the temperature difference between the uncovered skin and the environment:

$$\dot{Q}_C = K_C A (T_{skin} - T_{env}).\tag{4.27}$$

Here A is the total uncovered area of the skin, and K_C is a velocity-dependent constant with the unit $[J/m^2\,K\,s]$, which can be approximated by [14]

$$K_C = (10.5 - v + 10\sqrt{v}),\tag{4.28}$$

where v is the air flow velocity in units of m/s. Heat loss by convection can be substantial.

4.9.4 Sweating and shivering

When all other mechanisms for controlling body heat are exhausted, the body can still counteract overheating from sweating and hypothermia from shivering.

Sweating requires the ability to open pores in the skin where water or sweat can escape. In Fig. 1.9, sweat glands and pores are shown. The heat necessary to evaporate the sweat cools the skin. The heat of evaporation for water is exceptionally high. For the evaporation of 1 l of water at the condensation point, an energy amount of 2.26 MJ is required.

The loss of heat through the evaporation of water on the skin depends on the relative humidity. If the partial pressure of water vapor in air is greater than the vapor pressure generated on the skin at a temperature of 37 °C (6.3 kPa), water (sweat) cannot evaporate (sauna effect). Conversely, in calm and dry air with $T_{environ} > T_{body}$, evaporation is the only way to lose body heat.

The energy loss rate due to evaporation is given by

$$\dot{Q}_E = K_E A \Delta p,\tag{4.29}$$

where K_E is a heat coefficient, A is the surface area of the sweating body, and Δp is the partial pressure difference for water vapor in the air and at the skin surface. To estimate the heat loss by evaporation, consider the ½ l of water to be evaporated within 1 h. The energy loss rate is then $\dot{Q}_E = 2.26$ MJ$/2 \times 3600$ s $= 300$ W.

Shivering is the body's opposite response to hypothermia. Trembling means muscle activity that requires energy and gives off heat. The arrector pili muscles attached to the hair follicles in the skin (see Fig. 1.9) take part in the process of heat generation.

4.9.5 Equilibrium condition

To summarize, the combined heat loss of the body is given by:

$$\dot{Q}_{tot} = \dot{Q}_Q + \dot{Q}_R + \dot{Q}_C + \dot{Q}_E = \frac{1}{\lambda}\frac{A}{L}\Delta T + 4\varepsilon\sigma\, AT^3\Delta T + K_C A\Delta T + K_E A\Delta p. \qquad (4.30)$$

Referring to eq. (4.18), in equilibrium, the temperature of the body should be stable over time, implying $dT/dt = 0$. Hence

$$\frac{dT}{dt} = \frac{1}{C_{\text{body}}}\left(P_{\text{BMR}} - \dot{Q}_{\text{tot}}\right) = 0. \qquad (4.31)$$

The equilibrium condition $dT/dt = 0$ requires a mechanism of temperature control in the body that balances heat production and heat loss in a changing environment.

The body temperature is kept constant by a sophisticated control mechanism. For a healthy person, the core body temperature is 37 ± 0.5 °C. This temperature must be kept constant in the central part of the body, where the essential organs are located, and in the brain. The peripheral body parts belonging to the musculoskeletal system can have higher or lower temperatures depending on the environment. However, any temperature gradient between the core and the skin initiates thermal conductivity through the tissue. Isotherms for a body in environments of air temperatures of 20 and 30 °C are shown in Fig. 4.6.

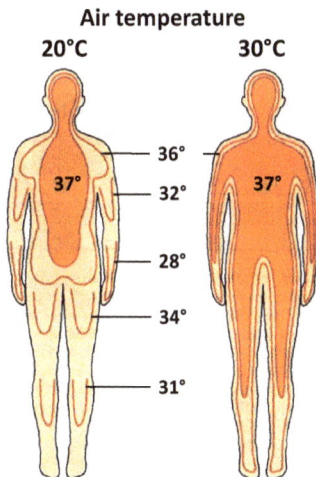

Fig. 4.6: Isotherms of the body temperature in different environments. All temperatures are in centigrade (reproduced from [13] with permission of Thieme Verlag).

Any temperature control system, including our body, consists of three major parts: (1) temperature sensor, (2) heat source, and (3) heat loss. The balance between heat source and heat loss keeps the temperature constant. The temperature sensors register the actual temperature and compare it with the target temperature. Deviations between the actual and set point temperature require a negative feedback system: if the actual temperature is higher than the set point, the heat generation is reduced and vice versa.

In the case of a resting adult human body under normal conditions, a heat generation of around 80–90 W is required to compensate for heat losses due to conduction, radiation, convection, and sweating. The metabolism generates heat. Heat sensors are distributed in the skin over the whole body, but with a particularly high density on the back. Excessive temperatures are compensated for by the expansion of the blood vessels in the skin, an increase in blood flow, and the opening of the pores for sweating. To counteract to low temperatures, blood vessels in the skin constrict, and blood is drawn from the skin into the body's inner parts, eventually causing tremors. In all cases, the thermostat, which compares the target temperature with the actual temperature and reacts to deviations, is located in the hypothalamus [15], which in turn is part of the cerebellum. The regulation of body temperature is known as body temperature homeostasis [16].

4.10 Summary

S4.1   Metabolism is the combustion of organic compounds with oxygen in living cells for the production of energy.

S4.2   Metabolism consists of two parts: catabolism and anabolism. Catabolism is the oxygen reactive part; anabolism is the synthetic part.

S4.3   ATP synthesis is the primary energy source of the body.

S4.4   ATP synthesis occurs in the mitochondria.

S4.5   COE measures the energy content of food via the oxygen consumption. About 1 l of oxygen is required for the production of 20 kJ of energy.

S4.6   The RER is the mole ratio of exhaled carbon dioxide to inhaled oxygen. The RER of glucose is 1.

S4.7   The ability to perform physical work depends on the oxygen uptake rate.

S4.8   The maximum oxygen consumption of a healthy body depends mainly on age.

S4.9   The BMR is the energy required to maintain the basic body functions at rest. The BMR depends on weight and age and is about 7–8 MJ/day.

S4.10 The ability to take up oxygen required for additional activity depends on the fitness of the body.

S4.11 The MHR is required for maintaining the body temperature. It is composed of the BMR and additional physical activity.

S4.12 Constant body temperature is achieved if the MHR and heat loss are in equilibrium.

S4.13   Heat loss is provided by heat conduction, radiation, convection or wind chill, evaporation, or sweating.

S4.14   To keep a constant body temperature, an elaborate temperature control system is activated, including temperature sensors everywhere in the skin and in the inner parts of the body. The information is transferred to the hypothalamus, acting as a thermostat.

Questions

?

Q4.1   What is the difference between anabolism and catabolism?
Q4.2   In which part of the cell is ATP synthesized?
Q4.3   What is ATP used for?
Q4.4   How many molecules of ATP are produced by a single molecule of glucose?
Q4.5   How is BMR defined?
Q4.6   What is the difference between BMR, RMR, and MHR?
Q4.7   How is the COE defined?
Q4.8   What is the daily BMR for an average person?
Q4.9   How many conventional and unconventional types of heat loss are important for the body to control the body temperature?
Q4.10   When does Newton's law for heat conductivity apply?
Q4.11   How is energy produced during anaerobic and aerobic exercise?
Q4.12   Is the homeostatic control of the body temperature a positive or negative feedback system?
Q4.13   What is cardiorespiratory fitness, and how can it be determined?
Q4.14   How is RER defined?

Attained competence checker   +   0   –

I know (roughly) what metabolism is and what it implies.

I know in which part of the cell the combustion with oxygen and the synthesis of high energy molecules takes place.

I know which molecule stores energy for cellular work.

I can calculate the COE for metabolic reactions.

I can determine the RER of a metabolic reaction.

I can distinguish between aerobic and anaerobic exercise.

I know how to measure the oxygen consumption of a person.

I can distinguish between MHR and BMR.

I know what the average MHR of an average adult is.

I can list the four processes by which the body can exchange heat with the environment.

I appreciate that the constancy of the body temperatures requires temperature control.

Suggestions for home experiments

H4.1 Using your porcelain coffee cup, how much time does it take before the outside becomes hot? Estimate the temperature difference and time it takes before the outside heats up. What is your guess for the thermal conductivity of the coffee cup? Do you have an idea why the handle does not become hot?

H4.2 During physical activity such as cycling, the transition from anaerobic to aerobic phase is clearly noticeable. Try to find out how long it takes you to make this transition.

Exercises

E4.1 **Caloric oxygen equivalent:** Consider the catabolic combustion of palmitic acid via the reaction:

$$CH_3(CH_2)_{14}COOH + 23O_2 + 129ADP + 29P_i \rightarrow 16CO_2 + 16H_2O + 129ATP + 3.76MJ.$$

Determine the (a) efficiency, (b) COE, (c) the energy density, and (d) RER for this reaction.

E4.2 **Burning fat:** How much fat is burned in a day with an O_2 inspirational consumption of 0.3 l/min if energy is only made available from fat stores?[11]

E4.3 **Heat conduction:** Determine the heat conduction of an undressed person with a core temperature of 37 °C in an environment that has an air temperature of 20 °C. The body's surface area shall be 1.6 m², and a tissue thickness of 5 cm forms a barrier between the core body and the skin.

E4.4 **Radiative heat loss:** Assume that an undressed person weighing 70 kg is standing in a room with a temperature of 20 °C. The resting metabolic rate is 80 W. The total surface area of the person is 2 m²; the emissivity $\varepsilon = 1$, and the body surface temperature is 32 °C.
a. Determine the rate of heat loss, assuming that only radiative heat plays any role.
b. If the rate of heat loss remains constant, how long does it take for the body temperature to drop by 5 °C?

E4.5 **Heat loss and gain:** A sunbather lies on the beach in the caribbean in the bright sun. The air temperature is 30 °C. In the afternoon, the sun at a polar angle of 45° emits radiation of 700 W/m². The skin absorbs 80% of this radiation. The person's skin temperature is 32 °C, the skin surface area exposed to the sum is 0.9 m², and the emissivity $\varepsilon = 1$.
a. With stagnant airflow, determine the total energy loss or gain per time due to thermal radiation.
b. Assume that there is a slight breeze of 4 m/s. How does the breeze change the heat loss due to convection?
c. If the person's metabolic rate is 90 W and breathing is still consuming 10 W, how much energy does the body have to lose per time through sweating for the body temperature to remain constant?

---

11 The tidal volume is much higher, about 8 l/min. But only a fraction is used for oxygen exchange in the catabolic process. For more information, see Section 9.4.

References

[1]   Campbell NA, Reece JB. Biology. 9th edition. New York Benjamin Cummings; 2009.
[2]   Moran LA, Horton RA, Scrimgeour G, Perry M. Principles of biochemistry. 5th edition. New York Pearson Education Limited; 2014.
[3]   Mifflin MD, St Jeor ST, Hill LA, Scott BJ, Daugherty SA, Koh YO. A new predictive equation for resting energy expenditure in healthy individuals. Am J Clin Nutr. 1990; 51: 241–247.
[4]   Kenney WL, Wilmore JH, Costill DL. Physiology of sport and exercise. 5th edition, Champaign, Il: Human Kinetics; 2012.
[5]   Åstrand PO, Rodahl K, Dahl HA, Stromme SB. Textbook of work physiology: Physiological bases of exercise. 4th edition. Human Kinetics; Champaign, Il; Windsor, Cananda; Leeds, UK 2003.
[6]   Dehn MM, Bruce RA. Longitudinal variations in maximal oxygen intake with age and activity. J Appl Physiol. 1972; 33: 805–807.
[7]   Hawkins SA, Wiswell RA. Rate and mechanism of maximal oxygen consumption decline with aging. Implications for exercise training. Sports Med. 2003; 33: 877–888.
[8]   http://www.sport-fitness-advisor.com/VO2max.html
[9]   Kvell K, Pongrácz J, Székely M, Balaskó M, Pétervári E, Bakó G. Molecular and clinical basics of gerontology. Hungary: University of Pécs; 2011.
[10] Eisemann JC, Pivarnik JM, Malina RM. Scaling peak VO_2 to body mass in young male and female distance runners. J Appl Physiol. 2001; 90: 2172–2180.
[11] Anderson HL. A physicist's desk reference. 2nd edition. American Institute of Physics; New York 1989.
[12] Eisberg R, Resnick R. Quantum physics of atoms, molecules, solids, nuclei, and particles. 3rd edition. Wiley & Sons; New York, London, Sydney, Toronto 2011.
[13] Tattersall GJ. Infrared thermography: A non-invasive window into thermal physiology. Comp Biochem Physiol A Mol Integr Physiol. 2016 Dec; 202: 78–98. doi: 10.1016/j.cbpa.2016.02.022. Epub 2016 Mar 2. PMID: 26945597.
[14] Cameron JR, Skofronick JG, Grant RM. Physics of the body. 2nd edition. Medical Physics Publishing; Madison, Wisconsin 1999.
[15] Pape H-C, Kurtz A, Silbernagel S, eds. Physiologie. 7th edition. Stuttgart, New York: Thieme Verlag; 2014.
[16] Boron WF, Boulpaep EL Medical physiology. 2nd edition. Saunders W.B. Elsevier; 2012.

Further reading

Atkins P. The laws of thermodynamics: A very short introduction. Oxford, New York, Athens: Oxford University Press; 2010.
Campbell NA, Reece JB. Biology. 11th edition, New York: Pearson; 2011.
Voet D, Voet JG, Pratt CW. Fundamentals of biochemistry. Life at the molecular level. New York, London, Sydney, Toronto: John Wiley & Sons Inc; 2016.
Moran LA, Horton HR, Scrimgeour KG, Perry MD. Principles of biochemistry. 5th edition. New York: Pearson; 2014.
Lehninger AL, Principles of Biochemistry: New York, W.H. Freeman 7th edition, 2017.

McArdle WD, Katch FI, Katch VL. Essentials of exercise physiology. 5th edition. Philadelphia, New York, London: Wolters Kluwer; 2015.

Brooks GA, Fahey TD, Baldwin KM. Exercise physiology: Human bioenergetics and its applications. Boston: McGraw-Hill Higher Education; 2004.

Herman IP. Physics of the human body. Berlin, Heidelberg: Springer; 2008.

Eisberg R, Resnick R. Quantum physics of atoms, molecules, solids, nuclei, and particles. 3rd edition. New York, London, Sydney, Toronto: Wiley & Sons; 2011.

Reif F. Fundamentals of statistical and thermal physics. Boston: McGraw-Hill; 1965.

5 Resting potential and action potential

Important physical properties of cells	
membrane thickness	4–5 nm
Resting potential	−75 mV
Cell capacitance	1 μF/cm^2
Single channel conductance	30 pS
Transmembrane specific resistance	2×10^7 Ωm
Time constant for depolarization	1 ms
Ion channel transport rate	10^6 ions/s

5.1 Introduction

In the previous chapters, we realized that all physical activities, whether voluntary or automatic, somatic or vegetative, require an initial electrical signal. In this chapter, we examine how cells generate such signals, and in the next chapter, we focus on the propagation of electrical signals from origin to destination. The key to signal generation is the electrochemical property of the cell. Here we consider a simple biological cell model to be sufficient to explain the conditions for electrical resting potentials and the generation of action potentials.

The body's electrical properties can be described by the principles of electrostatics and dc electric currents. However, the carriers of electric charges are not free electrons, and electric currents are not electron currents like in metals or semiconductors. Instead, charges are connected to ions (anions and cations), and currents are due to ion exchange and polarization propagation. The standard electrical devices: resistance, capacitance, and electromotive forces have their equivalents in ion channel resistance, cell capacitance, and cell potential. No magnetic fields are involved in the transport processes and there are essentially no electromagnetic fields generated by cells [1]. Ion diffusion and channel resistances determine the time constants for generating action potentials and polarization reversals. From a technical point of view, the body's electrical circuit is more or less treatable in terms of electrostatics since the fastest processes take place on a timescale of around 1 ms. So we are talking about sub-kHz circuits. Although such circuits appear unimpressive at first glance, the high performance of cognitive processing and body control via an ultra-responsive feedback system is extraordinary.

https://doi.org/10.1515/9783110756951-005

5.2 Cell membrane and ion channels

The main task of the cell membrane is the distinction between inside and outside
[2]. Figure 5.1 is a not-to-scale sketch of a cell, indicating a typical extension of a
nerve cell's soma. Everything inside is called *cytoplasm*; everything outside is the
extracellular space, also called *interstitial space*. The cell membrane consists of a
double lipid layer with hydrophobic tails inside and the hydrophilic heads in contact with the watery solution inside and outside (see Infobox I on cell membranes).
The thickness of the membrane, given by the chain length of the double lipid molecules, is about 4 nm. The watery solution in the cytoplasm and the extracellular
space is an electrolyte containing the salts NaCl and KCl. The cytoplasm exhibits an
excess of potassium cations (K^+), while the extracellular space has a sodium cation
(Na^+) surplus. The pH value of about 7.4 is the same on both sides of the membrane,
but the cation concentrations are drastically different [3]. This difference is the
main ingredient for the electrical potential of cells at rest, called *resting potential* or
resting membrane potential.

There is also a difference in the anion concentrations inside and outside of the
cell. Outside, the positive charge of the cations is mainly balanced by the negative
charge of the anion Cl^-. However, the Cl^- concentration is rather low in the cytoplasm, and the anions are mainly of organic type. The cation and anion concentrations of cells at rest are listed in Tab. 5.1. The anion disproportion does not play a
role for the resting potential nor for the cell's action potential as anions, in first approximation, cannot diffuse through the cell membrane and take no part in the ion
transport.

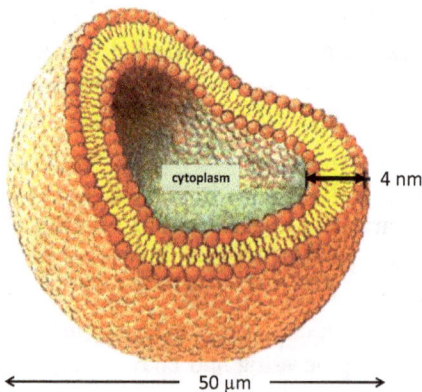

cytoplasm

4 nm

50 µm

Fig. 5.1: Not-to-scale sketch of a cell,
emphasizing the double lipid membrane
separating the cytoplasm from the extracellular
space.

Tab. 5.1: Second and third column: ion densities inside and outside of a cell at rest according to [3, 4]; fourth and fifth column: calculated and measured resting potentials derived from giant squid axons [13]. Potentials quoted are those between inside versus outside.

Ions	n_{inside} (mmol/l H_2O)	$n_{outside}$ (mmol/l H_2O)	Nernst potential (mV)	Measured resting potential (mV)
K^+	140	4	−92	−75
Na^+	12	145	65	+50
Cl^-	15	117	53	−54
Organic anions	140			

Infobox I: Cell membranes

Cell membranes are made up of phospholipid bilayers that form a dense packing (left panel) of chain-like double-stranded molecules (right panel). Each phospholipid molecule consists of a hydrophilic, polar, phosphate head that binds water, and a nonpolar, hydrophobic tail. The tails on both sides touch inside and eject water. The strands consist mainly of saturated fatty acids, and a few unsaturated fatty acid strands with a kink (cis bond) are distributed in the membrane and serve to increase flexibility. Proteins that cross membranes (transmembrane proteins) can form ion channels for charge transport.

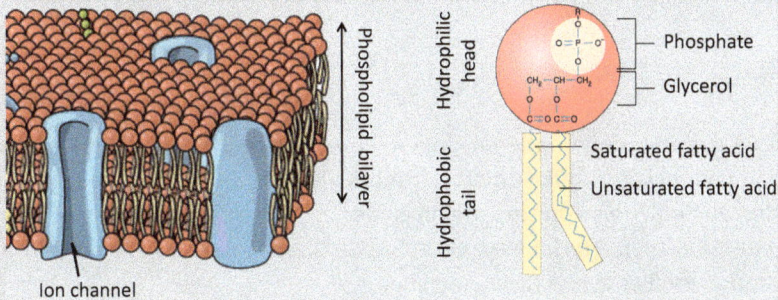

(Adapted from OpenStax Anatomy and Physiology, 2016, © creative commons)

Cell membranes are perforated by pores called *channels* that serve ion exchange between the cytoplasm and extracellular space (Fig. 5.2). The ion channels are cation-selective, and they are specialized for sodium (Na^+), potassium (K^+), calcium (Ca^{2+}), and other cations. Some channels are always open, and some are open or closed in response to a stimulus. The stimulus can be a voltage change, a ligand, or a mechanical strain. Accordingly, ion channels are classified as voltage-gated, ligand-gated, or mechanical-gated. We will consider in this chapter only voltage-gated ion channels. The ion channels are built from proteins that span the cell membrane and contain several subunits (transmembrane proteins). The protein structure and ion selectivity of the cation channels were investigated in great detail by MacKinnon[1] and coworkers, using mainly x-ray crystallography [5] and spectroscopic techniques [6]. There are also channels for chloride (Cl^-) and other anions. However, they appear to be less ion specific and their mechanism is less well understood [7]. The density of K^+ and Na^+ channels is still disputed in the literature [8], partly because it varies greatly depending on the location, such as soma or dendrites.

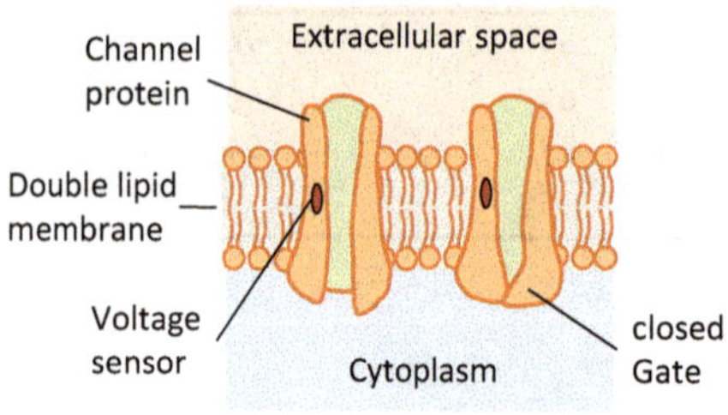

Fig. 5.2: Membrane with open and closed ion channels.

> Biological cells are functional units enveloped by a membrane that distinguishes between inside (cytoplasm) and outside (extracellular space). Ion channels provide chemical and electrical communication between interior and exterior.

5.3 Resting potential

Now we focus our attention on the Na^+ and K^+ ions. Figure 5.3 shows a model cell including ion channels for diffusion of Na^+ and K^+ ions in and out of the cell. The channels that allow K^+ ion diffusion are always open; the Na^+ channels are closed unless stimulated to open. The lower panel of Fig. 5.3 shows schematically the K^+/Na^+ ion concentration profiles across the membrane.

Because of the significant concentration difference of K^+ ions across the cell membrane, we expect a diffusive exchange following *Fick's first law*:

$$I_n = PA\Delta n. \tag{5.1}$$

Here I_n is the particle current, also called particle flow (unit: [mol/s]), i.e., the number of particles per unit time crossing a barrier (membrane) of total area A,

$$\Delta n = n_{inside} - n_{outside} \tag{5.2}$$

is the difference of the particle densities inside and outside (unit: [mol/l]), and

---

1 Roderick MacKinnon (*1956), American biophysicist, neuroscientist, and Nobel laureate in chemistry 2003.

$$P = D/\Delta x \qquad (5.3)$$

is the *permeability* of the membrane (unit: $[P]$ =m/s), where D is the diffusion constant (unit: $[\mathrm{m}^2/\mathrm{s}]$) and Δx is the membrane wall thickness.

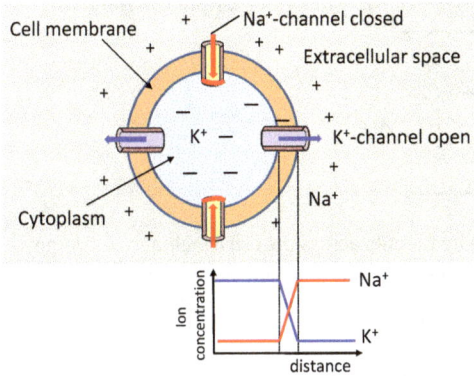

Fig. 5.3: Schematics of a biological cell, including specific ion channels for K^+ and Na^+ diffusion and ion concentration profiles in the lower panel.

Infobox II: Particle flux versus particle flow

Often the flux is considered instead of the flow. The particle flux is the particle flow per area, defined as follows:

$$j_n = \frac{I_n}{A} = P\Delta n = D\frac{\Delta n}{\Delta x}. \qquad (5.4)$$

Here P is the permeability of a barrier and D is the diffusion constant. Often the particle flow is represented in a form analogous to Ohm's law:

$$I_n = \frac{\Delta n}{R_D}. \qquad (5.5)$$

Here the density difference replaces the voltage difference and R_D is known as diffusional resistance. The reciprocal of R_D is called diffusional conductance. With the help of the expressions already stated in eqs. (5.1)–(5.3), one can find the following relationships for diffusion resistance R_D, diffusion constant D, and permeability P (Δt=diffusion time):

$$R_D = \frac{\Delta t}{\Delta V}; \quad D = \frac{1}{R_D}\frac{\Delta x}{A}; \quad P = \frac{1}{R_D A}; \quad \text{and } P = \frac{D}{\Delta x}. \qquad (5.6)$$

Figure 5.4 shows the ion density profiles before (blue) and after (red) ion exchange. A complete K^+-cation exchange, as indicated by the green line in the upper panel, is hindered by Coulomb attraction. Each K^+-cation that diffuses through the ion channel from the inside out leaves a negative unbalanced charge. In equilibrium,

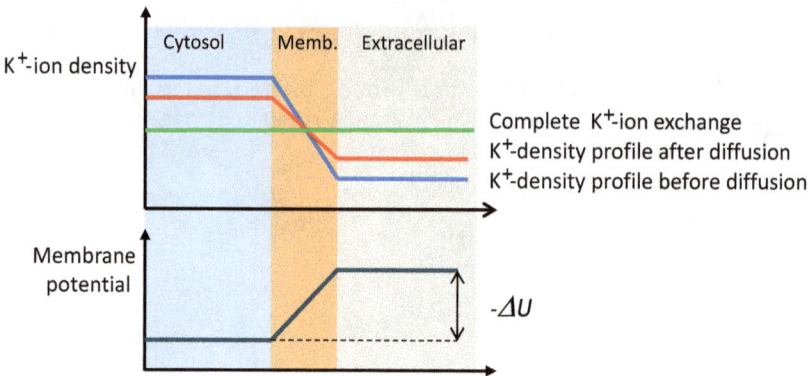

Fig. 5.4: Schematics of the potassium concentration profile across the cell membrane. The large concentration difference cannot be balanced completely because of Coulomb forces, hindering K$^+$ ions' outflow. The remaining concentration difference is responsible for the resting potential.

there must be a balance between thermal energy that promotes ion exchange by diffusion and electrostatic potential that imposes an increasing barrier against out-diffusion. Thus, the concentration ratio in equilibrium is given by

$$\frac{n_i}{n_o} = \exp\left(\frac{Q\Delta U}{RT}\right) = \exp\left(\frac{ZF\Delta U}{RT}\right). \tag{5.7}$$

Here Q is the total charge transported by diffusion, ΔU is the Coulomb potential difference (unit: volt (V)), R is the gas constant for 1 mol (8.314 J/K mol), T is the absolute temperature, and the subscripts i, o stand for inside and outside, respectively. The charge flow can be expressed in terms of the Faraday[2] constant $F = 96\,487$ C. The Faraday constant is the charge equivalent to 1 mol of singly charged ions, and Z is the ionicity, which is 1 for K$^+$. Equation (5.6) is called the Nernst[3] equation.

A similar but inverse density profile across the membrane exists for the Na$^+$ cations, and the same considerations would apply as for the K$^+$ cations. However, the Na$^+$ channels are closed, not perfectly, but well enough that we do not have to consider them here for determining the resting potential.

According to the density profile after K$^+$ ion exchange, the expected Coulomb (or electrochemical) potential difference is then

$$\Delta U = -\frac{RT}{ZF}\ln\frac{n_i}{n_o}. \tag{5.8}$$

The sign is negative because there is a negative gradient of the K$^+$ cations from the inside out. Substituting numbers we find at body temperature of 310 K that the

---

2 Michael Faraday (1791–1867), English physicist.
3 Walther Nernst (1864–1941), German physical chemist, Nobel Prize 1920.

prefactor yields a potential difference of −0.0267 J/C = −26.7 mV. Now we convert the natural logarithm to the decadal logarithm and obtain for the prefactor −61 mV:

$$\Delta U = -61 \text{ mV} \log \frac{n_i}{n_o}. \tag{5.9}$$

Here is a good opportunity to switch to the nomenclature preferred by physical chemists:

$$\Delta U = -\frac{RT}{ZF} \ln \frac{[K^+]_i}{[K^+]_o}, \tag{5.10}$$

where the square brackets indicate mole fractions.[4] We now take cation density ratios according to Tab. 5.1 and obtain a resting potential difference of −90 mV. We conclude that the resting potential is a result of the excess potassium concentration in the cytoplasm.

In a more advanced consideration, one may take into account the finite permeability of the cell membrane for Cl⁻ anions and reflux of Na⁺ cations. All this leads to the *Goldman[5]–Hodgkin[6]–Katz[7] equation* [10], which reads:

$$\Delta U = -\frac{RT}{ZF} \ln \frac{P_{K^+}[K^+]_i + P_{Na^+}[Na^+]_i + P_{Cl^-}[Cl^-]_i}{P_{K^+}[K^+]_o + P_{Na^+}[Na^+]_o + P_{Cl^-}[Cl^-]_o}. \tag{5.11}$$

Here P_i are the respective permeabilities. This equation yields a resting potential difference of about −75 mV, which is confirmed by experiments using microelectrodes [4]. Hence, at rest, there is a potential difference between the cytoplasm and the extracellular space maintained by a delicate gradient of the potassium cation concentration across the cell membrane. The cell is comparable to a battery or a capacitor, where cations and anions are spatially separated to create a voltage difference. The resting potential is constant over time (Fig. 5.5), aside from small voltage fluctuations. But this

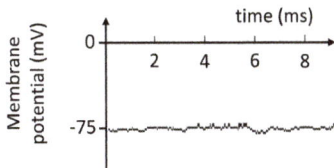

Fig. 5.5: Resting membrane potential as a function of time.

---

4 Mole fraction, also known as molarity, is the mole number of one component X_i divided by the sum of all mole numbers: $[X_i] = X_i / \sum_i X_i$.

5 David E. Goldman (1910–1998), American scientist.

6 Alan Lloyd Hodgkin (1914–1998), English biophysicist, Nobel Prize in Physiology 1963.

7 Bernard Katz (1911–2003), German-British biophysicist, Nobel Prize in Physiology 1970.

constancy is not fixed, instead it is controlled and maintained by a Na^+–K^+ ion exchange pump that always runs in the background. The resting membrane potential can be disturbed by changes of the pH value in the extracellular space. Extensive changes due to either too high or too low salt ion concentrations can be life-threatening.

> **!** The resting membrane potential is a consequence of a K^+-concentration gradient across the cell membrane combined with open K^+-channels for ion exchange. The dynamic equilibrium between ion diffusion and Coulomb potential results in a resting potential of about −75 mV.

5.4 Action potential

When triggered by an external stimulus, the cell reacts by depolarization, often called "triggering an action potential." The external stimulus can be anything from mosquito bites to the brain's decision to lift our bodies or listen to music.

How does activation of the action potential work? This is one of the most fascinating mechanisms that evolution has invented. It is comparable to the deep control of submarines. Triggering means opening Na^+ channels and changing the floating membrane potential. There is always some potential fluctuation and leakage of Na^+ ions from the outside to the inside. This is harmless as long as the fluctuations are smaller than the threshold potential. However, once the stimulus is strong enough to exceed a threshold, all of the voltage-gated Na^+ channels open at once. The Na^+ ions wash in and depolarize the cytoplasm in less than a millisecond. To this end, the Na^+ channels are much faster than the K^+ channels.

Each sense in the body has a different "knob" to activate an action potential. It can be a chemical stimulus to taste and smell, a mechanical tilt of tiny hair cells to hear, photoisomerization of the retinal to seeing, or a self-activated stimulus to the heartbeat. These different stimuli are discussed later in the respective chapters. The result is always the same. As soon as a cell is depolarized in response to a stimulus, it has to fulfill two tasks: First, communicate the state of depolarization to the neighboring cells, for example, through nerve conduction; second, repolarize as quickly as possible and return to the resting potential to be ready for the next stimulus. The refractory period from depolarization to repolarization is about 3 ms. The entire cycle from resting potential through activation to repolarization takes about 5 ms. It is not possible to fire any other action potential during the refractory period.

The time course during a depolarization cycle is sketched in Fig. 5.6. After crossing the threshold, the cell depolarization is limited by the automatic closing of the Na^+ channels as soon as the cell potential crosses the zero level. Then the K^+ channels open again, which have been temporarily closed to accelerate depolarization. And finally, the most important action is to start an ion exchange pump that exchanges 3 Na^+ ions from the interior (cytoplasm) for 2 K^+ ions from the extracellular

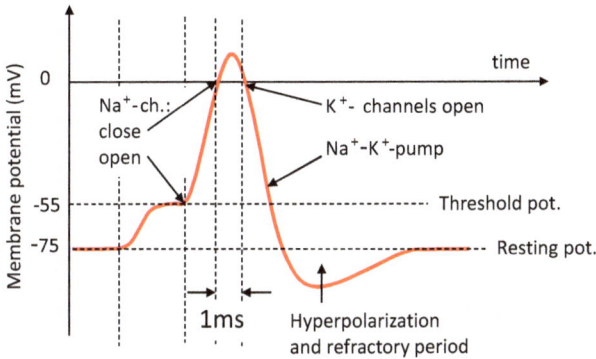

Fig. 5.6: Time development of the membrane potential.

space. This is an active ion transport against the ion concentration gradient, which requires energy. The energy is supplied by a molecule called *adenosine triphosphate* (ATP) . The mechanism of the ATP-assisted ion exchange pump is explained in Section 5.5. As we note, the depolarization is faster than the repolarization of the cell. This is because depolarization only requires the opening of fast Na^+ channels and diffusion does the rest. For repolarization, on the other hand, an active process against the concentration gradient requires more time. Nevertheless, the ion pump is so effective that brief hyperpolarization is possible during the refractory period.

> The action potential is initiated by opening of Na^+ channels, causing a depolarization of the cell. Repolarization requires the activation of an ion-pump (ATP). !

The response of a voltage-gated receptor to stimuli is shown in Fig. 5.7. With a subliminal stimulus, the membrane potential changes briefly, but no action potential is triggered. A single action potential occurs when the stimulus reaches the threshold and is brief. If the stimulus is strong and prolonged (superthreshold), a train of action potentials is activated whose frequency is proportional to the amplitude of the stimulus. The total number of action potentials represents the duration of the stimulus. The frequency is limited to about 200 Hz due to the refractory period, in which no new action potential can be triggered. This analog-to-digital conversion of stimuli is the basic working principle of all sensory receptor cells and is further explored in Chapter 6.

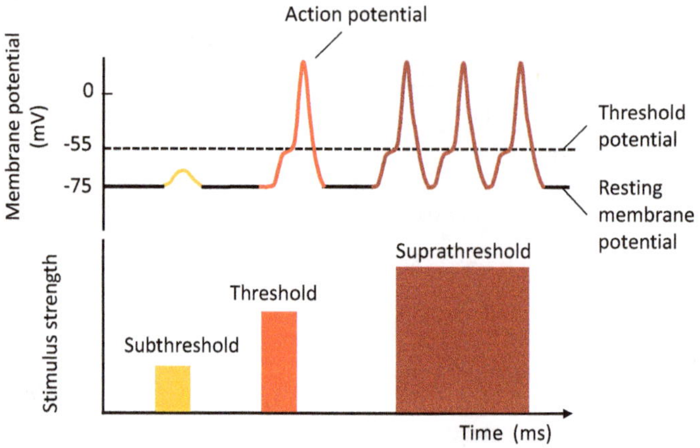

Fig. 5.7: Response of a voltage gated receptor cell for stimuli below and above threshold.

5.5 Electrochemical properties of cells

The cell may be considered a battery or a capacitor, loaded at rest and discharged upon depolarization. Using very simplified model assumptions, we can estimate some physical parameters of the cell's electrochemical properties and gain more insight into its mechanism.

5.5.1 Membrane capacitance

The Gauss law connects the electric field $\vec{E}$ with the charge Q enclosed in an area with surface A:

$$\oint \vec{E} \cdot d\vec{A} = \frac{Q}{\varepsilon \varepsilon_0}. \tag{5.12}$$

Here ε_0 (=8.854 × 10^{-12} As/Vm) is the dielectric constant of the vacuum, and ε is the static permittivity of the material enclosed by the capacitor plates. In our case, this is the permittivity of the cytoplasm, which is about 6, instead of 81 for pure water [9]. The enclosed total charge is, therefore:

$$Q = \varepsilon \varepsilon_0 EA. \tag{5.13}$$

Then the charge per area is

$$\frac{Q}{A} = \varepsilon \varepsilon_0 E. \tag{5.14}$$

The capacitance is defined as follows:

$$C = \frac{Q}{\Delta U_m},$$ (5.15)

and can be normalized to the surface area A:

$$\frac{C}{A} = \frac{Q}{A\Delta U_m} = \frac{\varepsilon\varepsilon_0 E}{\Delta U_m}.$$ (5.16)

The potential difference ΔU_m is the one across the membrane with thickness Δx: $\Delta U_m = E\Delta x$. Thus the transmembrane capacitance C_m is

$$C_m = \frac{\varepsilon\varepsilon_0 A}{\Delta x},$$ (5.17)

and normalized to the surface area:

$$\frac{C_m}{A} = C_{m,A} = \frac{\varepsilon\varepsilon_0}{\Delta x}.$$ (5.18)

Substituting numbers for the membrane thickness of typically $\Delta x = 5$ nm, we obtain:

$$C_{m,A} = 1\,\mu F/cm^2 = 10^{-2}\,F/m^2.$$ (5.19)

Various experiments have confirmed this average value for the membrane capacitance [10].

The standard capacitance of the cell per surface area A is $C_{m,A} = 1\,\mu F/cm^2$. !

5.5.2 Transmembrane ion exchange

Now the question arises as to how many ions have to pass through the membrane in order to generate a potential difference $\Delta U_m = -75$ mV. Using the capacitor equation normalized to the area $\Delta Q/A = C_{m,A}\Delta U_m$, a potential difference of 75 mV or about 0.1 V results in a charge difference of $\Delta Q/A = 0.1\,\mu C/cm^2$. This charge is equivalent to 0.1 $\mu C/e$ ($e = 1.6 \times 10^{-19}C$) = 6.25×10^{11} monovalent ions per cm^2 or only 6250 ions per μm^2. This is indeed a very small number of ions. With about 10 ion channels per μm^2, each channel would transport only 625 ions during an action potential. Rounding the numbers to 1000 ions transported by a single channel within a 1 ms depolarization time shows that the ion transport rate of ion channels is indeed rather high: 10^6 ions/s.

5.5.3 Transmembrane resistance

With the numbers derived so far, we can estimate the transmembrane specific resistance. First, we write for the ion current according to Ohm's law:

$$I_m = \frac{\Delta Q}{\Delta t} = \frac{\Delta U}{R_m}. \tag{5.20}$$

And normalized to the surface area A:

$$\frac{I_m}{A} = \frac{\Delta Q}{A\Delta t} = \frac{\Delta U_m}{AR_m}. \tag{5.21}$$

Now we recall that the resistance is given by

$$R_m = \rho_m \Delta x / A, \tag{5.22}$$

where ρ_m is the specific resistance, Δx and A have the same meaning as before. Substituting and solving for ρ_m yields

$$\rho_m = \frac{\Delta V \cdot \Delta t}{\Delta x \cdot \Delta Q / A}. \tag{5.23}$$

Substituting numbers: ΔU=100 mV; Δt=1 ms; Δx=5 nm; $\Delta Q/A$=0.1 μC/cm^2, we estimate the specific transmembrane resistance as ρ_m=1.5–2×10^7 Ωm. This value will be important for estimating the time constant for the depolarization.

> **!** During depolarization, about 1000 ions are transported per channel with a rate of 10^6 ions per second. The transmembrane specific resistance is about $\rho_m = 2 \times 10^7 \Omega$m.

5.5.4 Depolarization time constant

During depolarization, the cell discharges as a capacitor and transmembrane current flows through the channel resistance. The flow of current can be represented by an equivalent circuit diagram, shown in Fig. 5.8.

Taking the time derivative of the capacitor equation $\Delta Q = C_m \Delta V$, we find for the discharge current during depolarization:

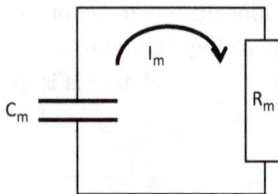

Fig. 5.8: Equivalent circuit diagram consisting of a membrane capacitance C_m and a transmembrane resistance R_m for describing the depolarization of a cell.

$$I_m = C_m \frac{d(\Delta U_m)}{dt}. \tag{5.24}$$

This current must be equal and opposite to the current through the resistor:

$$I_m = -\frac{\Delta U_m}{R_m}. \tag{5.25}$$

Therefore:

$$C_m \frac{d(\Delta U_m)}{dt} = -\frac{\Delta U_m}{R_m}. \tag{5.26}$$

Rearranging

$$\frac{d(\Delta U_m)}{\Delta U_m} = -\frac{dt}{C_m R_m} \tag{5.27}$$

and integrating both sides yields

$$\ln(\Delta U_m(t)) = -\frac{t}{C_m R_m} + \ln(U_0) \tag{5.28}$$

and

$$\Delta U_m(t) = U_0 e^{-t/C_m R_m} = U_0 e^{-t/\tau}, \tag{5.29}$$

where $\tau = C_m R_m$ is the time constant for the depolarization and $\ln(U_0)$ is an integration constant for $t = 0$. The time constant can be rephrased, using the expression for the transmembrane resistance $R_m = \rho_m \Delta x / A$:

$$\tau = C_m R_m = C_{m,A} A \rho_m \frac{\Delta x}{A} = C_{m,A} \rho_m \Delta x. \tag{5.30}$$

With the resistivity of a cell $\rho_m = 2 \times 10^7$ m, the membrane thickness $\Delta x = 5$ nm, and the capacitance per area $C_{m,A} = 10^{-2}$ F/m², the last equation contains only well-known values. Substituting numbers, we get for the time constant: $\tau = 1$ ms. This is a very reasonable estimate for the depolarization time constant and agrees well with experimental observations. As previously mentioned, repolarization is slower due to the active ion transport via the Na^+–K^+ ion pump.

> The time constant for depolarization of a cell is about 1 ms. !

5.5.5 Electrochemical potential

How can we experimentally probe the electrical resistance of a cell? In general, if we want to determine the electrical resistance of some material, we use a power supply and measure the current that goes through a well-defined piece of conductor versus voltage applied. The slope yields the conductance g by applying Ohm's law:

$$I_{ohm} = g \cdot \Delta U, \tag{5.31}$$

where $g = 1/R$ is the inverse of the resistance (unit ohm^{-1} = siemens, S). The linear relationship of Ohm's law implies that an external power supply provides a potential difference ΔU_{el}, such that $I_{ohm} = 0$ for $\Delta U_{el} = 0$. However, the cell is negatively biased by the resting (chemical) potential $\Delta U_K = -75$ mV. Therefore, the K$^+$ ion current is composed of the electrical and chemical potentials:

$$I_K = g_K(\Delta U_{el} - \Delta U_K). \tag{5.32}$$

The difference $(\Delta U_{el} - \Delta U_K)$ is known as the *electrochemical potential*. It can be described by an equivalent circuit diagram, which consists of a conductor that represents the K$^+$ ion channel in series with a battery representing the offset K$^+$ chemical potential at rest (Fig. 5.9(a)). In this form, the K$^+$ ion channel displays an ohm-like linear I_K~ΔU_{el} characteristic, with the conductance as the slope and offset by the resting potential $-\Delta U_K$. This linear relationship has been experimentally confirmed for isolated ion channels fixed in artificial lipid membranes [11, 12, 15] and is schematically shown in Fig. 5.9(b).

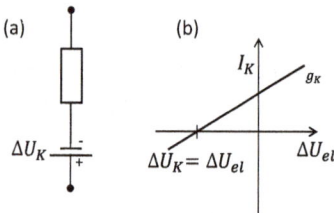

Fig. 5.9: Equivalent circuit for the K$^+$ channel conduction, and (b) K$^+$ channel current versus external potential difference.

5.5.6 Hodgkin–Huxley model

The Na$^+$ ion current I_{Na} and the leakage current I_L through the membrane can be treated in the same way as the K$^+$ ion current and they work in parallel with the K$^+$ channel (Fig. 5.10). All channels are characterized by their respective conductances g_K, g_{Na}, and g_L and resting potentials ΔU_K, ΔU_K, ΔU_L. Note that g_K and g_L have constant values, but the conductance of the Na$^+$ channel is voltage-gated and therefore varies greatly during depolarization. Also, note that the K$^+$ and Na$^+$ channels operate in opposite directions. The L and K$^+$ polarities are in the same direction, because the leakage current (mainly Cl$^-$) has the opposite charge and the opposite ion flow direction to the K$^+$ channel. In either case, the L conductance is much lower than that of the other two channels. To complete the circuit, a capacitance C_m per unit area representing the insulating properties of the membrane is included in parallel to the ion channels. The total membrane current per unit area is then expressed as follows:

$$I_{tot} = C_m \frac{dU_m}{dt} + I_K + I_{Na} + I_L, \tag{5.33}$$

or explicitly

$$I_{tot} = C_m \frac{dU_m}{dt} + g_K(\Delta U_{el} - \Delta U_K) + g_{Na}(\Delta U_{el} - \Delta U_{Na}) + g_L(\Delta U_{el} - \Delta U_L). \tag{5.34}$$

This expression is known as *Hodgkin–Huxley*[8] *equation* [13], according to the British biophysicists Hodgkin and Huxley, who modeled the membrane potential mathematically for the first time, long before discrete and specific ion channels were observed. The Hodgkin–Huxley equation can be refined by taking into account time and voltage-dependent coefficients for the channel conductances and also by considering the $Na^+ - K^+$ ion pump efficiency. Experimentally, the potentials and conductances have been determined using microelectrodes for measuring potential differences and currents between the cytoplasm and the extracellular space using the patch-clamp technique, discussed in Section 5.6.

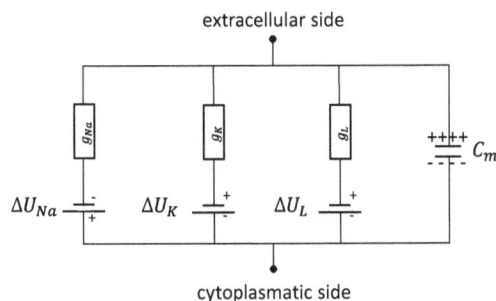

Fig. 5.10: Equivalent circuit of the cells' electrical properties. Resistors with conductances g_i represent the respective ion channel conductivities; capacitances symbolize resting potentials, and C_m denotes the global capacitance cell membrane.

Now we reconsider the cell condition at rest, characterized by a vanishing total current:

$$I_{tot} = 0 = C_m \frac{dU_m}{dt} + I_K + I_{Na} + I_L, \tag{5.35}$$

---

8 Andrew Fielding Huxley (1917–2012), British neurophysiologist and Nobel Prize laureate in physiology 1963.

and explicitly by the condition:

$$C_m \frac{dU_m}{dt} = 0. \tag{5.36}$$

Therefore at rest:

$$I_K + I_{Na} + I_L = 0,$$

or

$$g_K(\Delta U_{el} - \Delta U_K) + g_{Na}(\Delta U_{el} - \Delta U_{Na}) + g_L(\Delta U_{el} - \Delta U_L) = 0. \tag{5.37}$$

The last equation can be rearranged, yielding for the cells' electric potential at rest the expression:

$$\Delta U_{el} = \frac{g_K \Delta U_K + g_{Na}\Delta U_{Na} + g_L \Delta U_L}{g_K + g_{Na} + g_L}. \tag{5.38}$$

This equation is similar to the Goldman–Hodgkin–Katz equation (5.4), but it relates conductances and resting potentials instead of permeabilities and ion concentrations to the potential difference. In this form, experimental verification is easier as conductances can be better determined than permeabilities.

5.6 Single channel conductance

So far we have considered the electrochemical properties of cells as a whole and derived the global ion currents through their respective ion channels. We have also seen that there is a high density of such channels in the cell membrane. Therefore, the question arises as to whether it would be possible to determine the conductance of a single ion channel. The single channel ion current is

$$i_K = \gamma_K(\Delta U_{el} - \Delta U_K), \tag{5.39}$$

where

$$i_K = \frac{I_K}{N_K}, \quad \gamma_K = \frac{g_K}{N_K}; \tag{5.40}$$

and N_K is the number of K^+ ion channels per unit area. Measuring the conductivity of a single channel is a major technical and experimental challenge. In fact, this was achieved using the so-called patch-clamp technique, which Neher[9] and Sakmann[10] invented in 1976. The word "patch" refers to a piece of membrane that

---

9 Erwin Neher (*1944), German biophysicist, Nobel Prize laureate in physiology 1991.
10 Bert Sackmann (*1942), German physiologist, Nobel Prize laureate in physiology 1991.

contains an ion channel. The patch is selected by a very fine tapered and smooth glass pipette pressed against the membrane. With a tight seal, the electrical resistance is in the gigaohm range (10^9–10^{11} Ω) [15]. The term "clamp" does not refer to the clamping of the pipette on the membrane, as one might assume, but to the fixation of the membrane potential to a preselected value via a feedback control, i.e., a voltage clamp introduced by Hodgkin and Huxely [13]. Figure 5.11 shows a sketch of the experimental setup.

The high resistive seal is necessary to avoid leakage currents and to measure resistances, which are in the range of 100 GΩ. If a single channel has been located within the pipette's orifice, an electrolyte in the pipette connects the channel with an electrode, and a counter electrode connects the electrode to the extracellular fluid. A gate voltage triggers the voltage-sensitive channels to open or close. By varying the gate voltage, the current i_K as a function of ΔU_{el} can be measured as shown in Fig. 5.9 and the single-channel conductance γ_K follows from the slope.

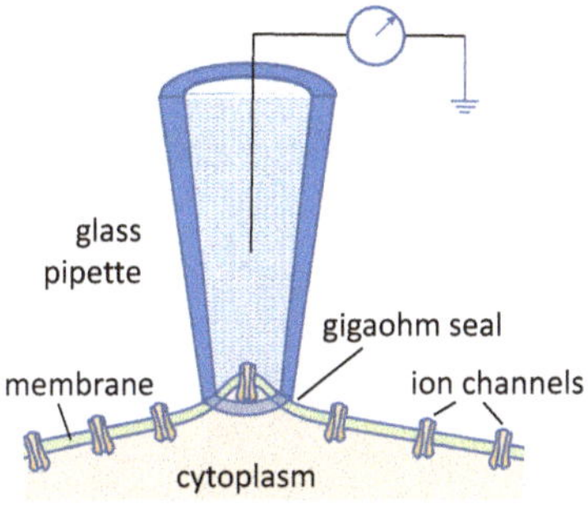

glass
pipette

gigaohm seal

membrane   ion channels

cytoplasm

Fig. 5.11: Sketch of the experimental setup for studying single-channel conductance with the patch-clamp method. A glass pipette with a smooth opening of 0.3–1 μm makes a tight seal with a membrane. The high-ohmic seal isolates the membrane from the ion channel and allows various electric measurements. (adapted from http://commons.wikimedia.org/wiki/File: Patchclamp.svg)

The patch-clamp technique not only enables in vivo studies of single-channel currents. It also allows membrane patches to be detached from the cell for isolated investigations [14]. Studying channels in isolation has the advantage that the channel conductivity can be determined from I–U curves without being offset by the resting potential. Furthermore, proximity effects from other nearby channels can be neglected. In addition, this technique enables various kinetic experiments that shed unprecedented light on the mechanism of individual ion channels [15]. An example is shown in Fig. 5.12. A membrane potential is applied stepwise to a single Na$^+$ channel, and the channel current is measured with an electrode. Instead of recording a continuous ion current, the Na$^+$ channel current occurs in a burst-like fashion for short periods of a few milliseconds. The burst has a fixed amplitude but varying time periods. The edges on either side of the pulse are very sharp, and the shape is completely reversible. The current amplitude increases linearly when changing the membrane potential, while the current bursts' stochastic nature remains. This behavior indicates that the Na$^+$ channel opens and closes with a certain probability p_{Na}, and that some conformational mechanism exists, which changes the channel conductivity abruptly and stochastically. Consequently, the description of the single-channel current should be modified to take the probability aspect into account:

$$i_{Na} = p_{Na}\, \gamma_{Na}(\Delta U_{el} - \Delta U_{Na}).\qquad(5.41)$$

The probability coefficient p_{Na} also expresses the time duration of the pulses. The ion current i_{Na} is plotted in Fig. 5.12(c) as a function of applied voltage. The slope yields the channel conductance of 30 pS for the Na$^+$ channel. The conductance of the K$^+$ channel turns out to be of similar magnitude, while the Cl$^-$ conductance appears to be somewhat lower.

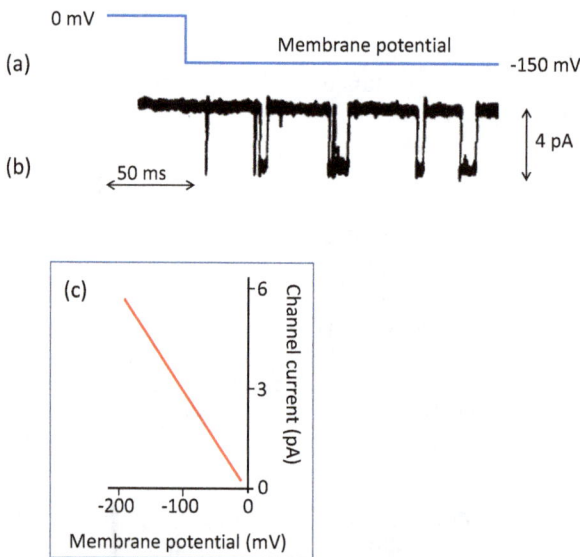

Fig. 5.12: Using the patch-clamp technique, the conductivity of a single Na$^+$ channel is probed. After application of a step-like membrane potential shown in the top panel (a), the Na$^+$ channel opens, and current pulses occur in stochastic sequence and over different time spans (b). The current amplitude is constant and proportional to the voltage applied. The bottom panel (c) shows a schematic plot of current amplitude versus membrane potential. The slope yields the channel conductance γ_{Na} (adapted from [15]).

! The patch-clamp technique allows to determine the conductivity of single ion channels. The single channel conductance is about 30 pS.

In view of the stochastic nature of the Na$^+$ channel conductance, we revisit the depolarization behavior of a cell by studying all Na$^+$ channels separately. The results are shown in Fig. 5.13 for a representative number of nine different depolarization events. After application of a membrane potential, current bursts with constant amplitude occur stochastically with different time lags. But in some cases, the channel remains completely silent. On average, the activity of the channels diminishes with time, even before turning off the potential.

Recording the response of a single channel N times is equivalent to recording N different channels a single time. Therefore, we may assume that these N channels belong to a single cell. Now we take the integral over all N sodium channels, which yields:

$$I_{Na} = Ni_{Na} = Np_{Na}\,\gamma_{Na}(\Delta U_{el} - \Delta U_{Na}). \tag{5.42}$$

The total depolarization current I_{Na} is plotted in the lower part of Fig. 5.13. This shows that the global depolarization current, such as the one plotted in Fig. 5.6, is – in fact – composed of millions of individual discrete and stochastically opening channel currents. The result is a smooth curve with no signs of the discrete nature of each individual ion channel conductivity.

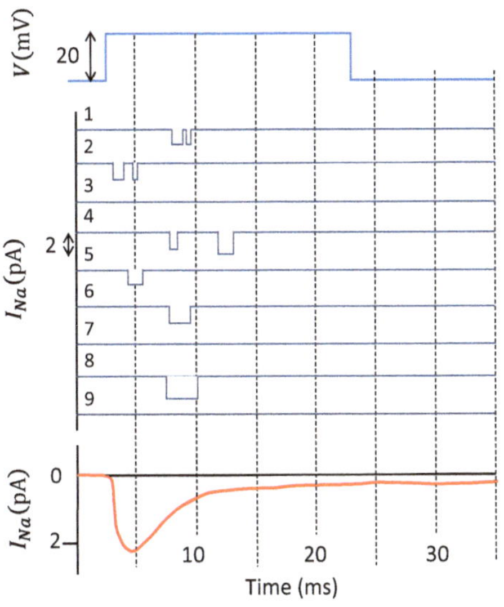

Fig. 5.13: After applying a membrane clamp-potential (blue line in the top panel) using the patch-clamp technique, the current response is recorded as a function of time. All current traces shown are schematic and smoothed. The red curve in the bottom plot is the result of averaging all recorded current traces (adapted from [16]).

Depolarization of a cell's membrane potential is the integrated result of many single ion channels that open burst like and stochastically over short time periods.

5.7 ATP-powered Na$^+$–K$^+$ pump

We have already realized that the Na$^+$–K$^+$ ion exchange pump for repolarization of the cell requires an energy source, which is ATP. This phosphate molecule consists of a carbon compound as a backbone (adenosine) and a phosphorous part depicted in Fig. 5.14. Three phosphorus groups $P_\alpha, P_\beta, P_\gamma$ are bound by oxygen ions, and they also have negatively charged side oxygen ions attached. These negative charges have a high repulsive potential. Removing one phosphate group from the end converts ATP to adenosine diphosphate (ADP), a process known as ATP hydrolysis. The reaction $ATP + H_2O \rightarrow ADP + PO_4{}^{3-} + \Delta G$ releases free energy of $\Delta G = -30.5$ kJ/mol. The split-off γ-phosphate group is often designated as Pi (inorganic phosphate). The body stores ATP as an energy source during the metabolic process (Chapter 4). About one-third to two-thirds of the total body energy consumption is used to synthesize fresh ATP by converting $ADP + PO_4{}^{3-} + \Delta G$ back to $ATP + H_2O$, a process referred to as *phosphorylation* (or cellular respiration). Indeed, ATP is the fuel that keeps the ion pumps running and, therefore, the entire body. ATP is the ubiquitous carrier of metabolic energy in all living cells.

The phosphorylation occurs in the cytoplasm when an enzyme attaches a third phosphate group to the ADP: $ADP + PO_4^{3-} + \Delta G \rightarrow ATP + H_2O$. An enzyme is required because the reaction is endothermic. The enzymatic mechanism underlying the synthesis of ATP was elucidated by Boyer[11] and Walker.[12]

The Na$^+$–K$^+$ ion exchange pump consists of a large transmembrane protein. On the cytoplasm side, it has binding sites for Na$^+$ and ATP, while on the extracellular surface it binds K$^+$. The action of the Na$^+$–K$^+$ ion exchange pump was discovered and described by Skou[13] [17].

The working principle of the Na$^+$–K$^+$ ion exchange pump powered by ATP is sketched in Fig. 5.15(A–F). First, we notice that the ion pump is asymmetric and is open to either one or the other side of the cell membrane. This configuration differs significantly from ion channels, which are always open or have a lock that opens for diffusive transport via a suitable gate potential. In the starting position (panel A), the ion pump is open to the cytoplasm so that Na$^+$ ions can penetrate and find three specific locking points. As soon as all three positive ions have settled in their pockets, their combined positive charge triggers the phosphate end group Pγ to dissociate and bind to a side group of the ion channel (panel B). The release of energy, in turn, changes the conformation of the ion pump, which now opens to the outside, releasing the Na$^+$ ions into the extracellular space (panel C). This conformational change allows

---

11 Paul D. Boyer (1918–2018), American biochemist, Nobel Prize recipient in chemistry 1997.
12 John Ernest Walker (*1941), British chemist. Nobel Prize recipient in chemistry 1997.
13 Christian Skou (1918–2018), Danish physiologist, Nobel Prize in Chemistry 1997.

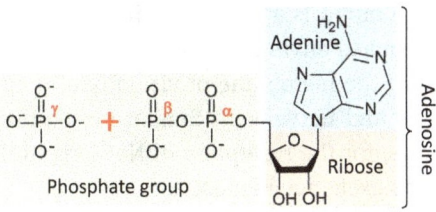

Fig. 5.14: Chemical structure of adenosine triphosphate (ATP).
(adapted from: https://commons.wikimedia.org/wiki/File:ATP_structure.svg © creative commons).

two K^+ ions to diffuse into the ion pump (panel D). Once they have found their sites, these two positive charges repel the phosphate group that leaves its binding site (panel E). As soon as the phosphate group leaves, the ion channel's shape returns to its original conformation, releasing the K^+ ions into the cytoplasm (panel F) and starting a new cycle until the cell is completely repolarized and returned to the resting potential. With each cycle, the $Na^+–K^+$ ion pump hydrolyzes one ATP molecule, from which it gains the energy for the ion transport against the electrochemical gradient.

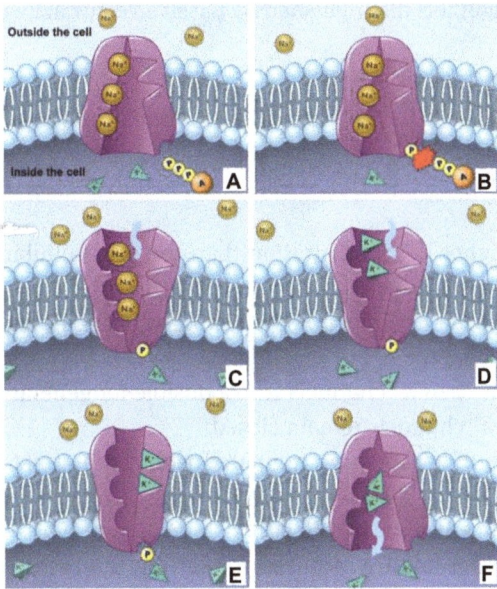

Fig. 5.15: The sodium–potassium exchange pump mechanism. The Na^+/K^+ pump moves two potassium ions (gold circles) from outside the cell to inside and three sodium ions (green triangles) from inside the cell to the outside by the breakdown of ATP molecules into ADP + P (yellow circles) (reproduced from https://openi.nlm.nih.gov/, open access).

The ATP-powered rocking ion exchange pump is a very powerful mechanism for active charge transport in wet electrochemical environments that resets the action potential to rest. The pump is found in the membrane of virtually all living cells, but is particularly common in muscle and nerve cells. Similar mechanisms apply to the H^+/K^+ exchange in the intestine, for the resorption of Na^+ ions in the urine of the kidneys, and for the Na^+/Ca^{2+} ion exchange in muscles.

> **!** The ATP-powered Na^+–K^+ ion exchange pump is a molecular motor that transports ions against the chemical gradient and is therefore responsible for repolarization of cells after foregone depolarization. Three Na^+ ions go out in exchange of two K^+ ions going in.

5.8 Summary

S5.1   Cell membranes define areas inside the cell (cytoplasm) and outside (extracellular space).

S5.2   The cell membrane consists of a double lipid layer.

S5.3   Without ion channels, the membrane is impermeable to ion exchange.

S5.4   There are two main types of channels distinguished by passive versus active ion transport.

S5.5   Passive ion channels are either open or closed.

S5.6   Passive ion channels are gated to open or to close either by voltage, ligands, or mechanical stress.

S5.7   Passive channels are specific for the type of ion that is allowed to permeate through.

S5.8   The cytoplasm has a much higher K^+ concentration than the extracellular space.

S5.9   Vice versa, the extracellular space has a much higher Na^+ ion concentration than the cytoplasm.

S5.10 The outflow of K^+ ions through open K^+ channels is responsible for generating a resting potential while Na^+ channels remain closed.

S5.11 The electrochemical properties of cells can be compared with those of batteries or capacitors.

S5.12 The capacitance of a cell is typically about $1\ \mu F/cm^2$.

S5.13 Only about 600 K^+ ions per channel are required for setting the resting potential.

S5.14 The ion exchange in equilibrium between the cytoplasm and extracellular space follows from the ratio of electric energy to thermal energy.

S5.15 The resting potential is about –75 mV and is described by the Nernst equation.

S5.16 Stimuli from receptors cause the Na^+ channels to open, which depolarizes the cell and causes an action potential.

S5.17 Depolarization requires about 1 ms.

S5.18 Repolarization requires active ion transport against the ion concentration gradients in the cell.

S5.19 Active transport is achieved by the ATP-assisted Na^+–K^+ pump, resetting the action potential to a resting potential.

S5.20 The total action potential lasts for about 5 ms.

S5.21 During the refractory period, cells cannot be activated.

S5.22 Using the patch-clamp technique, the single ion channel I–V characteristic and the channel conductance can be determined.

S5.23 With the patch-clamp technique, membrane patches holding single ion channels can be investigated in isolation.

S5.24 Typical values for single channel currents are a few picoampère (pA) and channel conductances of 10–30 pS.

Questions

Q5.1 What does the cell membrane consist of?

Q5.2 Which ion channels can open and close?

Q5.3 Which ion channels are always open?

Q5.4 Which cation is predominately in the cytoplasm, and which one is in the extracellular space?

Q5.5 What determines the resting potential?

Q5.6 What is the sign and value of the resting potential?

Q5.7 What causes an action potential?

Q5.8 How long does an action potential typically last before it returns to the resting potential?

Q5.9 What is the capacitance per unit area of a cell?

Q5.10 How is the strength of a stimulus translated into action potentials?

Q5.11 Explain the working principle of the Na^+–K^+ ion pump. What is pumped, and how is this achieved?

Q5.12 What is the energy source for the ion pump?

Q5.13 Which is correct: the ATP-powered Na^+–K^+ pump supports active ion transport, or it supports passive ion transport?

Q5.14 How many molecules of ATP are produced from one molecule of glucose?

Q5.15 Where does ATP synthase take place?

⚡ **Attained competence checker**	+	0	–
I can describe the main components of the cell acting as a rechargeable battery.			
I can derive the Nernst potential.			
I know what the Goldman–Hodkgin–Katz equation stands for.			
I know what the minimal requirement is to start an action potential.			
I realize that the cell membrane has a capacitance and the ion channels have a conductance.			
I can draw an equivalent circuit diagram for the cells' electric properties.			
I know what the Hodgkin–Huxley equation describes.			
I can describe the main idea of the patch-clamp technique.			
I realize that the single Na^+ ion channel current is a discrete and stochastic event for a short period of time.			
I can describe the different steps that the Na^+–K^+ pump uses for resetting the cell after an action potential.			

Suggestions for home experiments

HE5.1 **Salt in the extracellular space**
Place a single flower, such as a tulip, in a jam jar filled with water. Add a tablespoon of salt and stir it. After a day or two, observe the flower. How has it changed compared to a flower that has been standing in normal water for the same period of time. Can you explain the difference?

ℹ️ Exercises

E5.1 **The cell as a capacitor:** Consider the celle C, diameter 10 μm, plate spacing (thickness of the cell membrane) approx. $d = 5$ nm., dielectric constant $\varepsilon = 5$ for the cell membrane, and surface charge density $\Delta Q_{cell}/A = 0.1\,\mu C/cm^2$.
a. Calculate the electric field $E(Q, r)$ that drops across the cell membrane when a cell is at rest.
b. Compare your result in (a) with the electric field derived from the resting potential of the cell.
c. Determine the stored electric energy density in the resting state of the cell.

E5.2 **Time constant:** Figure 5.8 shows a resistor in series with a capacitor. The charging and discharging of the capacitor requires some time that depends on the capacitance C and the resistance R. Determine the resistance, knowing that the capacitance per unit area is $1\,\mu F/cm^2$.

E5.3 **Resting potential:** Knowing the resting potentials for the different ion channels according to Tab. 5.1 and the conductances (g_K = 10 μS, g_{Na} = 0.5 μS,g_L = 2.5 μS), calculate the resting potential of the cell.

E5.4 **ATP energy release:** With the hydrolysis of ATP, the calculated free energy release ΔG of 1 mol of ATP into ADP and Pi is −30.5 kJ/mol. How big is this energy in terms of electron volts per molecular bond?

References

[1] Funk RH, Monsees T, Ozkucur N. Electromagnetic effects – From cell biology to medicine. Prog Histochem Cytochem. 2009; 43: 177–264.
[2] Campbell NA, Reece JB. Biology. 11th edition. New York: Pearson; 2011.
[3] Alberts BE, Johnson A, Lewis J, Morgan D, Raff M, Roberts K, Walter P. Molecular biology of the cell. 6th edition. New York: Garland Science; 2014.
[4] Kandel ER, Schwartz JH, Jessell TM, Siegelbaum SA, Hudspeth AJ. Principles of neural science. 5th edition. New York, Chicago, San Francisco: McGraw Hill-Medical; 2013.
[5] MacKinnon R, Cohen SL, Kuo A, Lee A, Chait BT. Structural conservation in prokaryotic and eukaryotic potassium channels. Science. 1998; 280: 106–109.
[6] Bezanilla F. Ion channels: From conductance to structure. Neuron. 2008; 60: 456–468.
[7] Jentsch TJ, Stein V, Weinreich F, Zdebik AA. Molecular structure and physiological function of chloride channels. Physiol Rev. 2002; 82: 503–568.
[8] Kole MPH, Ilschner SU, Kampa BM, Williams SR, Ruben PC, Stuart GJ. Action potential generation requires a high sodium channel density in the axon initial segment. Nat Neurosci. 2008; 11: 178–186.
[9] Goldman DE. Potential, impedance, and rectification in membranes. J Gen Physiol. 1943; 27: 37–60.
[10] Gentet LJ, Stuart GJ, Clements JD. Direct measurement of specific membrane capacitance in neurons. Biophys J. 2000; 79: 314–320.
[11] Bezanilla F. Electrophysiology and the molecular level for excitability. The nerve impulse. Available at: http://nerve.bsd.uchicago.edu/
[12] Bezanilla F. Gating currents. J Gen Physiol. 2018; 150: 911–932.
[13] Hodgkin AL, Huxley AF. A quantitative description of membrane current and its application to conduction and excitation in nerve. J Physiol. 1952; 117: 500–544.
[14] Neher E, Sakmann B. The patch-clamp technique. Sci Am. 1992; 3: 44–51.
[15] Hamill OP, Marty A, Neher E, Sakmann B, Sigworth FJ. Improved patch-clamp techniques for high-resolution current recording from cells and cell-free membrane patches. Pflügers Arch. 1981; 391: 85–100.
[16] Pape H-C, Kurtz A, Silbernagel S, eds. Physiologie. 7th edition. Stuttgart, New York: Thieme Verlag; 2014.
[17] Skou JC. The influence of some cations on an adenosine triphosphatase from peripheral nerves. Biochim Biophys Acta. 1957; 23: 394–401.

Further reading

Boron WF, Boulpaep EL. Medical physiology. 2nd edition. Philadelphia, London, New York: Saunders W.B. Elsevier; 2012.
Malmivuo R, Plonsey R. Principles and applications of bioelectric and biomagnetic fields. Oxford, New York, Athens: Oxford University Press; 1995.
Waigh TA. Applied biophysics: A molecular approach for physical scientists. New York, London, Sydney, Toronto: Wiley & Sons; 2007.

6 Signal transmission in neurons

Nerve cells under the microscope

Important physical properties of nerve fibers

	Unmyelinated	Myelinated
Axon radius a	0.5–1 µm	
Membrane thickness d	5 nm	5–20 µm
Capacitance per unit length	6×10^{-8} F/m	2×10^{-10} F/m
Transverse resistance ρ_m	2×10^7 Ωm	
Parallel resistance ρ_p	0.5 Ωm	
Decay length λ	1 mm	10 mm
Signal velocity v	1 m/s	Up to 120 m/s
Signal time constant τ	1 ms	1 ms

6.1 Introduction

The previous chapter focused on action potentials in single cells at specific locations in the body, e.g., in the brain, muscles, or heart. Most cells can generate action potentials, but only nerve cells are specialized to transmit them over long distances. The body's nervous system has three functions:

(1) Perception of stimuli from the environment, conversion into action potentials and transmission of signals along afferent nerve fibers to the central nervous system (CNS);
(2) Integration and processing of received data in the CNS;
(3) Emission of signals along efferent nerve fibers from the center to the periphery for reactive movement.

The signal reception causes a graded receptor potential proportional to the stimulus strength. However, signal transmission requires action potentials with on-off

https://doi.org/10.1515/9783110756951-006

characteristics. How these two potentials work together to transmit signals over long distances is the topic of this chapter. Since neurons are neither metal wires for charge transport nor glass fibers for photon transport, another mechanism is needed to make the signal transmission along the nerve fibers responsive, fast, and reliable. Here we cover the transmission aspect of neurons. The sensory part of the nerves is exemplified for the visual perception in Chapter 11 and for the auditory and vestibular perception in Chapter 12. Standard human physiology and neurology textbooks cover all other aspects of the nervous system; some suggestions are listed under "Further reading." In particular, the book by Kandel and coauthors on *Principles of Neural Science* is highly recommended.

6.2 Overview on signal transmission

6.2.1 Components of neurons

Neurons are cells, i.e., one neuron is one single cell. Nerves, in contrast, may consist of a whole body of interconnected neurons. Neurons have widely different shapes, lengths, and tasks. However, all neurons share four characteristic components (Fig. 6.1):
(1) Cell body (*soma*): contains the cell nucleus and mitochondria
(2) Dendrites: fine fibers that connect the sensory part with the soma
(3) Axon: single fiber emanating from the soma and connecting to cells in the periphery
(4) Axon terminal: connection to a target cell

Dendrites and axons together are called nerve fibers. Dendrites are usually short; axons can be very long, stretching from brain to toe. Arrows in Fig. 6.1 indicate the information flow. Dendrites collect signals within the cell's areal reach, integrate those signals, and funnel them to the *axon hillock* (Fig. 6.1). The axon hillock is a specialized part of a neuron cell that connects to the axon. Receptor potentials that pass the axon hillock fire an action potential that travels down the axon toward the axon terminal. To speed up the signal transmission in long nerve fibers, the axon is coated by insulating sheaths, called *Schwann's cell*.[1] These cells are arranged like beads on a string and are separated by constrictions, called *Nodes of Ranvier*,[2] The purpose of Schwann cells and nodes of Ranvier is a jump-type (saltatory) conduction of the action potential from one node to the next, where at the same time the signal strength is boosted up and refreshed. Finally, the action potential reaches its

---

1 Theodor Schwann (1810–1882), German physiologist.
2 Louis-Antoine Ranvier (1835–1922), French anatomist.

destination, for instance, at a muscle cell, which may trigger a muscle contraction. Referring to the numbers in Fig. 6.1, the four essential steps of the signals' transduction from the sensory receptor to the target cell are as follows:

1. Signal perception: Incoming stimuli are received at the dendrites and change the membrane potential.
2. Signal integration: Changes of the membrane potential, which pass the axon hillock, initiate an action potential.
3. Signal conductance: Action potentials travel along axon fibers coated with Schwann cells up to the axon terminals.
4. Signal transmission: Neurotransmitters are released at synapses to activate target cells.

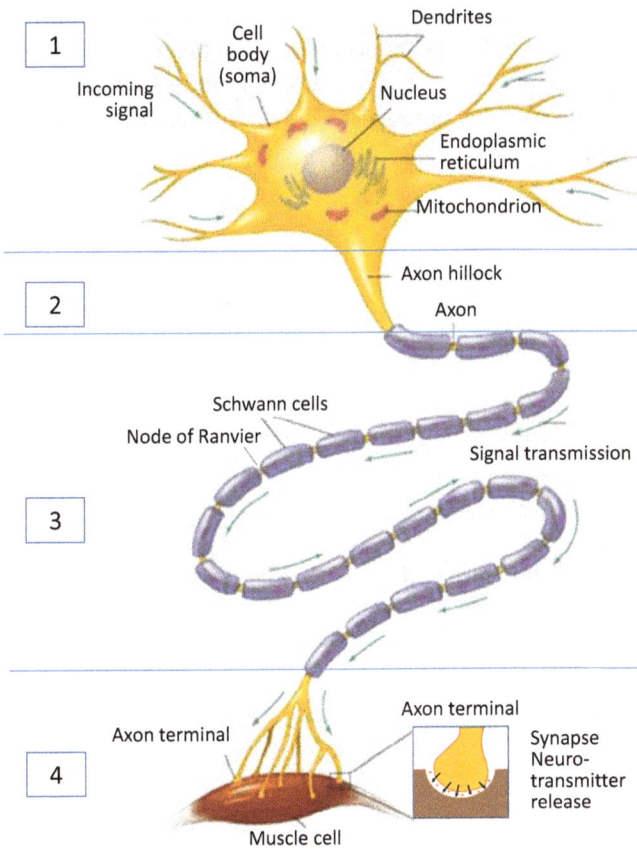

Fig. 6.1: Neuron consisting of dendrites, soma, axon, and axon terminal. The numbers refer to 1: signal reception, 2: integration, 3: conductance, and 4: transmission through an axon from the central nervous system to the target cell. At the axon terminal, the connection is established by neurotransmitters.

6.2.2 Afferent and efferent neurons

We distinguish between *afferent* neurons (input from the periphery guided to the CNS) and *efferent* neurons (leading away) *neurons*. Afferent neurons, also called *sensory neurons,* transmit signals from the sensory periphery to the center, i.e., to the spinal cord and the brain. Efferent connections, also called *motor neurons,* conduct signals from the center to the movable periphery to initiate actions (see Chapter 1 and Fig. 6.2). The neuron shown in Fig. 6.1 is an efferent neuron. The dendrite endings of this neuron are somewhere in the brain or in the spinal cord. The distal end of the axon is at a muscle fiber. There is no direct connection between a receptor cell and a muscle cell. All receptor signals first go to the CNS and from there back to the action center. There is a third type of neuron, called *interneuron* or *association neuron.* It is located within the CNS between sensory and motor neurons. In a few cases, the interneuron makes a direct connection between them. Note that the afferent and efferent neurons have different shapes. Afferent neurons are unipolar neurons, whereas efferent neurons are multipolar neurons. This distinction is presented in the next paragraph.

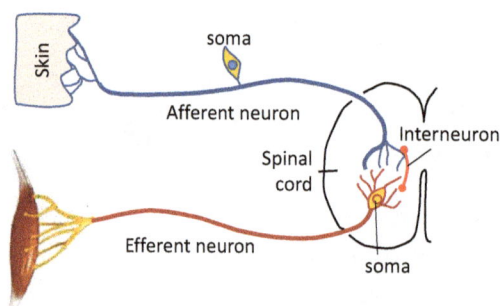

Fig. 6.2: Afferent neurons connect receptors with the CNS, efferent neurons conduct signals to action centers like a muscle. Afferent neurons are unipolar neurons; efferent neurons are multipolar neurons.

> **!** Two types of neurons connect the CNS with the periphery: afferent neurons transmit sensory information from the periphery to the CNS; efferent neurons transmit signals from the CNS to the periphery to initiate movements.

6.2.3 Unipolar and multipolar neurons

We distinguish neurons according to the number of excitable dendrites connected to an axon (Fig. 6.3): multipolar, bipolar, pseudo-unipolar, and unipolar neurons. Multipolar neurons have multiple dendrites connected to the soma and one axon coming out. Most neurons of the brain and the spinal cord are of this type. Bipolar neurons have one branched dendrite going into the soma and one axon coming out. They can be found in the retina of the eye (Chapter 11), in the inner ear (Chapter 12), and in the olfactory (smell) center of the brain. Pseudo-unipolar neurons have one branched dendrite going directly into an axon connecting to the terminal, whereas the cell body (soma) lies outside. Pseudo-unipolar neurons are in most cases afferent neurons, which connect a sensory receptor to the CNS. In unipolar neurons, the axon connects the soma directly with the terminal. Figure 6.2 shows the typical situation where a pseudo-unipolar afferent neuron connects from the periphery to the CNS and a multipolar efferent neuron conducts the action potential back to the target tissue.

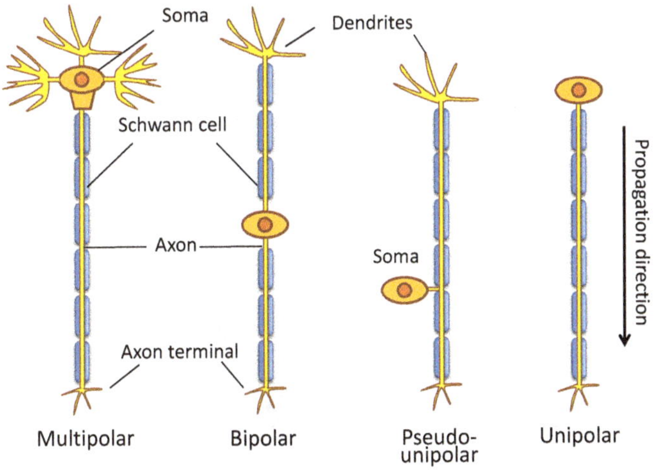

Fig. 6.3: Four types of neurons: multipolar, bipolar, pseudo-unipolar, and unipolar.

Infobox I: Charge transport

Transport of electric charges is possible in almost all kinds of media: gas, liquids, solids, polymers, and nerve fibers. However, the way charges are transported in various media is very different. Sometimes electrons carry the charge, sometimes ions, and sometimes only the polarization state of electric dipoles. Therefore, we distinguish between electron currents, ion currents, and polarization currents. Their differences are briefly presented further.

1. **Electron current (panel a):** In metals, electrons move relatively freely along a potential gradient. The electrons are not disturbed by the presence of other electrons but more by defects

in the metal lattice and by thermal movement of lattice ions. When a voltage is applied, the electrons do not accelerate but move through the metallic conductor with a drift velocity $\vec{v}_D$. The current density in a conductor is then $\vec{j}_e = \rho_e \vec{v}_D$, where ρ_e is the electron density. $\vec{v}_D$ depends on the electric field amplitude $\vec{E}$: $\vec{v}_D = \mu_e \vec{E}$, where μ_e is the electron mobility. Together this yields for the current density in metals: $\vec{j}_e = \rho_e \mu_e \vec{E} = \sigma \vec{E}$. Here, $\sigma = \rho_e \mu_e$ is the conductivity of metals.

2. **Ion current (panel b):** Pure water is an insulator. However, when adding salts, electrically charged ions (cations and anions) are created. The strong electric fields emanating from H_2O dipole moments lead to the dissociation of ionic bonds. NaCl dissolved in water splits into Na^+ and Cl^-. A solution with charged and mobile ions is called an electrolyte. If two metal electrodes are immersed in the solution at a fixed distance and a voltage source is applied to them, the positive ions (cations) migrate to the negative electrode (cathode) and the negative ions (anions) to the positive electrode (anode). The excess charge of the anions is delivered to the anode and passes through the circuit to the cathode, where the cations are neutralized by electron absorption. The electrolysis thus includes a charge transport and at the same time a mass transport. The positively charged metal and hydrogen ions are deposited on the cathode, while the negatively charged acid residues and oxygen ions are deposited on the cathode. The current density in the electrolyte is the product of the mobility μ, the charge density ρ, and the valence Z of the ions involved: $\vec{j}_{ion} = (\rho_+ \mu_+ Z^+ + \rho_- \mu_- Z^-)\vec{E}$.

3. **Polarization current (panel c):** Nerve fibers conduct electrical signals. However, neither electrons nor ions are transported in the direction of signal propagation, but a state of polarization. Ion exchange across the membrane initiates a local action potential that creates a local electrical dipole moment. While the electric dipole is directed from the inside (negative) to the outside (positive) at rest, the direction of the dipole is reversed during depolarization. When neighboring ion channels successively depolarize, the flipping of the dipole moment corresponds to a polarization current propagating along the nerve fiber. Nonmyelinated nerve fibers have a signal speed of approx. 2 m/s.

6.3 Sensory receptor potential

We have many types of sensory receptors in our body that react to various stimuli: mechanoreceptors sensing touch, pressure, vibrations, and sound (Chapter 12); thermoreceptors detecting temperature changes; nociceptors responding to pain from physical or chemical damage of the skin; photoreceptors sensing light that hits the retina (Chapter 11); chemoreceptors for smell and taste; and osmoreceptors for detecting osmotic pressure changes of body fluids [1]. Although this looks pretty complete, we lack receptors for electric or magnetic field gradients, unlike some animals.

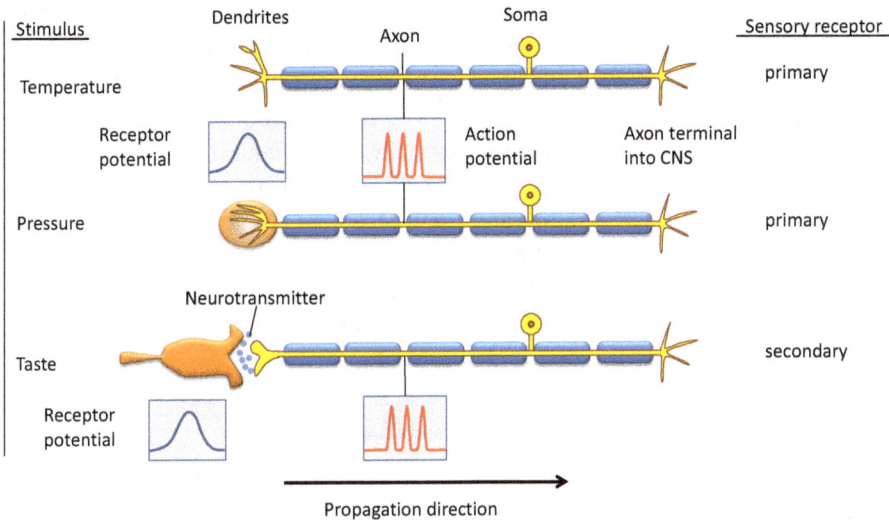

Fig. 6.4: Three types of sensory receptor neurons that are specialized for sensing temperature, pressure, and chemicals for taste and smell are illustrated.

Three main types of sensory receptor cells are sketched in Fig. 6.4. The receptors may directly be embedded in dendrites, such as for temperature sensitivity (top panel), or the dendrites may be encapsulated for pressure sensing (middle panel), or the primary sensor may be a chemical sensor for taste or smell (bottom panel). All three types of receptors connect to unipolar afferent neurons that transport the information to the CNS. The first two receptors are called *primary sensory receptors* because they translate the stimulus (pressure and temperature) directly into a receptor potential. In contrast, receptor potentials from taste and smell receptors are conveyed first into a release of neurotransmitters sensed by specific dendrites before being converted into action potentials. Sensors of this type are called *secondary sensory receptors*. The sensory part of primary and secondary receptors generates a change of the membrane potential in response to the stimulus, called the *receptor potential*.

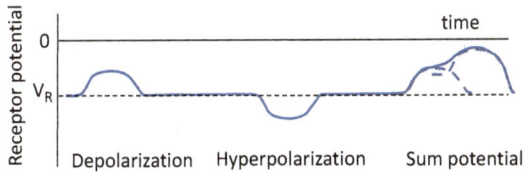

Fig. 6.5: Receptor potentials are graded potentials. The potential change may be positive (depolarization), or negative (hyperpolarization), or be summed up from two responses, which are timely not separated. V_R is the resting membrane potential.

The receptor potential responds to the stimulus continuously, without a threshold and without refractory period. Therefore these potential changes are called *graded potentials*. The graded potential change may be positive (*depolarization*) or negative (*hyperpolarization*), and can be added up to a sum potential (Fig. 6.5). Depending on the receptor type, they may respond proportionally to the stimulus (*p-receptors*) or emphasize changes of the stimulus called differential receptors (*d-receptors*). Pain receptors are of the p-type, temperature and smell receptors are of the d-type. Sense of smell is at the beginning strongest and dies off after a short time, so they are d-type. Most sensors are mixed pd-type. The distinction between receptor potentials and action potentials is essential and important for our subsequent discussion. Therefore, their main characteristics are compared in Tab. 6.1.

Tab. 6.1: Comparison between receptor potentials and action potentials.

	Receptor potential	Action potential
Threshold	No	Yes
Refractory time	None	2–3 ms
Summation	Yes	No
Polarization direction	Positive and negative	Only positive
Potential change	Graded	All or none
Propagation characteristics	Passive and damped	Active and regenerated at each node
Initiation	Receptors for specific stimuli	All membranes containing fast voltage-gated Na$^+$ channels

> ! Sensory receptors respond to stimuli by a graded potential. The graded potential change may be proportional to the stimulus or differential.

Infobox II: Types of ion channels

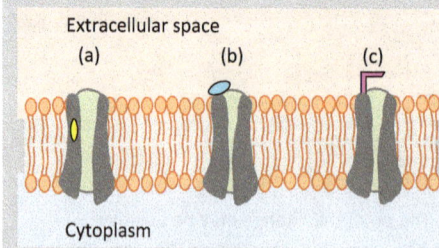

Extracellular space
(a)   (b)   (c)

Cytoplasm

In the body we find three different types of ion channels: (a) voltage-gated channels; (b) ligand-gated channels; and (c) strain-controlled channels.

Examples of voltage-gated channels are Na^+, K^+, and Ca^+ ion channels. They are the most common channels in the body. Ligand-gated channels open in the presence of neurotransmitters such as glutamate and can be found at all synapses between nerve fibers. The neurotransmitter acetylcholine (ACh) controls the ligand-gated ion channels on the motor endplates of muscle cells. The K^+ ion channels in the hair cells of the inner ear respond to mechanical stress, so do the hair cells in the vestibular organ.

6.4 Signal transduction

Figure 6.6(a) shows schematically the different steps from the stimulus to the receptor and action potential. In the example shown, the skin is exposed to a stimulus such as temperature or pressure that has different strengths for a short time. The dendrite reacts to the stimulus via a receptor potential of the pd type. The receptor potential is strong at the beginning of the stimulus and then decreases over time. With increasing stimulus strength, the receptor potential amplitude increases proportionally.

At Ranvier's first node, the signal observed is a mixture of a graded receptor potential and an action potential. The frequency of action potentials initiated depends on the amplitude of the receptor potential. A low receptor potential translates into a low-frequency action potential; a stronger receptor potential results in a higher frequency action potential. The conversion of a graded receptor potential into an action potential is made possible by the myelination of the nerve fiber, as explained below. The graded potential is filtered out a few nodes further down, and only the action potential propagates further along the axon. Therefore, the analog response to a stimulus at the nerve fiber's proximal end is successively converted into a discrete action potential.

The signal transduction from the graded potential to an action potential resembles an analog-to-digital conversion (ADC) in electronic circuits. In the case of nerve fibers, the strength of the stimulus and its duration are encoded in a sequence of action potentials. The stronger the stimulus, the higher the frequency of the action potentials. The rate may decrease over time, but action potentials will continue to be fired as long as the stimulus is "on." Technically speaking, the encoding of the receptor potential by action potentials corresponds to frequency modulation in telecommunication.

However, there is a mismatch between the dynamic range of the stimulus and the frequency bandwidth with which the action potentials can respond. Let us take the acoustic perception as an example. The acoustic intensity of a loudspeaker, for example, can extend over many orders of magnitude. On the other hand, there is only a bandwidth from 0 to a maximum of 200 Hz to encode the stimulus. The maximum frequency is 200 Hz, as it takes about 5 ms to refresh an action potential. To solve this problem, the receptor potential initially increases linearly with the intensity. However, as the intensity of the stimulus increases further, the amplitude of the receptor potential levels off (Fig. 6.6(b)). Accordingly, the frequency of the repetitive action potentials at the Ranvier node increases linearly at first, but then goes into saturation. A very intense stimulation causes an increasingly smaller incremental frequency change of the action potential. The nonlinear response is a fundamental principle of sensory perception that applies to almost all sensory receptors in the body, such as sound, light, color, pressure, temperature, and smell. In acoustics, the nonlinear perception of sound is the basis for the logarithmic scale

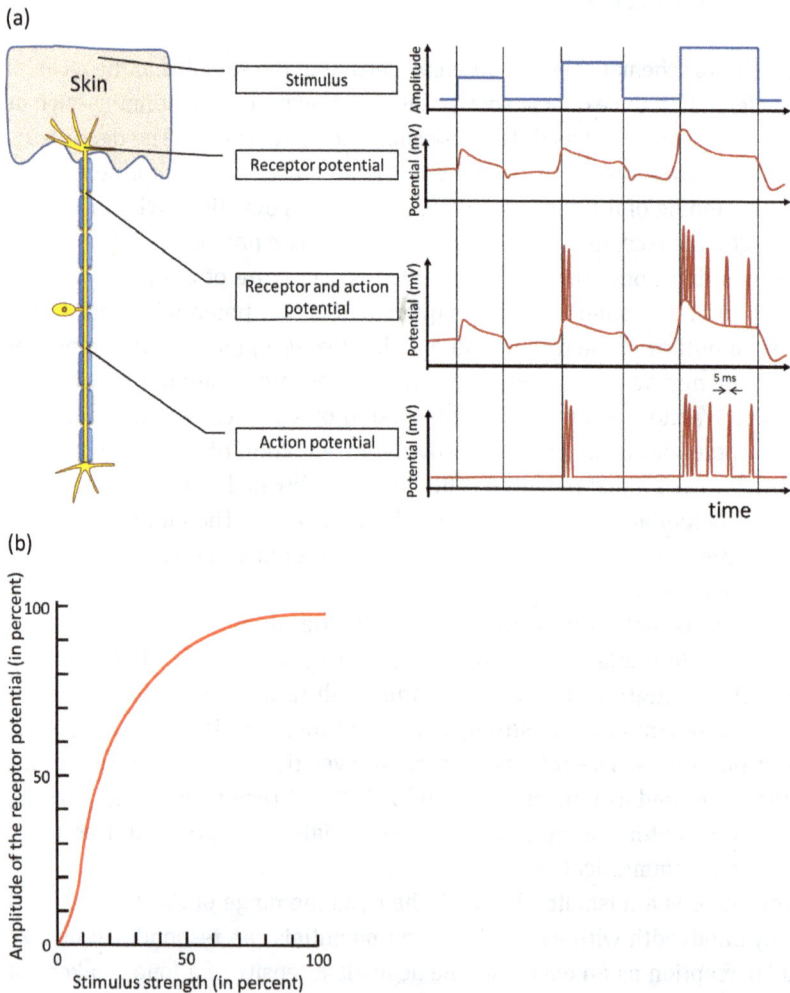

(a)

(b)

Fig. 6.6: (a) Transduction of receptor potentials with different amplitudes and duration to action potentials at the first and fifth node of Ranvier. (b) Amplitude of the receptor potential response to a stimulus of increasing strength (adapted from Pape et al. [2] and Guyton and Hall [3] with permission).

of the auditory level, also known as the Weber[3]–Fechner[4] law (see Section 12.4). This working principle has the advantage of a high sensitivity to weak stimuli while covering an extreme dynamic range of up to 12 orders of magnitude with just a limited bandwidth for the frequency modulation.

Now we have a closer look on the conversion of graded potential into an action potential at a receptor cell, as sketched in Fig. 6.7. Schwann cells, which coat the axon, have a high electrical resistance and block all ion channels within their range, such that the depolarization is suppressed. Transmembrane charges accumulate at the end of a receptor cell and generate an electric field that can jump from the axon hillock to the first Ranvier node. Voltage-gated Na^+ ion channels in the gap detect this strong electric field and depolarize. After a few more Ranvier nodes, the receptor potential is filtered out and only the digital action potential travels, jumping from node to node. The action potential is refreshed at each node and can continue forever or at least to the distal end of an axon. Suppose the electric field from the receptor is still present. Then another action potential is triggered to fire as soon as the refractory period of the previous one has expired. In this way, a sequence of action potentials can move along the axon as long as the stimulus is "on".

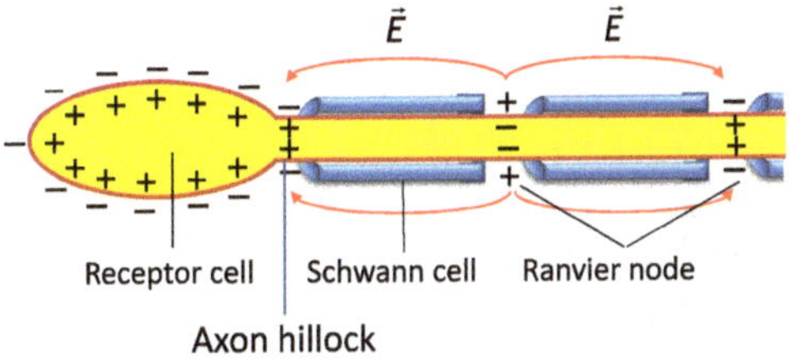

Fig. 6.7: Conversion of receptor potential into action potential.

6.5 Saltatory polarization current

At rest, axons have positive charges on the outside (Na^+ excess) and negative charges on the insides (K^+ deficit). The potential difference across the axon membrane has the expected value of about -75 mV. This potential is called the resting transmembrane potential. At the same time, the electrical dipole polarization $\vec{p}$ points from the inside out (Fig. 6.8(a)). When activated, Na^+ channels open, reverse the transmembrane potential and turn the membrane polarization from the outside in (Fig. 6.8(b)). In Fig. 6.8, each ion channel is symbolized by a pair of charges. The depolarization spreads successively over the axon (in Fig. 6.8 from left to right), which corresponds to a

---

3 Ernst Heinrich Weber (1795–1878), German physiologist and anatomist.
4 Gustav Theodor Fechner (1801–1887), German physician, physicist, and philosopher.

polarization current I_p that is oriented perpendicular to the polarization $\vec{p}$ (Fig. 6.8(c)). Since there is no potential gradient telling the polarization current which direction it should flow, there has to be some other mechanism deciding the directional preference. This is achieved via the refractory period, during which the polarization current can move forward into areas that have not yet been depolarized, but not backward (Fig. 6.8(d)). Repolarization during the refractory period is key to the directional preference of the polarization current. It should be noted that we are now talking about two different currents that are orthogonal to each other: the *transmembrane current* I_m and the *polarization current* I_p. The transmembrane current is local and responsible for depolarization and repolarization. The polarization current is nonlocal, propagative, and unidirectional. The distinction between these currents is very important for further analysis of axon's electrochemical properties.

> **!** We distinguish between local transmembrane ion current and nonlocal, propagative polarization current. The polarization current is saltatory in case that the nerve fibers are myelinated.

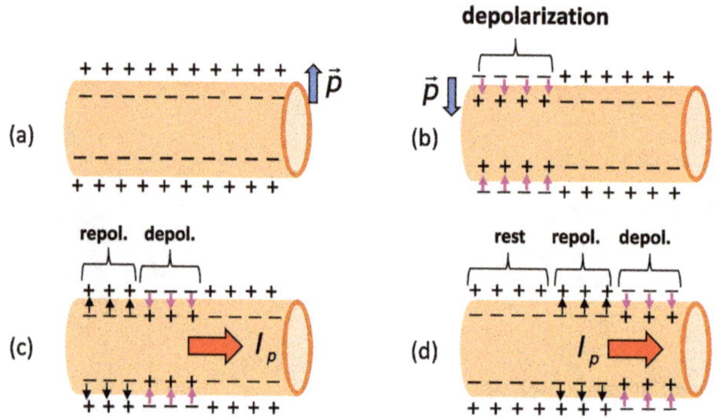

Fig. 6.8: Propagation of the depolarization along axon fibers. Areas that have been depolarized have a refractory period for repolarization, which hinders the action potential to move backward.

Nerve fibers that are not myelinated have a signal propagation speed of about 2 m/s [4]. This is pretty slow, but sufficient for short distances such as in the brain. Higher speeds are required when traveling long distances such as the distance from the brain to the lower limbs. Speeding up is achieved by myelination of the nerve fiber (Fig. 6.9(a)). Myelin is a dielectric material that forms sheaths that wrap around axon fibers. A myelin sheath can contain up to 300 bilipid membrane layers, each one increases the overall transmembrane resistance (Fig. 6.9(b)). Without myelination, the membrane potential decreases exponentially with the distance l from the receptor cell according to $\exp(-l/\lambda)$, where the extinction length λ is 2–5 mm. Under these circumstances, information transport and interneural communication over longer distances is impossible. However, myelination forces the transmembrane potential to jump from one Ranvier node to the next, as sketched in Figs. 6.7 and 6.9(a). The axonal membrane is not isolated in these nodes, allowing rapid depolarization over a high density of Na$^+$ channels. This depolarization refreshes the action potential and maintains the original signal level. The Ranvier nodes are all 0.2–2 mm apart and their length is about 1–5 μm. The saltatory propagation speed of the polarization current reaches values up to 120 m/s [4]. See also Section 6.7 for further discussion of this topic.

> We distinguish between local transmembrane ion current and nonlocal, propagative polariza-
> tion current. The polarization current is saltatory in case that the nerve fibers are myelinated.

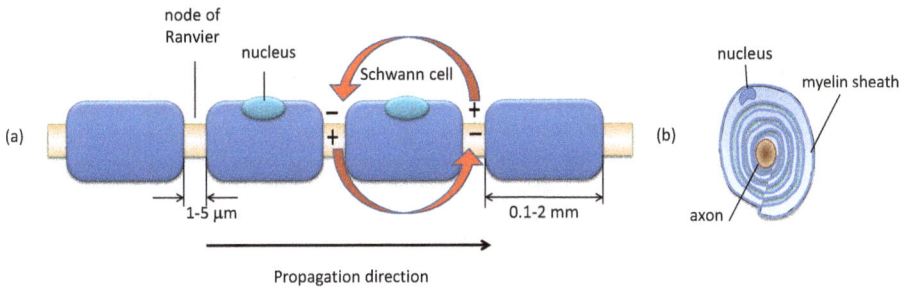

Fig. 6.9: (a) Schwann cells along an axon interrupted by nodes of Ranvier, arranged like beads on a string, allow jump-like fast signal propagation. (b) Cross section of a myelinated axon.

6.6 Electrochemical properties of axons

6.6.1 Transmembrane capacitance

Using the information provided, we can calculate some basic electrical properties of unmyelinated and myelinated axons. According to eqs. (5.15)–(5.18), we have for the cylindrical shape of axons the transmembrane capacitance:

$$C_m = \frac{\varepsilon\varepsilon_0 A}{\Delta x} = \frac{2\pi\varepsilon\varepsilon_0 aL}{\Delta x}. \tag{6.1}$$

Here a is the radius of an axon, L is the length, A is the surface are, and Δx is the membrane thickness. $\varepsilon(=6)$ and $\varepsilon_0(=8.854 \times 10^{-12}$ As/V m) are the permittivities of the cytoplasm and the dielectric constant of vacuum, respectively.

As an example, we calculate the capacitance for an unmyelinated axon per length L, radius a=1 μm, and membrane thickness $\Delta x = 5$ nm. The capacitance per length is $C_{m,umy}/L = 6 \times 10^{-8}$ F/m.

For a myelinated axon of the same length and radius, the membrane is effectively thicker. With 60 windings of 5 nm thickness, the effective membrane thickness becomes 300 nm instead of 5 nm. Therefore, the capacitance drops by a factor of 60 to the value: $C_{axon,my}/L = 1 \times 10^{-9}$ F/m.

It is easy to determine the total electric charge for the same axon length. Assuming a membrane resting transmembrane potential of $\Delta U_m = -75$ mV:

$$Q_{axon}/L = C_m \Delta U_m/L, \tag{6.2}$$

for the unmyelinated axon, the charge per length is

$$Q_{axon,\,umy}/L = 4.5 \times 10^{-9}\ \text{C/m} \text{ or } Q_{axon,\,umy}/L \cdot e = 2.8 \times 10^{10} \tag{6.2}$$

elementary charges per meter.

For the myelinated axon the respective values are:

$$Q_{axon,\,my}/L = 7.5 \times 10^{-11}\ \text{C/m} \text{ or } Q_{axon,\,my}/L \cdot e = 4.7 \times 10^{8}.$$

Thus, myelination has the effect of lowering the axon's capacitance and decreasing the surface charge.

> ❗ Myelination has the effect of lowering the axon's capacitance and decreasing the surface charge.

6.6.2 Resistance and time constant

Next we consider the electric resistance of myelinated and unmyelinated axons parallel and perpendicular to the axon axis. We start with the transmembrane resistance R_m experienced by the transmembrane current: $R_m = U_m/I_m$. In Chapter 5, we already discussed the ion channel resistance (or rather its conductance), which is adapted here to the cylindrical geometry. In general, the transmembrane ohm[5] resistance, which may be called transverse resistance, can be expressed as follows:

$$R_m = \rho_m \frac{\Delta x}{A} = \rho_m \frac{\Delta x}{2\pi a L}. \tag{6.3}$$

where ρ_m is the transmembrane specific resistance, Δx is the membrane thickness, a and L refer to axon diameter and length, respectively. The expression for R_m is fundamentally different from the parallel resistance R_p. The expression for the parallel or axial resistance of axons R_p is the same as for the ohm resistance of electric wires:

$$R_p = \rho_p \frac{L}{\pi a^2}. \tag{6.4}$$

---

5 Georg Simon Ohm (1789–1854), German physicist.

Here L is the wire (axon) length and πa^2 is the respective cross section.

The transmembrane resistance R_m scales with $\sim 1/L$. This is because the ion channels work in parallel, and the longer the axon, the lower the resistance. In contrast, the parallel resistance scales with $\sim L$ as usual for wires. The specific resistances are highly anisotropic. ρ_m has the value already calculated for a single cell according to eq. (5.23): $\rho_m = 2 \times 10^7$ Ωm. For the specific parallel resistance, a reasonable estimate is $\rho_p = 0.05$ Ωm [7].

According to eqs. (6.1) and (6.3), $C_m \sim L$ and $R_m \sim 1/L$. Therefore, the product, $C_m R_m$ must be a constant with the property of a time constant:

$$\tau = C_m R_m = \frac{\varepsilon \varepsilon_0 A}{\Delta x} \rho_m \frac{\Delta x}{A} = \varepsilon \varepsilon_0 \rho_m. \tag{6.5}$$

The time constant τ depends only on material parameters but not on the axon geometry. Substituting numbers, we find for an unmyelinated part of the axon a time constant of $\tau = 1.1$ ms. This is a typical time-constant for the depolarization of cells, which we have determined already in Section 5.4. In contrast, the myelinated part of the axon has an infinite time constant.

The time constant for the local depolarization of an axon is of the order of 1 ms.   **!**

6.6.3 Decay length of action potential

Now we consider the parallel ion current I_p and the potential drop over a length dz in an axon. This will give us some insight into the axon's signal propagation length and speed. Figure 6.10 shows the equivalent circuit diagram of an axon, underlying the calculation.

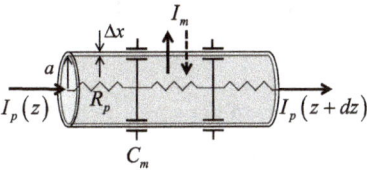

Fig. 6.10: Equivalent circuit diagram for the derivation of the telegraph equation that models the signal propagation of action potentials in axons.

The parallel current is given by the potential change ΔU_p in the parallel direction, divided by the parallel resistance:

$$I_P = \frac{\Delta U_p}{R_p} = \frac{\pi a^2}{\rho_p L} \Delta U_p = \frac{1}{r_p} \frac{dU_p}{dz}, \tag{6.6}$$

where R_p, is defined in eq. (6.4), $r_p = \rho_p/\pi a^2$ is the specific resistance normalized to the cross-sectional area πa^2 of the axon, a is the radius, the length L is replaced by dz, i.e., an incremental distance along the axon. The parallel current $I_P(z)$ at position z and at $z + dz$ are different, because some of the parallel current leaks out as transmembrane current I_m and changes the transmembrane potential (Fig. 6.10). Therefore, the transmembrane temporal potential change $dU(z,t)/dt = dU_p/dt$ of an axon that occurs when a depolarization signal travels between z and $z + dz$ can be expressed as follows:

$$C_m \frac{dU_p}{dt} = I_P(z) - I_P(z + dz) - I_m \tag{6.7}$$

or

$$C_m \frac{dU_p}{dt} = \Delta I_P - I_m.$$

Hence, the resulting parallel current is composed of two parts: parallel current difference minus leakage transmembrane current. The differential equation describing the propagation of the action potential $U(z,t)$ can be derived as follows (see mathbox).

$$\lambda^2 \frac{\partial^2 U(z,t)}{\partial z^2} - \tau \frac{\partial U(z,t)}{\partial t} + U(z,t) = \Delta U_m, \tag{6.8}$$

where ΔU_m is the resting potential and λ is the length constant defined by

$$\lambda = \sqrt{\frac{a \Delta x \rho_m}{2 \rho_p}} = 2 \times 10^5 \sqrt{\frac{a \Delta x}{2}}, \tag{6.9}$$

and Δx is the membrane thickness. The complete derivation of eq. (6.8) is provided in the mathbox for the experts. The partial differential equation (6.8) is known as the telegrapher's equation of transmission lines linking capacitive, resistive, and eventually also inductive elements. The equation was first used to describe electric pulses propagating in electric cables, such as the undersea cable between the UK and USA. Later, Hodgkin and Huxley used the telegrapher's or cable equation to analyze the axon conductance in squid mollusks [5].

Equation (6.8) can be solved numerically [6]. However, to get some insight, we will discuss two limiting cases. Assuming that the transmembrane capacitance $C_m = 0$ and therefore $\tau = 0$. Then eq. (6.9) has only a spatial dependence:

$$\lambda^2 \frac{d^2 U(z)}{dz} + U(z) = \Delta U_m \tag{6.10}$$

with the solution:

$$U(z) = \Delta U_m + U_0 e^{-z/\lambda}.$$

This shows that over the decay length λ, the signal potential drops to the trans-membrane resting potential ΔU_m, where λ is given by eq. (6.9). For an unmyelinated axon, the decay length is about 1 mm; for a myelinated axon, it is about 10 mm, spanning the distance between two nodes of Ranvier.

> The action potential decay length is estimated to be 1 mm for an unmyelinated axon and 10 mm for a myelinated axon.

Mathbox for the experts: telegraphic equation

Starting with eq. (6.7) and using eq. (6.6), we normalize the equation by the surface area $A = 2\pi a L$ of the axon and obtain:

$$\frac{C_m}{A}\frac{\partial U_p}{\partial t} = -\frac{I_m}{A} + \frac{\Delta I_p}{A} = -\frac{I_m}{A} + \frac{1}{A}\frac{\Delta(\partial U_p)}{A r_p\ \partial z} = -\frac{I_m}{A} + \frac{1}{2\pi a r_p}\frac{\partial^2 U_p}{\partial z^2}.$$

Here we have set $U_p = U(z,t)$. Multiplying with A/C_m:

$$\frac{\partial U_p}{\partial t} = -\frac{I_m}{C_m} + \frac{L}{C_m r_p}\frac{\partial^2 U_p}{\partial z^2}$$

$$= -\frac{\Delta U}{\tau} + \frac{L\Delta x}{2\pi\varepsilon\varepsilon_0 a L r_p}\frac{\partial^2 U_p}{\partial z^2}$$

$$= -\frac{\Delta U}{\tau} + \frac{a\Delta x}{2\varepsilon\varepsilon_0 \rho_p}\frac{\partial^2 U_p}{\partial z^2}.$$

Here $\Delta U = \Delta U_m - U_p$, where ΔU_m is the resting membrane potential. Normalization by A and multiplying with A appears to be cumbersome but makes the different steps more transparent. Using the relation for the time constant τ in eq. (6.5) we obtain

$$\frac{\partial U_p}{\partial t} = -\frac{\Delta U}{\tau} + \frac{a\Delta x\,\rho_m}{2\tau}\frac{\partial^2 U_p}{\rho_p\ \partial z^2}.$$

Now we introduce the length constant:

$$\lambda = \sqrt{\frac{a\Delta x\,\rho_m}{2}\frac{1}{\rho_p}} = 2\times 10^5\sqrt{\frac{a\Delta x}{2}}$$

and multiply once more with the time constant τ, which yields

$$\tau\frac{\partial U_p}{\partial t} = -\Delta U + \lambda^2\frac{\partial^2 U_p}{\partial z^2}.$$

ΔU is the difference between the resting potential ΔU_m and the amplitude of the action potential at position z and time t: $\Delta U = \Delta U_m - U(z,t)$. Rearranging the terms yields the partial differential equation of second order, also known as one form of the telegrapher's equation:

$$\lambda^2\frac{\partial^2 U(z,t)}{\partial z^2} - \tau\frac{\partial U(z,t)}{\partial t} + U(z,t) = \Delta U_m.$$

6.6.4 Propagation velocity of action potentials

Going back to eq. (6.9), and assuming that the transmembrane resistance is negligible ($\rho_m = 0$), then the differential equation has only a time dependence, expressed by

$$-\tau \frac{dU(t)}{dt} + U(t) = \Delta U_m, \tag{6.11}$$

with the solution:

$$U(t) = \Delta U_m + U_0 e^{-t/\tau}, \tag{6.12}$$

where τ is given by eq. (6.5). The time constant is about 1 ms, independent of myelination or not. Both constants, time constant and decay length, can be combined to yield the propagation velocity of the action potential:

$$v = \frac{\lambda}{\tau} = \frac{1}{\varepsilon\varepsilon_0} \sqrt{\frac{a\Delta x}{2\rho_m \rho_p}}. \tag{6.13}$$

We note that the propagation velocity is inversely proportional to the transmembrane resistivity and parallel resistivity. In the last equation, only the product $a\Delta x$ contains free variables; all other constants are fixed. Therefore, we can express the signal velocity as follows:

$$v = \frac{1}{\varepsilon\varepsilon_0} \sqrt{\frac{a\Delta x}{2\rho_m \rho_p}} = K\sqrt{a\Delta x}, \tag{6.14}$$

where $K = 4.7 \times 10^6$ s^{-1}. This shows that the propagation velocity depends on the axon's geometric properties, i.e., the axon radius a and the membrane thickness Δx. The radius a depends on the location, the thickness on whether the axon is myelinated or not. For myelination, the radius and the thickness are related according to $\Delta x = 0.4\ a$. Then

$$v_{my} = 0.6Ka \cong 3 \times 10^6 \times a\ \frac{1}{s}, \tag{6.15}$$

As an example, we assume for an unmyelinated axon: $a = 1\,\mu$m and for a myelinated axon: $a = 20\,\mu$m. This yields the respective speeds $v_{umy} = 3$ m/s and $v_{my} = 60$ m/s. Obviously, increasing the axon radius and the thickness of the myelinated sheath helps speed up the nerve signal. However, if each neuron were to be myelinated individually, it would take up too much space in the body. Therefore, the axons along the main nerve canals are packed in a bundle and wrapped in a common sheath.

> **!** The propagation velocity of action potentials in axons is proportional to their radius.

6.6.5 Experimental test of propagation velocity

The velocity of action potentials along axons can be tested and measured by using electromyographical methods explained in Section 6.10. In short, and referring to Fig. 6.11, electrodes are used to stimulate an action potential along an efferent axon such as the median nerve of the forearm. With another electrode on the wrist, the arrival time of the action potentials starting at points A and B is measured. Measuring the distance L_{AB} and the time lap between the arrival times $t_{BA} = t_B - t_A$ yields the velocity $v = L_{AB}/t_{BA}$. For the example shown, $L_{AB}=20$ cm, $t_{BA}= 4$ ms, and $v = 50$ m/s, in good agreement with the estimate made above. This is a typical velocity for the median nerve. Lower velocity would indicate injuries or some disease.

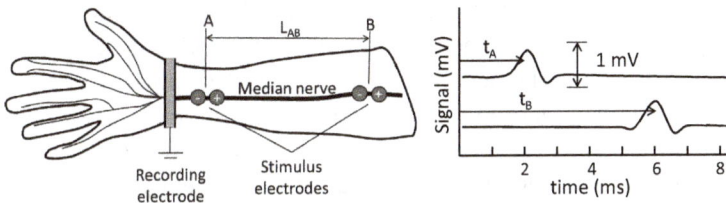

Fig. 6.11: Experimental determination of the action potential velocity along the median nerve fiber in the forearm.

Summarizing, the speed of action potentials along unmyelinated and myelinated axons is drastically different. For myelinated axons, the signal speed increases linearly with the axon radius. High speeds on the order of 100 m/s can only be reached with axons that have diameters on the order of tenths of micrometers.

6.7 Communication across axons

Neurons form networks with other neurons. A single axon may have as many as 15 000 connections with which it communicates [7]. Signal transmission across different axons is termed *synaptic transmission*. Synaptic transmission can be either of electrical or chemical nature.

In electrical junctions, also known as *gap junctions*, cells almost touch each other and are connected via common ion channels that allow electrical and ion transport (Fig. 6.12). The ion channels form pairs, one channel from either membrane. In order to function, the ion channel pairs have to be in a perfect registry. Gap junctions can be found in all body tissues, guaranteeing a very fast signal response. The channels of gap junctions are not continuously open, but Ca^{2+} ions regulate opening and closing. As such the permeability of gap junctions can change rapidly (within milliseconds) and reversibly, allowing efficient communication between cells [8].

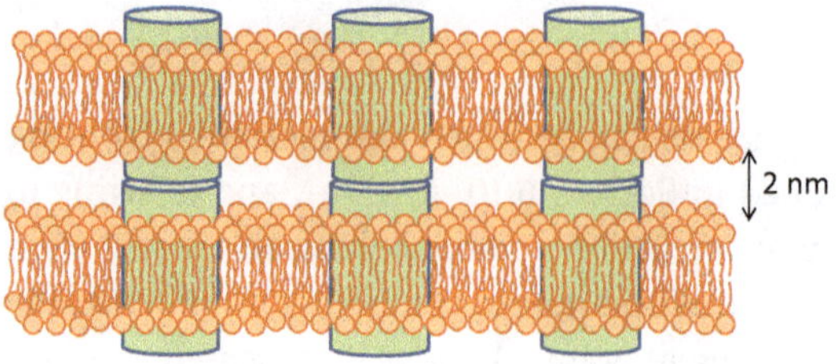

Fig. 6.12: Gap junctions between two membranes. Ion channels from both membranes pair up to a joined channel.

More common than gap junctions are chemical synapses that connect two cells by diffusing chemical neurotransmitters across a large gap. In contrast to gap junctions, which are only 2 nm apart, chemical synapses are about 30 nm apart [4, 8]. Neurotransmitters are stored in the presynaptic vesicles and released into the synaptic space (cleft) in response to incoming action potentials. This is shown schematically in Fig. 6.13(a). The most common neurotransmitter in the CNS is glutamate [9] (chemical formula: $C_5H_{10}N_2O_3$). The neurotransmitter molecules diffuse across the synaptic gap and activate the opposite cell by binding to certain receptor sites on the cell membrane. In this way, the incoming digital action potential is converted into an analog signal. Low-frequency action potentials release fewer chemical transmitters than high-frequency action potentials (Fig. 6.13(b)). After reaching the receptor of the postsynaptic cell, the analog signal again triggers a digital action potential (see Fig. 6.14). The frequency of the latter depends not only on the input of the

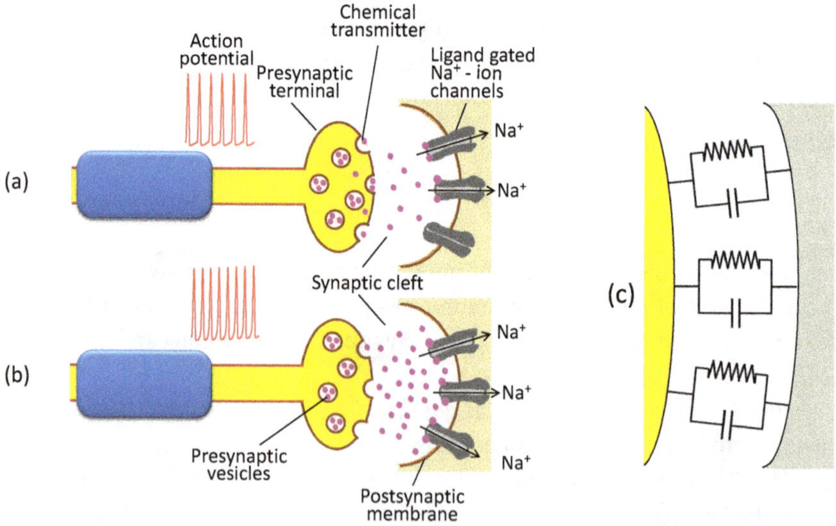

Fig. 6.13: (a) Chemical transmitter is released in response to action potentials, converting a digital signal into an analog signal. (b) Higher frequency of action potentials causes more neurotransmitters to be released. (c) Equivalent circuit diagram for the arrangement of parallel acting ion channels in the synaptic cleft.

presynaptic cell but also on the input of all the other neurons that bind to the synapse. This input can be both stimulating (reinforcing) and inhibiting (suppressing). When the sum of all analog input signals at the receptor site reaches the threshold potential of the postsynaptic cell, an action potential is triggered. Conversely, if the sum signal is too low, the action potential does not continue to propagate. Obviously, chemical synapses provide a variety of different signal transmissions, including stimulatory, inhibitory, and manipulative, through external chemicals such as anesthetics. The conductive properties of the parallel acting ion channels in the synaptic cleft can be modeled by an equivalent circuit diagram consisting of resistive and capacitive elements (Fig. 6.13(c)). Here the channel conductivity is controlled by the transmitter concentration. The complete sequence of events at a synaptic terminal is summarized in Fig. 6.14,

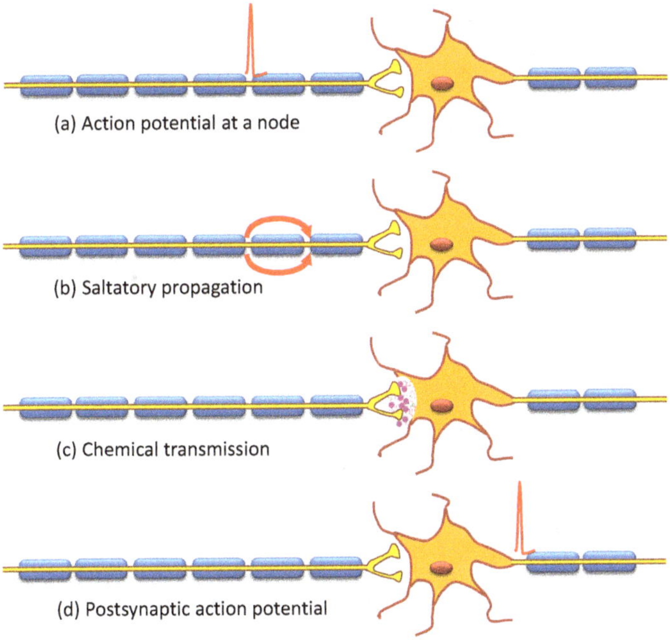

(a) Action potential at a node

(b) Saltatory propagation

(c) Chemical transmission

(d) Postsynaptic action potential

Fig. 6.14: Sequence of events at a chemical synapsis: (a) action potential at a node of Ranvier; (b) the polarization current jumps across a Schwann's cell to the next node of Ranvier; (c) release of neurotransmitters stored in presynaptic vesicles into the synaptic gap; and (d) the chemical signal is converted back into a digital action potential in the postsynaptic cell.

Signal transmission across different axons (synaptic transmission) can either be of electrical nature (gap junction) or based on the release of chemicals (neurotransmitter junction).

6.8 Neuromuscular junction – triggering muscle contraction

Efferent nerve-fibers (motor neurons) terminate at muscle fibers and bring them to contraction via action potentials. Each individual muscle fiber is innervated by only one motor neuron. However, a single motor neuron is usually split up in many branches and can innervate many muscle fibers, between 10 and 1000 depending on the muscle size. One motor unit is the combination of an individual motor neuron and all muscle fibers it innervates. Figure 6.15 illustrates one motor unit that innervates with only three muscle fibers. In reality, there may be up to a thousand motor units that connect to one muscle. For the distinction between muscles, bundles, fibers, myofibrils, and myofilaments, we refer to Section 2.4. Next, we inspect the processes occurring when muscles fibers are activated at one of the motor axon terminals.

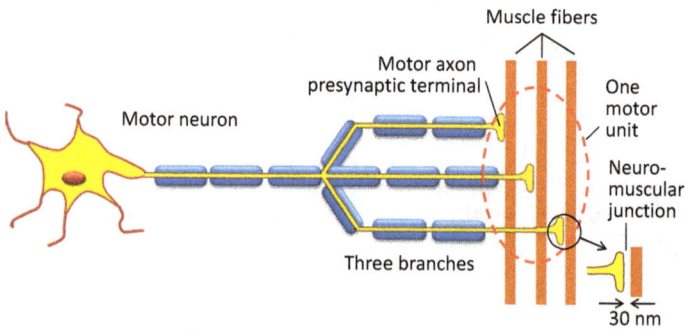

Fig. 6.15: Innervation of one motor neuron with fibers in a muscle.

The 30 nm wide gap between the motor axon terminal and a muscle fiber is the so-called *neuromuscular junction* (NMJ). After the action potential is transmitted across the NMJ, an action potential is stimulated in all the innervated muscle fibers of that particular motor unit. The sum of the electric activity leading to the action potential is called *motor unit action potential*. The NMJ is a chemical synapse, but differs in shape from synapses between neurons. The cross section of an NMJ is schematically shown in Fig. 6.16.

The presynaptic motor axon is not myelinated. Instead, it contains a large number of voltage-gated Ca^{2+} channels. Beyond the gap, the muscle fiber is folded up to increase the membrane's surface area exposed to the synaptic gap. These folds form the motor endplate of muscle cells hosting a huge number of receptors for the neurotransmitter ACh to bind on, as analyzed and identified by Dale.[6]

---

6 Henry Dale (1875–1968), British pharmacologist and physiologist, Nobel Prize 1936.

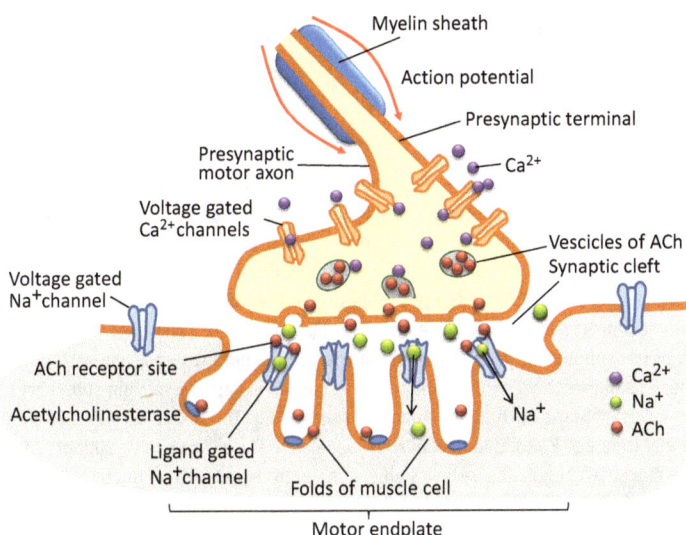

Fig. 6.16: Neuromuscular junction between an efferent presynaptic motor axon and a muscle fiber.

Referring to Fig. 6.16, we briefly sketch the sequence of events that occur at the NMJ after the arrival of an action potential at the distal end of a motor neuron [10]. First, voltage-gated calcium channels open in the membrane of the presynaptic terminal. When Ca^{2+} ions enter the neuron's cytoplasm, they bind to sensor proteins on vesicles that contain the neurotransmitter ACh. The neurotransmitter was previously synthesized in the terminal button and is now stored in the vesicles. Information on vesicles and liposomes is given in Infobox III. The binding of Ca^{2+} ions to the vesicles causes them to fuse with the active zone of the membrane and release ACh. ACh then diffuses through the synaptic cleft and binds to ligand-gated Na^+ channels in the motor endplate's folds. The binding of ACh to Na^+ channels triggers them to open. When Na^+ ions flow into the postsynaptic muscle cell, they depolarize to the threshold potential level and initiate an action potential. This depolarization is known as the end plate potential (EPP). As the action potential spreads over the muscle cell, the muscle cell contracts (see Chapter 2 for details). Then the ligand-gated Na^+ channels close and repolarize, allowing the muscle cell to relax. ACh, which was bound to the Na^+ receptor site, is released and cleaved into acetate and choline by the enzyme acetylcholinesterase located in the muscle cell membrane.

Meanwhile, acetate and choline are actively transported back into the presynaptic terminal. Here, acetate and choline are resynthesized in ACh and encapsulated in vesicles. Now the motor axon is ready to be activated again.

Some afferent connections are not controlled by the brain, but are connected to interneurons in the spinal cord that enable an automatic response. More information on spinal reflexes is given in Infobox IV.

> ❗ Muscle contraction is triggered by the reception of neurotransmitters that depolarize the motor endplate. The respective potential is the EPP.

ℹ Infobox III: Liposomes and vesicles

Cell membranes consist of phospholipid bilayers. Phospholipids are double-stranded chain-like molecules with hydrophilic heads and hydrophobic tails (see Infobox in Chapter 5). When a bilayer is formed, the hydrophilic heads are in contact with the cytoplasm and fluids in the extracellular space, while the hydrophobic tails hide inside. Phospholipids not only form cell membranes but also smaller spherical shells, so-called liposomes. Cells use liposomes to move around molecules, digest particles, and excrete molecular waste products. These carrier liposomes are called vesicles. Since vesicles are made up of phospholipids, they can break off the cell membranes and fuse with other parts of the membrane. This allows them to float around and act as small carriers, moving substances within cells and to the cell membrane such as the neurotransmitter ACh. The Nobel Prize in Medicine 2013 was awarded jointly to James E. Rothman,[7] Randy W. Schekman,[8] and Thomas C. Südhof[9] "for their discoveries of the machinery regulating vesicle traffic, a major transport system in our cells."

(Adapted from OpenStax Anatomy and Physiology, 2016, © creative commons)

ℹ Infobox IV: Spinal reflexes

Some afferent connections connect to the brain. Instead they are automatically controlled by the spinal cord, from where they go back to the periphery. These connections are made possible by interneurons, also called association neurons. Interneurons are suitable for verifying the connectivity as well as checking any possible injuries in the spinal cord. The best known automatic reflex is the patellar tendon or knee-jerk reflex shown in the following figure.

Striking the patellar tendon just below the patella with a hammer (stimulus) will stretch the femur muscle and receptors within the muscle spindle. The signal from the receptors travels along the afferent sensory neuron to the spinal cord completely independently without interference from higher centers. In the spinal cord, the afferent and efferent fibers are directly connected via an association neuron (interneuron). From the spinal cord, efferent motor neurons

---

7 James E. Rothman (*1950), US American biochemist.
8 Randy E. Schekman (*1948), US American biochemist.
9 Thomas Südhof (*1955), German-US-American biochemist.

conduct back the signal to the quadriceps and hamstring muscles, triggering a contraction of the quadriceps muscle and a relaxation of the hamstring muscle causing the leg to kick.

The coordination of muscle contraction on the one hand and interneuron inhibitory signal to the hamstring muscle on the other hand allows unconscious balance on our feet.

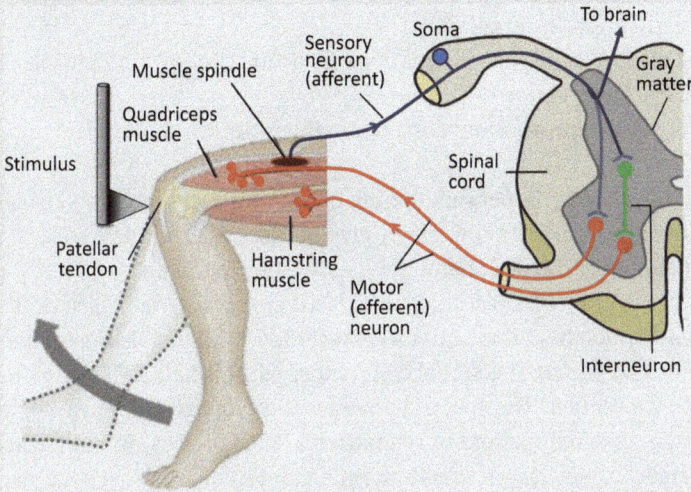

There are a couple of other automatic reflexes active in the body, such as the heat reflex and the eye blink reflex. In case of the heat reflex, the receptor afferent neuron connects to the spinal cord and automatically retracts the finger from the heat source by contracting the biceps brachii.

6.9 Electric and magnetic signals from the body

Several methods are available to investigate electrical signals in the body, providing rich information on functions and malfunctions of the body's nervous system. We distinguish between electromyography (EMG) for muscle activity, electrocardiography (ECG) for the heart activity, and electroencephalography (EEG) for the brain activity. This section introduces EMG and EEG, while the ECG is treated in Chapter 7 on the electrophysical aspects of the heart. Currents create magnetic fields and ionic currents generate an extremely weak magnetic signal that can be captured using a method known as magnetoencephalography (MEG). MEG is briefly described at the end of this section for completeness.

6.9.1 Electromyography (EMG)

EMG is a technique for evaluating and recording the electrical activity produced by skeletal muscles. EMG detects the electric potential generated by muscle cells when these cells contract and relax. Electrodes for EMG detection are either intramuscular needle electrodes or extramuscular surface electrodes. Fine needles are inserted into the muscle in the first case, and the transmembrane potential is measured. In the second case, the extracellular potential is determined over a larger area, such as taking an electrocardiogram (ECG) of the heart, as discussed in Chapter 7.

There are two main applications of EMG:

(1) Observation of electric potential fluctuations as a function of time and in response to a muscle contraction
(2) Measurement of signal propagation along nerve fibers after stimulation

Both types of tests concern the proper functioning of muscles and fibers versus malfunctioning due to various diseases that affect myelination, neurotransmitters, ion channels, muscle fibers, and the peripheral nervous system in general.

A typical EMG procedure is shown in Fig. 6.17. Two electrodes are placed on the muscle and extracellular potential fluctuations are recorded [11]. Possible fluctuations due to noise drop out by taking the signal difference, while small differences between positions 1 and 2 remain. The spectral power density of the difference signal is characteristic of the twitching pattern of the muscles (Section 2.3.5) in the relaxed state compared to the contracted state. The amplitude and frequency distribution of the recorded potential fluctuations can be used to assess normal versus diseased muscle contraction or nerve conductivity. Indeed, EMG is used to diagnose neurogenic or myogenic diseases. In addition, EMG is widely used in sports medicine to measure muscle contractility.

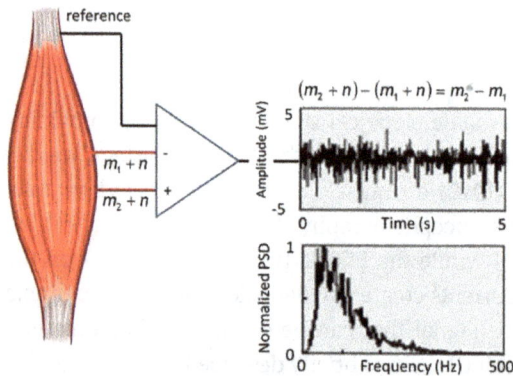

Fig. 6.17: Electrical potential fluctuations produced by muscle contraction can be tested by inserting electrodes into the muscle. n is the noise that cancels out by measuring at two positions. PSD is the power spectral density of the measured signal (adapted from [12]).

Another use of EMG is a measurement of the nerve conductivity, i.e., the ability to transmit action potentials to muscles. Degenerated or injured nerve fibers, or mechanically compressed nerves produce a numb feeling. In those cases, a diagnosis via EMG is indicated. An electrode is attached to a motor nerve fiber, and the motor nerve is stimulated by exposure to short voltage pulses to trigger an action potential, as shown schematically in Fig. 6.11. This procedure, already presented in Section 6.6.5, allows to measure the neuron signal velocity. For further information on EMG procedures, we refer to Ref. [12, 13].

EMG is useful for testing neuroconduction and muscle activity. !

6.9.2 Electroencephalography (EEG)

EEG was observed first on humans in 1924 by Berger.[10] Berger called his invention encephalography, which is a combined Greek word for brain and writing. EEG has become a widely used method for recording the brain's weak electrical potential fluctuations (several µV). The potential fluctuations are detected by electrodes placed on the scalp in form of a mesh and fixed with the help of a cap (Fig. 6.18). The counterelectrode is usually fixed on the pinna. The recorded electrical signals originate from nervous activities in the brain and are characteristic for the brain at rest, like sleeping, or for particular evoked reactions, like visual perception, muscle contraction, and cognitive activity.

An example is shown in Fig. 6.19. EEG is performed on patients to recognize normal brain behavior and diagnose abnormal activities indicative of brain tumors, epilepsy, or stroke. To this end, the recorded pattern is analyzed according to their frequency. Various waves are distinguished, such as the delta waves in the frequency band of 0.4–4 Hz for deep sleep, alpha waves in the 8–14 Hz frequency band for relaxed alertness, 20–40 Hz for concentrated learning, and the high-frequency γ-band (40–70 Hz) for very high concentration and information reception. A comprehensive review of EEG methods and applications can be found in Ref. [14].

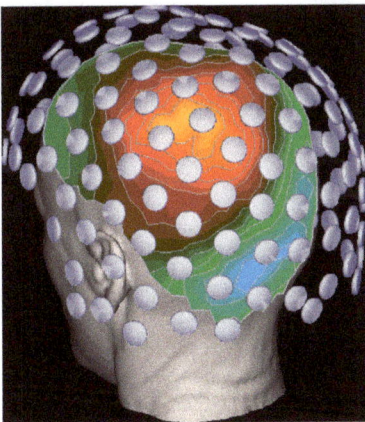

Fig. 6.18: EEG cap with electrodes attached (reproduced with courtesy of Los Alamos National Laboratory).

---

10 Hans Berger (1873–1941), German psychiatrist.

The EEG competes with functional magnetic resonance imaging (fMRI) of brain activity. The latter method, which is discussed in Chap. 3/Vol. 2, has the advantage of generating high spatial resolution images of brain activity in different layers of the brain. Compared to fMRI, the EEG has a much lower spatial resolution (several centimeters) to map potential or voltage fluctuations generated in the brain's upper part near the scalp. However, the EEG has a much higher time resolution of 1 ms or less. Another important difference to fMRI is the fact that the EEG shows brain activity directly, while fMRI signals result from an increased oxygen supply to the blood flowing in the vicinity of active neurons. Therefore, the EEG can also be used to study brain activity during sleep. In addition, the EEG is inexpensive, mobile, and noninvasive. Furthermore, the patients do not have to be immobilized, as is the case with fMRI. Therefore, the EEG is widely used in clinical practice and research, although the patterns are not easy to interpret. A typical example is shown in Fig. 6.19 of a person in a relaxed state. Frequency analysis of the EEG pattern shows that the dominant contribution is from the alpha wave band.

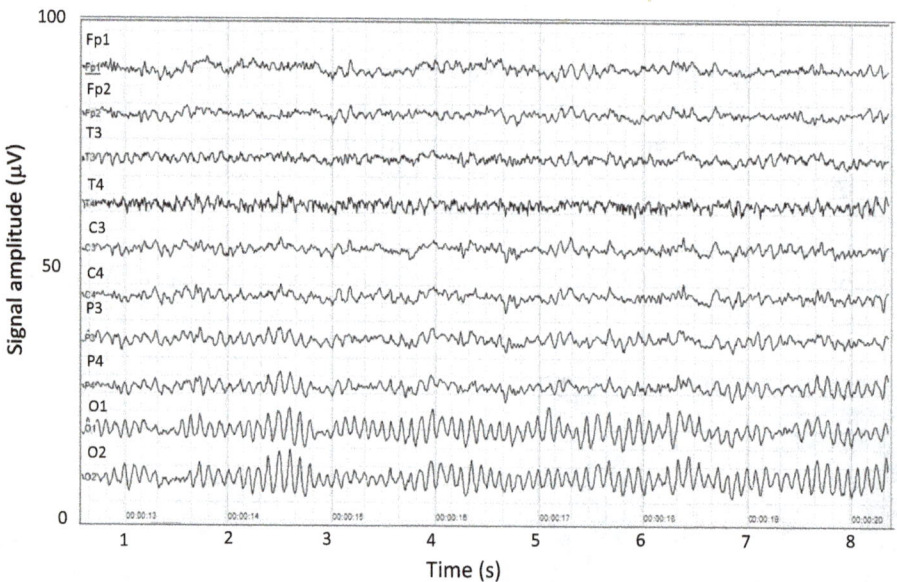

Fig. 6.19: EEG recording of brain activity with resting alpha rhythm. The 10 tracks recorded simultaneously come from different electrodes that are symmetrically arranged in pairs on the skull. A reference electrode is usually attached to the pinna. The O1 and O2 electrodes are placed on the back of the head near the right and left visual centers, Fp1 and Fp2 are positioned on the forehead near the eyebrows. Electrode pairs should have similar patterns when a person is healthy; asymmetries indicate diseases in the brain (adapted from https://en.wikipedia.org/wiki/Electroencephalography, © Creative Commons).

The body is generally charge-neutral despite all kinds of ionic currents constantly flowing in and out of the cells. However, transient electric dipoles occur throughout the body, particularly in the brain, where the frequency of action potentials is highest. Since the dipoles are randomly arranged in space, we expect the sum over all dipoles to give a zero potential. Nevertheless, EEG signals can be recorded in the upper part of the brain (see Fig. 6.19), suggesting that the brain activity derives in part from the synchronous activity of aligned dipoles in neurons. However, the noise level is quite high. Jittery potential fluctuations with variable latencies aggravate the problem. Therefore, precise phase-locking EEG signals to preceding activities achieve the best results.

6.9.3 Magnetoencephalography (MEG)

An alternative method of recording brain activity is MEG, which records magnetic Oersted fields generated by ionic currents in the brain (Fig. 6.20) instead of electrical dipole fields in EEG. These magnetic fields are extremely weak, the recorded amplitudes are only of the order of 50 pT. The earth's magnetic field is 50 µT, which is 10^6 times stronger than the brain's Oersted field. Therefore, the most sophisticated superconducting quantum interference device sensors are required to detect these weak fields on a person's scalp, which must be extremely well shielded against stray magnetic fields from the environment. It has been estimated that 50 000 neurons must fire at the same time in order to give a detectable signal. Time resolution and information gain are similar to the EEG, but at a much higher price. The advantage over the EEG is the better localization of the brain activity since stray magnetic fields have a shorter range compared to electric dipole fields. Due to the technical difficulties and the high investment costs, MEG is not often found in clinical practice but rather in research [15].

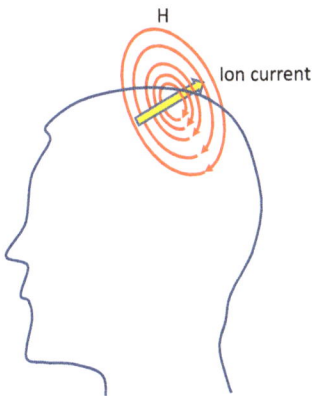

Fig. 6.20: Magnetoencephalography recording of magnetic fields generated by ion currents in the brain. The recording is performed with superconducting quantum interference device sensors distributed over the scalp in a magnetically well-shielded hutch.

6.10 Summary

S6.1   Neurons are single cells containing four parts: cell body, dendrites, axon, and axon terminal.

S6.2   Nerves consist of an ensemble of interconnected neurons.

S6.3   Long axons are myelinated by Schwann cells and gapped by nodes of Ranvier.

S6.4   The four essential steps of signal transduction are: 1. signal reception; 2. signal integration; 3. signal conductance; and 4. signal transmission.

S6.5   Afferent or sensory neurons conduct information from the periphery to the CNS.

S6.6   Efferent or motor neurons conduct action potentials to muscles for motion.

S6.7   Afferent and efferent neurons are not connected and have different pathways via the spinal cord to the brain and back.

S6.8   In a few exemptions, there are direct connections between afferent and efferent neurons in the spinal cord via association neurons.

S6.9   Stimuli from the environment are detected by specialized receptors.

S6.10   The receptor signal can be proportional or differential.

S6.11   The amplitude of the stimulus is encoded in the frequency of the action potential.

S6.12   The action potential propagates via a saltatory polarization current along the axon fiber.

S6.13   The propagation velocity of signals in myelinated axons can be up to 120 m/s.

S6.14   Different axons communicate via synaptic transmission.

S6.15   Synaptic transmissions can be electrical or chemical.

S6.16   Chemical synapses convert digital axon potentials into analog chemical transmitters and then back again into digital potentials in the postsynaptic axons.

S6.17   Efferent axons are connected to muscles fibers via NMJs.

S6.18   Muscle contraction is triggered by the reception of neurotransmitters that depolarize the motor end plate.

S6.19   One axon controls one motor unit of a muscle. One motor unit may contain up to 1000 NMJs.

S6.20   EMG is used for testing muscle activity and neuron conductance.

S6.21   EEG and MEG are used for the analysis of brain activities.

? Questions

Q6.1   What are the four main parts of a neuron?

Q6.2   What is the difference between efferent and afferent nerve fibers?

Q6.3   Why are some axons myelinated and others are not?

Q6.4   How are the three main receptor neurons distinguished?

Q6.5   What are the main differences between receptor and action potentials?

Q6.6   How does the ADC from receptor potential to action potential work?

Q6.7   Is there also a DAC in neuron signal transmission?
Q6.8   What is the difference between automatic neuron reflexes as compared to voluntary neuron action?
Q6.9   What can be tested with EMG, EEG, and MEG?
Q6.10 Describe how an action potential crosses a synaptic cleft.
Q6.11 What makes action potentials move in only one direction?
Q6.12 What do you understand saltatory conduction to mean?
Q6.13 What is the difference between electric conduction of a metal wire and a polarization current?

Attained competence checker	+	0	–
I can name the four parts of a neuron.			
I can list the four essential parts of signal transduction of neurons.			
I know how the ADC works in neurons.			
I can distinguish between graded potential and action potential.			
I know which physical parameter determines the speed of action potentials along axons with and without myelination.			
I recognize why Schwann cells speed up the propagation velocity of action potentials.			
I know how action potentials are transmitted across synaptic clefts.			
I can sketch the innervation of one motor neuron with fibers in a muscle.			
I know the basic idea of how to determine the signal speed by EMG.			
I can name the main advantages of EEG with respect to other methods such as fMRI.			

Suggestions for home experiments

HE6.1 **Patella reflex**
Test your patella reflex at both knees as described in Infobox IV. Which one is reacting more strongly? Which other automatic reflexes are you aware of?

HE6.2 **Reaction time**
Test your reaction time by using this visual-reaction tool:https://humanbenchmark.com/tests/reactiontime
Now test your reaction time in response to an audio signal:https://playback.fm/audio-reaction-time
Where do you score better, visual reaction or audio reaction? Note that both sensory systems have very different pathways in the brain.
You may repeat this test after drinking a glass of wine.

Exercises

E6.1 **Capacitance of axons:** Consider the axon as a cylindrical capacitor with a radius $a=1$ μm and membrane thickness $d = 5$ nm. Determine the capacitance per length L.
Confirm that the capacitance per length is $C_{m,\,umy}/L = 6.6 \times 10^{-8}$ F/m, using the dielectric constant of the vacuum $\varepsilon_0 (= 8.854 \times 10^{-12}$ As/V m) and the permittivity of the cytoplasm $\varepsilon(=6)$.

E6.2 **Propagation velocity:** How much time does it take for an action potential generated in the brain to arrive at the toe? Calculate the arrival time for unmyelinated and myelinated axons. Neglect the time loss at chemical synapsis and the exponential drop of the potential amplitude in unmyelinated axons.

E6.3 **Transmission lines:** Compare a technical transmission line with the model transmission line for the description of neurons, discussed in Section 6.7.

E6.4 **Analog-to-digital conversion:** Compare the electrotechnical ADC with the signal transduction from receptor potential to action potential. What is similar, and what are the differences?

E6.5 **Propagation equation:** Discuss why it is not possible to use the usual harmonic wave equation to describe the propagation of action potentials.

E6.6 **Piano virtuoso:** A piano virtuoso plays a piece of music with the tempo marking Prestissimo. With 600 beats per minute, the pianist plays particularly fast. Since you know that the fingers are controlled by nerve signals from the brain, compare the signal propagation time to the stroke frequency.

References

[1]   Marzvanyan A, Alhawaj AF. Physiology, sensory receptors. In: StatPearls [Internet]. Treasure Island (FL): StatPearls Publishing; 2021. https://www.ncbi.nlm.nih.gov/books/NBK539861/

[2]   Pape H-C, Kurtz A, Silbernagel S, eds. Physiologie. 7th edition. Stuttgart, New York: Thieme Verlag; 2014.

[3]   Guyton AC, Hall JE. Textbook of medical physiology. 11th edition. Philadelphia, Pennsylvania, USA: Elsevier Saunders; Elsevier Inc.,; 2006.

[4]   Siegel A, Sapru H. Essential neuroscience. Philadelphia, New York, London: Lippincott Williams & Wilkins; 2005.

[5]   Hodgkin AL, Huxley AF. A quantitative description of membrane current and its application to conduction and excitation in nerve. J Physiol. 1952; 117: 500–544.

[6]   Javidi M, Nyamoradi N. Numerical solution of telegraph equation by using LT inversion technique. Int J Adv Math Sci. 2013; 1: 64–77.

[7]   Kandel ER, Schwartz JH, Jessell TM, Siegelbaum SA, Hudspeth AJ. Principles of neural science. New York, Chicago, San Francisco: McGraw Hill; 2012.

[8]   Goodenough DA, Paul DL. Gap junctions. Cold Spring Harb Perspect Biol. 2009; 1: a002576.

[9]   Meldrum BS. Glutamate as a neurotransmitter in the brain: Review of physiology and pathology. J Nutr. 2000; 130(4S Suppl): 1007S–1015S.

[10] Boron WF, Boulpaep EL. Medical physiology. 2nd edition. Philadelphia, London, New York: Saunders W.B. Elsevier; 2012.

[11] Gohel V, Mehendale N. Review on electromyography signal acquisition and processing. Biophys Rev. 2020; 12: 1361–1367.

[12] Rodriguez-Falces J, Duchateau J, Muraoka Y, Baudry S. M-wave potentiation after voluntary contractions of different durations and intensities in the tibialis anterior. J Appl Physiol. 2015; 118: 953–964.

[13] Reaz MBI, Hussain MS, Mohd-Yasin F. Techniques of EMG signal analysis: Detection, processing, classification and applications. Biol Proced. 2006; 8: 11–35.

[14] Louis EK, Frey LC. Electroencephalography (EEG): An introductory text and atlas of normal and abnormal findings in adults, children, and infants. American Epilepsy Society; 2016.

[15] Zuo S, Nazarpour K, Bohnert T, Paz E, Freitas P, Ferreira R, Heidari H. Integrated pico-tesla resolution magnetoresistive sensors for miniaturised magnetomyography. Annu Int Conf IEEE Eng Med Biol Soc. 2020; 3415–3419. doi: 10.1109/EMBC44109.2020.9176266.

Further reading

Kandel ER, Schwartz JH, Jessell TM, Siegelbaum SA, Hudspeth AJ. Principles of neural science. New York, Chicago, San Francisco: McGraw Hill; 2012.

Purves D, Augustine GJ, Fitzpatrick D, Katz LC, LaMantia AS, McNamara JO, Williams SM, eds. Neuroscience. 2nd edition. Sunderland (MA): Sinauer Associates; 2001. Online textbook can be accessed but not browsed: www.ncbi.nlm.nih.gov/books/NBK11059/

Malmivuo J, Plonsey R. Bioelectromagnetism: Principles and applications of bioelectric and biomagnetic fields. Oxford, New York, Athens: Oxford University Press; 1995. Available at: www.bem.fi/book/

Siegel A, Sapru H. Essential neuroscience. Philadelphia, New York, London: Lippincott Williams & Wilkins; 2005.

Molecular biology of the cell. 4th edition. https://www.ncbi.nlm.nih.gov/books/NBK26871/

7 Electrophysical aspects of the heart

Physical properties of the heart	
Weight	300 g
Power consumption	6 W
Frequency of SA node	60–80 beats/min at rest
Pumping volume	7800 l/day and per ventricle
Absolute refractory period	250–300 ms
Highest potential difference in R point	4 mV
Leads recorded during an ECG	12

7.1 Introduction

The circulatory system of all vertebrates, including humans, consists of three main functional parts: pump, fluid, and tubings. The heart as a pump keeps the blood circulating as a liquid through a closed vascular system. In this chapter, we focus on the electrophysical aspects of the heart that allow the heart to contract and function as a pump. The mechanical aspects of the heart and the circulatory system are topics of Chapter 8. For further aspects, we refer to the standard textbooks of medical physiology and cardiology listed at the end.

The heart is undoubtedly a remarkable pump. With a power consumption of only 6 W, the 300 g light muscle pumps an average of 5–6 l/min with each ventricle, which amounts to approximately 7500 l/day, or 7.5 m^3 of blood, and twice that amount for both ventricles. The left and right ventricles of the heart work in series and must do so in a well-balanced and synchronized manner. The heart consists of four chambers with complete separation of oxygenated and deoxygenated blood. The right ventricle pumps blood to the lungs, while the left ventricle pumps blood to the rest of the body. Figure 7.1 shows a cross section of the heart, the different parts are presented further.

7.2 Cardiac action potential

7.2.1 Temporal evolution

The heart can be viewed as an elongated muscle that depolarizes when stimulated like normal skeletal muscle cells. The heart muscle, also called *myocard* or *myocardium*, differs from the skeletal muscles, however, in three essential ways [1, 2]:
1. Prolonged action potential (300 ms) and an extended refractory period (150 ms)
2. Self-excitatory pacemaker cells initiate the heart muscle contraction
3. Tetanic contraction of the heart muscle is excluded (see Fig. 2.17)

https://doi.org/10.1515/9783110756951-007

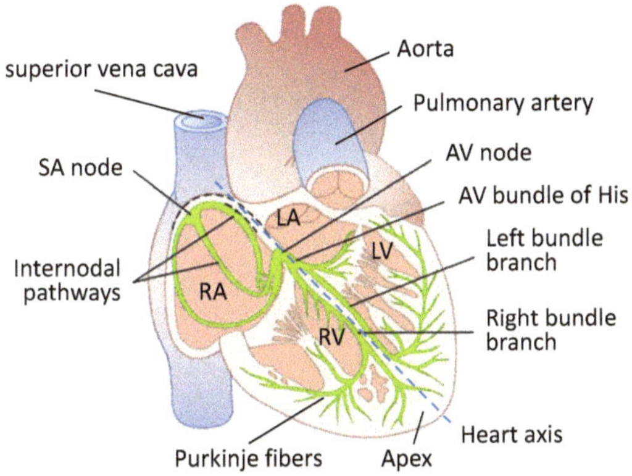

Fig. 7.1: Cross section of the heart showing main parts of the heart. Details are discussed in the text. AV node, atrioventricular node; SA node, sinoatrial node; RA, right atrium; LA, left atrium; RV, right ventricle; LV, left ventricle (adapted from http://www.medicinehack.com/).

The action potential of normal cells lasts about 2–3 ms. In contrast, the cardiac action potential (CAP) shown in Fig. 7.2 is characterized by a rapid depolarization and rapid but partial repolarization, followed by a prolonged plateau phase before full repolarization occurs. The entire CAP lasts about 200–400 ms, or 100 times longer than the standard action potential. The absolute refractory period, during which the heart is insensitive to new action potentials (blue area), ranges from depolarization to repolarization. The relative refractory period (yellow area) allows depolarization, but only with an increased threshold potential. The resting phase between two ventricular action potentials (green area) corresponds to the diastolic or filling phase. During the plateau phase, the action potential spreads from the sinoatrial (SA) node throughout the myocardium, causing the myocardium to contract. Therefore, the plateau phase is the ejection (systolic) phase.

The rapid depolarization is achieved by the opening of fast Na^+ ion channels. Complete depolarization is reached at a potential of approximately +40 mV. Early repolarization occurs when the Na^+ ion channels close, and a small number of K^+ ion channels open. Complete repolarization through K^+ ion channels is hindered by an influx of Ca^{2+} ions. Ca^{2+} ion control is typical for skeletal and cardiac muscle cells, but in the latter one, it is essential for maintaining a prolonged plateau phase. The Ca^{2+} ion concentration is rather low, but due to the double ionization, it very efficiently reduces the permeability of K^+ channels, which are otherwise open. Repolarization by K^+ ion channels is only possible once the Ca^{2+} concentration is exhausted. The time evolution of the ion-channel permeability for Na^+, Ca^{2+}, and K^+ is plotted in Fig. 7.3. This plot shows that slow Ca^{2+} channels control the long plateau phase, which is most characteristic of the CAP.

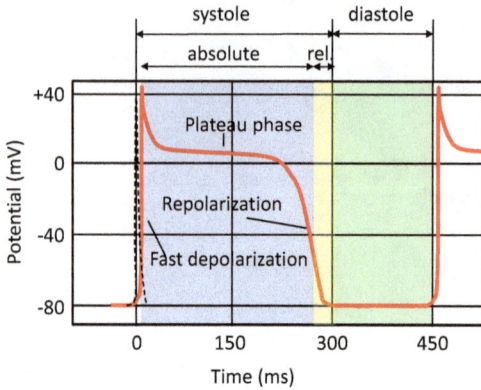

Fig. 7.2: The cardiac's action potential is characterized by very fast depolarization, extended plateau region, and repolarization after about 300 ms. The action potential is typical for a ventricular muscle fiber, as recorded by means of microelectrodes. The dashed black line shows the action potential of a skeletal muscle cell on the same timescale for comparison. Absolute and relative refractory periods, as well as the diastolic and systolic phase, are indicated by colored areas.

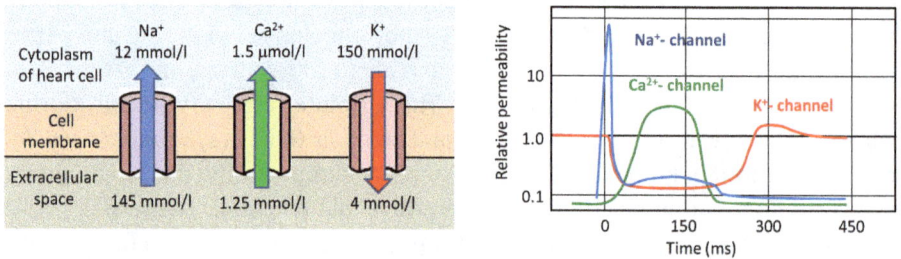

Fig. 7.3: Left panel: The cell membrane of the heart contains three main ion channels for Na^+, K^+, and Ca^{2+}. Cardiac cells express six different types of K^+ channels, which are required for maintaining the resting potential and for shaping the plateau phase. Right panel: The temporal opening of these channels increases their permeability during a cardiac action potential.

7.2.2 Spatial evolution

So far we have considered the heart as a whole and integrated the temporal evolution of the CAP over the entire myocardium. However, the CAP varies locally and evolves over time. Cardiac pacing begins in the SA node, which is located at the entrance to the right atrium (see Figs. 7.1 and 7.4). The SA node is made up of specialized heart cells called pacemaker cells that spontaneously and periodically generate action potentials (see Infobox I for further details). The action potential is "fired" at the SA node during the diastolic phase and then propagates in a highly organized manner via conductive fibers to other parts of the heart, opening Na^+ ion channels, resulting in further action potentials that stimulate the myocardium to contract.

As soon as the SA node has triggered an action potential, the heart's conduction system forwards the action potential to other parts of the heart muscle. This is possible because the muscle cells are strongly connected to one another, and electrical current flows between the muscle cells via *gap junctions*, enabling cell-to-cell coupling with low resistance (see Section 6.7). The sequence of depolarizations at different points in the heart is shown in Fig. 7.4. First, the atrium depolarizes, then the action potential jumps over to the *atrioventricular* (AV) *node*. The AV node is the only conductive connection between the atrium and the ventricle.

From the AV node, the action potential travels through the AV bundle parallel to the heart axis. The bundle then divides into left and right bundles up to the Purkinje[1] fibers. Finally, the action potential captures the ventricular muscle, which then contracts. The AV bundle and the Purkinje filaments are even more highly conductive fibers than the cardiac muscle cells, achieved by gap junctions between the fiber cells.

It is important to note that the CAP traveling through the bundle branches arrives at the ventricles at the same time. This allows both ventricular chambers to contract at the same time.[2] Unison contraction is important for in-phase systemic and pulmonary circulation.

The SA node is the heart's primary pacemaker at a rate of 60–80 beats/min at rest. The AV node also exhibits pacemaker capabilities, albeit at a lower rate of 0.6 to 1 Hz or 40–60 beats/min. The AV node acts as a low-pass filter: if the frequency of the SA node is too high, the AV node will not pass it on. Both the SA node and the AV node have efferent innervations via the sympathetic and parasympathetic nervous systems. They control the Ca^{2+} ion channel permeability and thus the beat frequency. In fact, when we get excited, our heart rate increases.

Infobox I: Action potential at the SA node

The action potential at the SA node shown in Fig. 7.4 (right panel) differs in several respects from the action potentials of the muscle cells [3]. It does not have a stable resting potential. Starting from a maximum diastolic potential of about −60 mV – much less than the surrounding −90 mV of the myocardial muscle – there is a slow spontaneous depolarization, which is referred to as diastolic depolarization or pacemaker potential. The threshold for starting an action potential is about −50 mV. Depolarization is mainly caused by a Ca^{2+} current carried by voltage-gated Ca^{2+}-selective ion channels. There is no contribution from voltage-gated Na^+ channels, as these are deactivated due to the diastolic membrane potential (−60 mV) which is less negative than myocardial cells at −90 mV. The action potential in the SA node lacks a pronounced plateau phase, and repolarization begins immediately after reaching the maximum. It takes about 100 ms from threshold to repolarization. The SA pacemaker generates a heart rate of approximately 70 to 80 beats/min or 1.2–1.3 Hz during resting.

---

1 Jan Evangelista Purkinje (1787–1869), Czech physiologist and histologist.
2 To be precise: the tricuspid valve in the right ventricle closes slightly before the mitral valve in the left ventricle by about 40 ms.

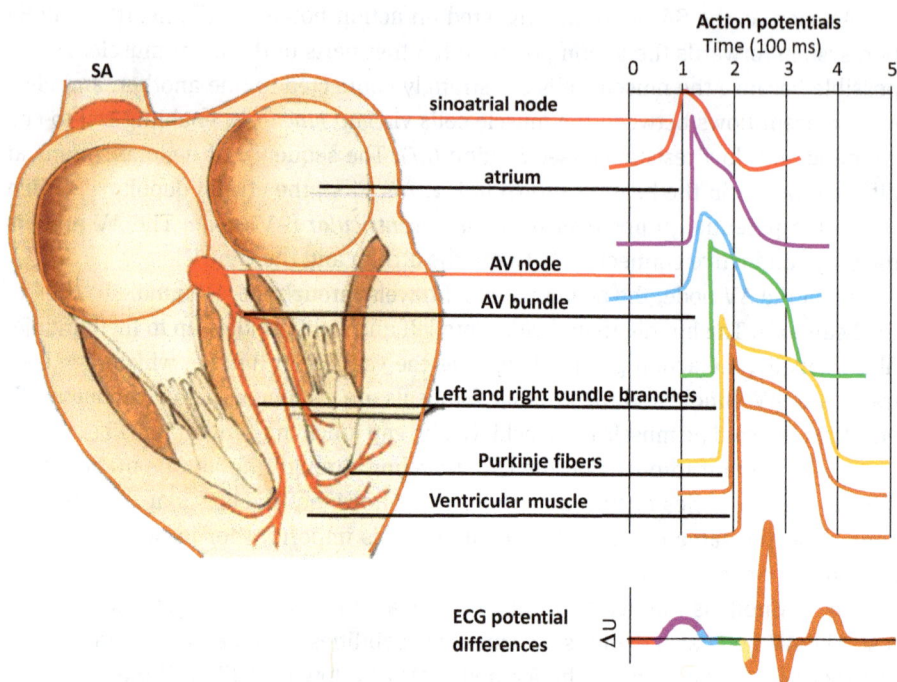

Fig. 7.4: Spreading of the action potential across the heart starting from the sinoatrial (SA) node as the self-organized pacemaker to the AV node, and from there down the Purkinje bundles to the ventricular muscle. The lower right trace shows the ECG potential difference as recorded by lead II with the respective color coding (compare Fig. 7.11) (adapted from Pape, Kurtz, and Silbernagel [2], with permission of Thieme Verlag, Stuttgart, New York, 7. Edition, 2014).

> **!** The CAP is self-excitatory, long-lasting, and without tetanic contraction.

7.3 Electric polarization of the heart

7.3.1 Transmembrane potential

Now we want to analyze the changes in the electrical potential over time that can be measured on the skin during a cardiac cycle. The action potentials traced in Fig. 7.4 are characterized as *transmembrane potentials*. They are measured with a pair of electrodes, one punching through the cell membrane into the cytoplasm and the other one placed outside. The *extracellular potential*, in contrast, is measured outside at different positions along a muscle or nerve fiber. The correspondence between transmembrane potential and extracellular potential during an action potential is plotted step by step in Fig. 7.5.

We start with the resting potential (panel (a)), which has the highest negative transmembrane potential difference (red line). On the other hand, no potential difference is measured outside (magenta line). When the cell is partially and locally depolarized (panel (b)), the transmembrane potential decreases and the extracellular potential shows a peak. At the same time, an electrical dipole moment occurs, which points from the negatively charged (depolarized) side to the positive side. The transmembrane potential is reversed with complete depolarization in the plateau phase (panel (c)), while the extracellular potential is again zero. On the way back to the resting potential, the electric dipole moment increases again and points in the opposite direction. In contrast, the extracellular potential has a negative amplitude (panel (d)) before it returns to zero in panel (e). As we shall see later, the extracellular potential recorded here corresponds to the potential changes observed during a cardiac cycle via an electrocardiogram. Note that there is a one-to-one correspondence between the extracellular potential amplitude and the existence of an electric dipole moment, including its orientation.

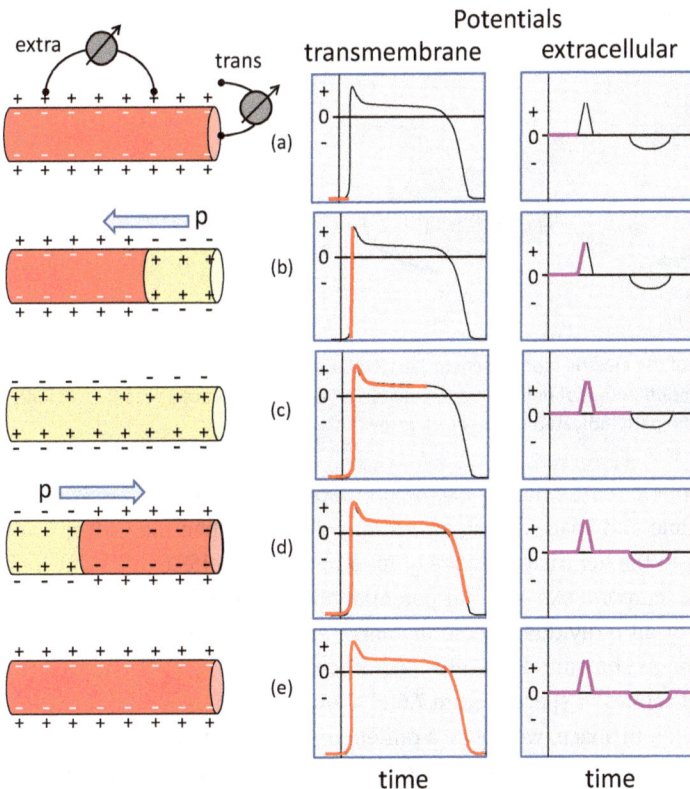

Fig. 7.5: Time evolution of a cardiac muscle cell's extracellular and transmembrane potential during an action potential. Arrows show the orientation of the effective electric dipole.

7.3.2 Dipole looping

Electrostatics informs us about electric monopoles, dipoles, quadrupoles, etc. [4]. Two monopoles of opposite sign combine to a dipole, two opposite dipoles combine to a quadrupole, etc. The dipole moment is defined as (Fig. 7.6(a))

$$\vec{p} = q\vec{d}, \tag{7.1}$$

where q is the charge (positive or negative) and the displacement vector $\vec{d}$ points from the negative charge to the positive charge. Therefore, $\vec{p}$ is also directed from minus to plus. The definition of the dipole moment is a mathematical construct. First, both charges must have the same magnitude, and second, the dipole moment is defined in the limit of $\vec{d}$ approaching zero, while the charge q goes to infinite and the product stays constant.

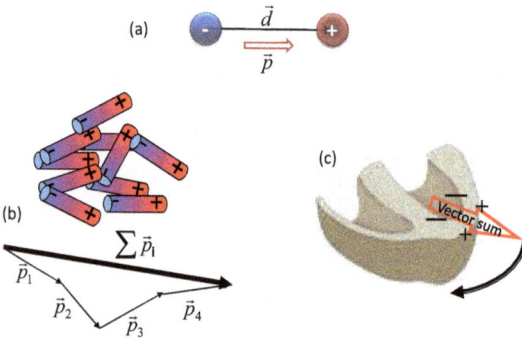

Fig. 7.6: (a) Definition of the electric dipole moment. (b) Resultant vector of dipole moments generated during an action potential in the muscle cells. (c) Snapshot of a cardiac dipole moment. The tip rotates along the path indicated by the black arrow.

In the body, each dipole represents the excitation (depolarization) of a muscle cell at a certain point in time and space during an action potential. When the CAP spreads through the atrium to the ventricle, billions of muscle cells are affected, all of which have their local and temporal extracellular potential difference and form electrical dipoles. The vectors of all individual dipole moments are superimposed and create a resultant vector sum, as shown in Fig. 7.6(b). Moreover, during a CAP, the sum vector changes length and rotates in space. Figure 7.6(c) shows one snapshot. When we follow this vector moving in space, we notice a rather complex threefold vector looping: a small loop for the atrial depolarization (P-loop), and a much bigger loop for the apical and ventricular depolarization (QRS-loop), and at the end another small loop during ventricular repolarization (T-loop). These three loops are shown with different projections in Fig. 7.7.

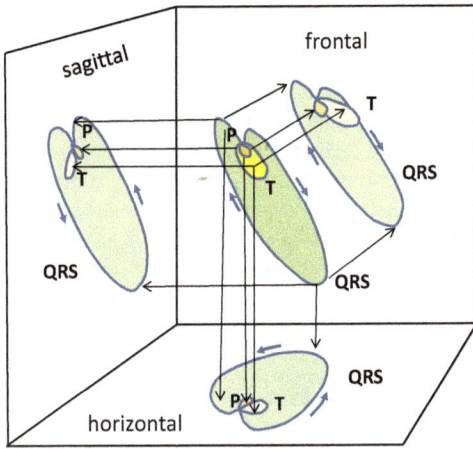

Fig. 7.7: Loops of the resultant dipole moment during a cardiac cycle and their projections on the sagittal, horizontal, and frontal plane. Three loops can be distinguished, labeled P, QRS, and T-loop (adapted from Ref. [5]).

7.3.3 Isopotential lines

According to electrostatics we know how to calculate the local electric field $\vec{E}(\vec{r})$ generated by electric charges and dipoles at distance $\vec{r}$ by using the principle of superposition (see Infobox II on isopotential lines). Knowing the electric field $\vec{E}(\vec{r})$, we can determine equal potential lines $\phi(r) = \text{const}$ that cross the electric field lines at right angles (see Infobox II for further details). The task of calculating $\vec{E}(\vec{r})$ and $\phi(\vec{r})$ for a known charge or dipole distribution is called *direct procedure*. The solution of this problem is exact and unique. However, in physiology, we have to solve the inverse problem: how can we determine the position and orientation of an electric dipole by measuring potential differences? The potential differences are measured outside on the skin, while we need to have information on the sum dipole vector inside that tells us about the heart condition. The direct problem has a unique solution. The indirect problem cannot be solved uniquely, as pointed out first by Helmholtz.[3] Electrocardiography (ECG) is an approximate approach to solve the inverse problem, as we will notice in the next sections.

---

3 Hermann von Helmholtz (1821–1894), German physicist and physiologist.

Infobox II: Isopotential lines

If we examine the electric field $\vec{E}$ of a single electric charge q with a tiny test charge, we will find that, if we follow the black arrows, $\vec{E}$ will decrease continuously and radially with distance $\vec{r}$ from the center: $\vec{E} = 1/4\pi\varepsilon_0 (q\vec{r}/r^3)$. Now, we take a path around the center charge and mark the position where the electric field has the same value. This leads to a circle highlighted in green. It does not cost any energy to move the charge along this line. Hence, it is called the isopotential line $\phi(r) = 1/4\pi\varepsilon_0 (q/r)$. We can measure the potential difference (indicated by blue boxes) between any points along the circle that always show a difference of zero. However, if we measure the potential difference between one circle and another one that is larger or smaller, the potential difference is finite. In this example, the isopotential lines are isotropic and have no directional preference.

The situation is different when we study electric fields and potential lines of an electric dipole $\vec{p} = q\vec{d}$, consisting of two opposite charges at distance $\vec{d}$. The dipole is a vector with a particular orientation. Therefore, the electric field lines are not radial, but start at the plus charge and end at the minus charge. The isopotential lines are distorted and a zero potential line occurs midway between the opposite charges and perpendicular to the dipole moment, according to $\phi(r) = 1/4\pi\varepsilon_0 (\vec{p} \cdot \vec{r}/r^3)$, where $\vec{r}$ is the distance from the dipole. The dependence of the dipole potential on the polar angle θ is $\phi(r) = 1/4\pi\varepsilon_0 (p\cos\theta/r^2)$, which allows to determine the orientation of the dipole in space. The blue boxes display the potential differences read by an electrometer: $\Delta U = 0$ V along an isopotential and $\Delta U = +2$ V or $+4$ V between isopotential lines of different charges.

7.4 Einthoven triangle

7.4.1 Coordinate system

The heart's electric dipole moment results from ion mobility and charge fluctuations taking place during the cardiac cycle. The dipole moment, in turn, generates electrical fields and potential differences that can be measured externally on the skin, illustrated in Fig. 7.8. Since the potential lines change over time, they provide information on the internal dynamics of the heart. Furthermore, when measuring the potential difference with an electrometer for different orientations on the skin, the heart's dipole moment can be determined, as indicated in Fig. 7.8. This is the essential point of ECG, invented by Einthoven[4] at Leiden University.

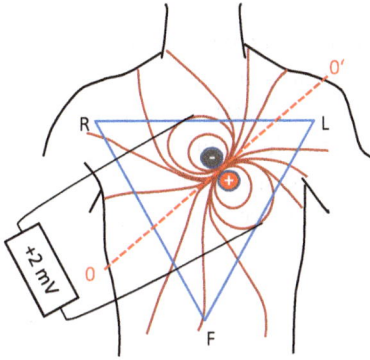

Fig. 7.8: Snapshot of an electric dipole moment during the cardiac cycle. The red loops around the plus and minus charges indicate isopotential lines. The red dashed line labeled (0, 0') is the zero volt isopotential line, oriented perpendicular to the dipole moment. Between different isopotential lines, a potential difference can be measured using a high-resolution electrometer. The blue triangle is the Einthoven triangle.

Einthoven introduced the non-Cartesian triangular coordinate system in the frontal plane, which is better adapted to the body's physiology than the Cartesian coordinate system. With reference to Fig. 7.9, the three sides of the equilateral *Einthoven triangle* are referred to as leads I, II, and III. The potential difference is measured between the limbs and projected onto these lines: between both arms (I), between the right arm

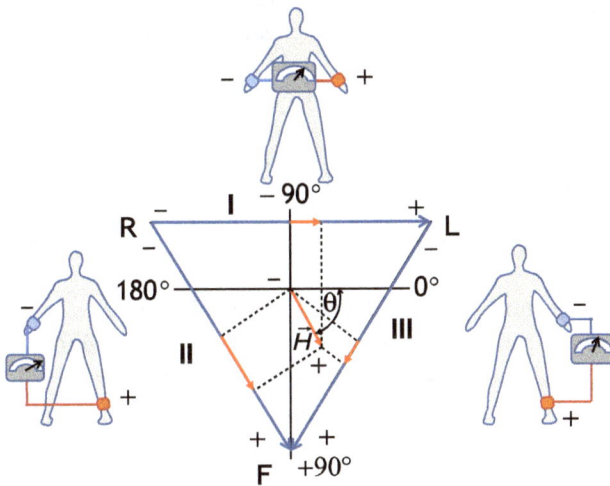

Fig. 7.9: The Einthoven triangle defines the projections of electric dipole moments and determines potential differences between extremities. I, II, III are the leads; R, L, and F stand for right, left, and foot, respectively.

---

4 Willem Einthoven (1860–1927), Dutch physiologist, 1925 Nobel Prize in Medicine.

and left foot (II), and between the left arm and left foot (III). In lead I, the negative electrode is connected to the right arm and the positive electrode to the left arm. Therefore, if isopotential lines of negative charges cross the right arm and the isopotential lines of positive charges cross the left arm, the reading on the electrometer is positive. The electric dipole is then oriented from the right to the left arm. The other two leads II and III have corresponding sign conventions. The blue arrows in Fig. 7.9 reflect the sign convention for positive potential differences: lead I from R to L, lead II from R to F, and lead III from L to F. The horizontal line parallel to lead I also defines the angular orientation θ of the dipole: $\theta = 0°$ for a horizontal dipole pointing from R to L, 180° for the opposite direction. Counter-clockwise rotation from horizontal orientation toward F is counted positive, otherwise negative.

7.4.2 Heart vector

In general, the electric potential ϕ_i generated by the dipole $\vec{p}$ at the limb i (i = R,L,F) is [4, 6]

$$\phi_i(r_i) = \frac{1}{4\pi\varepsilon_0}\frac{\vec{p}\cdot\vec{r}_i}{r_i^3},\tag{7.2}$$

where $r_i = |\vec{r}_i|$ is the distance between the dipole and the extremity's tip. Realizing that dipoles in the body are not static but the result of polarization currents, it is reasonable to express the dipole in eq. (7.2) in terms of a polarization current I_P [6, 7]:

$$\phi_i(r_i) = \frac{1}{4\pi\sigma}\frac{\vec{H}\cdot\vec{r}_i}{r_i^3}.\tag{7.3}$$

Here σ is the conductivity of the medium and

$$\vec{H} = I_P\vec{d}\tag{7.4}$$

is the heart vector. The heart vector has the same mathematical properties as an electric dipole: while I_P goes to infinite, d approaches zero and the product $I_P\vec{d}$ stays constant. A heuristic derivation of eq. (7.3) is provided in the Mathbox I.

With the help of the heart vector $\vec{H}$ and lead vector $\vec{l}$, we can now determine the respective potential differences at leads I, II and III:

$$\text{Lead I: } \Delta U_I = \phi_L - \phi_R = \vec{H}\cdot\vec{l}_{LR},\tag{7.5}$$

$$\text{Lead II: } \Delta U_{II} = \phi_F - \phi_R = \vec{H}\cdot\vec{l}_{FR},\tag{7.6}$$

$$\text{Lead III: } \Delta U_{III} = \phi_F - \phi_L = \vec{H}\cdot\vec{l}_{FL}.\tag{7.7}$$

Those are the potential differences determined by recording an electrocardiogram (ECG). Since the Einthoven triangle is overdetermined, the potential difference of two leads yields the third one. According to Kirchhoff's second law, the sum of all potential differences around any closed loop must be zero. Therefore,

$$\Delta U_{\mathrm{I}} + \Delta U_{\mathrm{III}} - \Delta U_{\mathrm{II}} = 0 \tag{7.8}$$

or

$$\Delta U_{\mathrm{II}} = \Delta U_{\mathrm{I}} + \Delta U_{\mathrm{III}}. \tag{7.9}$$

Example: we assume that the potential at the right arm is −0.2 mV with respect to the average potential of the body (reference point), at the left arm +0.3 mV, and at the left foot +1.0 mV. Then the potential difference ΔV_{I} measured in lead I between right and left arm is $\Delta V_{\mathrm{I}} = +0.5$ mV, in lead II it is $\Delta V_{\mathrm{II}} = +1.2$ mV, and in lead III, $\Delta V_{\mathrm{III}} = +0.7$ mV. Both potential differences add up to the third one: $\Delta V_{\mathrm{I}} + \Delta V_{\mathrm{III}} = \Delta V_{\mathrm{II}}$. The potential difference is positive and larger for lead II than for lead III. Hence, the dipole must be oriented downward close to parallel to lead II. Since $\Delta V_{\mathrm{I}} > 0$, we conclude that the angle $\theta < 90°$. A rough estimate of the angle θ is $\cos \theta \approx \Delta V_{\mathrm{I}}/\Delta V_{\mathrm{II}} = 0.5/1.2$, $\theta \approx 65°$.

The Einthoven triangle describes potential differences between the limbs evoked by the temporal heart dipole during a cardiac cycle.

!

Mathbox: Heart vector

i

According to Gauss law, integration of the electric field $\vec{E}$ over a closed surface $\vec{A}$ yields the enclosed charge q:

$$\oint \vec{E} \cdot \vec{A} = EA = \frac{q}{\varepsilon_0} \, .$$

On the other hand, the current density j is proportional to the electric field via $j = \sigma E$, where σ is the conductivity, here of the tissue. Therefore,

$$\frac{q}{\varepsilon_0} = EA = \frac{jA}{\sigma} = \frac{I_{\mathrm{P}}}{\sigma}.$$

Combining in scalar form, we have for the potential:

$$\phi_l(r) = \frac{1}{4\pi\,\varepsilon_0} \frac{q\,d}{r^2} = \frac{1}{4\pi\,\sigma} \frac{I_{\mathrm{P}}\,d}{r^2} = I_{\mathrm{P}} d \frac{1}{4\pi\sigma r^2} = Hl.$$

In vector form, we obtain for the potential of the heart vector:

$$\phi = \vec{H} \cdot \vec{l},$$

where $\vec{H} = I_{\mathrm{P}}\vec{d}$ is the heart vector, and $\vec{l} = \dfrac{\vec{r}}{4\pi\sigma r^3}$ is the lead vector. Remember that $\vec{r}/r$ is a unit vector in the direction of $\vec{l}$. The potential difference at two points separated by $|\vec{r}_1 - \vec{r}_2|$ is the dot product of the heart vector and the respective lead vectors:

$$\Delta U_{12} = \vec{H} \cdot \vec{l}_{1,2} = \vec{H} \cdot \frac{\vec{r}_1 - \vec{r}_2}{4\pi\sigma|\vec{r}_1 - \vec{r}_2|^3}.$$

7.5 Electrocardiogram

7.5.1 Einthoven leads

We now have collected all information to link the potential differences as determined by Einthoven leads I, II, and III, to the actual cardiac cycle. For the same phase of the cardiac cycle, each lead will show different but related potential variations from which the functioning of the heart, or in pathological cases, the malfunctioning, can be inferred. The recording of the potential differences versus time is known as *electrocardiography*; the plots of the potentials are *electrocardiograms*.

One example is shown in Fig. 7.9 and in more detail in Fig. 7.10(a), representing the beginning of the cardiac cycle when the depolarization starting at the SA node has spread over the atria. The depolarization causes a large dipole moment pointing parallel to the heart axis. The positive amplitude measured in all three leads is known as the *P-wave*. It occurs after about 100 ms and the amplitude is about 0.5 mV. Shortly after the depolarization of the atria is completed, the P-wave returns to zero potential, and the action potential crosses the AV node. The propagation through the AV node is very sluggish, resulting in a delay in the progress of activation. The delay allows the completion of ventricular filling.

After activating the AV node, the ventricular septum becomes depolarized, and a dipole occurs at an angle of about 170°. The projection of this dipole is negative in leads I and II, but positive in lead III. This point is referred to as the Q-point and is shown in Fig. 7.10(b). In lead III, the Q-point merges the R-wave and is therefore difficult to recognize. The time period between P- and Q-waves is called *PQ interval*, which takes about 160 ms. Sometimes it is also called *PR interval*, because the Q-wave is not much pronounced.

After passing the Q-spike, both ventricles and the apex become depolarized, which generates the largest dipole moment and the highest potential difference in the R peak of about 4 mV. Then follows a rapid change where only the left ventricle outside the wall shows a polarization, resulting in the S-spike. The potential variation from Q to S is known as *QRS complex*. Now all parts of both ventricles are depolarized and the potential returns to zero. Repolarization of the ventricles causes the T-wave. The time sequence of the polarization changes are plotted in Fig. 7.11 together with the potential changes according to lead II.

The QRS complex corresponds to the largest of the three loops shown in Fig. 7.7, and the other loops are connected to the P- and T-waves. It is interesting to note that the P-wave corresponds to depolarization of the atria, while the T-wave, although of the same sign, is due to repolarization of the ventricles. The repolarization of the atria is not noticeable as it is immersed in the large and dominant QRS complex.

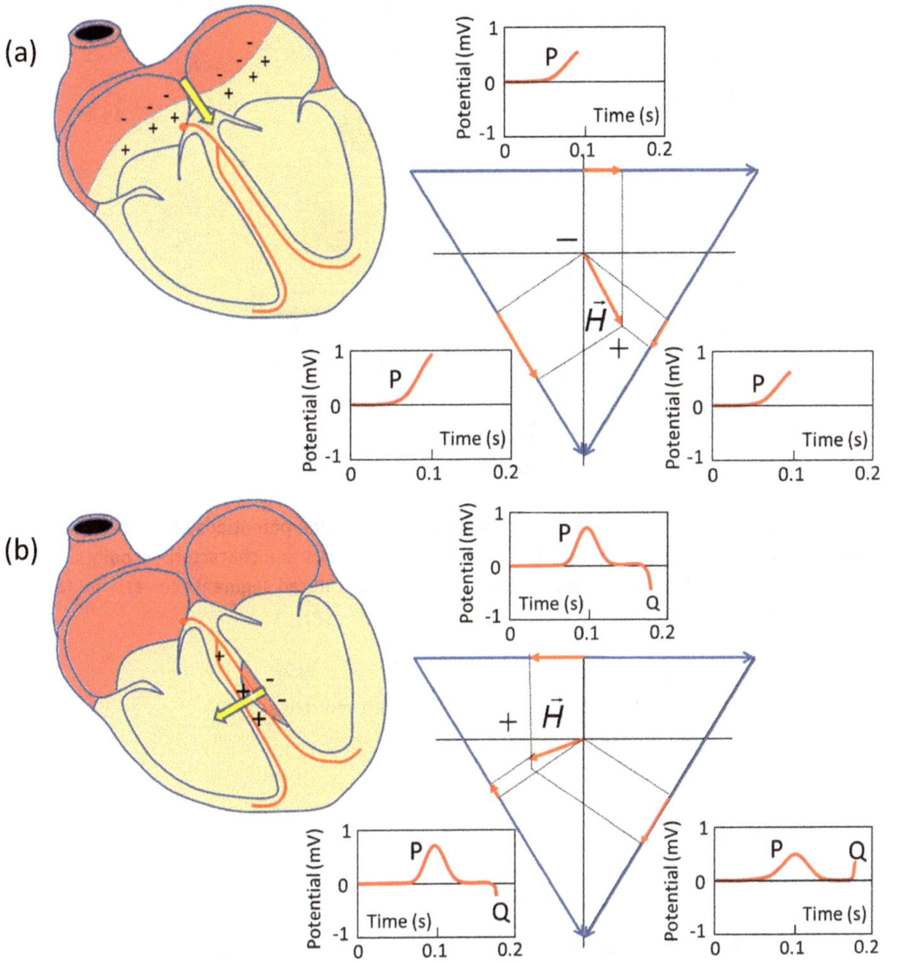

Fig. 7.10: Depolarization of the heart during two different phases: (a) depolarization of the atria generating the P-wave; (b) depolarization of the left side of the ventricular septum generating the Q-spike. The potential changes are shown for all three leads as a function of time. Red color in the heart contours corresponds to a depolarized state, and ocher color to a resting state.

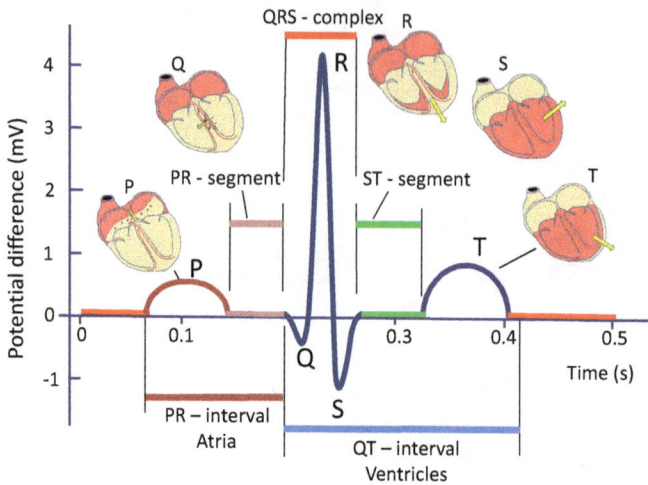

Fig. 7.11: Sequence of polarization changes during a cardiac action potential as recorded in lead II. The graph also gives the standard nomenclature for time intervals and characteristic points. Red color segment of the heart refer to depolarization, and ocher-colored segments correspond to a resting state or to repolarization.

> - The P-wave is due to the depolarization of the atria.
> - The QRS complex signifies the depolarization of both ventricles.
> - The T-wave reflects the repolarization of the ventricles.

7.5.2 Goldberger leads

The leads according to Einthoven are referred to as *bipolar leads.* Alternatively, *unipolar leads* can be defined according to the scheme of Goldberger.[5] The *Goldberger leads* define a neutral reference point, and the potential difference is measured between this neutral point and one of the extremities. The neutral point is defined to lie between two equally large resistances corresponding to the center of the Einthoven triangle. The leads indicated in Fig. 7.12 are labeled aVL, aVR, and aVF for augmented voltage left, augmented voltage right, and augmented voltage foot, respectively. Originally, these leads were measured without resistances defining the central point, but the signals were too small. The subsequent use of high ohmic resistances augmented the signal, therefore, the name "augmented voltage." The Goldberger leads are equivalent to the Einthoven leads I, II, and III. Although they yield redundant information, Goldberger leads allow more precise measurements of

---

5 Emanuel Goldberger (1913–1994), American cardiologist.

potential changes at the extremities than possible with Einthoven leads, and they confirm characteristic patterns in case of heart diseases.

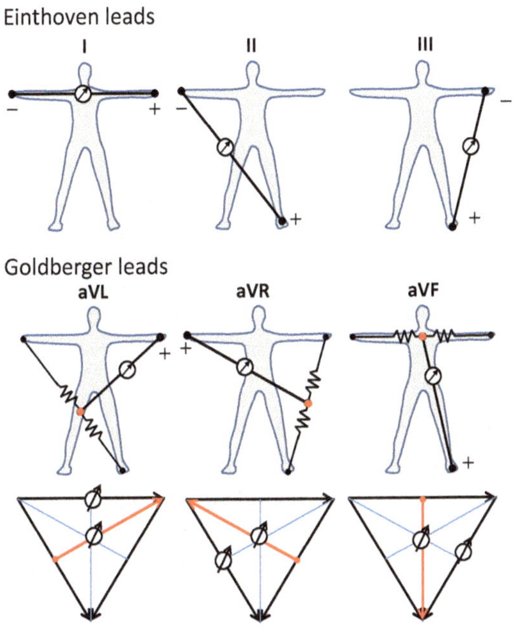

Einthoven leads

Goldberger leads

Fig. 7.12: Comparison of leads according to Einthoven and Goldberger. In the leads according to Goldberger, a neutral point is defined by two high and equal ohmic resistances between extremities. In the bottom panels, the equivalence of both lead systems (black: Einthoven; red: Goldberg) becomes evident.

The potential differences of the Goldberger leads are related to the Einthoven potentials by taking the difference between the potential at one of the limbs and the arithmetic mean of the two opposing potentials:

$$aVL = \phi_L - (\phi_R + \phi_F)/2, \tag{7.10}$$

$$aVR = \phi_R - (\phi_L + \phi_F)/2, \tag{7.11}$$

$$aVF = \phi_F - (\phi_L + \phi_R)/2. \tag{7.12}$$

Combining the Einthoven projections and the Goldberger projections, a vector projection is achieved in the frontal plane at increments of every 30°, as shown in Fig. 7.13. The circle is known as the *Cabrera circle*. The Einthoven projections are located every 60°, and the Goldberger projection lies in between. Note that with this system, orthogonal projections can be compared. For instance, if one of the leads shows a large amplitude in the QRS complex, the projection in the orthogonal direction should vanish. Using both projections, the precision of determining a vector orientation is about ±10°.

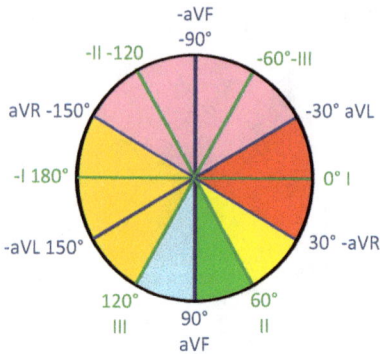

Fig. 7.13: The Cabrera circle combines Einthoven leads and Goldberger leads. The colors refer to the orientation of the heart axis. Yellow: standard indifferent; green: steep; blue: right; red: left. All other regions are extremes and usually not observed.

As an example, we discuss the heart axis (see Fig. 7.1). In the R-spike, the electrical dipole moment lies parallel to the *heart axis*. The standard orientation is about 60° with respect to the horizontal line (see Fig. 7.9 for the definitions of angles). Therefore, lead II should have the highest positive signal. Lead aVL is oriented at right angles to lead II and should display a zero-crossing passing through the perpendicular orientation. This is indeed the case when comparing all projections displayed in Fig. 7.15. At the same time, the projection onto aVR should be negative, which is actually observed. Heart axes that lie in the yellow angular region (30°–60°) of Fig. 7.13 are called *standard orientation* or *indifferent orientation*. The green region (60°–90°) refers to a "steep" heart axis, the blue region (90°–120°) to a right-oriented heart axis, whereas the red-colored region (30° to –30°) corresponds to a left-oriented heart. Extreme deviations from the standard orientation may indicate a heart disease or lung emboli, which pushes the heart to the right.

7.5.3 Wilson leads

Note that all projections taken so far according to Einthoven and Goldberger, yield a vector projection into the frontal plane. However, as we have seen in Fig. 7.7, the electrical sum dipole is a vector describing loops in three dimensions. In order to determine the third dimension in the horizontal plane, an additional system of leads was introduced according to Wilson.[6] These additional leads are placed directly on the left side of the chest on the left and right sides of the heart. The exact location of the electrodes, labeled V1–V6, is shown in Fig. 7.14.

---

6 Norman Wilson (1890–1952), American cardiologist.

The Wilson leads yield information on the horizontal vector projection. The potential is positive during periods when the dipole points in the direction of the electrode; the potential is negative whenever the vector points away from the electrode.

Fig. 7.14: Left panel: Placement of the Wilson leads on the chest. Right panel: Lead projections in the horizontal plane.

These 12 leads (3 Einthoven, 3 Goldberger, and 6 Wilson) are called *standard leads*. They are most frequently used in clinical practice. Figure 7.15 shows the recordings of electrocardiograms according to all 12 leads. In fact, they are highly redundant. Only three independent leads are required to determine all three vector components. Practical guidance to ECG diagnostics can be found in [8].

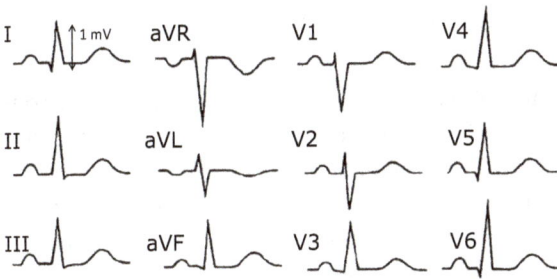

Fig. 7.15: Electrocardiograms recorded according to the leads of Einthoven (I, II, III), Goldberger (aVR, aVL, aVF), and Wilson (V1–V6).

7.6 Electrocardiography

7.6.1 Procedures

ECG[7] recordings can be performed at rest and during exercise to control long-term heart rhythms or abnormal and eventually life-threatening, arrhythmic perturbations (*cardiac arrhythmias*). ECG is also used to check heart functions during illnesses, strokes, and accidents. In all cases, the ECG provides invaluable and indispensable information on the heart's function. For instance, cardiac fibrillation is a condition of uncoordinated action potentials which leads to improper ventricular contraction. Figure 7.16 shows an example of ventricular fibrillation determined by lead II. Under these conditions, the pumping of blood is completely ineffective. As no blood is circulated, no oxygen is transported. This state is lethal unless stopped by an electroshock through the heart with the use of a *defibrillator*. The defibrillator provides a high dc voltage of 1000–4000 V for 3–40 ms to the chest on both sides of the heart, resetting the action potential. Then a new and fresh coordinated depolarization via the AV or SA node can be started. The total energy supplied by the defibrillator is limited by the charge of a capacitor which is usually 100–300 J. Ventricular fibrillation is the most commonly identified arrhythmia in cardiac arrest patients.

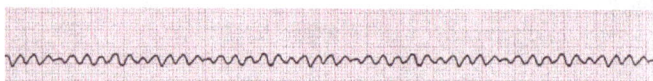

Fig. 7.16: Ventricular fibrillation measured at lead II.

In former days, ECGs were recorded with a single-channel strip chart recorder connected to an electrometer. This procedure required recording each lead one after the next rendering comparisons between different projections difficult if not impossible. More recently, the single-channel strip chart recorder was replaced by 12 channels in groups of 2 for all 12 leads; an example is shown in Fig. 7.17. Nowadays, the electrometer is connected to a digital oscilloscope after signal amplification and analog-digital conversion. For long-term ECG monitoring patients keep the electrodes attached to the body for a day or so, and the data are recorded in a small electronic storage device for later read-out. Alternatively, the data can be transmitted instantaneously to the clinic via the Internet, known as real-time telemedical (home) care.

---

7 In the literature one may also find the acronym EKG, referring to the German expression "Elektrokardiographie."

Fig. 7.17: ECG with 12 leads documented with a strip chart recorder.

Traditionally, electrodes for measuring ECGs are attached to the skin using Ag/AgCl electrolytic gels to reduce the ohmic resistance in the air gap between skin and electrode. A typical setup is shown in Fig. 7.18, including ergometer, recording unit, and attachment electrodes to the skin of a subject.

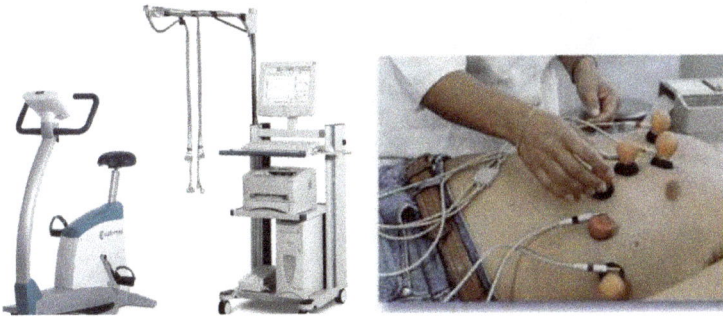

Fig. 7.18: Typical setup for ECG recording with ergometer, recording unit, and placement of electrodes on the chest.

7.6.2 New developments

In principle, electrical potentials can also be recorded by capacitive coupling. Capacitive sensors are well known in various industrial environments, but for measuring ECGs with very small millivolt potential changes, the sensitivity was insufficient so far. New capacitive sensors using high dielectric constant, so-called high k materials, have remedied this problem and provided the necessary sensitivity. The capacitive sensors are dry, do not need adhesives or electrolytes that could cause allergic reactions on the skin, and are easy to apply. Therefore, capacitive sensors will likely supersede ohmic contacts in the long run. They are already being used on wristwatches to monitor one-channel ECGs, usually applying the Einthoven lead I. One electrode on the backside of the watch is in permanent contact with the skin. The second front-facing electrode must be touched with a finger from the opposite hand to record an ECG. An electric potential integrated circuit (EPIC) sensor records the ECG signal and transmits it to the display of the wristwatch or to any other mobile device like a smartphone. Figure 7.19 shows an example of a handheld two thumb touchpad version with Bluetooth coupling to a smartphone reading out the cardiac signal from lead I [9].

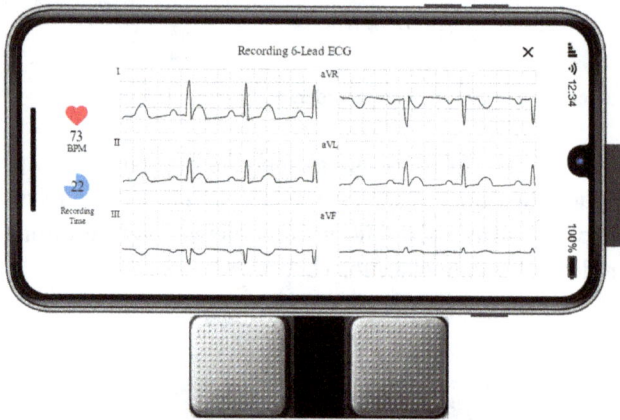

Fig. 7.19: Capacitive handheld device for recording cardiac signals (adapted from https://heartheroes.co.uk/portable-devices-for-ecg-recording-paediatric-cardiology-services/).

7.7 Magnetocardiography

The ion currents in the heart that cause myocardial contraction also create magnetic fields that are much weaker (~100 pT) than electric fields. Nevertheless, these weak magnetic fields can be detected with highly sensitive magnetic field sensors and a magnetocardiogram (MCG) similar to an ECG can be recorded. Two types of sensors are currently being tested and used: superconducting quantum interference devices (SQUID) based on high-temperature superconductors, which require liquid nitrogen instead of helium for cooling [10]; and laser-optically pumped Cs cells [11]. The advantage of MCG is the contactless recording of heart signals. In addition, a magnetic field mapping of cardiac activity of the entire chest can be performed with a number of sensors. Therefore, MCG can be carried forward from local signal acquisition to imaging. In principle, this should also be possible with the ECG electrodes and especially with the new EPIC sensor technology. However, the spatial resolution would be lower due to the larger spatial spreading of electric fields compared to magnetic fields, as already discussed in Chapter 6 with regard to MEG versus EEG. The downside to MCG is the weak signal that requires averaging over many cycles to get a decent signal-to-noise ratio (SNR). This works well for a healthy person. On the other hand, cardiac arrhythmia requires high time resolution, but MCG lacks them. Even so, this new technique is now available in some clinics. A schematic setup for MCG recording is shown in Fig. 7.20.

A promising application of MCG is the recording of the fetal heart function [12]. ECG has been used in the past, but with mixed success. This is due to a low SNR resulting from an interfering maternal ECG and the small size of the fetal heart. On the other hand, fetal MCG, although very weak, has been shown to give results at an adequate noise level, making it possible to analyze fetal arrhythmias for possible treatment before birth.

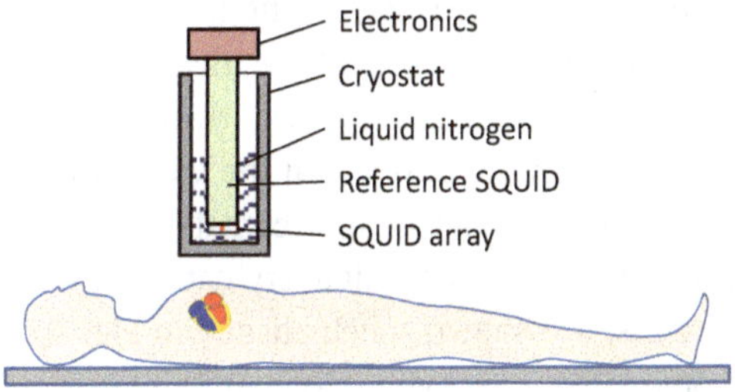

Fig. 7.20: Magnetocardiography uses SQUID sensors for recording weak magnetic fields from cardiac activity.

The analytical methods of cardiac function are by no means exhausted with ECG and MCG. In the next chapter, we will get to know additional analytical methods. Furthermore, numerous imaging methods are used that can represent the heart function in real time. These include ultrasound echocardiography (Chapter 1, Vol. 2), turbomagnetic resonance tomography (Chapter 3, Vol. 2), and myocardial single-photon emission computed tomography, discussed in Chapter 9, Vol. 2.

7.8 Artificial pacemaker

An artificial pacemaker is a small electronic device that can be used to control the heart rate when the natural heartbeat is too slow or irregular (arrhythmias). Artificial pacemakers consist of three main parts: battery, pulse generator, and one or more leads with electrodes on each lead. A flat, thin metal box contains the battery and electronics implanted under the skin through an incision in the chest just below the collarbone. The electrodes are passed through the vein that goes into the right ventricle. The electrodes have a double purpose: They record the ECG from the heart and signal the impulses back to the electronics. The electronics compares the actual heart rate with the standard settings. The pacemaker can also monitor temperature, breathing rate, and other factors and adjust the heart rate for changes in activity. If the rate is too low or arrhythmic, the electrodes stimulate the heart muscle with electrical discharges. A Li-ion battery lasts about 8–10 years, and when it needs replacement, the metal case has to be opened by a small surgical intervention. Modern pacemakers have Bluetooth connections to the outside world so that the patient or a doctor can monitor the heart rate via the Internet (telemedicine). Depending on the design, cardiac pacemakers can also contain a defibrillator that delivers stronger impulses than a standard pacemaker. Developments for charging batteries via microwave radiation have not been successful. Efforts to extend battery life are more promising.

Recent developments have the potential to revolutionize the use of artificial pacemakers [13]. A tiny pen-shaped cordless pacemaker was developed that contains all the parts, including the battery, electronics, sensors, and pulse generator. The dimensions of 6 mm wide and 42 mm long are such that it can be inserted directly into the heart without surgery. It is introduced by a steerable catheter delivery system funneled from a leg vein through the inferior vena cava to the right ventricle of the heart and anchored in the lower portion of the right ventricle. After fixation and electrical approval, the catheter is removed. Since no wires are inserted, there are significantly fewer complications to be expected, such as infection of the veins into which the wires are inserted and venous injuries. The battery, which has a lifetime of approximately 12 years, can be replaced by removing the pacemaker using the same delivery catheter. Both types of pacemakers, conventional and wireless, are shown schematically in Fig. 7.21. In any case, all electrodes and parts that come into contact with blood and tissue must be made biocompatible to prevent blood clotting.

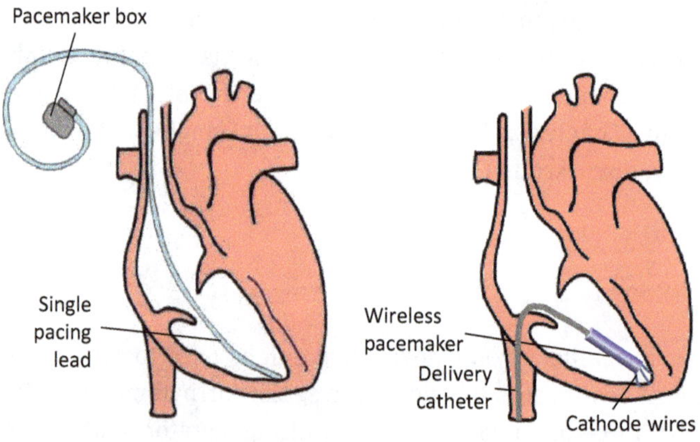

Fig. 7.21: Comparison of a single lead pacemaker (left) and a wireless pacemaker (right). The pacemaker box contains the battery, electronics, and pulse generator. The left panel shows only a single lead in the right ventricle, but up to three leads are used depending on the design. The wireless pacemaker consists of a metal tube that contains all parts in miniature and is inserted into the lower part of the right ventricle. Cathode wires are anchored on the inside wall [13].

7.9 Summary

S7.1   The CAP is much longer than the action potential of normal cells.

S7.2   The CAP is characterized by a very fast depolarization, fast but partial repolarization, followed by an extended plateau region before complete repolarization takes place.

S7.3   The myocardium does not show tetanic contraction.

S7.4   The action potential is self-excitatory and starts at the SA node.

S7.5   The AV node acts as low-pass filter.

S7.6   The sum vector of electric dipoles describes three loops during one cardiac cycle, characterized by a P-wave, PR interval, Q-spike, QRS complex, S-spike, ST interval, and T-wave.

S7.7   The P-wave reflects the depolarization of the atria, the QRS complex reflects the depolarization of both ventricles while the atria recovers, and the T-wave reflects the repolarization of the ventricles.

S7.8   The Einthoven triangle describes projections of the temporal electric dipole onto three directions, called leads.

S7.9   The potential differences recorded by the three Einthoven leads reflect the depolarization and repolarization of the heart during one cardiac cycle.

S7.10 Goldberger leads are taken between the center and the extremities.

S7.11 The Cabrera circle combines Einthoven leads and Goldberger leads.

S7.12 Combination of Einthoven and Goldberger leads allow a determination of the heart axis within ±10°.

S7.13 Wilson leads are placed on the chest and add vector information on the third dimension.

S7.14 Standard ECG recordings use 12 leads: 3 according to Einthoven, 3 according to Goldberger, and 6 according to Wilson.

S7.15 New capacitive sensors allow ECG recording by simply touching contacts with fingers from both hands.

S7.16 Magnetocardiography shows promising applications for recording fetal heart activity.

S7.17 Artificial pacemakers record the heart rate and stimulate the ventricular contraction if the heart rate drops below normal.

S7.18 Miniaturized pacemakers are inserted directly into the right ventricle and controlled by Bluetooth technology.

Questions

Q7.1   What do you understand by an absolute refractory time and a relative refractory time?

Q7.2   What is the role of Ca^{2+} ions for the CAP?

Q7.3   How is the action potential of the SA node different from that of the myocardium?

Q7.4   What keeps the heart beating, and what is the natural pacemaker?

Q7.5   How is the action potential distributed over the myocardium?

Q7.6   How is the Einthoven triangle arranged?

Q7.7   What is the purpose of the Einthoven triangle?

Q7.8   What is the difference between the Einthoven leads and the leads according to Goldberger?

Q7.9   Name the three characteristic intervals of the heart cycle.

Q7.10 Describe the potential variations measured at lead I with respect to the cardiac cycle.

Q7.11 If lead II reads positive values, but aVL reads negative during the QRS complex, which orientation does the heart axis have?

Q7.12 What is recorded during an ECG examination?

Q7.13 What type of artificial pacemakers are available?

Q7.14 When should a defibrillator be used?

Attained competence checker	**+**	**0**	**–**
I know the reasons why the CAP is much longer-lasting than a usual action potential.			
I can describe the spreading of the CAP across the heart.			
I know the function of the SA node and the AV node.			
I can distinguish between the transmembrane potential and extracellular potential.			
I can sketch the electric field lines of an electric monopole and electric dipole.			
I can draw the equipotential lines for an electric monopole and dipole.			
I realize that the Einthoven procedure measures potential differences between the extremities.			
I know the difference between the Einthoven procedure and Goldberger procedure.			
I understand that the Einthoven and Goldberger leads are projections of the heart vector into the frontal plane.			
I know how many loops the heart vector makes during one CAP.			
I know how many leads are measured during an ECG examination.			
I am aware of different types of pacemakers.			

Suggestions of home experiments

HE7.1 **Potential difference**
Use a laboratory multimeter with a mV scale and measure the potential difference between your left and right thumb. What is you reading? Change the polarity and confirm your result. Do the same between your left hand and right toe. If the contact is weak, you may wet your fingers and toe with saltwater. Note that the voltage reading is a time average and therefore lower than the spikes observed in a time resolved measurement.

HE7.2 **Heartbeat**
Do 10 standups and measure your heart frequency before and after. Is the heart frequency already at the limit, or can you do more?

Exercises

E7.1 **Dipole orientation:** When the potential difference ΔV is strong and negative in lead III, which direction does the dipole point? What phase in the cardiac cycle does this situation correspond to?

E7.2 **Potential difference and dipole orientation:** Measurements of potential differences yields the following values: $\Delta V_I = -0.6$ mV and $\Delta V_{II} = -0.3$ mV. What is ΔV_{III} and which direction does the dipole point?

E7.3 **QRS complex:** The QRS complex shows a maximum in lead II and a zero potential in aVL. Explain why.

E7.4 **ST segment:** During the ST segment, the recording in lead II shows a zero potential difference. Explain how this is possible.

E7.5 **Goldberger leads:** Show that the Goldberger leads are rotated by 30° with respect to the Einthoven leads.

E7.6 **Cartesian coordinates:** Express the potential differences according to Einthoven in terms of Cartesian coordinates, setting $|\vec{l}| = 1$.

E7.7 **AV block:** Use sketches to describe how you can identify an AV block of the first grade using an ECG recording.

References

[1] Guyton AC, Hall JE. Textbook of medical physiology. 11th edition. Philadelphia, Pennsylvania, USA: Elsevier Saunders; 2006.

[2] Pape HC, Kurtz A, Silbernagel S, eds. Physiologie. 7th edition. Stuttgart, New York: Thieme Verlag; 2014.

[3] Boron WF, Boulpaep EL. Medical physiology. 2nd edition. Philadelphia, London, New York: Saunders W.B. Elsevier; 2012.

[4] Jackson JD. Introduction to electrodynamics. 3rd edition. New York, London, Sydney, Toronto: Wiley; 1998.

[5] Zabel M, Acar B, Klingenheben T, Franz MR, Hohnloser SH, Malik M. Analysis of 12-lead T-wave morphology for risk stratification after myocardial infarction. Circulation. 2000; 102: 1252–1257.

[6]   Malmivuo J, Plonsey R. Bioelectromagnetism: Principles and applications of bioelectric and
      biomagnetic fields. Oxford University Press; 1995. Available at: www.bem.fi/book/
[7]   Tendera J. Linear Functionals in ECG and VCG. Master thesis, Germany: University Münster;
      2013.
[8]   Vecht R, Gatzoulis MA, Peters NS. ECG diagnosis in clinical practice. 2nd edition. Berlin
      Heidelberg: Springer-Verlag; 2009.
[9]   https://heartheroes.co.uk/portable-devices-for-ecg-recording-paediatric-cardiology-services/
[10] Zhang Y, Wolters N, Lomparski D, Zander W, Banzet M, Schubert J, Krause HJ, van Leeuwen
      P. Multi-channel HTS rf SQUID gradiometer system recording fetal and adult
      magnetocardiograms. IEEE Trans Appl Supercond. 2005; 15: 631–634.
[11] Bison G, Castagna N, Hofer A, Knowles P, Schenker JL, Kasprzak M, Saudan H, Weis A. A room
      temperature 19-channel magnetic field mapping device for cardiac signals. Appl Phys Lett.
      2009; 95: 173701.
[12] Sameni R, Clifford GD. A review of fetal ECG signal processing: Issues and promising
      directions. Open Pacing Electrophysiol Ther J. 2010; 3: 4–20.
[13] Reynolds D, et al. A leadless intracardiac transcatheter pacing system. N Engl J Med. 2016;
      374: 533–541.

Further reading

Crawford MH, DiMarco JP, Paulus WJ. Cardiology. 3rd edition. Philadelphia, Pennsylvania, USA:
      Elsevier Saunders; 2010.
Seeley R, Vanputte C, Russo A. Seeley's anatomy and physiology. Boston: McGraw Hill Book Co.;
      2016.
Guyton AC, Hall JE. Textbook of medical physiology. 11th edition. Philadelphia, Pennsylvania, USA:
      Elsevier Saunders; 2006.
Pape HC, Kurtz A, Silbernagel S, eds. Physiologie. 7th edition. Stuttgart, New York: Thieme Verlag;
      2014.
Malmivuo J, Plonsey R. Bioelectromagnetism: Principles and applications of bioelectric and
      biomagnetic fields. Oxford, New York, Athens: Oxford University Press; 1995. Available at:
      www.bem.fi/book/

8 The physical aspects of the circulatory system

Red blood cells

Physical properties of the circulatory system

Total peripheral flow resistance (R_{TPFR})	1.4×10^5 Pa · s/l
End-diastolic volume	120 ml
End-systolic volume	50 ml
Stroke volume	70 ml
Ejection fraction	0.6
Cardiac output per ventricle	88 ml/s or 5.3 l/min or 7600 l/day
Power consumption of the heart	5.6 W
Mechanical efficiency of the heart	25%
End systolic pressure	140 hPa
Pressure at hydrodynamic indifference level	130 hPa
End diastolic pressure	3 hPa
Mean systemic filling pressure	10 hPa
Flow velocity in the capillary bed	0.005 m/s
Flow velocity at aorta	1 m/s
Pulse wave velocity	6 m/s
Blood volume in the body	5–6 l
Total oxygen content of blood	0.03 mol oxygen
Number of red blood cells per 1 µl blood	5×10^6 RBCs
Life time of RBCs	100–120 days
Hematocrit value	0.6
Mean corpuscular volume of erythrocyte	100 fl
Mean corpuscular hemoglobin (MCH)	27–34 pg
Mean corpuscular hemoglobin concentration (MCHC)	32–36 g/dl
Viscosity of blood at 37 °C	4 mPa s
Osmotic pressure of erythrocytes	7.5 bar = 7500 hPa
Number of hemoglobin molecules per erythrocyte	2.5×10^8
Reproduction rate of erythrocytes	3.5×10^6 erythrocytes/s
Viscosity of blood	4 mPa s

https://doi.org/10.1515/9783110756951-008

8.1 Introduction

Circulation is the circular flow of a medium (gas, liquid) driven by a pump. In the physiological sense, the bloodstream is a circulatory system in which the heart serves as a pump. In this chapter, we describe the physical aspects of the circulatory system, also known as the *cardiovascular system*. We pay special attention to the mechanical work of the heart as a pump and the hemostatics and hemodynamics of the blood vessels. This chapter requires the information provided in Chapter 7. It is therefore advisable to work through the previous chapter first or become familiar with the basics of cardiac-electrophysiology before proceeding with this chapter.

The cardiovascular system is reponsible for supplying all body parts with oxygen and nutrients and for disposing of metabolic waste products such as carbon dioxide, urea, water, and heat. There are further functions of blood flow associated with fighting disease via leukocytes, transporting hormones, supplying thrombocytes to injured blood vessels, etc. Oxygen is taken up in the lungs during inhalation, and carbon dioxide is disposed of during exhalation. How this occurs is the subject of Chapter 9 on the respiratory system. Oxygen molecules (O_2) bind to the heme complex in hemoglobin proteins located in erythrocytes. These small disc-shaped oxygen carriers float in the bloodstream to the most distant parts of the body and through vessels from wide to extremely narrow.

Figure 8.1 shows schematically the circulatory system consisting actually of two pumps. The left ventricle is the pump for oxygenated blood, and the right ventricle is the pump for deoxygenated and carbon dioxide-rich blood. Oxygen arriving from the lungs is pumped to the capillary system and delivered to organs and muscles, while carbon dioxide is resorbed from the tissue and pumped back to the lungs for expiration. In the following, we will describe the circulatory system in more detail from a physical point of view. For physiological aspects, literature is provided at the end of this chapter.

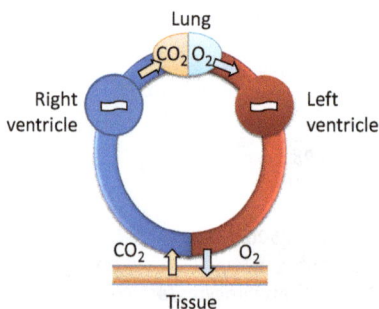

Fig. 8.1: Schematics of the circulatory system. O_2 is taken up in the lungs and pumped by the left ventricle to the organs and muscles, whereas metabolic CO_2 is pumped back through the right ventricle to the lung for expiration. Note that both pumps are arranged in series.

8.2 Two circuits

We may also look at the circulatory system from another side as consisting of two circulations: a lung circuit or *pulmonary circuit* and a body circuit *or systemic circuit*. Both circulations are displayed in Fig. 8.2 and require the action of both ventricles, which act in a series:

1. The pulmonary circuit pumps deoxygenated blood from the right ventricle to the lung. On the return, it delivers oxygenated blood to the left atrium.
2. The systemic circuit pumps oxygenated blood through the left ventricle to the body. On the return, it delivers deoxygenated blood to the right atrium.

Sometimes these circuits are also referred to as *small circuit* and *large circuit*. Both circuits run synchronously and in-phase, and they pump the exact same blood volume of about 70 ml per systolic ejection, called *stroke volume (SV)*.

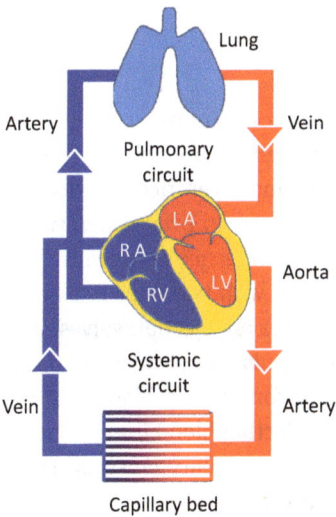

Fig. 8.2: The circulatory system consists of two circuits: the pulmonary and systemic circuits. RA = right atria, LA = left atria, RV = right ventricle, LV = left ventricle.

> The cardiovascular system consists of two circuits: a lung circuit or *pulmonary circuit* and a body circuit *or systemic circuit.*

There are three types of blood vessels in the body that are schematically distinguished in Fig. 8.3:

1. *Arteries* and *arterioles* carry blood away from the heart.
2. *Veins* and *venules* carry blood to the heart.
3. *Capillaries* allow the diffusion of oxygen and nutrients from the blood to the tissue and the diffusion of waste products to the blood. The *capillary bed* is the counterpart to the alveoli in the lung. Both serve the purpose of gas exchange in and out of the blood.

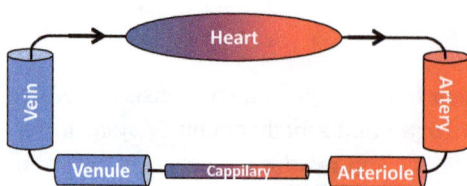

Fig. 8.3: Blood vessels in the circulatory system.

There is still another way to categorize the circulatory system according to the blood pressure. Arteries carry high-pressure blood, whereas veins support the low-pressure blood system. These differences are also expressed in the structure of the blood vessels, as we will discuss later.

8.3 The heart as a pump

8.3.1 Four phases

The heart consists of two pumps and two chambers each, one pump for the pulmonary circuit and one for the systemic circuit. These two pumps are colored in blue and red in Fig. 8.4. Each pump has an atrium for filling and a ventricle for ejection of blood, and each chamber is equipped with a valve. The two *atrioventricular valves* are filling valves: *tricuspid valve* and *mitral valve* (MV). The *tricuspid valve* with three flaps fills blood from the body in the right atrium. The *MV* with two flaps fills blood from the lung into the left atrium. The two *ventricular valves* are ejection valves, also called *semilunar valves*, because of their crescent shape: *pulmonary valve* and *aortic valve*. The pulmonary valve in the right ventricle opens for ejecting blood into the lung, and the aortic valve in the left ventricle opens for ejecting blood into the body. All valves open in the flow direction by the contraction of papillary muscles. There are three of them in the right ventricle and two in the left ventricle.

The cardiac pumping cycle can be divided into four distinct phases. They are shown schematically in Fig. 8.5 (1–4) for the left atrium and ventricle. The red color indicates oxygenated blood flowing through the left ventricle. The right ventricle operates in precisely the same manner for deoxygenated blood. In Fig. 8.6, the corresponding pressures in the atrium and in the ventricle, the cardiac volume, and the ECG's relation are plotted. Also indicated at the bottom are the opening and closing phases of the valves and the time span for the diastole and systole.

The four phases of the pumping cycle are characterized as follows [1]:
1. *Inflow phase*: The MV is open, and the semilunar aorta valve (AV) is closed. This phase is initiated by the P-wave of the ECG (see Chapter 7), leading to the atrium's contraction. As the atrium contracts, pressure increases and more blood flows through the open MV, leading to a rapid filling of the ventricles.

2. *Isovolumetric contraction phase*: All valves are closed. This phase begins with the QRS complex of the ECG, which represents ventricular depolarization. The depolarization triggers excitation–contraction of the ventricle and a rapid increase in intraventricular pressure. When the atrium valves close, a heart sound can be heard. As the tricuspid valve in the right ventricle closes slightly before the MV in the left ventricle by about 40 ms, a double sound ("lub" and "dub") is actually picked up with the help of a stethoscope.

3. *Outflow phase*: The semilunar valves (aortic and pulmonic) are open; the MV valve remains closed. This phase represents a rapid ejection of blood into the artery (aorta and pulmonary arteries). The ejection begins as soon as the intraventricular pressures exceed the pressures within the artery, causing the aortic and pulmonic valves to open. This is the case immediately after the S-wave of the ECG. The highest outflow velocity is reached early in the ejection phase, while the systolic pressure reaches a maximum.

4. *Isovolumetric relaxation phase*: All valves are closed, and the pressure decreases. Late in the ejection phase, the blood flow drops back to low values until it may reverse. At this point, the semilunar valves close to hinder backflow. The closing defines the onset of the diastole at the end of the T-wave.

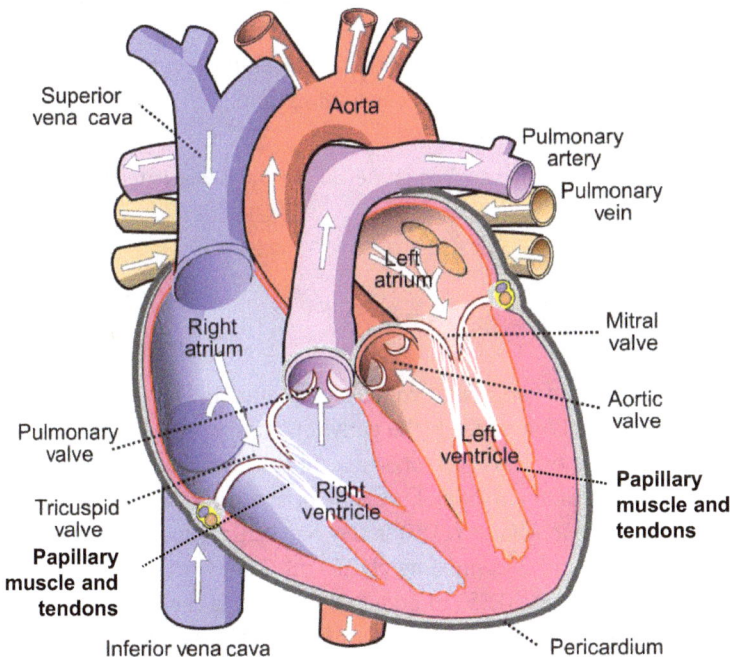

Fig. 8.4: Valves in the heart and blood flow directions (reproduced from Wikimedia, Wapcaplet, CC BY-SA 3.0, https://commons.wikimedia.org/w/index.php?curid=830253).

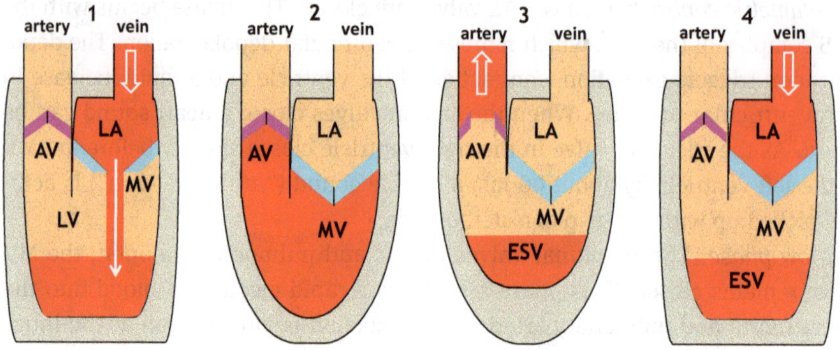

Fig. 8.5: Four phases of the left part of the heart: (1) filling phase; (2) contraction phase; (3) ejection phase; (4) relaxation phase. LA, left atrium; LV, left ventricle; AV, aortic valve; MV, mitral valve; ESV, end systolic volume.

These four phases are commonly divided into two sequences: systole, including phases 2 and 3, and diastole, comprising phases 1 and 4. The four phases can also be identified with the help of ECG, where phase 1 corresponds to the P-wave, phases 2 and 3 to the QRS complex, and phase 4 to the T-wave and to the subsequent resting phase (compare Figs. 7.11 and 8.6). In a normal heart cycle at rest with a rate of 75 cycles/min, corresponding to a cycle duration of 800 ms, the systole occupies about 300 ms and the diastole about 500 ms. With increasing heart rate, the diastole shortens, whereas the systole remains roughly constant in time.

> **!** The systole phase is the compression and ejection phase: the diastole phase is the filling and relaxation phase.

8.3.2 Stroke volume

The pressure dependence and the ventricular volume are plotted in Fig. 8.6 in relation to the electrocardiogram. The filling phase is the early low-pressure phase during the diastole with a blood pressure on the order of 3 hPa. In the succeeding contraction phase, the pressure steeply rises to a maximal pressure of about 140 hPa immediately before ejection. The *end-diastolic volume* (EDV) is the ventricular blood volume after filling and just before ejection and is approximately 120 ml. After ejection, the volume decreases to the *end-systolic volume* (ESV) of about 50 ml. This is the rest volume remaining in the ventricle. The *SV* is defined as the difference between EDV and ESV. Therefore the SV is typically

$$V_{SV} = V_{EDV} - V_{ESV} = 120\,\text{ml} - 50\,\text{ml} = 70\,\text{ml}. \tag{8.1}$$

The *ejection fraction* (EF) is defined as

$$\text{EF} = \frac{V_{\text{SV}}}{V_{\text{EDV}}} = \frac{70\,\text{ml}}{120\,\text{ml}} \cong 0.6. \tag{8.2}$$

An EF of 0.6 is typical for a healthy person. The SV and the ejection fraction are independent of the heart frequency up to about 180 beats per minute. Beyond this frequency, the SV and EF decrease rapidly because there is not sufficient diastolic time for refilling the ventricles.

We can also determine the *cardiac output*, which is defined as the product of heart rate and SV. Assume that the heart makes 75 beats per minute at rest, which is a heart rate of 1.25 Hz. Then we find the cardiac output ($\dot{V}_{\text{CO}}$), defined as volume rate (volume per time unit):

$$\dot{V}_{\text{CO}} = V_{\text{SV}} \times f_{\text{heart}} = 70\,\text{ml} \times 1.25\,\text{Hz} = 88\,\text{ml/s}. \tag{8.3}$$

The $\dot{V}_{\text{CO}}$ amounts to 5.3 l/min and 7600 l/day for the left ventricle and 15 200 l/day for both ventricles at rest.

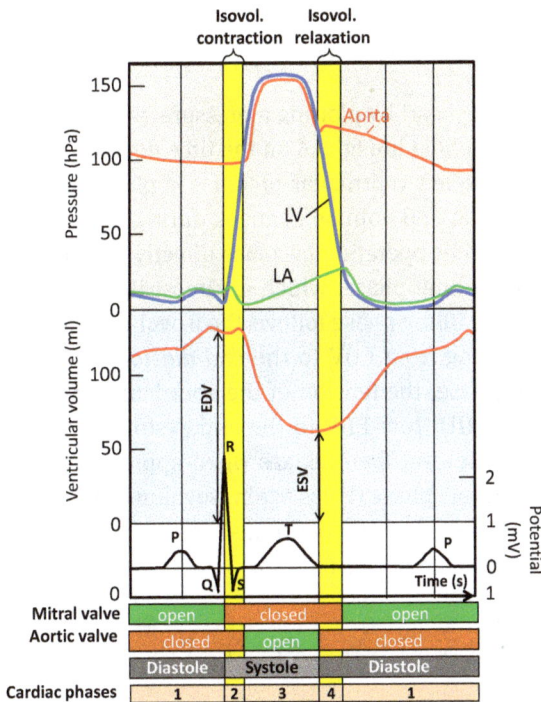

Fig. 8.6: Pressure changes during a cardiac cyle in the left ventricle, aorta, and atrium. Also shown are the ventricular volume and the cardiac action potential. The time axis spans about 1 s. EDV = end diastolic volume, ESV = end systolic volume; pressures: LA = left atrium (green line), LV = left ventricle (blue line), aorta (red line). PQRST have their usual meaning.

> ❗ Systole = high-pressure contraction phase. Diastole = low-pressure filling phase.

The heart has a built-in mechanism that allows adapting to fluctuations of the blood volume arriving at the venous side. Whenever the blood volume increases, for instance during exercise, the cardiac output will increase at the same heart frequency. This feedback system is known as the *Frank[1]–Starling[2] law* of the heart [1, 2]. Increased blood volume stretches the cardiac muscle, and during the systole, the myocardium is required to contract more strongly to eject the extra volume. Important is the fact that the left and right ventricles always pump precisely the same amount at the same frequency and in phase.

8.3.3 Energy, power, and efficiency

We consider now how much energy the heart requires to keep it running. The total energy of the heart is composed of volume or potential energy, ejection or kinetic energy, and tension or thermal energy:

$$E_{tot} = E_{pot} + E_{kin} + E_{tension}. \tag{8.4}$$

We determine the potential energy of the heart by plotting a pressure-volume phase diagram and using the information presented in Fig. 8.6 on the time dependence of the volume and pressure. In Fig. 8.7 the left ventricular pressure is plotted versus volume. The red line shows the pressure and volume changes during the cardiac cycle. In this pV-diagram, the heart's cyclic operation becomes directly visible. The first phase (I) from ESV to EDV is the filling phase, which is almost isobaric with only a slight pressure increase. After the filling phase follows the isovolumetric (isochoric) contraction phase (II), stretching from EDV to the *end-diastolic pressure* (EDP) point. During the contraction phase, the tension of the ventricular muscles increases. During the ejection phase (III) from EDP to the end systolic pressure (ESP), the pressure first slightly increases but then relaxes when approaching the relaxation phase from ESP to ESV. This last phase (IV) is again isovolumetric.

The mechanical volume work (potential energy) performed by the left ventricle equals the enclosed area, which can be estimated as follows:

$$E_{pot} = \int_{ESP}^{EDP} p_{ejection} dV - \int_{ESV}^{EDV} p_{filling} dV \approx \Delta p \int_{ESV}^{EDV} dV = \Delta p \times V_{SV}. \tag{8.5}$$

---

1 Otto Frank (1865–1944), German physiologist.
2 Ernst Henry Starling (1866–1927), British physiologist.

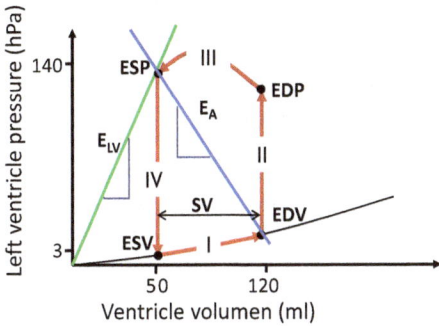

Fig. 8.7: pV-diagram of the cardiac cycle for the left ventricle. I: isobaric filling phase; II: isovolumetric contraction; III: ejection phase; IV: isovolumetric relaxation. ESV, end systolic volume; EDV, end diastolic volume; ESP, end systolic pressure; EDP, end diastolic pressure; SV, stroke volume; E_A, arterial elastance; E_{LV}, end systolic elastance.

According to Figs. 8.6 and 8.7, the ejection pressure is about 140 hPa and the pressure during filling is about 3 hPa. Therefore we obtain the potential energy of the left ventricle:

$$E_{pot} = \Delta p_{LV} \times V_{SV} = 137\,\text{hPa} \times 70\,\text{ml} = 0.96\,\text{J}. \tag{8.6}$$

The kinetic energy is equal to the ejection energy of the blood into the aorta with a flow velocity of about 1 m/s. Assuming for the blood a density of 1 g/cm^3, we obtain

$$E_{kin} = \frac{1}{2}m_{SV}v^2 = 0.5 \times 0.07\,\text{kg} \times 1\,\text{m/s} = 0.035\,\text{J}. \tag{8.7}$$

The total mechanical energy $E_{mech} = E_{pot} + E_{kin}$ is therefore about 1.0 J for one stroke. With a heart rate of 1.25 Hz and for a person at rest, we find a mechanical power consumption of

$$P_{mech} = E_{mech} \times f_{heart} = 1.25\,\text{W}. \tag{8.8}$$

The right ventricle operates at a much lower pressure difference Δp_{RV} of about 20 hPa. Accordingly, the potential energy and the mechanical power are only one-seventh of the left ventricle. The power of both chambers together is roughly $P_{mech} = 1.4$ W. This difference is directly expressed in the wall thicknesses of the left and right ventricles as shown in a cross section in Fig. 8.8. During exercise, f_{heart} increases while V_{SV} remains constant. Therefore, the power consumption will also go up, but never much beyond 2.5 W.

It is slightly more challenging to estimate the tension energy. During isometric tension, the muscle length stays constant. But to keep the tension constant, energy is required; otherwise, the muscle would stretch and relax (see Fig. 2.19). The aerobic energy required is supplied by ATP and oxygen and converted into heat. Therefore

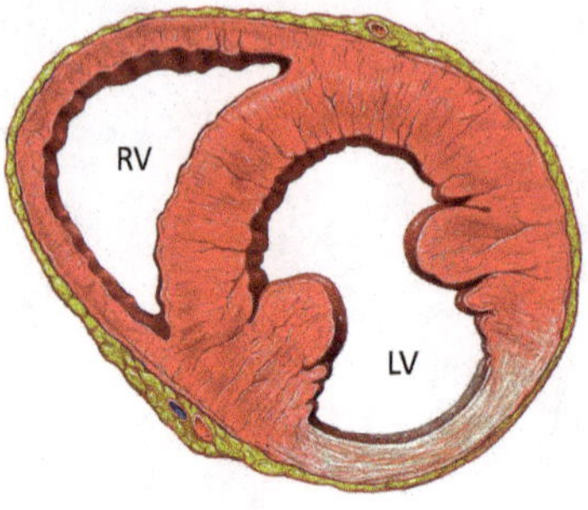

Fig. 8.8: Cross section through the heart showing the myocardial wall thicknesses of the eft and right ventricle. The left ventricle (LV) works at a much higher pressure level, and therefore, the tension of the walls needs to be much larger than for the lower pressure right ventricle (RV) https://commons.wikimedia.org/wiki/File:Heart_in ferior_wall_scar.jpg#/media/File:Heart_inferior_wall_ scar.jpg.

this thermal heat is also referred to as tension heat. It is proportional to the tension of the myocardial muscles T_{muscle} and the time span Δt of the stress:

$$E_{\text{tension}} = \alpha T_{\text{muscle}} \Delta t, \tag{8.9}$$

where α is a conversion factor to SI units. It is difficult to give numbers for E_{tension}. Therefore, we refer to the power consumption and recall from Chapter 4 that a person at rest has a basal metabolic rate of 80 W. Seven percent of the metabolic rate goes to the heart, which makes $P_{\text{heart, tot}}$= 5.6 W. As $P_{\text{heart, tot}} = P_{\text{mech}} + P_{\text{heat}}$, we find for $P_{\text{heat}} = 4.2$ W. This is 75% of the total energy requirement, and only 25% are needed for the mechanical work. The *efficiency* of the heart is defined as

$$\epsilon = \frac{P_{\text{mech}}}{P_{\text{heart, tot}}}. \tag{8.10}$$

Substituting numbers yields an efficiency of 25%. In comparison, the heart is still reasonably efficient, and as we have already seen, the heart's cardiac output is amazingly high, albeit exactly the amount required for.

> **!** The power consumption of the heart is approximately 6 W. The mechanical efficiency is about 25%.

The slope of the blue line in Fig. 8.7 is $E_A = \text{ESP/SV}$. This ratio characterizes the *arterial elastance*. The arterial elastance is a measure of the heart's total compliance and the vascular system's impedance. The slope of the green line is $E_{\text{LV}} = \text{ESP/ESV}$ referring to the *end-systolic elastance*. The ratio E_{LV}/E_A is proportional to the ejection fraction: $\text{EF} = E_{\text{LV}}/E_A$ (compare with eq. (8.2)). Both slopes and their ratio are indicators of the heart and arterial functionality. Deviations from normal behavior are early signs of potential heart failure.

All numbers derived so far are average numbers, not considering the gender or the size of a person. However, personal conditions can be adapted by normalizing the SV with the *body surface area*, which is often done in cardiovascular physiology.

For estimating the energetics and efficiency of the myocardium, we have mainly used information on the SV and left ventricular pressure difference. In the past, this information could only be gained invasively using catheters inserted into the heart. However, with the availability of modern imaging methods, in particular, Doppler sonography (Chapter 1, Vol. 2) and MRI (Chapter 3, Vol. 2), the myocardial efficiency can be evaluated by noninvasive methods. Another important myocardial stress test is performed with spingle photon emission computer tomography, using the radioisotope [99m]Tc. The method is described in detail in Chapter 9/Vol. 2. An overview on present-day methods for determining cardiac efficiencies is given in [3].

8.4 Hemostatics of the systemic circuit

When describing the physical properties of liquids, a distinction is made between hydrostatics and hydrodynamics, whereby the general aspects and associated equations do not only apply to water. Blood is a special liquid, as we shall see. Therefore, it is justified to coin the terms hemostatics and hemodynamics to describe their physical properties. We start here with hemostatics and cover hemodynamics in the next section.

8.4.1 Hemostatic pressures

When we consider the static properties of liquids, the focus is on the liquid itself rather than on tubes and vessels that hold the liquid. It does not matter whether the liquid is contained in glass, ceramic, or metal vessels. In contrast, the vessels play an important role in hemostatics, and so we need to look at both the blood as a liquid and the flexible vessels as a containment. The combination is called a *systemic circuit*.

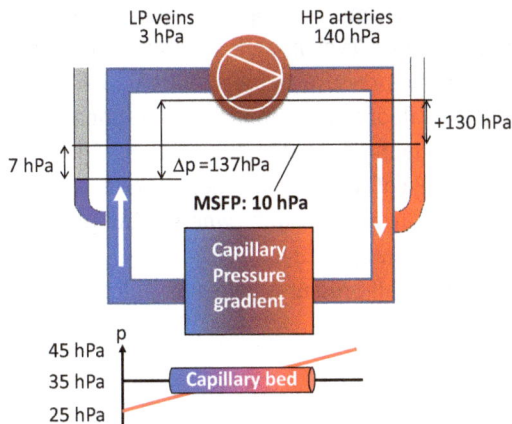

Fig. 8.9: The heart as a mechanical pump, maintaining a pressure difference. MSFP = mean systemic filling pressure is the remaining pressure when the heart stands still. LP, low pressure; HP, high pressure.

Without heart activity, the pressure in the blood vessels is the same everywhere. This pressure is known as the mean systemic filling pressure (MSFP). The MSFP is approximately 10 hPa for a subject lying flat. The MSFP pressure depends on the filling capacity of all blood vessels, including the heart, and on the compliance of the blood vessels. As soon as the heart is turned on, the arterial side's pressure rises from 10 to 140 hPa and lowers the venous side's pressure to around 3 hPa. Thus, the pressure difference between arterial and venous side is about 137 hPa, as already stated in the previous section. Figure 8.9 schematically shows the circulatory system with the relevant pressures. High and low pressures are symbolized by Pitot[3] tubes. The high pressure drops continuously with increasing distance from the aorta, but mainly in the capillary bed to a low level on the venous side.

Furthermore, we have to distinguish between a person's pressure distribution in a horizontal and a vertical posture, as demonstrated in Fig. 8.10. In the horizontal posture, the aortic pressure is high and constant across the body. The blood pressure drops off across the capillary bed, which represents the main flow resistance. On the venous side, the pressure is also constant, albeit on a much lower level. If Pitot tubes were inserted into the vessels, the blood would rise in the tubes to the indicated levels. The open Pitot tubes measure the fluid's static pressure with respect to the standard atmospheric pressure of 1000 hPa. A pressure of 130 hPa then implies that the pressure is 130 hPa higher than the atmospheric pressure.

When standing up, the circulatory system has to adapt to the superimposed pressure gradient due to gravitation:

$$p(h) = \rho g h. \tag{8.11}$$

Here ρ is the blood density, which is essentially the same as water, g is the gravitational acceleration, and h is the height of blood vessels above the ground. $p(h)$ is called the gravitational pressure or hydrostatic pressure. At a certain height, the blood system's pressure is independent of the person's position, horizontal or vertical. This height is known as the *hydrostatic indifference level* h_{ind}. Its position is below the heart at about 120 cm above the ground and the pressure p_{ind} is about 130 hPa. When standing up, the 120 cm high blood tube from floor to h_{ind} causes an additional pressure of 120 hPa at the feet. Therefore the arterial pressure of 130 hPa at h_{ind} goes up to 250 hPa at the toe and drops down to 70 hPa at the pinna. Hence the arterial pressure in the head drops by roughly half in an upright posture, whereas the pressure in the feet doubles in comparison to the horizontal posture.

In general, for an upright standing posture, all pressures below h_{ind} follow from

$$p(h \leq h_{ind}) = p_{ind} + \rho g h, \tag{8.12}$$

---

3 Henri de Pitot (1695–1771), French hydraulic engineer.

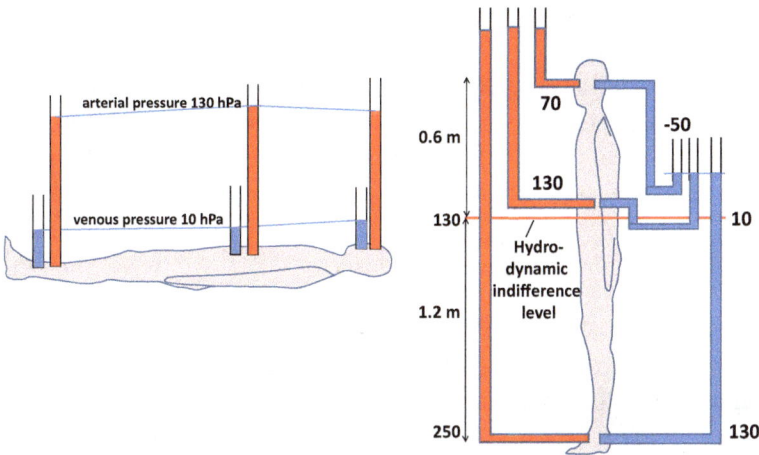

Fig. 8.10: Hydrostatic pressure of the blood system when lying horizontally and for standing upright. The vertical length of the Pitot tubes schematically indicate the various pressures. All numbers in the right panel are in units of hectopascal, but the heights are in meters.

and all pressures above h_{ind} are accordingly

$$p(h \geq h_{ind}) = p_{ind} - \rho g h. \tag{8.13}$$

On the venous side, we find similar pressure relationships: the pressure of 10 hPa for the horizontal position increases to 130 hPa at the feet and drops to −50 hPa in the head by standing upright. What does it mean to have a negative pressure of −50 hPa in the head? It means that the venous pressure in the head is about 50 hPa below the atmospheric pressure. This lower pressure is very small and nothing to worry about getting a brain drain. Overall the pressure difference Δp between arterial and venous blood vessels remains constant at a level of 120 hPa.

8.4.2 Elastic properties of blood vessels

The structure of blood vessels reflects the pressure difference between arteries and veins. The veins have large inner diameters and thin walls, whereas the arteries have small inner diameters and thick walls. Cross sections of both vessel types are shown in Fig. 8.11 at the level just above the capillary bed. Because of the larger

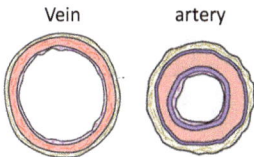

Fig. 8.11: Cross sections of blood vessels. Note the different inner diameters and wall thicknesses for veins and arteries (reproduced from Ref. [22] with permission of Thieme Verlag).

diameter, most of the blood volume resides in the venous system (75%), which is considered a blood reservoir.

Blood vessels have to be regarded as elastic tubes responding to pressure changes. The diameter depends on the pressure difference between the internal pressure p_i and the external pressure p_e. The pressure difference across the vessel wall is called *transmural pressure* difference (Fig. 8.12):

$$\Delta p_{tm} = p_i - p_e. \tag{8.14}$$

As we increase the transmural pressure difference, the radius and the tension T_{wall} in the walls increase, similar to inflating a balloon. The wall tension is equivalent to the surface tension in a fluid. The transmural pressure difference exerts an expanding force on the walls:

$$F_1 = \Delta p_{tm}\, 2\pi r \Delta r, \tag{8.15}$$

while the wall tension has a constricting effect, expressed by the constricting force:

$$F_2 = T_{wall} 2\pi \Delta r. \tag{8.16}$$

In the last equation, T_{wall} has the role of a force constant and Δr is the expansion of the inside radius r. The linear dependence in eq. (8.16) corresponds to the well-known equation for the force when pulling on a spring: $F = k\Delta x$, where k is the spring's force constant and Δx is the extension. The unit of T is $[T] = \mathrm{N/m}$.[4]

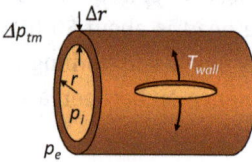

Fig. 8.12: Blood vessel as an elastic tube with radius r, wall thickness Δr, and wall tension T_{wall}. Δp_{tm} is the transmural pressure difference between inside and outside pressures p_i and p_e, respectively.

In equilibrium, both forces F_1 and F_2 must balance, and therefore we have

$$\Delta p_{tm}\, 2\pi r \Delta r = T_{wall} 2\pi \Delta r. \tag{8.17a}$$

For the wall tension we obtain

$$T_{wall} = r\Delta p_{tm}. \tag{8.17b}$$

For low-pressure differences, the relationship between T_{wall} and Δp_{tm} is linear. This is the elastic reversible region. The vessel wall can rupture if the tension increases

---

4 In older textbooks one can find the unit dyn/cm. The conversion is 1000 dyn/cm = 1 N/m.

beyond a critical value, as illustrated in Fig. 8.12. Equation (8.17b) is known as the Laplace[5] equation (see also Chapter 9, eq. (9.32)).

Veins and arteries both react elastically to pressure changes. However, their elastic modulus is very different. This can be best characterized by defining the elastic compliance. According to Chapter 3, eq. (3.5), the volume change and the pressure difference are related as

$$\frac{\Delta V}{V} = \frac{\Delta p_{tm}}{Y},$$
(8.18)

where Y is the elastic modulus or Young's modulus. Rearranging, we have

$$C = \frac{\Delta V}{\Delta p_{tm}} = \frac{V}{Y}.$$
(8.19)

where C is called the *elastic compliance*, here specifically of the blood vessels. The unit is $[C] = m^3/Pa$. C is basically the inverse of the elastic modulus. The higher the compliance, the more responsive the elastic medium is. A small pressure increase already results in a large volume expansion. Contrary, an infinitely stiff material does not deform under load and therefore shows no compliance. Copper tubings have a compliance $C=0$. When applied to blood vessels, we find that veins and arteries have very different compliances. In Fig. 8.13 the volume change as a function of transmural pressure difference is plotted. A steep slope indicates high compliance. Veins have very high compliance, but starting from 15 hPa, further volume expansion stops or rupture would follow. This implies that veins have a low yield point. In contrast, the compliance of the arteries is much lower, but the pressure can rise up to 250 hPa without rupture, i.e., the yield point of arteries is much higher than that of veins.

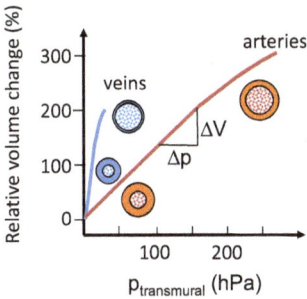

Fig. 8.13: Volume–pressure relationship for blood vessels. Blue: veins, red: arteries. The slope $\Delta V/\Delta p$ marks the compliance of the vessels. Note that the plot is rotated by 90° compared to standard plots of stress–strain relationships, for instance, in Figs. 3.1 and 3.3.

---

5 Pierre-Simon Laplace (1749–1827), French mathematician and physicist.

> **!** Veins have a high compliance to pressure changes and a low yield point. Arteries have lower compliance, but a much higher yield point.

8.5 Hemodynamics of the systemic circuit

8.5.1 Basic equations and assumptions

Fluid dynamics of the circulatory systems is called *hemodynamics* [4]. This is because blood circulation has special characteristics as a non-Newton-type fluid in a circuit of elastically compliant vessels. Nevertheless, we start with classical fluid dynamics, which is governed by four fundamental laws:

1. *Continuity equation*, expressing the conservation of mass of a fluid in a closed system:

$$\rho \vec{\nabla} \cdot \vec{v} + \frac{\partial \rho}{\partial t} = \vec{\nabla} \cdot \vec{j} + \frac{\partial \rho}{\partial t} = 0. \tag{8.20}$$

Here ρ is the density of the fluid, $\vec{v}$ is the velocity, and the product $\vec{j} = \rho \vec{v}$ is the flux density. The continuity equation expresses that any spatial change in the flux density results in a temporal change in the fluid density, or vice versa. Any temporal change in the fluid density results in a spatial change in the flux density. If there is no temporal change of the density, i.e., no sources or sinks, then $\partial \rho / \partial t = 0$. Consequentially, the divergence of the flux density must also be zero: $\vec{\nabla} \cdot \vec{j} = 0$. If we now take the integral over different cross-sectional areas A_i that the fluid passes through, then the product $A_i j_i$ stays constant. This shows that the volume flow rate is constant, i.e., the product of cross section A and velocity v is constant in time and space:

$$I_V = \frac{dV}{dt} = Av = \text{const.} \tag{8.21}$$

2. *Bernoulli*[6] *equation*, stating the conservation of energy for ideal incompressible fluids:

$$\Delta p = \frac{1}{2}\rho v^2 + \rho g h, \tag{8.22}$$

where Δp is the pressure difference in the fluid maintaining flow. The first part on the right side is the kinetic energy density and second is the potential energy density.

---

6 Daniel Bernoulli (1700–1782), Swiss mathematician and physicist.

3. *Ohm's law,*[7] recognizing the dissipation of energy by viscous flow, expressed by

$$I_V = \frac{\Delta p}{R_{\text{flow}}},$$

(8.23)

where R_{flow} is the flow resistance. The flow resistance is discussed later in Section 8.5.3.

4. *Kirchhoff's laws*[8] describe the total flow and branching in circuits with parallel and serial resistances. For serial resistances R_i, the total resistance is

$$R_{\text{tot}} = \sum_i R_i$$

(8.24)

and for parallel resistances R_i the conductances $G_i = 1/R_i$ add up to the total conductance:

$$G_{\text{tot}} = \sum_i G_i.$$

(8.25)

These equations[9] are based on a number of inherent assumptions:
(a) incompressibility of fluids;
(b) independence of viscosity on shear rate;
(c) rigid tubes with zero compliance;
(d) continuous laminar flow;
(e) closed circuit without fluid volume gain or drain;
(f) one component, simple fluids.

All these assumptions are more or less violated for the blood flow in the body's circulatory system. The compressibility of blood is slightly higher than that of water due to the suspension of squeezable red blood cells (RBCs). The viscosity of the blood depends on the shear rate. The flow is pulsatile rather than steady. Blood vessels respond to changes in transmural pressure; blood volume varies; and blood is a rather complex, multicomponent fluid. Even so, we will apply these laws of fluid-dynamics to get some first useful insights into the blood flow. Later we will refine our conclusions by taking into account nonlinear effects.

---

7 Georg Simon Ohm (1789–1854), German physicist.
8 Gustav Robert Kirchhoff (1824–1887), German physicist.
9 Ohm's law and Kirchhoff's law were originally applied to electric currents, i.e., the flow of electric charges. These laws apply in an analogous way also to viscous fluids, as long as the flow is laminar.

Infobox I: Laminar flow and turbulent flow

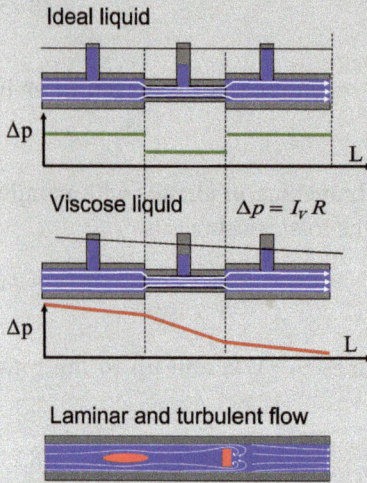

Ideal liquid

Δp

L

Viscose liquid $\Delta p = I_V R$

Δp

L

Laminar and turbulent flow

In the sketches on the left, the flow of liquids through a pipe is symbolized by flux lines. The density of the flux lines is proportional to the flow velocity, and the continuity/discontinuity of the lines indicates a laminar flow versus a turbulent flow. Pitot tubes indicate the static internal pressure, which is also shown in the respective diagrams. In ideal liquids, the velocity increases in the area of a constriction (continuity equation), the static pressure drops (Bernoulli equation), and behind the constriction the pressure is the same as before. In viscous liquids the flow velocity also increases in the region of the constriction, but the pressure drops more drastically and does not recover after the constriction (Hagen–Poisseuille equation). Streamline obstructions in the tube do not interfere with laminar flow, but severe clogging can cause turbulence, which is characterized by disruption of the flux lines and the formation of swirls.

8.5.2 Pulsatile flow

The pressure of the left ventricle sends a periodic pulse wave down the bloodstream through the artery. Due to this, the blood flow is not continuous but pulsatile, as indicated in Fig. 8.14. If the blood flow were steady, there would not be much of a difference between a rigid tube and an elastic rubber-like vessel. However, because of the pulsing flow combined with compliant blood vessels, the flow resistance is no longer constant but changes with the pulse amplitude. Altogether, the flow resistance has to be replaced by a complex fluid impedance that properly describes dissipation and phase retardation between source and resistor response [5]. The further away from the aorta, the more the pulsing blood flow is damped out and replaced by more continuous flow. Ohm's law is only applicable to the time average mean pressure, indicated by the red line in Fig. 8.14. At distances beyond the arterioles, the pulsatile pressure oscillations are diminished, and hence the local pressure is also the time average pressure.

The mean arterial pressure $p_{m,a}$ (red line in Fig. 8.14) can be estimated by determining the systolic pressure p_{sys} and diastolic pressure p_{dia} at the brachial artery (main arterial supply to the arm) and using the expression:

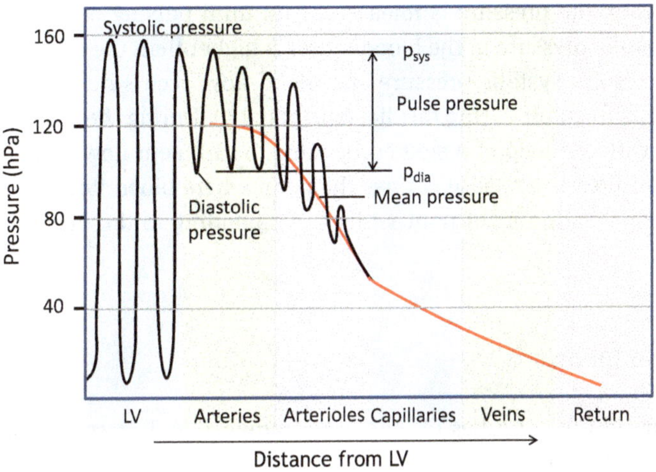

Fig. 8.14: Arterial pressure: time and distance relation at different lengths from the left ventricle (LV).

$$p_{m,a} = p_{dia} + \frac{p_{sys} - p_{dia}}{3}. \tag{8.26}$$

On the other hand, at the brachial artery, the systolic versus diastolic pressure can be measured directly using the method of Riva-Rocci[10] (Fig. 8.15). This method employs an inflatable cuff wrapped around the upper arm or the wrist and a tube connecting the cuff with a pressure manometer. A handheld pump first increases the pressure in the cuff to a level that blood flow is stopped and pulsing is no more

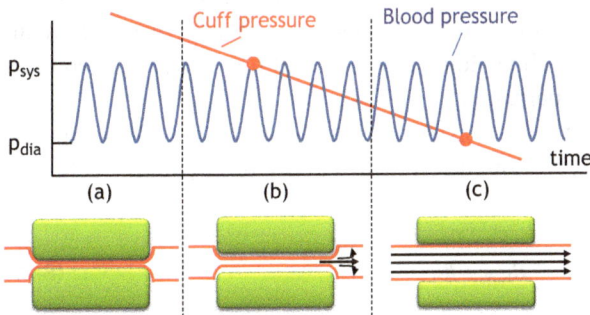

Fig. 8.15: Blood pressure measurement according to the method of Riva-Rocci. The blue line indicates the pressure amplitude oscillating between systolic and diastolic pressure. The red line is the pressure applied by a cuff around the upper arm. When releasing pressure in the cuff, first the systolic pressure level and later the diastolic pressure level is passed, indicated by red dots. The corresponding squeezed and expanded blood vessel by the cuff (in green) is shown in the lower panel.

---

10 Scipione Riva-Rocci (1863–1937), Italian internist, pathologist, and pediatrician.

noticeable (panel a). Then the pressure is released again until pulsing occurs as soon as the internal systolic pressure in the blood vessel is higher than the external pressure. This point fixes the systolic pressure and blood flow occurs during the systole while the flow is still cutoff during the diastole (panel b). During the systole, a sound can be heard with the help of a stethoscope due to turbulent flow behind the constriction. Further pressure release defines the point where blood flows in a laminar fashion and noise is no longer noticeable, corresponding to the diastolic pressure (panel c).

8.5.3 Perfusion and flow resistance

The flow resistance for cylindrical tubes of length ΔL, radius r, and cross section $A = \pi r^2$ is according to Hagen[11]–Poiseuille:[12]

$$R_{\text{flow}} = \frac{8}{\pi} \eta \frac{\Delta L}{r^4} = 8\pi\eta \frac{\Delta L}{A^2}, \tag{8.27}$$

where η is the viscosity of the fluid (units: $[R_{\text{flow}}] = \text{Pa} \cdot \text{s/m}^3$; $[\eta] = \text{Pa} \cdot \text{s}$). In analogy to electric resistors, the flow resistance has a geometric dependence (length and radius of the tube) and a material-specific dependence (viscosity). Since

$$I_V = \frac{\Delta p}{R_{\text{flow}}} \propto r^4 \propto A^2, \tag{8.28}$$

the volume flow increases with the radius's fourth power or the cross section's square power. This has severe consequences if constrictions of blood vessels occur (stenosis), which may lead to a stroke unless removed in time, for instance, by inserting a stent (see Chapter 13). The volume flow of blood I_V is often referred to as perfusion, more precisely hemoperfusion.

Although the Hagen–Poiseuille resistance depends on the cylindrical geometry of tubes, the essential proportionalities always apply to any cross section of tubes:

$$R_{\text{flow}} = k\eta \frac{\Delta L}{r^\alpha}, \tag{8.29}$$

where k is a constant and α is an exponent that depends on the geometry.

---

11 Gotthilf H. L. Hagen (1797–1884), German engineer.
12 Jean Léonard Marie Poiseuille (1797–1869), French physicist and physiologist.

8.5.4 Total peripheral flow resistance

The capillary bed connects the artery with the veins and the porosity of the thin-walled capillaries allows exchange of oxygen and nutrients (Fig. 8.16). These blood vessels have the smallest radius and therefore the highest flow resistance. Using the Hagen–Poisseuille equation, we can estimate the flow resistance for a short capillary tube. As an example we consider a tube of 10 mm length, a radius of 5 μm, and a viscosity of 4 mPa s. For such a capillary, the flow resistance is about 1.6×10^{12} Pa · s/l. The blood flow in the capillaries runs in parallel circuits, which reduces the overall resistance according to Kirchhoff's law (eq. 8.23):

$$\frac{1}{R_{\text{tot}}} = \frac{1}{R_1} + \frac{1}{R_2} + \frac{1}{R_3} + \cdots = \frac{N}{\langle R_i \rangle}. \tag{8.30}$$

Thus the total resistance R_{tot} decreases with the number of parallel branches N, where $\langle R_i \rangle$ is the average resistance of a single branch. On the other hand, the *total peripheral flow resistance* (R_{TPFR}), which is the sum of the resistances of all peripheral vasculature in the systemic circulation, can be estimated from the ratio of the known mean arterial pressure $p_{\text{m,a}}$ and the cardiac output I_{CO}:

$$R_{\text{TPFR}} = \frac{p_{\text{m,a}}}{I_{\text{CO}}}, \tag{8.31}$$

where $p_{\text{m,a}}$ is about 120 hPa, and the cardiac output I_{CO} is about 5 l/min. Then we have for R_{TPFR}

$$R_{\text{TPFR}} = \frac{120 \text{ hPa}}{5 \text{ l/min}} = 24 \frac{\text{hPa min}}{1} = 1.4 \times 10^5 \frac{\text{Pa} \cdot \text{s}}{1}.$$

This is a much lower value than the resistance of a short capillary tube. Therefore we conclude that about $N \approx 10^7$ capillaries run in parallel in order to lower the overall flow resistance to the TPFR value determined. Cross section, local velocity, and pressure drop across the capillary bed are plotted in Fig. 8.16. We notice that the blood flow velocity decreases as the total cross section in the capillary bed increases according to the continuity equation $A \times v = \text{const}$. The low velocity of about $v_{\text{capillary}} = 0.005$ m/s in the capillary bed is physiologically essential since it provides sufficient time for exchange of oxygen and nutrients.

The total peripheral flow resistance is lower than expected because of high branching of blood vessels in the capillary bed. !

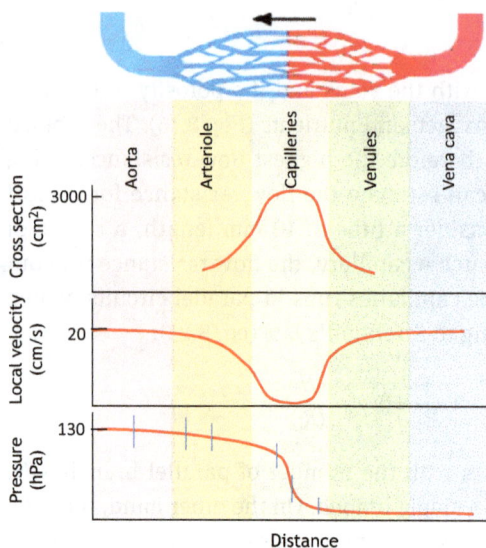

Fig. 8.16: Top panel: Branching of capillaries in the capillary bed connecting arteries and arterioles with veins and venules. Lower panels: Dependence of cross section, local velocity, and blood pressure as a function of position across the capillary bed. The blue vertical bars in the lowest panel indicate the pressure pulse between diastole and systole.

8.5.5 Turbulent flow and Windkessel function

Low velocity and narrow cross section are beneficial for assuring *laminar flow*. However, in the aorta, the conditions are just the opposite: high ejection velocity combined with a large cross section. With the dimensionless *Reynolds*[13] *number* (Re), defined as

$$\text{Re} = \frac{\rho v d}{\eta},\tag{8.32}$$

where $d = 2r$ is the diameter of a tube, one can estimate whether a *turbulent flow* is to be expected. The flow is laminar for Re numbers below 1000, but for Re numbers beyond 2000, turbulent flow is very likely. Turbulent flow is characterized by a disruption of flow lines and the formation of curls. Turbulent flow is not only chaotic but also destructive and must be avoided. See also the Infobox I on laminar and turbulent flow.

In the case of the aorta, the ejection velocity is about 1 m/s, the diameter is 20 mm, and the viscosity of blood on the average is $\eta = 3$–4 mPa s. Therefore we obtain a Reynolds number for blood flow after ejection into the aorta:

---

13 Osborne Reynolds (1842–1912), Irish-British phyiscist and engineer.

Re(aorta)

$$= \frac{\left(10^3 \, \text{kg/m}^3\right) \times (1 \, \text{m/s}) \times \left(20 \times 10^{-3} \text{m}\right)}{4 \times 10^{-3} \, \text{Ns/m}^2}$$

$$= 5000.$$

This number is critically high. Nature has invented a scheme known in engineering as *Windkessel* to reduce the danger of turbulent blood flow. The Windkessel is an air chamber or an elastic reservoir that temporarily stores kinetic energy. The aorta is an elastic piece of the artery with particularly high compliance that absorbs part of the kinetic energy and temporarily converts it into potential energy. The potential energy is converted back into kinetic energy to maintain organ perfusion during the diastole when the cardiac output has ceased. Figure 8.17(a) is a sketch of the left ventricle together with the aorta including the Windkessel. The Windkessel is therefore an energy storage system that suppresses turbulence and regulates blood pressure fluctuations. With age, the compliance of the arteries decreases, resulting in a less effective Windkessel conversion and increased systolic pulse pressure.

8.5.6 Mathematically modeling

The first mathematically modeling of the Windkessel is attributed to Otto Frank, the basics we want briefly discuss here with reference to Fig. 8.17(b). The blood flow starts from the aortic valve (AV) and delivers a blood volume V_{in} at an end-systolic pulsed pressure p_{ESP}. The blood volume is then partly stored in the Windkessel with the mechanoelastic volume capacity V_C and compliance C and finally merges into the vascular circulation with a peripheral resistance R [5].

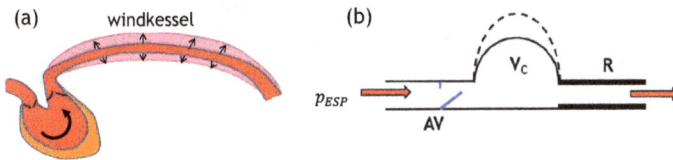

Fig. 8.17: (a) Left ventricle and Windkessel function of the aorta; (b) equivalent circuit diagram for modeling the pulsed flow from the aortic valve (AV) with ESP p_{ESP}, passing by the Windkessel with volume capacity V_C and compliance C to the periphery with resistance R.

Mass conservation requires that the mass flow into the Windkessel $\dot{V}_{in}$ equals the mass flow stored temporarily in the Windkessel $\dot{V}_C$ plus the amount flowing out into the peripheral circuit $\dot{V}_{out}$:

$$\dot{V}_{in} = \dot{V}_C + \dot{V}_{out}.$$ (8.33)

Rephrasing this equation:

$$\frac{\partial V_{in}}{\partial t} = \frac{\partial V_C}{\partial p_{ESP}}\frac{\partial p_{ESP}}{\partial t} + \frac{p_{ESP}}{R} = C\frac{\partial p_{ESP}}{\partial t} + \frac{p_{ESP}}{R},$$ (8.34)

and rearranging yields the differential equation:

$$\frac{\partial p_{ESP}}{\partial t} + \frac{p_{ESP}}{CR} = \frac{1}{C}\frac{\partial V_{in}}{\partial t}.$$ (8.35)

Here $C = \partial V_C/\partial p_{ESP}$ is the compliance of the Windkessel and R is the flow resistance. This inhomogeneous differential equation of first order can be solved with particular model assumptions. We make the most simplifying assumption and set the right side to zero:

$$C\frac{\partial p_{ESP}}{\partial t} + \frac{p_{ESP}}{R} = 0.$$ (8.36)

The consequences of this assumption are the subject of Exercise E8.7. The differential equation is then solvable by integration, yielding the solution:

$$p_{ESP} = p_0 \exp\left(\frac{t - t_0}{\tau}\right),$$ (8.37)

where $\tau = RC$ is the time constant for the pulse pressure decay after maximum expansion of the Windkessel. Substituting numbers for the flow resistance and the compliance is not so easy. The flow resistance of the aorta can be estimated according to eq. (8.23) to be about $R_{aorta} = 100$ Pa·s/l.

The compliance is even more difficult to estimate. It should be higher than general arteries but lower than veins. A value of $C_{aorta} = 8$ ml/Pa appears reasonable. Then the time constant is estimated to be about 0.8 s. This simple two-component RC model cannot describe the initial pressure rise in the Windkessel, but reproduces well the subsequential exponential pressure drop with a time constant that is very reasonable, as seen in Fig. 8.18.

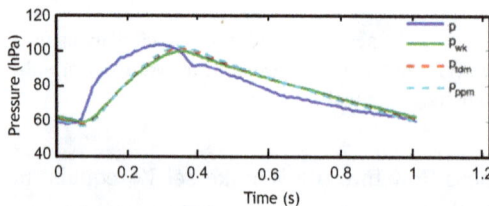

Fig. 8.18: Measured aortic pressure (solid dark blue line) and time dependence of pressure according to different model assumptions (dashed lines) (adapted from [5]).

In the Mathbox below alternative models are discussed and the generic character of the models is emphasized that allows to describe many other but similar time dependent events.

Mathbox: Windkessel: alternative models

The model which we used for describing the Windkessel function can also be described by a circuit indicated in panel (a) below. The pump (heart) provides a pulsatile pressure $p(t)$, R is the flow resistance, and the capacitance C corresponds to the mass storage capacity of the aorta. We can solve the corresponding differential equation with respect to the time dependence of the pressure or the mass flow. Depending on circumstances, the circuit shown in panel (a) can also be slightly modified to the circuits shown in panel (b) and (c). Circuit (a) describes situations where the flow ceases after charging (discharging) the capacitor. Circuit (b) yields an immediate response to the source due to the resistance R and a time-delayed response via the capacitance C. Model (c) is a combination of (a) and (b).

These circuits are generic models for the descriptions of reoccurring events such as the pumping cycle of the heart, the breathing cycle, and the depolarization of an action potential. In all cases, the analysis yields a generic first-order differential equation of the type:

$$\frac{dO(t)}{O(t)} = -\frac{dt}{RC}, \tag{8.38}$$

where $O(t)$ is a time-dependent observable, and $\tau = RC$ is the time constant. C may be anything from membrane capacitance to lung compliance. Similarly, R may be the ion channel resistance or the airway resistance, depending on the circumstances. If τ is known, R or C can be determined.

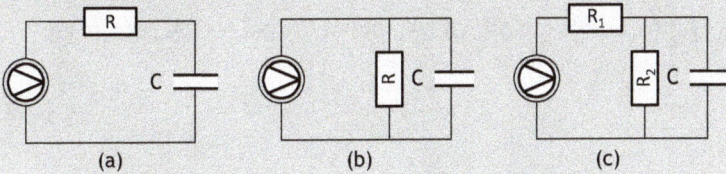

(a) (b) (c)

8.5.7 Flow velocity and pulse wave velocity

We return to the pulsatile nature of the blood flow. The left ventricle ejects a SV at high pressures into the artery. The systole–diastole pressure wave propagates down the aorta and into the carotid artery. The pressure wave results in two effects: at any distance Δx from the aortic valve and after a propagation time Δt we measure a local flow velocity $v(\Delta x, \Delta t)$ in the fluid and a local pressure amplitude $\Delta p(\Delta x, \Delta t)$ of the vessels, both having pulsing and propagating character, as plotted in Fig. 8.19.

Hence we distinguish between a time-independent *mean flow velocity*:

$$\langle v_{\text{mean}}(\Delta x) \rangle = \frac{v_{\text{sys}} t_{\text{sys}} + v_{\text{dia}} t_{\text{dia}}}{T_{\text{cardiac}}},$$ (8.39)

where $T_{\text{cardiac}} = t_{\text{sys}} + t_{\text{dia}}$, and a time-dependent *pulse wave velocity (PWV)* v_{PWV}. The mean velocity is measured at a fixed distance Δx from the aorta and averaged over time, which can be done, for instance, via Doppler shift sonography (see Section 1.8/ Vol. 2). For this purpose, laminar flow is assumed. The velocity is measured in the middle of the velocity profile at the center of a cylindrical tube, as indicated in Fig. 8.19. In Tab. 8.1 the typical values for the mean flow velocity at different locations are listed. As the distance from the aorta increases, the pulsatile nature of the blood flow ceases, and the flow takes on a more continuous character.

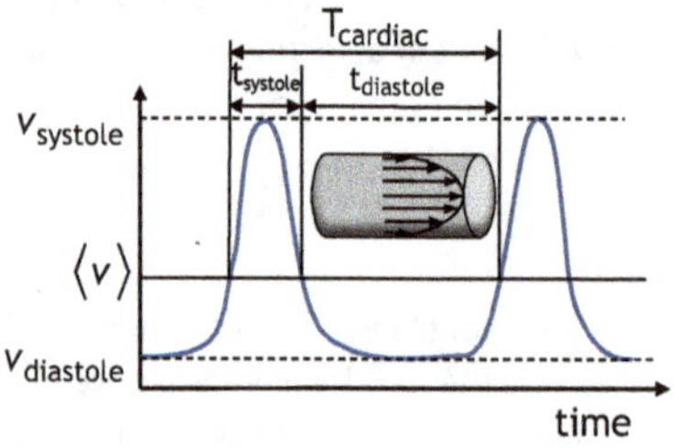

Fig. 8.19: Velocity versus time for the pulsatile blood flow determined at a fixed distance Δx from the aorta. The same plot can be made for the time structure of the pressure amplitude.

The pulsatility of the flow is quantified by a *pulsatility index* (PI), defined as

$$\text{PI} = \frac{v_{\text{sys}} - v_{\text{dia}}}{v_{\text{mean}}}.$$ (8.40)

For $v_{\text{sys}} \rightarrow v_{\text{dia}}$, PI $\rightarrow 0$, and $v_{\text{PWV}} = \langle v_{\text{mean}} \rangle$. Note that in general, v_{PWV} and $\langle v_{\text{mean}} \rangle$ are very different. $\langle v_{\text{mean}} \rangle$ is the mean flow velocity of the blood, whereas v_{PWV} is the pulse propagation speed along the blood vessels. As detailed below, the latter one depends on the elastic properties of the blood vessels and is usually much faster than $\langle v_{\text{mean}} \rangle$. Only in the limiting case PI $\rightarrow 0$, the difference between v_{PWV} and $\langle v_{\text{mean}} \rangle$ is negligible.

In contrast to the mean flow velocity, the PWV is the propagating velocity of the pressure wave as it travels along the artery. The puls propagation resembles a soliton-like pulse propagation along a stretched rope illustrated in Fig. 8.20.

In general terms, the pulse propagation can be expressed by the equations:

$$t = 0: \quad y(x, 0) = f(x),$$ (8.41)

$$t > 0: \quad y(x, t) = f(x - v_{\text{PWV}} t),$$ (8.42)

where $f(x)$ is a function describing the shape of the pulse, and v_{PWV} is the PWV.

Tab. 8.1: Mean flow velocities and cross sections of blood vessels (compiled from [5, 6]).

Blood vessel	Mean flow velocity (cm/s)	Total cross section (cm²)
Aorta	92	3–5
Main diastolic pulmonary artery	65	3–4
Vena cava	15	14
Capillaries	0.05	4500–6000

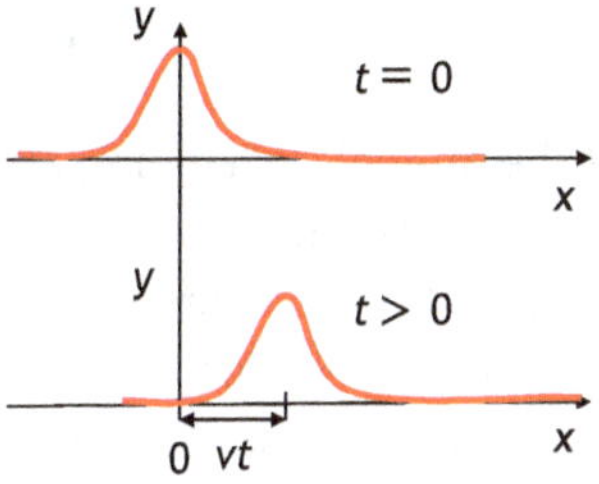

Fig. 8.20: Pulse propagation in one direction.

Because of the compliance of the blood vessels, the propagating pressure wave can directly be observed by a bulging of the arteries (Fig. 8.21), which is similar to the Windkessel effect discussed before. A pressure wave in a rigid tube would have pure longitudinal character. However, in a blood vessel due to the compliance of the walls, the pressure wave has both longitudinal and transverse components.

Fig. 8.21: Local bulging of blood vessels due to propagating pressure wave.

The bulging effect is used for determining the blood pressure according to the method of Riva-Rocci, as already presented before. It can also be used for detecting the arrival time of a pulse wave with a pressure-sensitive device such as a tonometer or a piezo-electric transducer. The amplitude of the pressure wave is a question of vessel compliance. The softer the vessel wall, the bigger is the amplitude. Since elastic deformation such as bulging costs energy, the PWV decreases with increasing compliance C. This dependence is expressed in the Moens[14]–Korteweg[15] equation for the PWV [5]:

$$v_{PWV} = \sqrt{\frac{1}{C}\frac{A\Delta r}{\rho}} = \sqrt{\frac{Y}{\rho}\frac{\Delta r}{2r}}, \tag{8.43}$$

---

14 Adriaan Isebree Moens (1846–1891), Dutch physician and physiologist.
15 Diederik Johannes Korteweg (1848–1941), Dutch mathematician.

where A is the cross-sectional area of the vessel, Δr is the wall thickness of the blood vessel, Y is the elastic modulus, $2r$ the diameter, and ρ is the density of the fluid. The PWV thus provides information about the compliance of the blood vessels.

> **!** We distinguish between a *mean blood flow velocity* and a *PWV*. The mean flow velocity is a longitudinal velocity of the fluid governed by the laws of hemodynamics. The PWV is the velocity of a transverse pressure wave and depends on the vessels' elastic properties.

8.5.8 Methods for determining PWV

PWV provides information about the elastic properties of the arterial system, in particular, about the compliance and stiffness of the arterial vessels. The mechanical properties of the arterial walls change along the arterial system. From the large arteries to the periphery, the stiffness of the walls increases. Therefore the PWV is not a constant, but increases with the distance from the heart. In addition, the wall stiffness also changes with age: the vessels become less compliant, which leads to increased PWV and increased blood pressure. Therefore, the PWV measurement is also used for the early indication of cardiovascular diseases that stiffen the blood vessels.

There are several methods for determining the PWV: (1) direct contact methods (tonometer, piezoelectric transducer), (2) imaging methods (sonography, magnetic resonance tomography), and (3) optical interferometry methods. We will only discuss nonlocal and direct contact methods. For the other procedures, we refer to the respective chapters on imaging techniques in Vol. 2.

In contact mode, PWV is determined by measuring the distance Δx from AV over travel time Δt. Favorable points for measuring the pulse amplitudes are those where the pulse can be easily detected because of its proximity to the skin, like the carotid artery and the femoral artery. In one-point measurements, the travel time Δt is taken as the time difference between the R point of the systole, which is the time when the isovolumetric contraction starts (see Fig. 8.6), and the arrival time of the pulse at the carotid artery (t_1) or femoral artery (t_2). The arrival time is defined as the ascending foot of the pressure pulse, indicated by a red dot in Fig. 8.22. Thus one-point measurements require simultaneous recording of ECG and pressure amplitudes at specific points on the skin. The distance Δx between the AV and points C or F is usually measured with a meter stick on the skin. This is not a precise measurement but usually sufficient for the purpose.

In a two-point measurement, ECG recording can be avoided. In this case, the *transfer time* (TT), which is the difference of the arrival times t_1 for the carotid pulse and t_2 for the femoral pulse, is measured: $TT = t_2 - t_1$. This time difference corresponds to the length difference of the femoral point $\Delta x_2 = F$ compared to the carotid point $\Delta x_1 = C$, as indicated in Fig. 8.22.

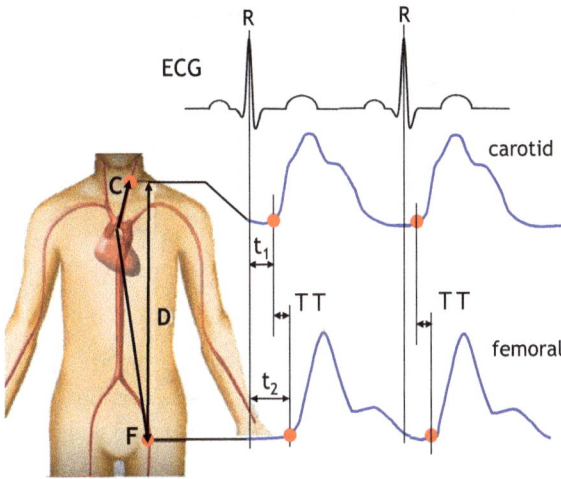

Fig. 8.22: Determination of the pulse wave velocity using either a one-point measurement together with recording of ECG or a two-point measurement by measuring the time difference TT and the distance D between the carotid and femoral artery.

Therefore the PWV is determined by taking the ratio

$$v_{PWV} = \frac{F - C}{TT}. \tag{8.44}$$

F and C are again measured with a meter stick on the skin that may bear some systematic errors. A much easier but less accurate method is obtained by taking the distance D between the carotid and femoral artery and calculating the velocity according to

$$v_{PWV} = \frac{D}{TT}. \tag{8.45}$$

The oversimplification can be rectified by taking a correction factor into account. As D is easily measured, and the ratio $\alpha = C/F$ is roughly constant independent of the body size, a better approximation is

$$v_{PWV} = \frac{D}{TT}\left(\frac{1-\alpha}{1+\alpha}\right). \tag{8.46}$$

Typical values for C and F are 100 and 560 mm, respectively, yielding $\alpha = 0.18$ and a ratio $(1-\alpha)/(1+\alpha) = 0.7$. PWV determined in such a manner is in the order of 6–8 m/s. When the PWV goes up to 10–12 m/s, a cardiovascular disease is likely. It should be noted that the PWV is much higher than the mean blood flow velocity in this pulsatile part of the artery. In addition, the pulse shape shown in Fig. 8.22 changes drastically from point to point. This is due to the varying stiffness of the

blood vessels with distance and partial reflection of pressure waves as the impedance changes along the pathway. Furthermore, pulses slightly vary from one systole to the next. Therefore, the foot of pulses or the initial slope is a better indicator
for pulse positions than the peak.

8.6 Physical properties of blood

8.6.1 Composition of blood

Blood is a complex fluid. It consists of extracellular fluid called plasma and blood
cells in suspension. The plasma is composed of water (92%), plasma proteins (7%),
and solutes (1%). The blood cells are mostly RBCs (*erythrocytes*, 95%), and to a much
smaller fraction thrombocytes (4.8%) and leucocytes called white blood cells (0.2%).
A 1 µl (=1 mm^3) of blood contains about 5×10^6 RBCs. Male adults have 5–6 l blood,
female adults 4–5 l blood in the body.

  The lifetime of erythrocytes is 100–120 days. About 2–4 million new erythrocytes are synthesized any second from special stem cells (hemocyctoblast) located
in the red bone marrows at the heads of long bones. The newly formed RBCs use
the bone marrow vasculature as channel to the systemic circulation of the body.

  The fraction of erythrocytes to the total blood volume is called *hematocrit value*
(Hct) or *packed cell volume*. The Hct is determined by centrifugation of a test tube
filled with a blood volume $V1$ (Fig. 8.23). Centrifugation and ultracentrifuges were
introduced by Svedberg[16], and is explained in Infobox II. After centrifugation, three
areas are recognized: yellowish plasma on top, buffy coat-containing thrombocytes
and leucocytes, and the rest volume $V2$ containing erythrocytes. The ratio

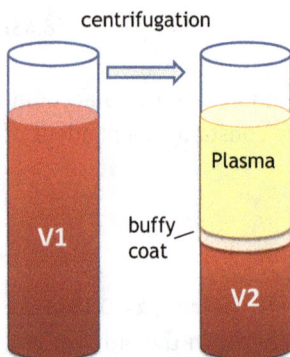

Fig. 8.23: Test tube filled with blood separates into three zones
after centrifugation: plasma, buffy coat, and erythrocyte
volume.

---

16 Theodor Svedberg (1884–1971), Swedish chemist and Nobel laureate in chemistry 1926.

$$\text{Hct} = \frac{V2}{V1} \tag{8.47}$$

is the Hct. Typically Hct = 0.4–0.5.

RBCs contain the red pigment hemoglobin, which binds and transports O_2. Each RBC is a biconcave disc with a diameter of 7–8 μm and disk thickness of 2.5 μm (Fig. 8.24).

7-8 μm   ↕ 2 μm

Fig. 8.24: Shape and size of red blood cells.

Using these values, the *mean corpuscular volume* (MCV) of an erythrocyte is 100 fl. With this value, we can determine the number of erythrocytes in one liter of blood: 5×10^{12}. Assuming that the density of blood is similar to water (1 g/cm³), the mass of an erythrocyte can be estimated to be about 100 pg. This mass value is called in medical textbooks *mean corpuscular hemoglobin* (MCH) and its accepted value is 27–34 pg. Compared to our rough estimate, the lower actual mass indicates that either the density or the number of erythrocytes per liter of blood was underestimated. In any case, the ratio of the mass to volume of a single erythrocyte yields the *mean corpuscular hemoglobin concentration* (MCHC): MCHC = MCH/MCV = 32–36 g/dl. More precisely, this value is a volume density rather than a concentration.

Infobox II: Ultracentrifuge

The centrifuge is a spinning apparatus about a single central axis. For simplicity you may imagine a spinning disk. On each mass point in the disk acts a centrifugal force:

$$F_c = m\omega^2 r.$$

Therefore, the centrifugal force on each mass point in the disk increases with the mass m, with the square of the rotational frequency ω, and the radius r.

Now consider two mass points m_1 and m_2 fixed at the same radius r from the center and spinning with the same angular frequency ω. If $m_1 > m_2$, the centrifugal force $F_{c,1}$ on m_1 will be higher than $F_{c,2}$ on m_2. Now we assume that these two mass points are embedded in a viscous fluid. Then we expect that they will move radially outward in the fluid with a constant velocity v:

$$v = \frac{m\omega^2}{6\pi\eta},$$

where η is the viscosity of the fluid. m_1 with the higher mass will move faster than mass m_2. After some time, these two mass points will have separated: mass m_1 will be at a larger distance r_1 from the center of the centrifuge than mass m_2. This shows that centrifuges are suited for separating different masses. Mass separation by centrifuges has been used by biologist to separate molecules such as proteins and by physicist to separate isotope such as ^{235}U and ^{238}U.

The effect is small and therefore high angular frequencies are required. The method gained increased interest among biologist after Svedberg built a centrifuge in 1924 with 12 000 rpm that produced a centrifugal force 7000 times the earth gravitational acceleration g. He called it the ultracentrifuge. Nowadays accelerations of 100 000 up to 750 000 times g are achieved with ultracentrifuges. Typically 1000–2000 rpm are used to separate blood plasma and blood cells. See also exercise E8.15 for further information.

8.6.2 Viscosity of blood

The viscosity of fluids is an important intrinsic and material-specific parameter determining the flow resistance. There are many experimental methods to determine the viscosity of fluids, which are discussed in standard textbooks. For practical reasons, we mention the gliding plate method (Fig. 8.25): a fluid of constant thickness h separates two plates, one fixed and another movable. The top free plate is moved with a constant speed v_1 by a tangential shear stress $\tau = F/A$, where F is the force and A is the area of the plate. Assuming that the fluid layers next to the surface of the plates stick to the plates by adhesion and do not slip, there will be a velocity gradient of the fluid from the lower plate ($v = 0$) to the upper plate ($v = v_1$). The shear rate $\dot{y}$ is defined as $\dot{y} = v_1/h$ (units $[\dot{y}] = s^{-1}$). Then the viscosity η is defined by the shear stress divided by the shear rate:

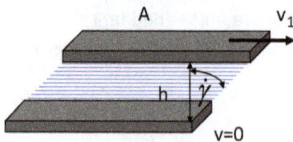

Fig. 8.25: Gliding plate method for determining the viscosity of a fluid.

$$\eta = \frac{\tau}{\dot{y}} = \frac{F/A}{v_1/h}. \tag{8.48}$$

The unit for the viscosity is $[\eta] = \text{N} \cdot \text{s}/\text{m}^2 = \text{Pa} \cdot \text{s}$. The cgs unit Poise[17] is often used in medical textbooks. The conversion to SI units is: 1 Poise (P) = 0.1 Pa · s.

The viscosity of fluids depends on temperature and sometimes also on the shear rate. Usually, the viscosity of fluids decreases with increasing temperature due to the increasing mobility of molecules in fluids. Typical examples are glycerin or honey, which stick at low temperatures but flow at higher temperatures. Table 8.2 lists a few viscosity values for water and blood. From these values, we notice that at the body temperature of 37 °C, the viscosity of blood is five times higher than that of water. In fact, the viscosity of blood depends on the Hct – value as shown in Fig. 8.26. The viscosity quoted in Tab. 8.2 refers to an average Hct of 45% typical for a healthy person.

Tab. 8.2: Viscosity values for water and blood at different temperatures.

Viscosity (mPa s)	0 °C	20 °C	37 °C
Water	18	1	0.8
Blood			4

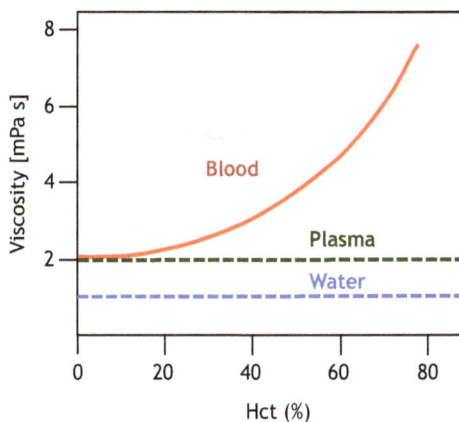

Fig. 8.26: Dependence of the viscosity on the hematocrit value.

Fluids, having a viscosity independent of the shear rate $\dot{y}$, are called *Newton fluids*. Fluids, whose viscosity depends on $\dot{y}$, are called *non-Newton fluids*. Blood is a non-Newton fluid. The viscosity of blood depends on the shear rate: the viscosity decreases with increasing shear rate (see Fig. 8.27). Because of this characteristic, the flow resistance decreases with increasing flow velocity. A Newton liquid has a parabolic velocity profile when flowing through a cylindrical pipe, In the center the fluid velocity is

---

17 Poise stands for Poiseuille.

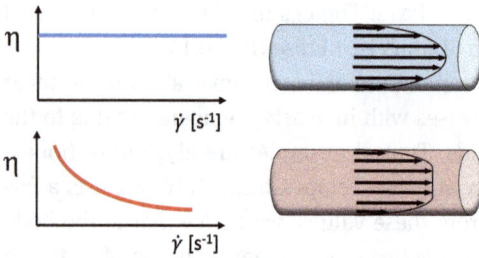

Fig. 8.27: Velocity profiles for a Newton fluid with constant viscosity (top panel) and a non-Newton fluid with shear rate-dependent viscosity (bottom panel). The Newton fluid shows a parabolic velocity profile, the non-Newton fluid, a flattened velocity profile in the center.

highest and decreases to zero at the walls. In contrast, for non-Newton fluids the velocity profile is flattened in the center.

The special properties of blood are due to the shape of the RBCs. Due to the biconcave disc shape, RBCs can randomly arrange themselves in streams of wide vessels. However, when pushed through narrow vessels, the plates rearrange in parallel to the blood flow and on top of each other. Three regions can be distinguished by increasing the shear rate (Fig. 8.28(a)). In Region I, the orientation of the erythrocytes is random. In Region II, the erythrocytes are arranged parallel to one another for better slip. In Region III, the RBCs have to form chains and only one at a time can pass through narrow capillaries. In fact, they can squeeze through arterioles in the capillary bed as small as 4–5 μm in diameter without getting caught by rolling up and bending over to reduce their size, as shown in Fig. 8.28(b). The high deformability of the RBCs is based on a network of fibers inside the double lipid cell membrane of the erythrocytes, which are known as the *cytoskeleton* of RBCs.

> **!** From a physical point of view, blood is a multicomponent non-Newton fluid.

(a)   (b)

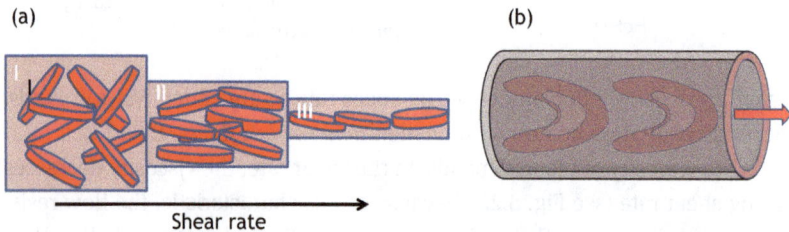

Shear rate

Fig. 8.28: (a) Three regions of blood cell organization can be distinguished, depending on the diameter of blood vessels. With increasing shear rate, the RBCs organize themselves in plates or chains. (b) Curled up shape of erythrocytes in narrow capillary tubes.

8.6.3 Osmotic pressure

RBCs contain no nucleus, no mitochondria, and no ion channels. Therefore RBCs cannot perform action potentials and they cannot reproduce themselves. They are synthesized from stem cells in the red bone marrows, as already mentioned. Because of the lack of ion channels, the ion concentration in the intracellular space of the RBC remains constant and consists mainly of K^+-ions. However, the cell membrane is permeable to water, which is essential for controlling the *osmotic pressure* of RBCs.

To understand the origin of osmotic pressure, we first consider the partial pressure p_i of one gas component i in a gas mixture such as air

$$p_i = \frac{RT}{V_{gas}} n_i, \tag{8.49}$$

where R is the gas constant, T is the absolute temperature, V is the gas volume, and n_i is the mol number of gas component i in the mixture. This equation shows, as explained in more detail in Section 9.3.1, that the partial pressure p_i is proportional to the mole number of gas component i in the mixture. Analogous to the partial pressure of a gas component in a gas mixture, any substance (or solute) dissolved in a solvent generates a partial pressure in the solvent:

$$\pi_i = \frac{RT}{V_{sol}} n_i. \tag{8.50}$$

Equation (8.50) is known as the van't Hoff law,[18] and the pressure π_i is the *osmotic pressure* of component i. We define the solubility S_i of substance i as

$$S_i = \frac{n_i}{V_{sol}} \tag{8.51}$$

where n_i is the mole number of the solute and V_{sol} is the volume of the solvent. Then the osmotic pressure π_i of component i is

$$\pi_i = S_i RT. \tag{8.52}$$

Equation (8.52) is known as Henry's law.[19] Note that the osmotic pressure depends only on the mole number, not on the specifics of the ions dissolved.

If two components with different mole numbers $n_1 \neq n_2$ are dissolved in the same solvent, such as water, the osmotic pressure will be different: $\pi_1 \neq \pi_2$. Like in gas, if these solvents are in contact, the system tries to equalize the pressures. Let

---

18 Jacobus Henricus van t'Hoff (1852–1911), Dutch physical-chemist, 1901 Nobel Prize in Chemistry.
19 William Henry (1774–1836), British chemist.

us assume that a membrane separates these two components. Then equalization of the pressures can be achieved in only two different ways:

1. By diffusion of the solutes i across the membrane up until the mole numbers equalize: $n_1 = n_2$;
2. Or by a shift of the solvent volume V_{sol} across the membrane so that the solubilities equalize on both sides: $S_1 = S_2$.

The second option is known as *osmotic pressure equalization* or simply as *water displacement*. Water displacement always occurs if the membrane is permeable to the solvent (water) but not to the solute. Both options, diffusion and water displacement, are compared in the Infobox III. The RBCs have a water-permeable membrane but no ion channels. Therefore, water displacement is the only option to equalize osmotic pressures from the different solutes: K^+ in the cell plasma and Na^+ in the cytoplasm, schematically shown in Fig. 8.29.

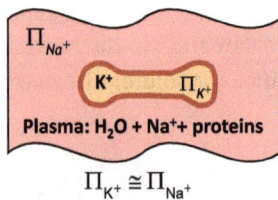

Π_{Na^+}

K^+ Π_{K^+}

Plasma: H_2O + Na^+ + proteins

$\Pi_{K^+} \cong \Pi_{Na^+}$

Fig. 8.29: Osmotic pressures in the plasma and inside of the blood cell must be well balanced.

The K^+-solubility determines the osmotic pressure in RBCs, and the Na^+-solubility determines the osmotic pressure in the plasma. The osmotic pressures are high on both sides of the membrane (~7.5 bar). If the pressures are about equal inside and outside ($\Pi_{K^+} \cong \Pi_{Na^+}$), the condition is called *isotonic*. Then the blood cells have their regular shape as shown in Fig. 8.30. However, any imbalance of the osmotic pressure results in a movement of water across the cell membrane. For $\Pi_{K^+} > \Pi_{Na^+}$ the condition is *hypotonic*. Additional water penetrates into RBCs, the cells expand and may eventually burst. Hypotonic conditions occur due to lack of Na^+ in the plasma. Vice versa, *hypertonic* conditions are present if the Na^+-concentration in the plasma is too high. Then $\Pi_{K^+} < \Pi_{Na^+}$ and water moves from RBCs into the plasma to balance the pressures. The RBCs will shrink and lose their ability as gas exchange carriers. Moreover, the changing shape under hypotonic and hypertonic conditions alters the elasticity and deformability of RBCs and, therefore, the viscosity of blood.

In case of high loss of blood volume and lack of blood plasma, an *isotonic saline solution* can be injected as immediate and temporary first help. The isotonic fluid consists of a 0.9% solution, which is 9 g of NaCl per 1000 ml of H_2O (see Exercise E8.10).

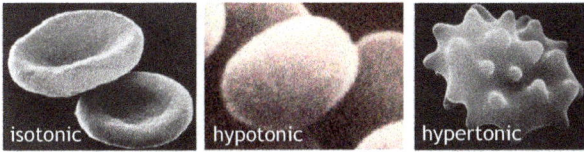

Fig. 8.30: Red blood cells under isotonic, hypotonic, and hypertonic conditions. The spiky appearance of the hypertonic erythrocyte is due to the internal scaffolding by the cytoskeletal fiber proteins actin and spectrin.

Infobox III: Osmotic versus diffusive particle exchange, what is the difference?

Let us consider a vessel with a solvent (water), which is divided by an impermeable wall into two chambers of equal size. On both sides of the wall there is a different number of molecules $N_1 = 10$ and $N_2 = 20$ dissolved in water, causing different osmotic pressures π_1 and π_2. In the shown example $\pi_2 = 2 \times \pi_1$. This is the starting situation in the left panel below. Now we replace the impermeable wall by one which allows diffusion of the solute across the wall. From Fick's law we know that a diffusion current will take care of equalizing the number of molecules on both sides, in our cases: $N_1 = N_2 = 15$. At the same time, the osmotic pressures will be identical: $\pi_1 = \pi_2$. This case is shown in the right panel.

Now we go back to the starting situation and we replace the wall with one that allows water to move across but not the dissolved molecules (middle panel). Then water will diffuse from left to right until the osmotic pressures are equal: $\pi_1 = \pi_2$. Now we notice that one water level is higher than the other. The difference in height exerts a hydrostatic pressure $\Delta p = \rho g h$ that is equal to the osmotic pressure difference before water displacement: $\Delta p = \pi_2 - \pi_1$. Osmotic exchange costs potential energy $\Delta E = mgh$. Diffusion is free of energy cost.

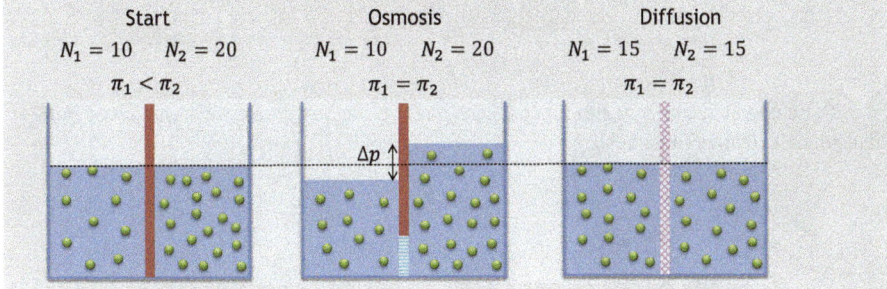

8.7 Blood as an oxygen carrier

RBCs are carriers of oxygen (O_2) and carbon dioxide (CO_2). The color ranges from bright red (O_2 – rich blood, oxygenated blood) to dark red (O_2 – poor blood, deoxygenated blood). Gas transport via RBCs requires permeability of cell membranes for gases in question, binding during transport, and release at targeted destinations. RBCs are predestined for these tasks as they contain the protein hemoglobin. Each erythrocyte hosts about 2.5×10^8 *hemoglobin* molecules. The mean hemoglobin concentration in a RBC is about 32–36%, the rest of the cell is filled with fibers, intracellular fluid, and other proteins.

8.7.1 Structure of hemoglobin

Each hemoglobin molecule consists of four helical chains, two α-chains, and two β-chains (Fig. 8.31). The polypeptide chains have slightly different lengths: 141 amino acids in α-chains and 146 amino acids in β-chains. More importantly, all chains form a pouch containing one heme molecule in the middle and a channel for oxygen access. Each flat heme molecule (Fig. 8.32), also called Fe(II)-protoporphyrin, has one Fe^{2+} ion at its center for binding oxygen molecules during inspiration. The Fe^{2+} ion in the heme molecule is bound to four nitrogen atoms in the center. The structural description of the protein hemoglobin by Max Perutz[20] published in 1959 was an early milestone

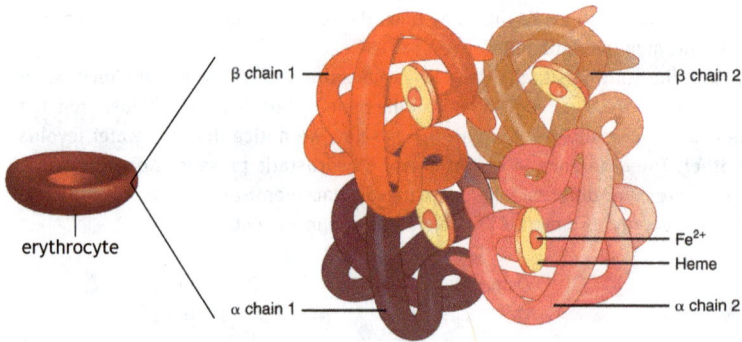

Fig. 8.31: Hemoglobin with four side chains, each hosting one heme molecule (reproduced from Wikimedia, © creative commons).

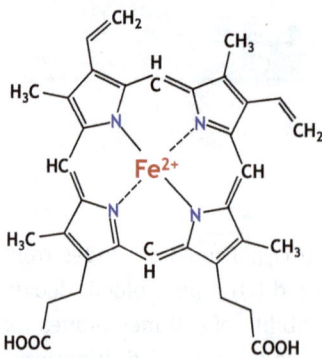

Fig. 8.32: Heme molecule with a Fe^{2+} ion at the center.

---

20 Max Perutz (1914–2002), Austrian-British biochemist, Nobel Prize in Chemistry 1962.

in biochemistry and protein crystallography [7]. Nowadays, with advanced and partially automated techniques using synchrotron radiation and cryogenic electronmicroscopy (cryo-EM), thousands of protein structures are determined every year.

8.7.2 Saturation curve

The reversible binding of oxygen to hemoglobin in erythrocytes or myglobin in muscle cells can be represented in oxygen-binding curves, where the fractional saturation of oxygen is plotted versus the partial pressure of oxygen during inspiration. The *fractional saturation of oxygen* is defined as

$$Y = \frac{[\text{HbO}_2]}{[\text{HbO}_2] + [\text{Hb}]},\tag{8.53}$$

where $[\text{HbO}_2]$ and $[\text{Hb}]$ are the concentrations of hemoglobin with and without oxygen bonded, respectively. According to Henry's law (eq. 8.52), the oxygen concentration in the blood is proportional to the partial oxygen pressure p_{O_2} in the alveolar space where the oxygen exchange takes place: $[\text{O}_2] \sim p_{\text{O}_2}$. The rate of O_2 bonding to Hb depends on the concentration of O_2 in the plasma and the concentration of vacant Hb sites:

$$\dot{Y}_{\text{bond}} = k_b [\text{O}_2] N (1 - Y).\tag{8.54}$$

Here N is the total number of hemoglobin molecules, $N(1 - Y)$, the number of sites that O_2 can still occupy, and k_b is a bonding coefficient. Conversely, the rate of dissociation depends on the oxygen concentration that is bonded to Hb, where k_d is a dissociation constant:

$$\dot{Y}_{\text{diss}} = k_d N Y.\tag{8.55}$$

In equilibrium, bonding and dissociation rates must equilibrate at constant pressure and temperature, and therefore we have

$$k_b [\text{O}_2] N (1 - Y) = k_d N Y.\tag{8.56}$$

Solving for the fractional occupation Y, we obtain

$$Y = \frac{[O_2]}{K + [O_2]},\tag{8.57}$$

where $K = k_a/k_b$. Equation (8.57) is the standard *Langmuir equation*[21] [8]. In Fig. 8.33, Y is plotted as a function of the oxygen concentration $[O_2]$. It shows a linear and steep rise at low $[O_2]$ concentration and a saturation at higher concentrations. The relation (8.57) holds for the bonding of oxygen to myoglobin, but not for hemoglobin. Hemoglobin has an S-shaped saturation curve at low oxygen pressures before going into saturation (Fig. 8.33), described by the equation

$$ Y = \frac{[O_2]^n}{K + [O_2]^n}. \tag{8.58} $$

In this form the equation is known as the *Hill equation*,[22] n is called the *Hill coefficient* [9, 10]. n is a measure of the cooperativity of different binding sites. If $n > 1$, the binding is cooperative, meaning the affinity to O_2 increases with the number of O_2 already bonded. If $n < 1$, the binding is counter-cooperative. For $n = 1$ the bonding–dissociation rates follow the Langmuir rate equation.

For hemoglobin, the Hill coefficient n is about 2.5. Myoglobin is another protein that contains a heme molecule for reversible oxygen binding and transport in muscles. In the case of myoglobin, the S-shape is missing and the saturation follows the more simple Langmuir curve with $n = 1$.

> **!** The reversible chemical binding of oxygen to the hemoglobin molecule is the key to supplying cells with oxygen for cellular respiration.

From a physiological point of view, the saturation curves for myglobin and hemoglobin show some remarkable features. First, they signal that oxygen molecules are chemically bound to hemoglobin (myoglobin), as both curves display a steep rise at low pressures and saturation at higher pressures. A physical solution of gaseous O_2 is also present, but at a much lower level and following a linear dependence as a function of the oxygen pressure p (O_2) (see Fig. 9.4). In addition, the S-form for hemoglobin indicates that the binding of O_2 is weaker than for myglobin at low pressures and increases with increasing fractional saturation. This trend implies that the affinity for oxygen increases with each oxygen molecule that is already bound by positive feedback. Indeed, in response to oxygen uptake, the α- and β-chains rearrange from a tense form (T) without oxygen to a relaxed form (R) with oxygen, schematically shown in Fig. 8.34. This cooperation is typical for hemoglobin and requires the cooperative effect of four chains. A similar cooperation does not exist in myoglobin since myoglobin has only a single chain. As a result, the saturation curve for myoglobin is not S-shaped.

---

21 Irving Langmuir (1881–1957), US chemist, Nobel Prize in Chemistry 1932.
22 Archibald Hill (1886–1977), British physiologist, Nobel Prize in Medicine 1922.

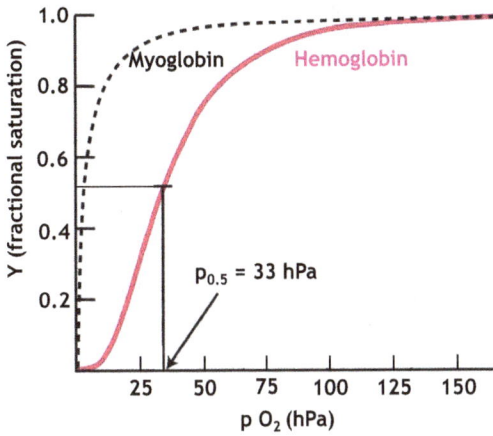

Fig. 8.33: Fractional saturation of oxygen in hemoglobin and myoglobin as a function of partial oxygen pressure.

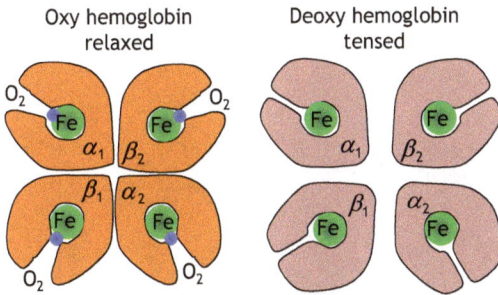

Fig. 8.34: Tense and relaxed states of hemoglobin consisting of four chains (α_1, α_2, β_1, β_2) that cooperate positively for increased oxygen affinity: the more oxygen is bound, the more relaxed the four chains become, and the affinity to more oxygen-binding increases.

In saturation, each heme molecule carries one O_2 molecule. Therefore, one erythrocyte carries $4 \times 2.5 \times 10^8 = 10^9$ O_2 molecules, and 1 l of blood conveys 5×10^{21} O_2 molecules. Assuming that 2 l are fully oxygenated on average, and the remaining 3 l of the total blood volume have an oxygen content of 75% after expiration, then the total oxygen content of blood is 0.03 mol oxygen.

The partial pressure at half-saturation $p_{0.5}$ is a measure of the affinity for oxygen binding. In the case of hemoglobin, $p_{0.5}$ is about 33 mbar, whereas saturation is reached beyond 100 mbar. Myoglobin has a much higher oxygen affinity: $p_{0.5}$ is 2–5 mbar, and saturation is already reached at 50 mbar. The partial pressure of atmospheric O_2 is 210 mbar = 210 hPa, sufficient for blood oxygen saturation at sea level. However, at higher altitudes, such as on top of Mt. Everest, the partial O_2 pressure drops to 70 mbar, not sufficient for sustainable life without an oxygen mask.

Oxygen is not the only molecule with a high affinity to heme. Carbon monoxide can also bind to Hb with an affinity that is by a factor of 200 higher than for O_2. Carbon monoxide poisoning results from two factors: CO blocks sites in Hb for O_2 uptake and hinders O_2 transport. And second, the presence of CO increases the bonding strength in the remaining HbO_2, thus impeding O_2 release at target sites in tissues.

8.7.3 Ferritin

From the foregoing, we appreciate the importance of iron for oxygen transport. We also noticed that many erythrocytes, including hemoglobin, are being synthesized any second in the bone marrow. Each hemoglobin molecule synthesized takes up four Fe atoms. Therefore the body must run a Fe recycling system and a Fe storage for buffering Fe concentration fluctuations. About 3.7 g of Fe are in the body at any time. From these, 2.5 g are bound in hemoglobin for oxygen transport, some 0.2 g are bound in myoglobin, and another 0.02 g are distributed over various proteins. The remaining 1 g is the reserve and stored in ferritin.

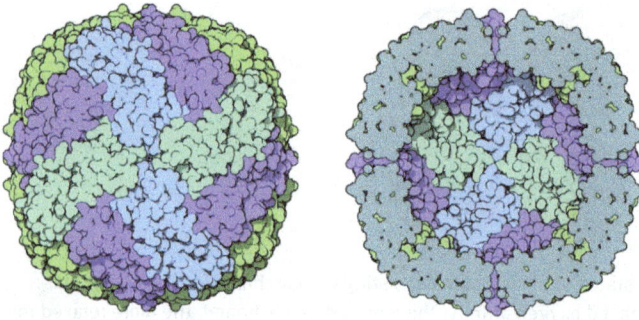

Fig. 8.35: Ferritin is composed of 24 proteins forming a hollow sphere that houses Fe in small crystals. Left side: closed structure; right side: cross section of the hollow ferritin sphere (from http://www.rcsb.org/pdb/molecules/pdb35_1.html).

Ferritin is a highly interesting biomaterial [11]. It consists of 24 identical proteins forming a hollow sphere, shown in Fig. 8.35. Only a few pores allow access to the interior. Once Fe ions have penetrated through the pores, they are bound in ferritic molecules, forming small crystals. Each ferritin stores about 4500 Fe ions. Ferritin can be found mainly in the liver and spleen.

In ferritin, Fe is bonded in the oxidation state Fe^{3+} within small crystals with the stoichiometry $[FeO(OH)]_8[FeO(H_2PO_4)]$. If needed for hemoglobin synthesis, Fe^{3+} must first be reduced to Fe^{2+}. Then $Fe(H_2O)_6^{2+}$ goes into the solution and can leave ferritin through the pores. The entire metabolism of iron is described in textbooks on physiology. Here we just mention that ferritin is used in nanomedicine for transporting targeted drugs and magnetic nanoparticles protected by a biocompatible shell. These aspects are discussed in more detail in Chapter 7/Vol. 3. The fascinating magnetic properties of Fe^{2+} in hemoglobin are presented in the Infobox IV, including the high-spin–low-spin transition upon oxygen bonding to the heme molecule.

Infobox IV: High-spin–low-spin transition (advanced topic)

Fe^{2+} in the heme molecule has two important properties: it is highly reactive and likely to bind oxygen, and it is magnetic. The electronic configuration of Fe^{2+} is $3d^6$. According to Hund's rule, the spin quantum number is $S = 2$, the orbital quantum number is $L = 2$, and the total angular momentum quantum number is $J = 4$. The Landé factor g_J is 3/2, and therefore, the expected magnetic moment is $m_J = g_J \sqrt{J(J+1)}\mu_B = (3/2) \times \sqrt{20}\mu_B = 6.7\mu_B$, where μ_B is the Bohr magneton. However, Fe^{2+} is surrounded by a local *crystal electric field* E_{CF} and therefore the electronic considerations need to be modified, depending on whether the local crystal electric field E_{CF} is stronger ($E_{CF} > \vec{L} \cdot \vec{S}$) or weaker ($E_{CF} \leq \vec{L} \cdot \vec{S}$) than the *spin-orbit coupling* $\vec{L} \cdot \vec{S}$. In any case, the local electric field E_{CF} lifts the degeneracy of the atomic orbital moments m_L into two subsets of orbitals, e_g and t_{2g}, as is shown in the left panel for local octahedral symmetry.

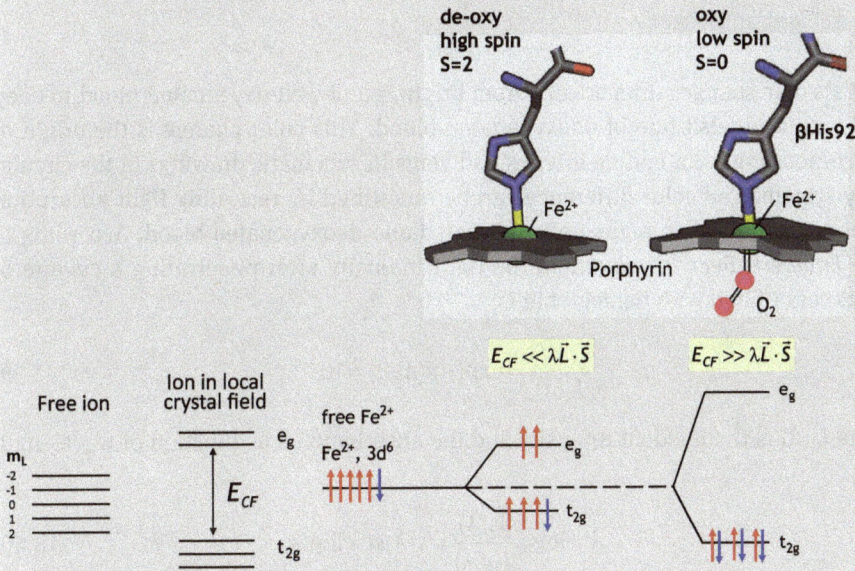

In the heme molecule, the- Fe^{2+} ion is well-protected by a so-called βHis92 molecule. Without this protection, O_2 would immediately bind to Fe^{2+} and form a complex $Fe^{3+}(O_2)^-$, i.e., Fe^{2+} would be oxidiced by electron transfer to O_2. This bond is so strong that it would be impossible to recycle O_2 during respiration. Therefore, with the help of the globin crevice and in particular the histidine chain βHis92, the electron transfer is only partial and much weaker, permitting the Fe-O_2 bond to be reversible. The binding of O_2 to hemoglobin is usually written: $Hb + O_2 \rightarrow HbO_2$.

When O_2 binds to the heme group, the Fe^{2+} ion is coplanar within the heme molecule and experiences a strong *octahedral crystal field* (oxy-state, right panel). Then only the t_{2g}-states are occupied. According to the Pauli principle, the spins have to pair up and therefore the total spin moment cancels to zero: $S = 0$. Upon oxygen release, Fe^{2+} moves slightly out of the heme plane by 0.04 nm pushing the βHis92 molecule up. This weakens the crystal field and the splitting between e_g and t_{2g} states become much reduced. Then also the energetically higher e_g

states can be occupied, resulting in a high spin $S = 2$ state. The schematics of the spin occupancies is shown in the de-oxy state of the right panel. Hence, the reversible oxygen binding in hemoglobin simultaneously drives a *high-spin–low-spin* (HSLS) *transition* in Fe^{2+}. At the same time, the HSLS transition is associated with a magnetic-non magnetic transition. For the physiology of the respiratory system, this spin transition is not important. However, for functional imaging of brain activity via magnetic resonance tomography, the HSLS transition is essential because the proton relaxation time $T2$ depends on the local Fe magnetic moment. Using scans, which are sensitive to the $T2$ relaxation time, one can map out those parts of the brain, which consume more oxygen (low magnetic moment, slow $T2$ relaxation) in response to an external stimulus, like listening or reading, as compared to other nonactivated parts (high magnetic moment, short $T2$ relaxation) of the brain. This type of brain mapping is referred to as functional MRI (fMRI).

8.7.4 Light absorbance

RBCs' color changes dramatically from bright red of well-oxygenated blood to deep red with a blueish hue of deoxygenated blood. This color change is the origin of corresponding color coding arteries and veins in schematic drawings of the circulatory system. The color difference can be quantified by recording light absorption spectra in the visible regime of oxygenated and deoxygenated blood. According to the *Lambert–Beer*[23,24] *equation,* the light intensity after penetrating a cuvette of thickness t filled with hemoglobin is

$$\frac{I(\lambda, t)}{I_0(\lambda)} = \exp(-\varepsilon(\lambda)st) = 10^{-A}. \tag{8.59}$$

Here $I_0(\lambda)$ is the incident intensity and the absorbance A as function of wavelength λ is

$$A = \log_{10}\frac{I_0(\lambda)}{I(\lambda, t)} = \varepsilon(\lambda)st \times \log(e). \tag{8.60}$$

$\varepsilon(\lambda)$ is the wavelength-dependent extinction coefficient; $s = f[\text{Hb}] + (1\text{-}f)[\text{HbO}_2]$ is the combined concentration of oxygenated and deoxygenated hemoglobin in the sample with fraction $(1\text{-}f)$ and f, respectively. The extinction coefficient has the units $[1\,\text{mol}^{-1}\,\text{m}^{-1}]$. The absorbance of oxygenated hemoglobin (red curve) and deoxygenated hemoglobin (blue durve) is plotted in Fig. 8.36. The color that we observe is the one that is less absorbed and more reflected. In oxygenated hemoglobin, absorption bands are visible at 535 and 575 nm, whereas little absorption occurs at 560 nm and above 600 nm. An absorption band exists at 560 nm for deoxygenated

---

23 Johann Heinrich Lambert (1728–1777), Swiss-French mathematician and physicist.
24 August Beer (1825–1863), German physicist, chemist, and mathematician.

hemoglobin, but low absorption is seen beyond 650 nm. The absorption spectra reproduced in Fig. 8.36 are characteristic and can be quantified with respect to the oxygen concentration bond to hemoglobin [12]. Absorbance measurements are usually
performed in vitro.

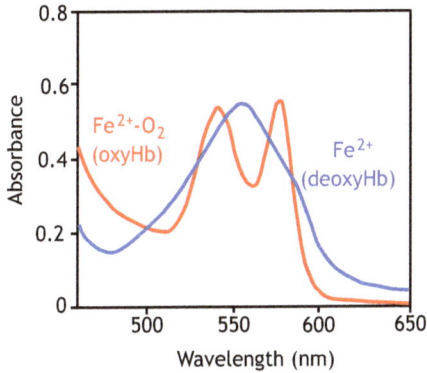

Fig. 8.36: Absorbance spectra of oxygenated and
deoxygenated hemoglobin.

In vivo absorbance measurements have become very popular recently with the
availability of health trackers, smartwatches, and especially pulse oximeters. The
latter is important for early warning of low arterial blood oxygen levels in diseases,
including viral infections such as COVID-19 [13]. Consumer market oximeters may
not provide the precision required for clinical applications. However, the measuring
principle of consumer and occupational units is identical. In contrast to the in vitro
measurements described above, in vivo pulse oximeters use two wavelengths in the
red and infrared part of the spectrum provided by two LEDs. The exact wavelength
is not important, but one LED should emit photons at around 700 nm and the other
at around 900 nm. A photodiode detects both wavelengths either in transmission
mode or in reflection mode. The finger oximeter determines the light transmitted by
the finger, while trackers carried on the wrist work in reflection mode or, more precisely, in scattering mode. Figure 8.37 reproduces the wavelength dependence of
the extinction coefficient in the near-infrared range. According to this plot, oxygen-
containing blood (HbO_2) absorbs more infrared light at 900 nm and less red light at
700 nm. Conversely, oxygen-free blood allows more infrared light to pass through
at 900 nm and absorbs more red light at 700 nm. By taking the transmitted intensity ratio $I(700\,\text{nm})/I(900\,\text{nm})$, the peripheral oxygen concentration can be determined empirically. For a healthy person, a ratio of 0.5 corresponds to 100% oxygen
saturation, a ratio of 1 corresponds to about 82% saturation, and a ratio of 2 corresponds to nearly zero oxygen concentration [14, 15].

The breakthrough in oximetry technology came with the introduction of a lock-in
detection scheme: the photodiode detects a temporally oscillating signal caused by the
pulsating arterial blood flow. As soon as the heartbeat frequency is determined and

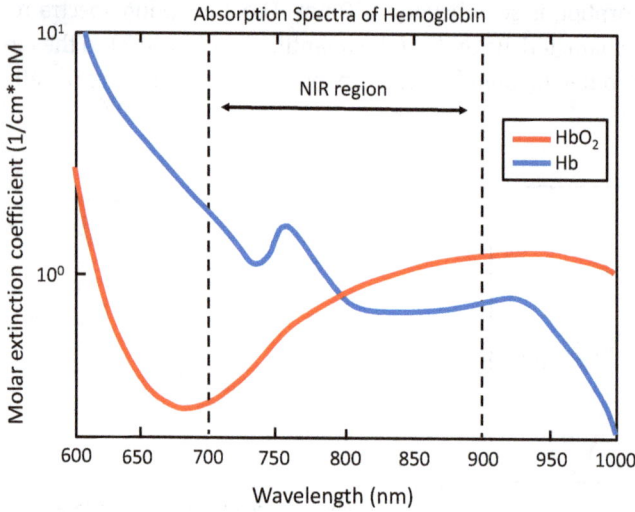

Fig. 8.37: Extinction coefficient of oxygenated (red) and deoxygenated (blue) hemoglobin plotted as a function of wavelengths in the near infrared regime from 600 to 1000 nm. The vertical dashed lines indicate the wavelength at which LEDs in oximeters determine the transmitted light.

stable, the red LED measures the intensity ratio at the top and bottom of the pulsed intensity. The same is repeated with the infrared LED. The bottom intensity level, indicated by the dashed line in Fig. 8.38, is due to the nonpulsating arterial and venous blood volume and to the absorption in the tissue. Only the pulsatile fraction is representative of the arterial oxygen level. Therefore, the intensity below the dashed line is subtracted and only the peak intensities are used to determine the oxygen concentration. The measurements are repeated at high frequency, and an average is taken. With this pulse-locked technique, previous problems with weak signals and noise could be eliminated, which is why this method is called "pulse oximeter."

Fig. 8.38: Schematics of finger positioning, light-emitting diodes, and photodetectors in a pulse oximeter. Pulsatile arterial blood flow arrives from the left. Red dots indicate oxygen-saturated hemoglobin; blue dots refer to deoxygenated hemoglobin and venous blood, gray dots indicate absorption in other parts of the finger, like skin and tissue.

> **!** The absorption spectrum of blood in the visible and infrared regime is characteristically different for oxygenated and deoxygenated hemoglobin. Phase-locking to the pulsatile blood flow allows determining the blood oxygen concentration.

Infobox V: Absorption or absorbance: what is the difference?

In the previous paragraph we used both terms equally. Absorption is clearly defined as the elimination of particles by transmission through some material. For example, photons can be absorbed (eliminated) by the photoelectric effect and converted into electrons and energy. The term "absorbance" is synonymous with the term "attenuation." Attenuation implies that the incident beam is weakened by passing through a material either by absorption or scattering, or both. The terms "attenuation" or "absorption" are collective terms for both effects and are always used when the cause of the attenuation is not analyzed further. In the biophysical literature, the term "absorbance" is more popular than "attenuation," which is more commonly used in the physics literature.

8.8 Cardiopulmonary bypass intervention

8.8.1 Cardiopulmonary bypass

Surgery on the heart, pulmonary arteries, or large vessels is usually impossible while the heart is beating. The heart and lungs are removed from the circulatory system during such surgical interventions and brought to rest. Their functions are then temporarily taken over by an extracorporeal circuit, also known as *cardiopulmonary bypass* (CPB) or extrapulmonary ventilation [16].

Surgery on the surface or inside the heart requires extracorporeal circulation, especially for the repair or replacement of heart valves, the correction of congenital heart failure, the repair of bypasses of the coronary arteries, heart replacement, the removal of pulmonary thromboses, and any surgery on the aorta. Open heart surgery is performed only if no other minimally invasive procedure promises success.

For the duration of the operation, which can take several hours, the heart and lung activity is intentionally and temporarily immobilized. This artificial cardiopulmonary arrest is achieved by injecting cardioplegic substances such as K^+, Mg^+ acetylcholine, and neostigmine and injecting a sodium-free solution. The Na^+ ion current is the largest ion current in the heart. By reducing the Na^+ ion concentration, the rapid depolarization from the sinoatrial node is stopped immediately.

CPB is a method that bypasses either all or part of the bloodstream and redirects the blood in a machine called a *heart–lung machine* (HLM). The HLM takes over the circulation at the correct frequency and blood pressure, provides gas exchange (O_2/CO_2), and filters the blood [17]. In the end, the blood is returned to the arteries at body temperature and volume rate. The HLM was first used by Gibbon[25] in 1953 and has since been used successfully in numerous heart surgeries, at least half a million a year worldwide [18].

---

25 John Heysham Gibbon (1903–1973), American surgeon and inventor of the heart-lung machine.

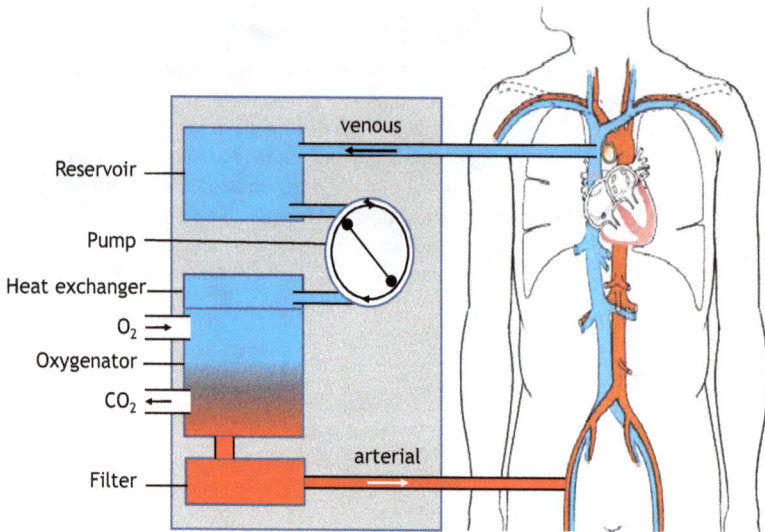

Fig. 8.39: Schematic overview of the main components of a heart–lung machine.

The main components of the HLM are indicated in Fig. 8.39 and consist of:
1. Reservoir for collecting the blood from the veins
2. Roller pump taking over the function of the heart
3. Heat exchanger controlling the blood temperature
4. Oxygenator taking over the function of the lung
5. Filter for the blood

These components are described in more detail as follows.

Before connecting the patient to a CPB circuit, the entire circuit must be expelled from the air in all parts, particularly the arterial cannula that are reconnected to the patient.

A reservoir is used to collect the venous blood from the vena cava by gravity or by suctioning the blood. Blood that occurs during the operation is also sucked out and collected in the reservoir. Various filters and a defoamer are integrated into the reservoir to remove debris and gas bubbles. Medicines such as anticoagulants are also added to prevent blood clotting in the circulatory system. Cardioplegia suppress heartbeats and lower myocardial metabolism to protect heart tissue.

The pump in the HLM is used to maintain the circuit at the correct pressure and flow rate and to circulate medication. The requirements for pumps in the HLM are strict. They have to work reliably and as simply as possible, without damaging the blood cells, and at the same time deliver an exact pump performance and an exact pressure. Two types of pumps are generally used: roller pumps and centrifugal pumps. Roller pumps push the blood in pulsed fashion through flexible hoses.

Centrifugal pumps create blood flow using centrifugal forces instead of mechanically squeezing the blood. Hence, it is believed to be superior to roller pumps and cause less blood damage. Both working principles are shown schematically in Fig. 8.40.

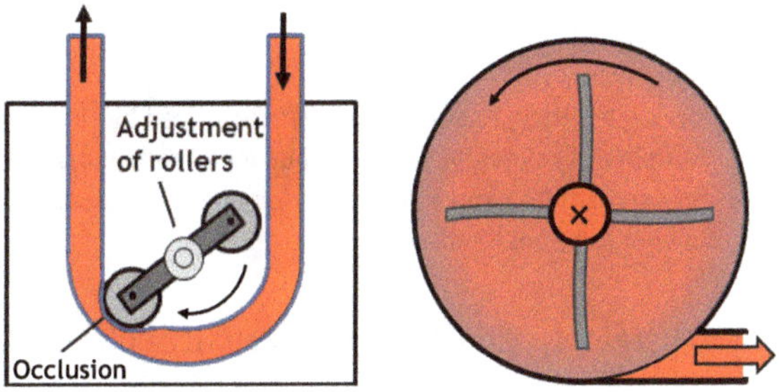

Fig. 8.40: Working principle of roller pump (left) and centrifugal pump (right). In the case of the centrifugal pump, blood flows in from the top and leaves on the side after being circulated by a wheel in a round box.

The task of the heat exchanger is to lower the patient's temperature during the operation and to warm up the body after the hypothermia has ended. The body cooling allows more time for the operation without causing brain damage. The heat exchanger is also used for operations at normal body temperatures to compensate for heat losses.

As the blood is warmed up, the gaseous solution decreases and microbubbles form. Therefore, in the CPB, the heat exchanger is installed before the oxygenator; otherwise, microbubbles' formation during warming up would be even greater.

8.8.2 Oxygenator

The oxygenator is the most important part of the HLM, the schematics is shown in Fig. 8.41. The oxygenator takes over the task of the lungs: oxygen supply and CO_2 removal. In the oxygenator, the blood flows in the opposite direction to the gas flow separated by microporous fibers. The gas is a mixture of air and oxygen. The gas exchange takes place at a membrane separating gas and blood. According to the same principle as in the lungs, the gas exchange follows the concentration gradients for O_2 and CO_2. The deoxygenated dark red blood enters the oxygenator and leaves the oxygenator in a bright red color after being due to the oxygen enrichment. The filter membrane is made of microporous polypropylene carbon fibers. The lower limit of the pore size in the filter is determined by the size of the erythrocytes and by the maximum flow resistance that can be allowed

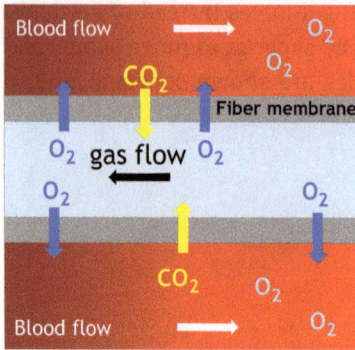

Fig. 8.41: Reoxygenation of blood is achieved by diffusion of oxygen across a porous membrane located between a tube separating O_2 enriched air and blood, flowing in opposite directions.

to maintain the required flow rate. Therefore, the pore size is usually chosen to be 20–40 μm.

Reanimation of the heart occurs when the heart comes back into contact with blood after opening the clamp on the aorta. It takes another 30–60 min for the heart to recover and before it can resume its self-stimulated pumping ability.

Operations that require CPB are the most serious surgical interventions that can cause a number of related problems, such as hemolysis, blood clotting in the circulatory system, air embolism, and breathing problems. Problems with the biocompatibility of tubes and walls within the HLM can lead to blood clotting. This issue may partially be overcome by using rough inner walls in all parts in contact with blood. In the case of rough walls, the adhesion and adherence of blood to the walls is improved so that the circulating blood only comes into contact with a sacrificial layer of blood adsorbed on the rough surface.

In recent years, so-called miniaturized heart-lung devices (mini-CPB) or miniaturized extracorporeal circulation have been developed. They are mainly used for emergency applications and postoperative treatments. Studies have shown that these miniaturized devices also positively affect reducing inflammatory, coagulopathic, and hemodilution side effects compared to normal size HLMs in CPB surgery [19]. Based on the positive experience to date, mCPBs have the potential to replace HLMs of normal size in the long term. A comparative study of both techniques is published in Ref. [20, 21].

8.9 Summary

S8.1 The heart contains two separate but not independent pumps.

S8.2 Right and left ventricle work synchronously and have to pump the exact same amount of blood, which is about 15 000 l/day.

S8.3 The right ventricle pumps deoxygenated blood from the body to the lungs at low pressure. The left ventricle pumps oxygenated blood from the lungs to the body at high pressure.

S8.4 The pumping cycle consists of four phases: inflow phase or isobaric filling phase, isovolumetric contraction phase, outflow phase or isobaric ejection phase, and isovolumetric relaxation.

S8.5 · The mechanical power of the heart is about 1.25 W with a cardiac output of approximately 90 ml/s. The total power consumption of the heart is about 5–6 W.

S8.6 The efficiency of the heart is about 25%.

S8.7 The pressure difference between arterial and venous side during a systole is about 140 hPa.

S8.8 The Windkessel suppresses turbulence of the aorta and stores potential energy during ejection. The potential energy is converted back into kinetic energy during the diastole phase.

S8.9 The elastic properties of blood vessels for high-pressure (arteries) circulation are distinctly different from blood vessels for low-pressure (veins) circulation. Veins have a high elastic compliance but low yield; arteries have a low compliance but high yield.

S8.10 Blood flow into the arteries is pulsatile.

S8.11 The pulsatile blood flow can be used for measuring the blood pressure according to the method of Riva-Rocci.

S8.12 The PWV is higher than the mean blood flow velocity in the arteries.

S8.13 The PWV is proportional to the elastic modulus of blood vessels.

S8.14 The mean blood flow resistance is about 100 kPa s/l, which is rather low and due to the large branching in the capillary bed.

S8.15 Blood consists of plasma and mostly RBC. The volume fraction of RBCs is called hematocrit and is about 40%–45%.

S8.16 Blood is a non-Newton fluid. The viscosity of blood decreases with increasing shear rate.

S8.17 RBCs have no ion channels but have pores for water exchange. The osmotic pressure inside and outside must be balanced (isotonic conditions).

S8.18 RBCs are carriers of oxygen and of carbon dioxide. Oxygen binds to the Fe^{2+} in heme molecules, sitting in a pouch surrounded by amino acid chains.

S8.19 Upon binding of oxygen, the Fe^{2+} ion undergoes a high spin–low spin transition, which is the basis for functional MRI.

S8.20 Fe^{3+} is stored in ferritin. Ferritin is an important biocompatible hollow nanoparticle formed by proteins.

S8.21 The absorption spectrum in the visible regime is characteristically different for oxygenated and deoxygenated hemoglobin explaining the color of blood.

S8.22 The pulsatile blood flow combined with light absorption in the infrared regime can be used to determine the oxygen concentration with the help of a pulse oximeter.

S8.23 Heart–lung machines take over the pumping activity of the heart and the gas exchange of the lung during surgery at the open heart or at large vessels.

? Questions

Q8.1 The circulatory system can be subdivided in which main blood circulations?

Q8.2 What is the main task of the circulatory system?

Q8.3 How can veins and arteries be distinguished?

Q8.4 Can veins and arteries be distinguished by their oxygen level?

Q8.5 How many valves does a heart contain?

Q8.6 How can the four main phases of the heart cycle be characterized?

Q8.7 What is the ejection fraction in percent?

Q8.8 What do you understand by cardiac output?

Q8.9 What is the total power consumption of the heart and its efficiency?

Q8.10 How big is the pressure difference between the arterial and venous side of the circulatory system?

Q8.11 Describe the potential variations measured at the Einthoven lead I (see Chapter 7) with respect to the cardiac cycle.

Q8.12 Which blood vessels have the higher compliance, veins, or arteries?

Q8.13 Is the circulatory system a good example of classical hydrodynamics? If not, why?

Q8.14 How can the rather low total peripheral flow resistance be understood if the capillaries deliver a high flow resistance?

Q8.15 What is the purpose of the Windkessel?

Q8.16 Which information can be gained from the PWV?

Q8.17 Why is it important to keep an isotonic osmotic pressure in blood?

Q8.18 What is the main structure of hemoglobin?

Q8.19 How does the high-spin–low-spin transition come about and what is it important for?

Q8.20 Why do myglobin and hemoglobin have different oxygen binding curves?

Q8.21 What is the task of ferritin?

Q8.22 What are the main absorption bands of hemoglobin?

Q8.23 Which spectral region determines the color of blood?

Q8.24 What is the task of a CPB.

Attained competence checker	+	0	–	⚡
I can name the four phases of the heart cycle.				
I know what the cardiac output of the heart is per day.				
I know what the mechanical power and the efficiency of the heart is.				
It is clear to me why the cross section of veins and arteries is different.				
I can distinguish between laminar flow and turbulent flow.				
I realize that in viscous liquids, the resistance increases with the length of the tube and is inversely proportional to the fourth power of its radius.				
I know that the total resistance of a branching tubular system can be lowered with increasing the number of branches.				
I know what the limit is for the onset of turbulent flow according to the Reynolds number.				
I realize that blood is a non-Newtonian liquid.				
I know the reasons why the saturation of oxygen in hemoglobin is S-shaped.				
I know that the magnetic properties of the Fe^{2+} ion can be used for functional MRI.				
I know that with the help of a pulse oximeters the oxygen saturation can be determined in vivo.				

Suggestions for home experiments

HE8.1   Try to observe the erythrocytes in an optical transmission microscope. You need at least a magnification of 400× or better.

Exercises

E8.1   **Number of blood cells**: Blood cells make up about 20% of the total number of cells in the body. How many blood cells do we have and how many cells in total are in the average body with a blood volume of 5 l.

E8.2   **Compliance of the aorta**: The aorta's "wind chamber function" is based on its compliance (passive elasticity of the aortic wall). Among other things, it is responsible for dampening the cardially generated pulse wave. With 150 ml of blood in the "air chamber" there is a transmural pressure of 10 kPa, with 180 ml it is 15 kPa. How big is the compliance of the aorta in this pressure range?

E8.3   **Flow rate**: By selecting a proper stent for a patient it turned out that the flow velocity in the stent has to be twice as big as compared to the velocity without stent in order to keep the volume flow rate I_V constant. What is the ratio of the radii of the vessels with and without stent?

E8.4   **Bypass**: A patient with an acute heart attack is given a bypass to relieve the constricted blood vessel. The bypass has the same length but twice the radius as the diseased blood vessel. How many times greater is the volume flow rate in both vessels together compared to the case without bypass, assuming the same pressure difference?

E8.5 **Giraffe:** What is the minimum systolic blood pressure that a giraffe with a head 2.5 m above the shoulder has to build up so that its head is supplied with blood?

E8.6 **Critical velocity:** What is the critical velocity at which turbulence can occur in the aorta? You either know all the necessary numbers needed or you can find them in this chapter.

E8.7 **Frank's differential equation:** In eq. (8.33), the right side is set to zero. What is the consequence of this assumption and what can/cannot be described with the remaining differential equation.

E8.8 **Number of RBCs:** The RBC volume is 100 fl. How many RBC are in 1 l of blood?

E8.9 **Foramen oval:** The heart of a fetus has an opening between the left and right ventricles that does not close until a few months after birth, called a foramen oval.

a. Schematically draw the functional units of the heart of a fetus with the location of the foramen oval;

b. Provide a reason why foramen oval makes biological sense.

E8.10 **Isotonic solution:** Explain how one can produce an isotonic osmotic pressure with a 0.9% saline solution.

E8.11 **Osmosis:** To care for a patient, 6 g of glucose $C_6H_{12}O_6$ are added to a physical isotonic saline solution of one liter of water. The molecule does not dissociate in water, but is osmotically effective. How much water still has to be added to achieve again an isotonic solution.

E8.12 **Reproduction rate:** Erythrocytes have life cycle of about 100 days. Assume that a person has 6 l of blood with a number density of $5 \times 10^6/\mu l$. How large must the reproduction rate be (number of erythrocytes per second) so that the total number of erythrocytes remains constant?

E8.13 **Total oxygen content of blood:** Verify the statement: in saturation, each heme molecule carries one O_2 molecule. This makes $4 \times 2.5 \times 10^8 = 10^9$ per erythrocyte and $5 \times 10^{21}/l$ of blood. Assuming that 2 l are fully oxygenated on average, and the remaining 3 l have an oxygen content of 75% after expiration, then the total oxygen content of blood is 0.03 mol oxygen or 6 mmol/l blood.

E8.14 **Venous blood flow:** The venous blood vessels in the lower limbs have a mechanical problem. The blood must rise to the heart and not flow back down due to gravity. Do you know how nature has solved this problem? Please write a short essay on this topic and suggest a solution to this problem.

Veins with right and wrong flow directions.

E8.15 **Ultracentrifuge:** In his experiment, Sbedberg used 12 000 rpm and generated a centrifugal force 7000 times the earth gravitational acceleration g. What was the radius during the centrifugation?

References

[1]   Guyton AC, Hall JE. Textbook of medical physiology. 11th edition. Philadelphia: Elsevier
      Saunders; 2006.
[2]   Jacob R, Dierberger B, Kissling G. Functional significance of the Frank-Starling mechanism
      under physiological and pathophysiological conditions. Eur Heart J. 1992; 13: 7–14.
[3]   Knaapen P, Germans T, Knuuti J, Paulus WJ, Dijkmans PA, Allaart CP, Lammertsma AA, Visser
      FC. Contemporary reviews in cardiovascular medicine, myocardial energetics and efficiency:
      Current status of the noninvasive approach. Circulation. 2007; 116: 434–448.
[4]   Milnor WR. Hemodynamics. Baltimore: Williams & Wilkins; 1982.
[5]   Hellevik LR. Cardiovascular biomechanics. Script NTNU; 2018. https://folk.ntnu.no/leifh/
      teaching/tkt4150/._main026.html
[6]   Burman ED, Keegan J, Kilner PJ. Pulmonary artery diameters, cross sectional areas and area
      changes measured by cine cardiovascular magnetic resonance in healthy volunteers. J
      Cardiovasc Magn Reson. 2016; 18: 12(1)–12(10).
[7]   Perutz MF, Rossmann MG, Cullis AF, Muirhead H, Will G, North ACT. Structure of
      haemoglobin: A three-dimensional Fourier synthesis at 5.5-Å resolution, obtained by X-ray
      analysis. Nature.1960; 185 (4711): 416–422.
[8]   Langmuir, I. The adsorption of gases on plane surfaces of glass, mica and platinum. J Am
      Chem Soc. 1918; 40: 1361–1403.
[9]   Hill AV. The possible effects of the aggregation of the molecules of hæmoglobin on its
      dissociation curves. J Physiol 1910; 40(Suppl): iv–vii.
[10] Weiss JN. The Hill equation revisited: Uses and misuses. FASEB J. 1997; 11: 835–841
[11] Theil EC. Ferritin: Structure, gen regulation, and cellular function in animals, plants and
      microorganisms. Annu Rev Biochem. 1987; 56: 289–315.
[12] Pittman RN. Regulation of Tissue Oxygenation. San Rafael (CA): 2nd Edition. San Rafael, CA,
      USA: Morgan & Claypool Life Sciences; 2016.
[13] Quaresima V, Ferrari M. COVID-19: Efficacy of prehospital pulse oximetry for early detection
      of silent hypoxemia. Crit Care. 2020; 24: 501.
[14] Cannesson M, Desebbe O, Rosamel P, Delannoy B, Robin J, Bastien O, Lehot JJ. Pleth
      variability index to monitor the respiratory variations in the pulse oximeter plethysmographic
      waveform amplitude and predict fluid responsiveness in the operating theatre. Br J Anaesth.
      2008; 101: 200–206.
[15] https://web.archive.org/web/20150318054934/http://www.oximetry.org/pulseox/principles.htm
[16] Sarkar M, Prabhu V. Basics of cardiopulmonary bypass. Indian J Anaesth. 2017; 61:760–767.
[17] https://en.wikipedia.org/wiki/Cardiopulmonary_bypass
[18] Edmunds LH Jr. Advances in the heart-lung machine after John and Mary Gibbon. Ann Thorac
      Surg. 2003; 76:S2220–S2223.
[19] Harling L, Punjabi PP, Athanasiou T. Miniaturized extracorporeal circulation vs. off-pump
      coronary artery bypass grafting: What the evidence shows? Perfusion. 2011; 26(Suppl):
      40–47.
[20] Pereira SN, Balta Zumba I, Sulzbacher Batista M, Da Pieve D, dos Santos E, Stuermer R,
      Pereira de Oliveira G, Senger R. Comparison of two technics of cardiopulmonary bypass
      (conventional and mini CPB) in the trans- and postoperative periods of cardiac surgery. Braz J
      Cardiovasc Surg. 2015; 30: 433–442.
[21] Liebold A, Albrecht G. Minimized extracorporeal circulation in non-coronary surgery. J Thorac
      Dis. 2019; 11(Suppl 10): S1498–S1506.
[22] Faller A, Schünke M, Schünke G. The human body. An introduction to structure and function.
      Stuttgart, New York: Thieme Verlag; 2004.

Further reading

Feher J. Quantitative human physiology: An introduction. Amsterdam, Boston, Heidelberg, etc:
    Academic Press imprint of Elsevier; 2012.
Rhodes RA, Bell DR. Medical physiology. 3rd edition. Philadelphia, New York, London: Wolters
    Kluwer, Lippincott Williams and Wilkins; 2009.
Boron WF, Boulpaep EL. Medical physiology. 2nd edition. Philadelphia, London, New York:
    Saunders W.B. Elsevier; 2012.
Blundel S. Magnetism in condensed matter. Oxford, New York, Athens: Oxford University Press;
    2001.
Klabunde RE. Cardiovascular physiology concepts. 3rd edition. Philadelphia, Baltimore, New York:
    Wolters Kluwer; 2021

9 Physical aspects of the respiratory system

Physical properties of the respiratory system	
Diameter of alveoli	75–300 µm
Branching generations	23
Number of alveoli in both lungs	3×10^8
Average surface area of all alveoli	80 m^2
Total air volume of alveoli	5–6 l
Weight of one lung	500–600 g
Oxygen uptake by the lungs	200–250 ml O_2/min or 360 l/day.
Oxygen concentration in blood	6–10 mmol O_2/l blood
Minute ventilation	8 l/min
Tidal volume	0.5 l
Vital capacity	3–4.5 l
Compliance of lung/thorax	1–2 l/kPa
Airway flow resistance	600 Pa · s/l
Alveolar pressure difference	−200 Pa to 200 Pa
Breathing frequency at rest	16/min
Time constant for inspiration-expiration	0.75 s

9.1 Introduction

Heartbeat and breathing are the most visible vital signs of the body. Both the heart and the lungs have their own rhythm and both are self-stimulating. A comparison of the pulsatile arterial blood pressure and the pleural cavity pressure variation is shown in Fig. 9.1 as a function of time. The frequency ratio is roughly 4:1, or 80 heartbeats per minute correspond to roughly 20 respiratory cycles per minute at rest. One might have the impression that both organs work independently. However, the respiratory and cardiovascular systems are closely related. Breathing supplies the circulation with oxygen from the environment. Hemoglobin in the erythrocytes is the oxygen carrier to the tissue wherever oxygen is needed to generate energy. During the recirculation process, CO_2 is absorbed as a metabolic end product and exhaled into the environment. This loop path is necessary because direct diffusion of oxygen to the cells in the organs is not possible in our body, in contrast to small animals such as insects. In small animals, the gas exchange may occur through the skin without special breathing membranes, known as cutaneous respiration. In larger animals, oxygen diffusion is not fast enough for sufficient oxygen supply. Therefore an oxygen rich conveyer fluid is necessary to reach all body parts.

Considering the respiratory system in humans from a physical point of view, the gas exchange, the elastomechanics of the chest, and the airways' aerodynamics are the central points of interest. First, we introduce some basic anatomical and physiological facts before discussing the physical mechanism of the respiratory system.

https://doi.org/10.1515/9783110756951-009

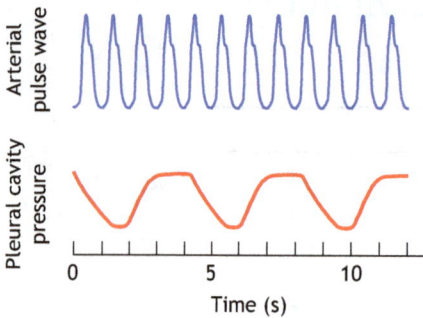

Fig. 9.1: Comparison of the heart (blue) and lung (red) rhythms according to the arterial pulse pressure and the pleural cavity pressure, respectively, plotted over time. The frequency ratio f_{heart}/f_{lung} is roughly four.

9.2 Organs of the respiratory system (overview)

The respiratory system consists of the following organs from top to bottom [1–5], which are sketched in Fig. 9.2:

1. *Oral cavity*: The nose and *oral cavity* allow for inhalation of air.
2. *Pharynx:* The pharynx is the common pathway above the epiglottis of the digestive and respiratory systems. It receives air from the nasal passage and air, food, and liquids from the mouth. The pharynx is connected to the respiratory system at the larynx and to the digestive system at the esophagus.
3. *Epiglottis:* Air and food, both passing through the pharynx, are separated at the *epiglottis*. The epiglottis is like a two-way switch: it either opens for air to enter the trachea or closes the trachea entrance and opens the esophagus for swallowing food. Muscles control the folding of the epiglottis: either upright for breathing or touching the base of the tongue, thereby blocking the glottis and directing food to the esophagus.
4. *Larynx:* The *larynx* is part of the trachea and is a gatekeeper for the lower part of the respiratory ducts and the vocal cords. The larynx consists of an assembly of cartilage connected to joints, ligaments, and membranes. The larynx, containing the voice box (glottis), can regulate the position of the cartilage and the tension of the vocal cords. It is the organ that makes the sound. Larynx and glottis are again topics of Section 12.10 on the origin of the sound.
5. *Trachea, bronchi, alveoli:* The *trachea* branches off in the left and right *bronchus*, conducting air to the left and right lungs. The primary bronchi branches further out in many more secondary and tertiary bronchi. Air is channeled through bronchioles into terminal alveoli, where the O_2-CO_2 gas exchange takes place. About 23 generations of branching from the trachea to the final bronchioles exist. The final branching and terminating alveoli are indicated in the inset of Fig. 9.2. The diameter of the alveoli ranges from 75 to 300 μm. There are about 3×10^8 alveoli in both lungs with an average surface area of approximately 80 m² and a total volume of

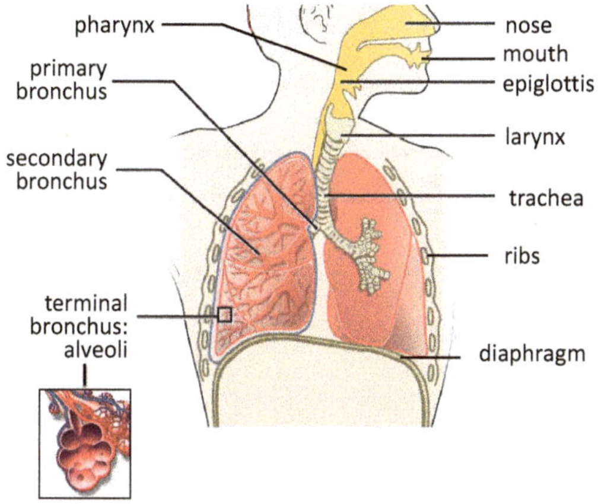

pharynx — nose
— mouth
primary bronchus — — epiglottis
— larynx
secondary bronchus —
— trachea
— ribs
terminal bronchus: alveoli —
— diaphragm

Fig. 9.2: Overview of the respiratory system, indicating their main organs from the upper parts (mouth, nose, pharynx, larynx) to the lower parts (bronchus, lung, diaphragm). Bronchioles terminate in spherical air sacs called alveoli, where O_2-CO_2 gas exchange occurs. Capillaries surround the alveoli.

roughly 5–6 l. These numbers are mean values; actual numbers depend on body size and sex [1, 3].

6. *Thorax cavity:* The lungs are the central organ of the respiratory system. Concerning volume and weight, they are the largest organ in the body. Each lung weighs about 500–600 g. The lungs are contained in the *thorax cavity* consisting of the *thoracic wall* and the *diaphragm* at the lower end. The thoracic wall is made up of thoracic vertebrae, ribs, intercostal muscles, and breast bone (sternum). The diaphragm is a large dome-like muscle separating the thoracic cavity from the abdominal cavity. The muscles associated with the thorax cavity are responsible for respiration, as we will see later.

7. *Pleural cavity*: The lungs would not be ventilated without interaction of the thorax with the pleural cavity surrounding the lungs and maintaining a reduced pressure of about − 500 Pa to the environment. Each lung is enclosed by a separate pleural cavity formed by a membrane, as shown in Fig. 9.3. The membrane is folded back onto itself, forming a pleural sac with an inner wall covering the surface of the lung (visceral pleura) and an outer wall surrounding the inner thoracic wall (parietal pleura). The pleural cavity is filled with a pleural fluid that serves as a lubricant between the gliding inner and outer walls when the lungs change shape during respiration. The pleural fluid also holds the two membranes together when the thoracic volume changes. In case of a rapture of the pleural cavity, the negative pressure is lost, the lungs will collapse, and respiration is no longer possible. The condition is known as pneumothorax. Since the pleural cavities of the left and right lungs are not connected, the rapture of one cavity will not affect the other, and therefore, limited breathing with one lung will still be possible.

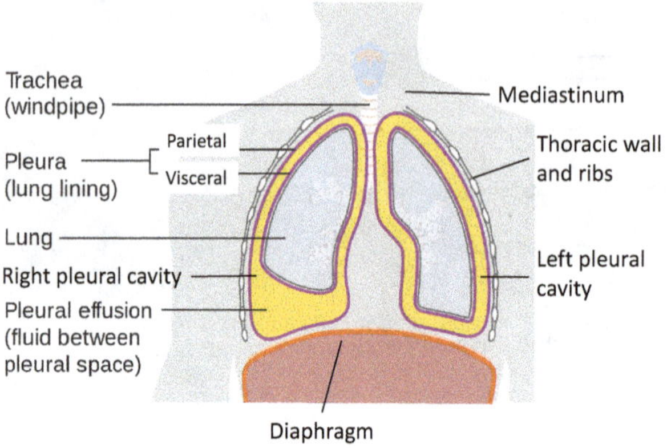

Fig. 9.3: The thoracic cavity consists of a thoracic wall on the sides and a diaphragm at the bottom. A pleural cavity encloses each lung. Left and right pleural cavities are separated by the mediastinum formed by the heart, trachea, esophagus, and associated structures. Adapted from Wikipedia. https://en.wikipedia.org/wiki/Pleural_cavity.

Within the respiratory system, the lungs are responsible for gas exchange and the blood's pH level, heat exchange, moistening the inflow of air, control of airflow through the vocal cords for articulation, and cleaning inhaled air from dust and contaminations. However, the dust and smoke filter is quickly exhausted by overload from industry, traffic, and tobacco smoking, resulting in hindered gas exchange in the alveoli. Severe diseases such as silicosis and bronchial carcinoma may result from such adverse effects. A concise and up-to-date overview of the lungs and thorax anatomy is given in [4].

9.3 Gas exchange

Blood is pumped from the right atrium into the pulmonary circuit at low pressures of about 30 hPa. The body holds about 5–6 l of blood; about 1 l is always present in the lung. This blood volume spreads out over the capillary bed in the lung. In the capillary bed, three actions take place:

1. **Perfusion:** blood flows from the pulmonary circuit to the capillary bed and back again; hemodynamics controls perfusion.
2. **Ventilation:** air reaches the alveolar surface in the periodic rhythm of inspiration and expiration; aerodynamics controls ventilation.
3. **Diffusion:** gas exchange across the membrane walls of the blood vessels and the alveoli walls occurs by diffusion, following a concentration gradient and pressure difference.

Perfusion is discussed in Chapter 8 on the circulatory system, ventilation is presented in Section 9.4, and gas exchange by diffusion is the topic of this section.

Gas exchange in the lungs operates by means of three physical methods: perfusion, ventilation, and diffusion. !

9.3.1 Partial pressures of the atmosphere

First, we analyze the gases and their respective partial pressures on both sides of the membrane. The pressure difference will then control the gas exchange by diffusion across the barrier.

We start with our standard atmospheric conditions. If n_i is the mole number of one of the gas components i, then the sum of all components yields the mole number n of the air:

$$n = n_1 + n_2 + n_3 + \ldots = \sum_i n_i. \tag{9.1}$$

According to the general gas law, also known as Dalton's law,[1] these mole components generate a total pressure:

$$p_{tot} = \frac{RT}{V_{gas}} n = \frac{RT}{V}(n_1 + n_2 + n_3 + \ldots) = p_1 + p_2 + p_3 + \ldots. \tag{9.2}$$

Here R $(= 8.314\,\text{J/K}\cdot\text{mol})$ is the gas constant, T is the absolute temperature, and V is the volume filled with gas. Therefore, each component has the partial pressure:

$$p_i = \frac{RT}{V_{gas}} n_i = \frac{n_i}{(n_1 + n_2 + n_3 + \ldots)} p_{tot} = y_i p_{tot}, \tag{9.3}$$

and y_i is the *mole fraction* of the component i. Our atmosphere has the following mole fractions for the respective gases:

$$y_i = 78\% \ N_2, \ 20.95\% \ O_2, \ 0.93\% \ Ar, \ 0.03\% \ CO_2.$$

Assuming *standard ambient temperature and pressure* conditions (SATP), defined by 293.15 K (20 °C) and absolute pressure of 1000 hPa, the partial pressures of the gases are accordingly:

$$p_{O_2} = 210\,\text{hPa}, \ p_{N_2} = 780\,\text{hPa}, \ p_{Ar} = 930\,\text{Pa}, \ p_{CO_2} = 30\,\text{Pa}.$$

So far, we have neglected the pressure produced by water vapor. At 20 °C and relative humidity of 50%, the water vapor pressure is 12 hPa. All mole fractions and partial pressures of the air are listed in Tab. 9.1.

---

1 John Dalton (1766–1844), English chemist.

Tab. 9.1: Mole fractions and partial pressures of the air components at normal SATP conditions.

Gas component	Mole fraction [%]	Pressure [hPa]
N_2	78.00	780.00
O_2	20.95	209.5
Ar	0.93	9.3
CO_2	0.03	0.3
H_2O	1.20	12.0

9.3.2 Alveolar pressures

Now we consider the pressure conditions for inspiration and expiration. During inspiration, the air volume inhaled, called the *tidal volume*, is about 0.5 l of fresh air at a partial oxygen pressure of $p_{i,O_2} = 210$ hPa. In the alveolar space, the fresh air is mixed with 2 l of alveolar residual air. Due to this mixing, the oxygen partial pressure drops to $p_{A_{O2}} = 130$ hPa. At the same time, the CO_2 partial pressure in the alveolar space is essentially zero from inspiration but increases to 50 hPa during expiration. This CO_2 pressure rise is due to the metabolic CO_2/O_2 conversion. Recalling the respiratory exchange rate *RER* discussed in Section 4.4, the CO_2/O_2 conversion is 1 for metabolizing carbon hydrates and about 0.8 for burning fat and proteins. Therefore, the expirational CO_2 pressure is higher for metabolizing carbon hydrates than for other food items. After a meal, carbon hydrates are digested first, followed by the other foods. Correspondingly, the expirational CO_2 pressure drops gradually from 50 to about 40 hPa. Hence, the expirational CO_2 pressure is about 100 times higher than the environmental CO_2 pressure.

The mole fractions and partial pressures during inspiration and expiration, together with the conditions within the alveoli are listed in Tab. 9.2. According to this table, the pressure difference between the inspirational oxygen gas and the oxygen pressure in the alveolar space is: $p_{i,O_2} - p_{A_{O2}} = 210 - 130\,\text{hPa} = 80\,\text{hPa}$. This pressure difference should equal the partial pressure of CO_2 in the alveoli air space. However, $p_{A_{CO_2}}$ is only 53 hPa. The difference is due to the partial pressure of the moisture in the air.

Tab. 9.2: Mole fractions and partial pressures during respiration. Values from [1].

	%O_2	p_{O2} [hPa]	%CO_2	p_{CO2} [hPa]
Inspirational air in the trachea	21	210	0.04	0.4
Alveoli air space	14	130	5.6	53
Expirational air in the trachea	16	150	4.5	43

9.3.3 Solubility of gases in water

The pressure difference on both sides of the membrane between alveoli and arterioles controls the direction of gas diffusion across the barriers. Furthermore, the partial pressure of oxygen in the alveoli governs the solubility of oxygen gas in the blood according to *Henry's law*:[2]

$$S_i = \frac{n_i}{V_{liquid}} = K_H p_i. \tag{9.4}$$

Here K_H is Henry's constant and p_i is the partial external pressure of the gas component *i*. This equation shows that the partial gas pressure p_i controls the solubility $S_i = n_i / V_{liquid}$ of this specific gas component *i* in the liquid. The unit of K_H is [K_H] = mol/1 × Pa. Tab. 9.3 gives an overview on some solubilities of gases in water [6]. Tab. 9.3 also shows that the solubility of oxygen in water is comparatively low, about a factor of 26 less than the solubility of CO_2 in water. These considerations also apply to oxygen solubility in lakes, rivers, and the ocean. When the oxygen solubility in water drops from a normal value of 2.7×10^{-4} mol/l to less than half, life in water, which depends on oxygen, ceases. Conversely, the high solubility of CO_2 shows that

Tab. 9.3: Henry constants and solubility of gases in water according to their partial pressures in the atmosphere. K_H values are reproduced from [6].

	K_H [10^{-6} mol/l × hPa]	S [10^{-4} mol/l]
Ar	1.5	0.14
CO_2	34	0.07
N_2	0.6	4.7
O_2	1.3	2.7

---

2 William Henry (1774–1836), English chemist.

water is an effective store for carbon dioxide, which leads to a lower CO_2 concentration in the atmosphere than would otherwise be the case.

> **!** The partial pressure of O_2 in the alveolar space is only 130 hPa, 35% less than in the atmosphere. The solubility of CO_2 in water is much higher than of O_2 by a factor of 26.

i **Infobox I: Deep water diving**

Deep water diving requires careful preparation and a number of safety precautions. At a depth of up to 30 meters, the hydrostatic pressure is 4 bar. Three bars from the water level above and another bar from the atmospheric pressure. Therefore, the air for inspiration must be offered at a pressure of 4 bar. At 4 bar, however, the solubility of N_2 in the blood increases by a factor of 4. During the ascent, N_2 bubbles form, which burst and destroy the vessels. Therefore, helium is added to the air, which has a lower solubility in the blood and reduces the risk of blistering. In addition, at a pressure of 4 bar, the inspiratory gas volume is four times higher than at normal atmospheric pressures. Exhaling is more difficult, which may lead to increased levels of carbon dioxide in the blood and eventually blood poisoning.

9.3.4 Solubility of oxygen in blood

Let us assume that the solubility of oxygen in blood plasma is the same as in normal water. Then at a partial pressure of 130 hPa, corresponding to the alveolar partial pressure of oxygen during inspiration, the oxygen concentration in the blood plasma would be 2.7×10^{-4} mol/l, which is only 4% of what is needed to support the body's oxygen consumption. If physical solubility were the only mechanism for supplying oxygen to the cells, the heart would have to pump about 150 l/min of blood instead of the actual 6 l/min. From this estimate, we conclude that there must be some other active mechanism to increase the oxygen level in the blood. The other mechanism is the chemical binding of oxygen to hemoglobin according to the reaction:

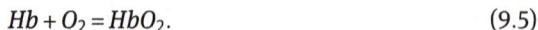

$$Hb + O_2 = HbO_2. \tag{9.5}$$

The bonding process of oxygen to hemoglobin is described in Section 8.6.2. With chemical bonding of oxygen to hemoglobin in the erythrocytes (four O_2 molecules per hemoglobin), 98% of saturation is already reached at an alveolar pressure of

130 hPa. Physical solution and chemical bonding of O_2 in hemoglobin is compared in Fig. 9.4 as a function of oxygen pressure. This graph shows that chemical bonding and transport by hemoglobin is by far the dominating process of oxygen supply in the body. The oxygen transport through the protein hemoglobin was first discovered by Hünefeld[3] and later confirmed by Hoppe-Seyler.[4]

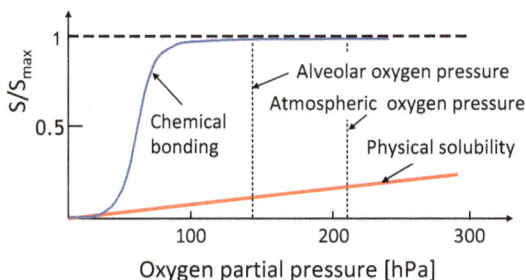

Fig. 9.4: Oxygen solution in blood and oxygen bonding in red blood cells.

Each erythrocyte carries about 10^9 O_2 – molecules in saturation. This makes about 5×10^{21} oxygen molecules per liter of blood, or roughly 0.01 mole O_2/l, much more than the physical solution would ever deliver. At rest, by far not all oxygen that is bonded to hemoglobin is delivered to the cells. The venous blood still contains about 75% of the original oxygen uptake. More oxygen is taken out of the venous reservoir and exchanged if the oxygen concentration drops in the muscles due to activity. Only in the coronary vessels the $O_2 \leftrightarrow CO_2$ exchange is complete.

9.3.5 Carbon dioxide transport in blood

CO_2 transport in the blood is much different from O_2 transport. There are three main pathways for CO_2 from the cells, where metabolic CO_2 is produced, to the lungs. First: according to Tab. 9.3, about 7–8% of CO_2 are physically dissolved in the blood plasma. Second: another 5–8% of CO_2 diffuses across the erythrocytes' membrane and becomes reversibly bonded to the β-chains of hemoglobin forming carbaminohemoglobin, while CO_2 is not bonded to the heme molecule itself. Third: most of the CO_2 (85%) reacts with water in the cytoplasm of the erythrocytes to form carbonic acid (H_2CO_3) with the help of the enzyme carbonic anhydrase. Carbonic acid, also called respiratory acid, is volatile and breaks down into bicarbonate (HCO_3^-) and protons (H^+). These two reactions are as follows:

---

3 Friedrich Ludwig Hünefeld (1799–1882), German medical scientist.
4 Felix Hoppe-Seyler (1825–1895), German medical scientist and chemist.

$$CO_2 + H_2O \leftrightarrow H_2CO_3$$

$$H_2CO_3 \leftrightarrow HCO_3^- + H^+$$

Two thirds of the bicarbonate diffuse from the cytoplasm back into the blood plasma and are transported to the lungs via the bloodstream.

In the capillaries of the lungs, all reactions run backward: bicarbonate migrates back to the erythrocytes, where the molecules are split into CO_2 and water. Together with CO_2 from carbaminohemoglobin and physically dissolved O_2, CO_2 finally diffuses outwards through the air-blood barrier, following the pressure gradient of 53 hPa inside versus 43 hPa in the alveolar space. As a side note, carbon monoxide, unlike carbon dioxide, is bound to the same location in hemoglobin as O_2, but with a much higher affinity. Therefore, CO is poisonous while CO_2 is exhaled.

> **!** CO_2 transport in the blood differs greatly from O_2 transport. O_2 chemically binds to hemoglobin, while CO_2 reacts with water in the cytoplasm of the erythrocytes.

9.3.6 Gas exchange in the alveoli

Now we are primed to re-examine the $O_2 \leftrightarrow CO_2$ gas exchange in the alveoli and pulmonary capillary. Referring to Fig. 9.5, blood arrives at the pulmonary arterial side of the lungs with an O_2 concentration corresponding to a partial pressure of about 50 hPa. The pressure difference across the joint membrane is $\Delta p_{O_2} = 130$ hPa $-$ 50 hPa $= +80$ hPa.

The pulmonary arterial CO_2 concentration corresponds to a partial pressure of about 60 hPa, while the CO_2 pressure in the alveoli is lower at about 50 hPa. Therefore, the pressure difference is $\Delta p_{CO_2} = -10$ hPa.

Thus, on the pulmonary arterial side, the O_2 partial pressure in the alveoli is higher than in the capillaries. In comparison, the CO_2 partial pressure in the alveoli is lower than in blood plasma. Therefore, oxygen diffuses from the alveoli to the erythrocytes and carbon dioxide diffuses in the opposite direction. Consequentially, on the pulmonary venous side of the capillary bed, the erythrocytes are oxygen rich while carbon dioxide is depleted.

The joint membrane acts as blood-air barrier and is called *alveolar-capillary barrier*. The barrier consists of four layers: a surfactant layer, an alveolar epithelium, a basal lamina, and endothelium, adding up to a total thickness of about 1 µm. All together the oxygen molecules have to pass nine membrane walls on their way from the alveolar airspace to the inside of the erythrocytes. Nevertheless, the gas exchange rate, controlled by the diffusivity and permeability of the blood-air barrier, is high enough for sufficient O_2 supply to the body. The rate of oxygen uptake by the lungs is 200–250 ml O_2/min or 360 l/day. This agrees well with our daily oxygen consumption according to the metabolic rate estimated in Chapter 4.

The diffusion depends on pressure differences and concentration gradients. If more oxygen is required during hard work or sport, the partial pressure differences rise across the blood-air barrier and thus also the oxygen uptake.

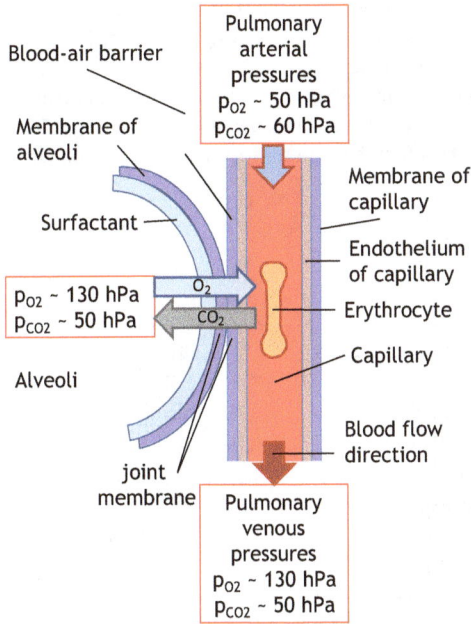

Fig. 9.5: Partial pressures on both sides of the blood-air barrier. Details are discussed in the text.

The gas exchange across the blood-air barrier can, in principle, be modeled by Fick's first law of diffusion:

$$j_D = D\frac{dn}{dx}. \tag{9.6}$$

In this equation, j_D is the flux density (number of particles per area and time), D is the diffusion constant for the gases through the membranes (units: area per time), and dn/dx is the concentration gradient across the barrier (units: 1/Volume × 1/length). However, there are some problems with applying this law to the blood-air barrier. First, there are different media on either side of the barrier: gas on one side and liquid on the other side of the barrier. Therefore, it is better to consider the pressure gradient across the barrier rather than the concentration difference. Second, Fick's first law describes a constant diffusion flow in which the concentration profile does not change over time (see Mathbox on Fick's law). However, this condition is not met during breathing. During inspiration, the area and thickness of the membranes change as the lungs expand. Similarly, the gas pressures of O_2 and CO_2 change in time and space during inhalation and exhalation. Therefore Fick's first equation can only be applied locally and instantly. For a complete picture of the respiratory process, all snapshots for all locations must be added up. It is more appropriate to use Fick's second law of diffusion to model the breathing cycle in these circumstances.

ⓘ **Mathbox: Fick's first and second law**

Adolf Fick (1829–1901), a German mathematician and physiologist, has among many other discoveries, formulated two mathematical expressions for the gas exchange across barriers like membranes. According to Fick, we have to distinguish between particle flows that come to a standstill over time due to concentration equalization (right), and those that are constantly maintained by infinite reservoirs on both sides and which are independent of time (left). The constant diffusion currents are described by Fick's first law, the time-dependent diffusion currents by Fick's second law.

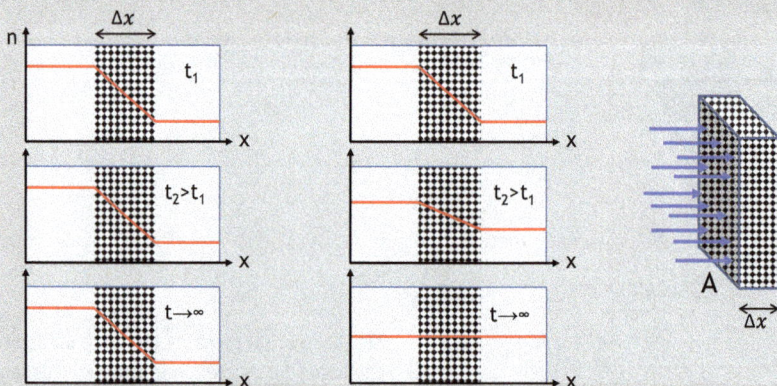

We consider first the steady state condition and define the particle flow $I_N = \Delta N/\Delta t$, which is the number of particles N that cross the barrier with thickness Δx and area A in units of time Δt. The particle flux density $j_n = I_N/A = \Delta n(\Delta x/\Delta t)$ is the flow per area and Δn is the difference of the particle densities on both sides of the barrier. The particle flux is proportional to the density gradient $j_n = -D(\Delta n/\Delta x)$, where the proportionality factor is the diffusion constant $D = (\Delta x)^2/\Delta t$. Expressed in vector form for application in three dimensions, Fick's first law reads:

$$\vec{j}_n = -D\vec{\nabla}n(\vec{r}). \tag{9.7}$$

The direction of the net current vector $\vec{j}$ points from large to small concentrations. The direction of the gradient $\vec{\nabla}n(\vec{r})$ points from small to large concentrations. In order for the two directions to be identical, a minus sign is introduced in the proportionality.

In the nonstationary case of diffusion, the density is not only a function of the location but also of the time. In this case the continuity equation applies (here in one dimension):

$$\frac{\partial n(x)}{\partial t} = -\frac{\partial}{\partial x}(j_N). \tag{9.8}$$

Inserting the equation of Fick's first law, we obtain in case that the diffusion constant D does not depend on location:

$$\frac{\partial n(x)}{\partial t} = -\frac{\partial}{\partial x}\left(D\frac{\partial n}{\partial x}\right) = D\frac{\partial^2 n}{\partial x^2} \tag{9.9}$$

or in vector form:

$$\frac{\partial n(\vec{r})}{\partial t} = -D\nabla^2 n(\vec{r}, t). \tag{9.10}$$

The last equation is Fick's second law, which applies to diffusion currents that depend on time. For instance, during inspiration, the O_2 pressure difference in the alveolar space and the pulmonary veins is different at the beginning and at the end. Therefore, modeling the inspiration with the second law is more appropriate than with the first law.

9.3.7 Respiratory coefficient and exchange rate

Alternatively, we may describe the gas transport across the blood-air barrier by the volume flow equation, according to Ohms' eq. (8.23):

$$I_V = \dot{V} = G_R \Delta p. \tag{9.11}$$

Here the proportionality constant G_R is the particle flow conductance, which is the reciprocal of the particle flow resistance, the subscript R stands for respiratory, and Δp is the pressure difference across the barrier. Using this equation, we face similar problems concerning varying pressures and conductances during the respiratory cycle. This is discussed in much detail in physiological books such as the one of Boron and Boulpeap [1]. Nevertheless, we can make a couple of interesting conclusions. For the oxygen diffusion during inspiration, the pressure difference is highest at the beginning (80 hPa) and is almost zero at the end. For the CO_2 out-diffusion during expiration, the pressure difference is much smaller and almost constant, only 10 hPa instead of 80 hPa for O_2. Furthermore, CO_2 is a more heavy and bigger molecule than O_2, and therefore, we expect CO_2 to be slower than O_2 diffusion. Despite all these differences, the gas exchange rate for O_2 and CO_2 are about equal, such that the *respiratory quotient* (RQ) is:

$$RQ = \dot{V}_{CO_2}/\dot{V}_{O_2} \leq 1, \tag{9.12}$$

Equation (9.12) implies that the flow conductance of CO_2 across the membrane must be eight times higher than that of O_2, contrary to our expectation. Equation (9.12) also implies that the flow of O_2 and CO_2 corresponds to the *respiratory exchange ratio* (see eq. (4.6)):

$$RQ = RER. \tag{9.13}$$

The *respiratory exchange ratio* (RER) tells us that the number of CO_2 moles exhaled equals the number of O_2 moles inhaled. The RER is thus a statement about gas volumes but not about volume rates. In contrast, RQ is a statement about volume rates. Therefore, eq. (9.13) assures that the gas exchange proceeds via the same gas volumes and over the same time span. This was not unexpected but intriguing to see it in terms of an equation. Equation (9.13) also holds when RQ drops to lower values when metabolizing proteins or fat.

> **!** Upon $O_2 \leftrightarrow CO_2$ gas exchange, the respiratory quotient is 1.
> The number of CO_2 molecules exhaled equals the number of O_2 molecules inhaled.
> The respiratory quotient equals the respiratory exchange rate.

9.4 Tidal volume and vital capacity

9.4.1 Inspiratory and expiratory air volumes

The inspiratory and expiratory air volumes can be determined with the help of a spirometer (see Infobox II). The spirometer records gas volumes as a function of time. The test person is asked to breathe normally at rest, inhale as much as possible, and exhale to the maximum. The corresponding air volumes are collected and measured. Figure 9.6 shows a schematic chart of such a recording.

At rest, the inspiration-expiration volume is called the *tidal volume* (TV). The tidal volume is on average about 0.5 l, independent of sex. When inhaling to the absolute maximum capacity, air fills the *inspiratory rest volume* (IRV). This is the volume beyond the tidal volume, which is about 3 l for men and 2 l for women. Conversely, upon expiration to the absolute minimum, the *expiratory reserve volume* (ERV) is reached (1.2 l for men, 0.7 l for women). The residual volume (RV) cannot be measured by inspiration-expiration. However, it must be either estimated by mixing air with a nontoxic gas such as He or by imaging via special MRI methods discussed in Chapter 3 in Vol. 2. Typically, RV is about 1.5–2 l. It is important to maintain a residual gas volume to prevent the lungs from collapsing, and it facilitates the expansion of the lungs during inspiration. These four main volumes, TV, IRV, ERV, and RV do not overlap. The remaining volume definitions follow from addition and subtraction, as can be seen by the colored chart in Fig. 9.6. Those are as follows:

VC = vital capacity = ERV + TV + IRV is the capacity from maximum expiration to the maximum inspiration. For a healthy adult person, VC is about 3 l (women) to 4.5 l (men).

$$FRC = \text{fractional residual capacity} = ERV + RV.$$

$$TLC = \text{total lung capacity} = VC + RV.$$

These lung capacities are listed in Tab. 9.4 for men and women.

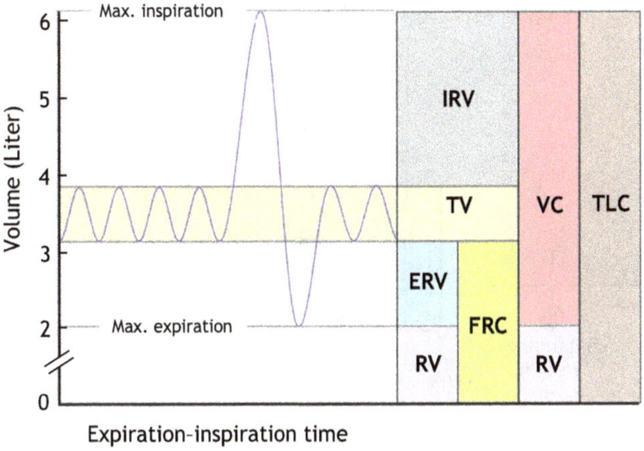

Fig. 9.6: Various volume capacities of the lung. TV = tidal volume; IRV = inspiratory rest volume; VC = vital capacity; ERV = expiratory rest volume; RV = residual volume; FRC = fractional rest capacity; TLC = total lung capacity.

Tab. 9.4: Typical inspiratory and expiratory air volumes of the lungs. All numbers are in units of liter.

Term	Abbreviation	women	men
Tidal volume	TV	0.5	0.5
Inspiratory rest volume	IRV	1.8	2.8
Expiratory rest volume	ERV	0.7	1.2
Vital capacity	VC	3.0	4.5
Residual volume	RV	1.5	2.0
Total lung capacity	TLC	4.5	6.5

Infobox II: Spirometer

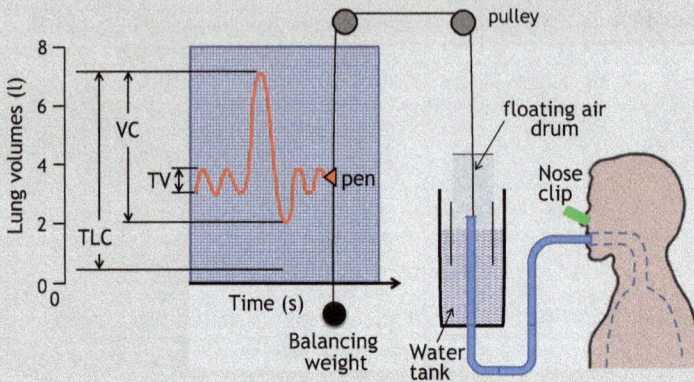

A spirometer is a simple device for examining lung ventilation and measuring the volume of air during inhalation and exhalation. A drum is filled with air or oxygen and floats over a water tank so that no air can escape. A hose connects the air drum with the mouth of a test subject. The drum is balanced by a counterweight and moves up and down according to the subject's breathing. The movement of the drum as function of time was monitored by a strip chart recorder in the past, but is now measured electronically. A nose clip prevents the airflow from bypassing. The acronyms of the different volumes are explained in the main text.

A healthy person can exhale 70% of the vital capacity within 0.5 s, another 15% in 1 s, and a total of 97% within 3 s. During this exercise, the volume flow rate is 6–8 l/s, a value that we will need later on. The flow velocity can be as high as 0.5–1 in units of the Mach number (1 Mach = 340 m/s).

The *breathing frequency* (BF) of an adult is about 15–16 breathes per minute. For an infant, the breathing rate is about 40 per minute. Taking a tidal volume (TV) at rest of about 0.5 l, we determine a *minute ventilation* ($\dot{V}_{air}$):

$$\dot{V}_{air} = BF \times TV = 8\,l/\mathrm{min}. \tag{9.14}$$

This makes 11 500 l of inhaled air per day. It turns out that this is about the same amount of inhaled air in liters as blood is pumped by the heart per day.

From the 11 500 l air/day only 13% arrive as oxygen in the alveoli, which amounts to about 1500 l/day. However, only about 30% of the oxygen is really exchanged, which corresponds to 430 l oxygen/day.

According to Chapter 4, the *caloric oxygen equivalent* (COE) is 1 l of O_2 for generating 20 kJ of energy. The energy requirement per day is about 8 MJ at rest, or 400 l of oxygen. These estimates show that the volumes of oxygen supply and metabolic oxygen requirement agree very well. With increasing activity levels, there is a sufficient oxygen reservoir to respond to the needs.

Oxygen supply by inspiration and oxygen consumption during metabolism agrees very well, amounting to about 400 l of oxygen per day. !

9.4.2 Inspirational oxygen flow and cardiac output

Inspiration and expiration flows[5] are connected to the gas fractions in the following way [7]. When $\dot{V}_{O_2}$ is the inspiration flow of O_2 and $\dot{V}_{air}$ is the minute ventilation, then

$$\dot{V}_{O_2} = \dot{V}_{air}\left(v_{i,O2} - v_{e,O_2}\right), \tag{9.15}$$

where $v_{i,O2}$, $v_{e,O2}$ are the fractional oxygen gas volumes during inspiration and expiration, respectively. The inspirational oxygen volume is 13% of the inhaled air, and the expirational oxygen volume is 70% of the inhaled volume. Therefore, $\dot{V}_{O_2} = \dot{V}_{air} \times 0.13 \times 0.3 = \dot{V}_{air} \times 0.04 = 0.3\,l/min$.

The same also holds for the CO_2 flow with the exception that there is no inspirational gas fraction of CO_2:

$$\dot{V}_{CO_2} = \dot{V}_{air} \times v_{e,CO_2}. \tag{9.16}$$

These relations are quite clear, and with the help of a spirometer, the air volumes and fractional gas volumes can be determined.

Next, we ask whether the inspirational oxygen flow $\dot{V}_{O_2}$ can also be determined via the oxygen concentrations in the pulmonary veins and arteries. If so, it would require an analysis of the blood composition with respect to the oxygen concentration. Because of the O_2 uptake in the capillary bed of the lungs, the O_2 concentration in the pulmonary veins is higher than in the pulmonary arteries. In the pulmonary arteries, the O_2 concentration is similar to the one in the right ventricle, which is a mixture of venous blood from different organs, denoted (c_m). In the pulmonary veins, the blood composition is the same as in peripheral arteries, designated as (c_a). Analysis of the blood concentration yields for $c_m \cong 150$ ml per 1 l of blood, and $c_m \cong 200$ ml/l.

Let $\dot{V}_L$ be the blood flow through the lungs. Then the inspirational O_2 flow can be determined from the product of $\dot{V}_L$ and the blood oxygen concentration difference [7]:

$$\dot{V}_{O_2} = \dot{V}_L(c_a - c_m). \tag{9.17}$$

As the perfusion of the lung amounts to the same volume rate as the cardiac output: $\dot{V}_L = \dot{V}_{CO}$ (see eq. 8.3), we can rephrase eq. (9.17):

---

5 This section can be skipped on a first reading.

$$\dot{V}_{O_2} = \dot{V}_{CO}(c_a - c_m).\tag{9.18}$$

Note that the volumes V_L and V_{CO} are different, but the respective volume rates must be identical. The last equation yields a method to determine the cardiac output:

$$\dot{V}_{CO} = \frac{\dot{V}_{O_2}}{(c_a - c_m)}.\tag{9.19}$$

Hence, the cardiac output can be determined by measuring the blood's oxygen flow and the arteriovenous concentration difference. If we now take the ratio of cardiac output and the beat frequency f_{heart}, we obtain the stroke volume V_{SV}:

$$V_{SV} = \frac{\dot{V}_{CO}}{f_{heart}} = \frac{\dot{V}_{O_2}}{f_{heart}(c_a - c_m)}.\tag{9.20}$$

With this approach, Fick succeeded in 1872 in estimating the heart's stroke volume for the first time, a milestone in medical sciences. With the numbers at hand, we find for the stroke volume:

$$V_{SV} = \frac{\dot{V}_{O_2}}{f_{heart}(c_a - c_m)} = \frac{0.3\,\mathrm{l\,min}^{-1}}{70\,\mathrm{min}^{-1} \times 50\,\mathrm{ml\,l}^{-1}} = 90\,\mathrm{ml}.$$

This result compares well with the quoted value in eq. 8.1 of $V_{SV} = 70$ ml.

9.5 Pulmonary volume and pressure changes

Lung, diaphragm, and thorax form a unity during breathing. The lung sits in the *thoracic cage* separated from the chest wall by an intrapleural space called the *pleural cavity* (see Fig. 9.3). This cavity consists of a folded membrane, where the visceral pleura covers each lung and the parietal pleura covers the inner thoracic wall. The interaction of the chest and the lung across the intrapleural space determine the lung volume. The volume change during respiration is dramatic, as we notice from the schematics in Fig. 9.7. During inspiration, the diaphragm contracts and flattens, and the intercostal muscles elevate the ribs and the sternum. The largest volume expansion of the thoracic cage results from movement of the diaphragm. During expiration, the diaphragm and the intercostal muscles relax, and the elastic properties of the thorax and lungs allow a passive reduction of the thoracic cage volume. Also, the abdominal muscles support the expiratory contraction.

As the thorax volume varies during respiration, the gas pressure also changes in the lungs. Figure 9.8 displays a simple mechanical model for the lungs that represents all relevant pressures and forces. The pressures include the total pressures and pressure differences, irrespective of the partial pressures that constitute the atmospheric pressure. We start with the resting position, defined as the position where the gas pressure in the lungs is the same as in the atmosphere, referred to as barometric pressure p_B. The barometric pressure is the one present in the environment irrespective of the actual altitude above sea level. In the resting position, the lungs are expanded and would normally contract and eventually collapse if not attached to the thoracic wall. The thorax is compressed and would normally extend if not linked to the lungs. The force F_L pulls the lungs together, and the force F_T expands the thorax. Both forces point in opposite directions and keep in balance (red arrows in Fig. 9.8). Moreover, the force pair pulls on either side of the membrane, separating thorax and lungs. This pull increases the intrapleural space slightly and causes a reduction of the pleural cavity pressure Δp_{PB} to -500 Pa with respect to the barometric pressure. The transmural pressure difference between alveolar space in the lungs and pleural cavity, Δp_{TM}, follows from:

$$\Delta p_{TM} = \Delta p_{BA} - \Delta p_{PB} = +500 \text{ Pa}, \tag{9.21}$$

where Δp_{BA} is the pressure difference between alveolar space and the environment. This pressure acts on the visceral wall from the inside.

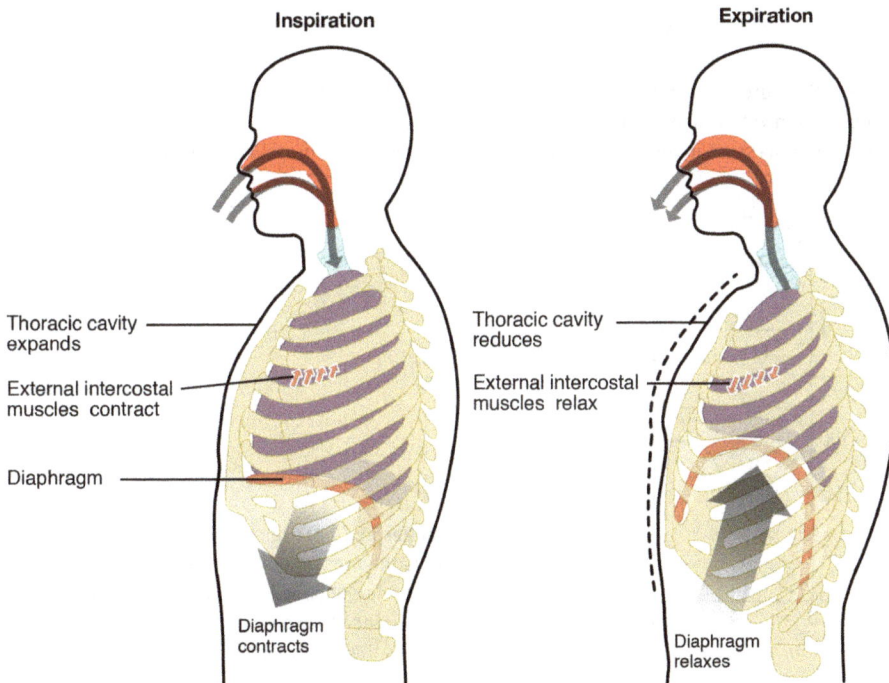

Fig. 9.7: Volume change of the thoracic cavity during respiration (reproduced from OpenStax Anatomy and Physiology, © Creative Commons).

During inspiration, the thorax expands due to the action of the intercostal muscles and due to the diaphragm pulling downwards. This pull increases the thoracic cage volume. As the lungs adhere to the thoracic wall, the lung volume will also expand. Furthermore, the elastic forces between lungs and thorax increase slightly the intrapleural space such that the pleural pressure difference Δp_{PB} drops from −500 to −750 Pa with respect to the barometric pressure. As the pleural pressure decreases and the lungs expand, so do the alveoli inside the lungs. This expansion causes a pressure drop in the alveoli with respect to the environment to about $\Delta p_{BA} = -150$ Pa. The transmural pressure difference $\Delta p_{TM} = \Delta p_{BA} - \Delta p_{PB}$ increases to about +600 Pa during inspiration. Because of this inspiratory pressure drop in the alveolar space and increased volume, air will flow into the lungs until the pressure difference is balanced.

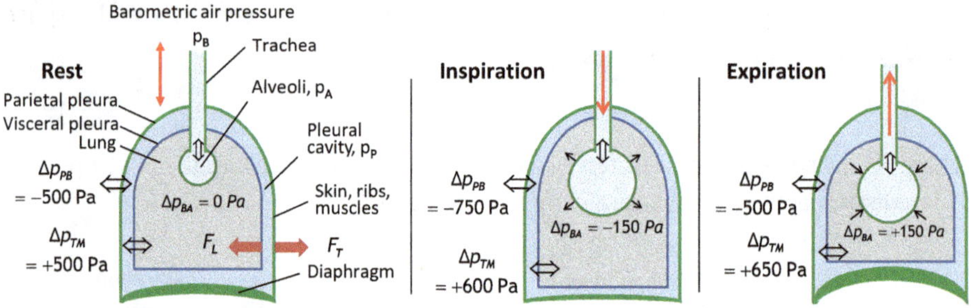

Fig. 9.8: Pressures and pressure differences at rest and during inspiration and expiration, demonstrated with a simple mechanical model. p_B = barometric pressure, p_A = gas pressure in the alveoli, p_P = pressure in the pleural cavity. Δp are the respective pressure differences.

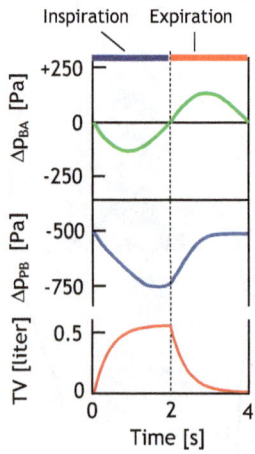

Fig. 9.9: Pressure changes and tidal volume during one inspiration-expiration cycle. Δp_{BA} is the pressure difference between the barometric pressure outside the body and alveolar air in the lungs; Δp_{PB} is the pressure difference between the pleural cavity and the barometric pressure. TV is the tidal volume.

During an inspiration-expiration cycle, the thorax expands and contracts due to the action of the intercostal muscles and due to the diaphragm pulling downwards and upwards.

During expiration, the volume of the thoracic cage decreases beyond the resting position. Therefore, in the alveoli, an expiratory overpressure Δp_{BA} = +150 Pa develops, and air flows out of the lungs through the trachea into the environment. The other corresponding pressure differences are indicated in Fig. 9.8.

The time dependences of the alveolar pressure difference Δp_{BA}, the interpleural pressure difference Δp_{PB}, and the tidal volume (TV) are plotted in Fig. 9.9. Shown is one inspiration – expiration cycle, assuming that the breathing frequency is 0.25 s^{-1} or 15 breathings per minute. Tab. 9.5 provides an overview of the various pressure differences introduced before.

Tab. 9.5: Overview on different pressure differences which are active during respiration.

Pressure difference	Main function
Δp_{BA}	Alveolar pressure difference, responsible for the transport of gas in and out of the lung through the airways.
Δp_{PB}	Intrapleural pressure difference, responsible for maintaining an underpressure in the intrapleural space.
Δp_{TM}	Transmural pressure difference, responsible for inflating the alveolar sacs during inhaling.

9.6 Volume work, compliance, and surface tension

9.6.1 Volume work

The volume work done by the thorax during respiration can be represented in a pV-diagram as indicated in Fig. 9.10. The enclosed area corresponds to the volume work W_L during respiration at rest. This volume is equivalent to the tidal volume and the pressure refers to the intrapleural pressure difference Δp_{PB}. The volume work performed can be estimated as:

$$W_T = \frac{1}{2}(\Delta p_{PB} \times TV).\qquad(9.22)$$

However, during expiration, the thorax recoils elastically from the expanded to the relaxed volume, which costs essentially no energy. Therefore, the volume work done by the thorax should be estimated as:

$$W_T = \frac{1}{4}(\Delta p_{PB} \times TV). \tag{9.23}$$

This estimate amounts to $W_T = 0.25 \times 200 \text{ Pa} \times 0.5\,l = 0.025\,\text{J}$ per breathing cycle. With a breathing frequency $BF = 15/\text{minute}$, the mechanical power consumption is only 6 mW. Even if we consider a mechanical efficiency of 20%, the power consumption of the respiratory system is still negligibly low, thanks to the elastic recoil properties of the thorax.

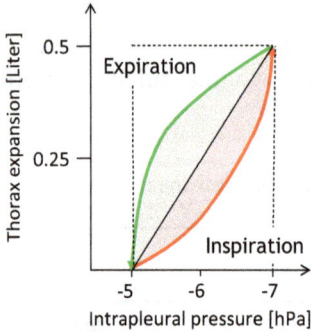

Fig. 9.10: Volume work of the lung during inspiration (red area) and expiration (green area).

9.6.2 Compliance

The ease of expanding the lung can also be recognized by the high compliance of the lungs C_L. The lungs' and the thorax compliances are defined as the volume change ΔV by which the lungs and the thorax expand for each increment of pressure change Δp_{BA} in the alveoli. The relation between pressure change and volume change of elastic bodies is usually expressed in terms of Hooke's law (see eq. (3.5)):

$$\Delta p = B\frac{\Delta V}{V}. \tag{9.24}$$

Here B is the elastic or bulk modulus of the expandable medium. Rephrasing this equation, we find:

$$\Delta V = \frac{V}{B}\Delta p = C\Delta p. \tag{9.25}$$

The ratio $C = V/B = \Delta V/\Delta p$ is called the *elastic compliance*. It can be considered as a mechanical susceptibility, where ΔV is the extensive property and Δp is the conjugated field to the volume change ΔV. The reciprocal value $E = 1/C$ is called the *elastance*.

Figure 9.11 shows a plot of lung volume versus *transmural pressure difference* Δp_{TM}. The local first order derivative refers to the compliance of the lung, C_L. Obviously, the compliance is not a constant. C_L is large in the beginning when for a small pressure change, the volume change is large. As the volume increases, it becomes more difficult to expand it further, i.e. the compliance decreases with inspiration beyond the *TV* region into the *IRV* region. The drop of the compliance is to be expected for a system that levels off.

Figure 9.11 also shows the volume-pressure relationship for two frequent lung diseases. *Emphysema* is a disease where surfactants and cell membranes of alveoli are irreversibly destroyed, for instance, by smoking (see next section). This leads to a saggy type of the lungs that lacks substantial elastic recoil (blue line). On the contrary, *fibrosis* is a condition where the cell walls of the alveoli harden. Then the lungs are much less elastic and the compliance is strongly decreased (red line).

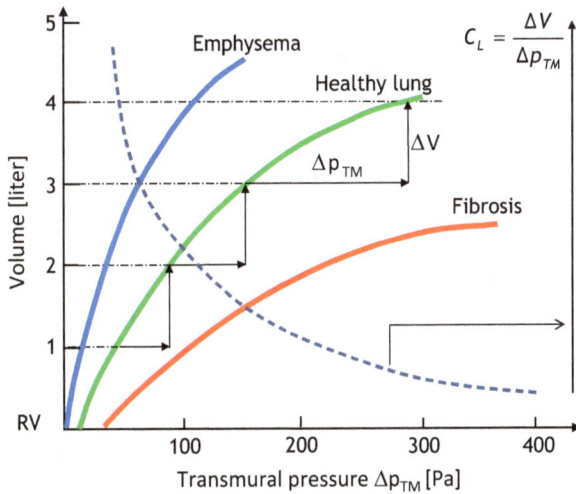

Fig. 9.11: Volume-pressure dependence of the lung for a healthy person (green line), for a patient with emphysema (blue line), and for a patient with fibrosis (red line). The dashed blue line shows the decreasing compliance with increasing transmural pressure. The arbitrary scale is to be read on the right side.

The volume-pressure relationship is different for the thorax and the lung alone, as shown in Fig. 9.12. Thorax and lung have opposite elastic properties. The compliance of the thorax C_T is small at the beginning and increases with expansion, similar to the compliance of an air balloon. In contrast, the compliance of the lung C_L is large at the beginning but drops to zero when approaching saturation, as we have just seen.

The green line shows the combined compliance of thorax and lung in Fig. 9.12. In a mechanical model shown in Fig. 9.13 both lung and thorax can be represented by coupled springs with an equilibrium rest position corresponding to the fractional rest volume (*FRC*). When released from their coupling, the spring representing the lung will contract to a volume less than the residual volume *RV*, whereas the spring representing the thorax will expand to a volume below the total lung capacity.

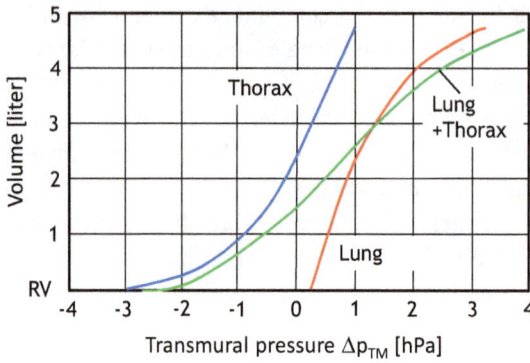

Fig. 9.12: Compliance of thorax and lung alone and of the combined organs.

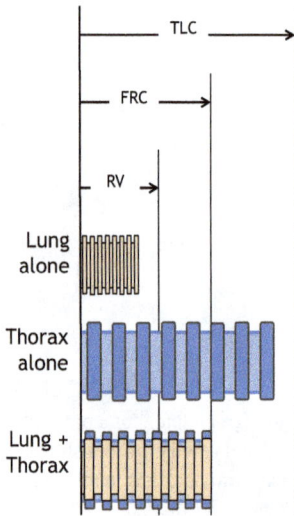

Fig. 9.13: Mechanical spring model representing lung and thorax.

Because the lung and thorax are arranged in parallel, the reciprocal values of the compliances, i.e., the *elastances*, add up similar to parallel ohmic resistances:

$$\frac{1}{C_{tot}} = \frac{1}{C_L} + \frac{1}{C_T}.$$ (9.26)

Now we try to estimate both compliances independently. The compliance of the lung is defined by:

$$C_L = \frac{\Delta V}{\Delta \hat{p}_{TM}} = \frac{TV}{\Delta \hat{p}_{TM}}.$$ (9.27)

Here we set $\Delta V = TV$. The transmural pressure difference was defined before as:

$$\Delta p_{TM} = \Delta p_{BA} - \Delta p_{PB}.$$

However, for estimating the lungs' compliance, we need to take the modified pressure difference $\Delta \hat{p}_{TM}$:

$$\Delta \hat{p}_{TM} = \Delta p_{TM,f} - \Delta p_{TM,i}. \tag{9.28}$$

$\Delta \hat{p}_{TM}$ is the difference between the transmural pressure difference $\Delta p_{TM,f}$ at the plateau region after inhaling is completed, and the transmural pressure difference $\Delta p_{TM,i}$ at the beginning of inhaling. This transmural pressure difference from the beginning to the end of inhaling is 150 Pa. Therefore, the compliance of the lung can be estimated as:

$$C_L = \frac{\Delta V}{\Delta \hat{p}_{TM}} = \frac{TV}{\Delta \hat{p}_{TM}} = \frac{0.5\,l}{150\,\text{Pa}} = 3.3\,\frac{l}{\text{kPa}}.$$

The compliance of the thorax, defined by

$$C_T = \frac{\Delta V}{\Delta \hat{p}_{PB}}, \tag{9.29}$$

can be estimated similarly by considering the intrapleural pressure difference at the start and end of inhaling. The pressure difference is 250 Pa:

$$C_T = \frac{\Delta V}{\Delta \hat{p}_{PB}} = \frac{0.5\,l}{250\,\text{Pa}} = 2\,\frac{l}{\text{kPa}}$$

The total compliance is therefore:

$$C_{tot} = 1.25\,l/\text{kPa}.$$

Often the total compliance is estimated by considering the thorax alone, setting:

$$C_{tot} = C_T = \frac{\Delta V}{\Delta \hat{p}_{PB}} = 2\,l/\text{kPa}.$$

Typical values for adults are 1–2 l/kPa. Measurements of the compliance may indicate diseases of the lung, as we have already noticed for the emphysema and fibrosis diseases.

9.6.3 Surface tension

Now we discuss the physical properties of the alveoli in terms of surface tension of soap bubbles (Fig. 9.14(a)). The force F_1 on the rim of a soap bubble tends to increase the bubble's surface (factor of two because of two surfaces, inside and outside):

$$F_1 = 2y2\pi r. \tag{9.30}$$

Here r is the radius and the proportionality factor y is the surface tension. This equation holds if the force required to increase the surface is constant and independent of the radius. On the other hand, the force F_2 acts on the equatorial plane with area πr^2 and is due to the inside gas pressure p_i:

$$F_2 = p_i \pi r^2. \tag{9.31}$$

In equilibrium, both forces must balance, and therefore we obtain the internal pressure p_i:

$$p_i = \frac{4y}{r}, \tag{9.32}$$

known as Laplace[6] law.

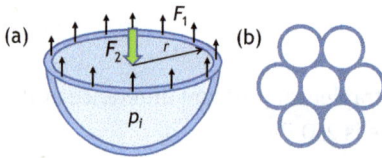

Fig. 9.14: (a) Model of a soap bubble; (b) Model of alveolar sacs. Alveoli surrounded by other alveoli have only one inside surface.

The lung of adults contains about $3-6 \times 10^8$ alveoli [2]. Each alveoli has a diameter of 50 μm at expiration, which expands to about 250 μm at inspiration. The wall thickness is about 0.5 μm. These thicknesses can be obtained by microscopy of histological cuts through the lung tissue [7].

The surface tension of the alveoli y tends to reduce the diameter and to contract the lung. During inspiration, the pressure in the alveoli must equilibrate the pressure due to the wall tension:

$$p_{wall} = \frac{2y}{r}. \tag{9.33}$$

A factor of 2 is missing because the alveoli have only one inside surface unlike free and unclustered soap bubbles, as indicated in Fig. 9.14(b).

The surface tension of water is $y = 0.072$ N/m. For the transmural pressure difference, we then obtain:

$$\Delta p_{TM} = \frac{2 \times 0.072\,\text{N/m}}{25\,\mu\text{m}} = 5.75\,\text{kPa}.$$

---

6 Pierre-Simon Laplace (1749–1827), French mathematician and physicist.

This pressure is by far bigger than is observed. The actual transmural pressure difference is only a tenth of the calculated one: $p_{TM} = 500$ Pa instead of 5000 Pa. The surface tension reduction in the alveoli is due to surfactants in the wall membrane. The main component of this surfactant is the molecule dipalmitoylphosphatidylcholine (DPPtdCho), which can lower the surface tension from 72 mN/m to almost zero. Any damage to the pulmonary surfactant increases the transmural pressure and leads to dyspnea. In newborns, the pulmonary surfactant is not completely developed, which may cause *infant respiratory distress syndrome.*

Alveoli with smaller radii have higher internal pressure than larger ones because the internal pressure is proportional to the reciprocal radius: $p \sim 1/r$. Air may then be pushed out of the smaller alveoli, which eventually completely collapse (Fig. 9.15). However, the surfactants not only lower the surface tension but simultaneously hinder gas exchange. In case the surfactants are depleted or damaged, entire areas in the lungs may collapse, which is known as *atelectasis.*

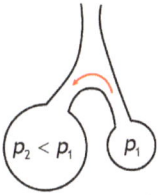

Fig. 9.15: Pressure in small bubbles is bigger than in larger bubbles because the pressure scales with p ~ 1/r.

One of the major dangers to surfactants is tobacco smoke. Tobacco smoke not only destroys the surfactants but also clogs up the airways and increases the airway resistance. The result is potential emphysema, chronic bronchitis, and lung cancer.

Infobox III: The effect of surfactants on surface tension

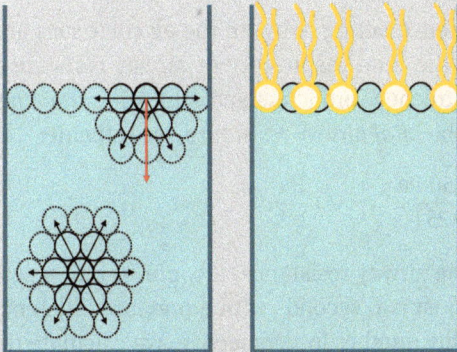

The term "surfactant" is an abbreviation for surface active agent. Surfactants are those that lower the surface tension of water. The surface tension of water is quite high due to the strong dipolar interaction of the water molecules. Within the liquid, the resulting force on a water molecule is canceled out. On the surface, however, the resulting force points into the liquid. Therefore, water forms a molecular film on the surface, which prevents, for example, paper clips from sinking and enables water fleas to walk on water. The high surface tension of water can be reduced by using hydrophilic molecules. The best known are phospholipids. These are chain-like molecules with hydrophilic heads and hydrophobic tails. Upon contact with water, the hydrophilic heads squeeze between the water molecules and lower the surface tension, while the tails protrude from the water surface.

9.7 Airway resistance

9.7.1 Total airway resistance

When breathing, air flows through the airways in and out of the lung. The volume flow rate of air is defined as (eq. 9.14)):

$$I_V = \dot{V}_{air} = \frac{dV_{air}}{dt} = TV \times BF, \tag{9.34}$$

where TV is the tidal volume and BF the breathing frequency at rest. During breathing, we inhale and exhale the same amount. Therefore the tidal volume of 0.5 l has to be taken twice for determining the flow rate. Assuming BF to be 15/min, the flow rate is about 0.25 l/s or 15 l/min (recall that the minute ventilation MV is half this amount). Knowing the flow rate, we can determine the total airway resistance by applying Ohm's law:

$$\dot{V}_{air} = \frac{\Delta p_{BA}}{R_R} \tag{9.35}$$

Here Δp_{BA} is the alveolar pressure difference acting between the alveolar sacs and the barometric pressure outside. R_R is the total respiratory or airway resistance, composed of the resistances due to the trachea, bronchus and all further branches before reaching the alveolar space. For the *total airway resistance*, we estimate:

$$R_R = \frac{\Delta p_{BA}}{\dot{V}_{air}} = \frac{150\,\text{Pa}}{0.25\,\text{l/s}} = 600\,\frac{\text{Pa}}{\text{l/s}}.$$

This is indeed a rather small value for the airway resistance. It confirms our experience that breathing half a liter of air in just two seconds with a pressure difference of a mere 150 Pa is rather easy. To set this number in perspective, we calculate the flow resistance of different tubes, representing the trachea, bronchus, and alveoli. Assuming laminar flow and cylindrically shaped tubes, the flow resistance according to the expression by Hagen–Poisseuille is (see eq. (8.27)):

$$R_{tube} = \frac{8}{\pi}\eta\frac{\Delta L}{r^4} = 8\pi\eta\frac{\Delta L}{A^2}.$$

The viscosity of air at 20 °C is roughly 17×10^{-6} Pa · s [8]. Typical values for radius, cross sections, and lengths of the trachea, bronchial tube, and alveoli are given in Fig. 9.16 and in Tab. 9.6. With this information, we can calculate the flow resistance of different parts in the airway; the values are also listed in Tab. 9.6.

According to Tab. 9.6, the flow resistance of the trachea is very low. The bronchial tubes do not contribute much to the flow resistance either. The main contribution to the resistance is due to the alveoli. However, these numbers are misleading, and it turns out that the interpretation of the flow resistance is not proper.

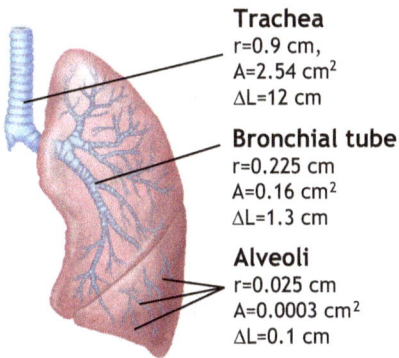

Trachea
r=0.9 cm,
A=2.54 cm²
ΔL=12 cm

Bronchial tube
r=0.225 cm
A=0.16 cm²
ΔL=1.3 cm

Alveoli
r=0.025 cm
A=0.0003 cm²
ΔL=0.1 cm

Fig. 9.16: Airways and left lung with typical values for radius, cross-section, and length of trachea, bronchial tube, and alveoli, respectively.

Tab. 9.6: Geometric data for airways in the respiratory system and respective flow resistances.

	Radius [cm]	Cross section [cm²]	Length [cm]	R [Pa · s l⁻¹]
Trachea	0.9	2.5	12	0.8
Bronchi	0.225	0.16	1.3	17
Alveoli	0.025	0.002	0.1	10^4

9.7.2 Turbulent flow

Airflow can easily become turbulent because of the low gas viscosity. We can experience turbulent flow by watching candlelight. The trachea and even the bronchial tree have radii that could make the airflow turbulent. Therefore we calculate the Reynolds number for the airflow according to (see eq. (8.30)):

$$Re = \frac{\rho v d}{\eta},$$

Here v is the flow velocity, ρ is the density, d the diameter of the conducting system, η is the viscosity. The dimensionless number should be smaller than 1000 to warrant laminar flow. A value above 2000 is indicative for turbulent flow. The density of air is roughly 1.2 kg/m^3, the flow velocity is v = 1.9 m/s and follows from the continuity eq. (8.17):

$$v = \frac{\dot{V}_{air}}{A},$$

where A is the cross-section of the tube. If we substitute numbers listed above, we get a Reynolds number of 2400 for the trachea, 1000 for the bronchial tree, and about 100 for the alveoli. From these values, we conclude that the flow in the trachea may be turbulent, while in the bronchi and In the bronchioles, the flow is most likely laminar. Turbulent flow increases the resistance to airflow dramatically.

Second, we must consider that the total cross-section from the trachea to the bronchioles increases significantly due to the branching. The increasing flow resistance due to the decreasing tube diameter is more than compensated for by an increasing number of divisions, which reach 100 million at the terminal bronchioles with an area of approx. 80 m^2. Accordingly, the flow resistance decreases due to the parallel conductance of all branches. The situation is similar to the blood flow resistance in the capillary bed described in Chapter 8.

Figure 9.17 shows a graph of airway resistance as a function of distance starting from the trachea. In contrast to our first impression, the resistance is highest in the region of the trachea and then decreases continuously with increasing cross-sectional area. The high resistance at the beginning is due to a turbulent flow in the trachea and even increases in the direction of the first division into the primary bronchi. At this junction, the turbulent flow dominates. The resistance drops and calms down as the flow becomes laminar. Therefore, this area is called the "quiet zone" of the lungs. The quiet and noisy zones can easily be distinguished with the help of a stethoscope placed at different locations on the chest.

In summary, we have found that the airway resistance of the lungs is mainly due to turbulent flow in the trachea. The bronchioles and their terminal branches account for less than 20% of the total resistance. Thus the true distribution of airway resistance is the opposite of what we initially anticipated.

9.7.3 Time constant

Having derived the total respiratory resistance R_R and the compliance C_L of the lungs, the product of both yields the time constant for breathing [9, 10]:

$$\tau = R_R C_L \tag{9.36}$$

Substituting numbers, we get for the inspiration a time constant of:

$$\tau = R_R C_L = 600 \, \frac{Pa \cdot s}{l} \times 1.25 \cdot 10^{-3} \, \frac{1}{Pa} = 0.75 \, s$$

and the same time constant for expiration. Inspiration (or expiration) is complete after about three time constants or approximately 2 s, and 4 s for a complete cycle. This estimate agrees well with the respiratory rate of 15 per minute at rest. The exponential increase of the tidal volume (TV) during inspiration and the corresponding decrease during expirations are schematically plotted in the lower panel of Fig. 9.9.

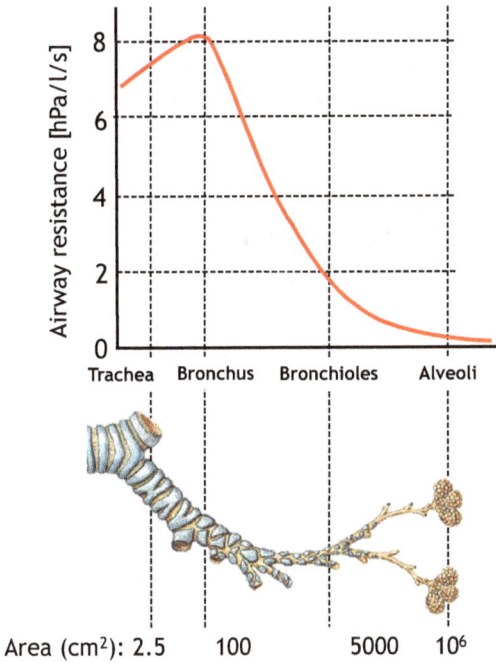

Fig. 9.17: Airway resistance in different parts of the respiratory system. At the bottom of the panel, the cross-sectional area of the airways is indicated as it follows from the cross-section of individual branches times the number of branches.

> The airway resistance is highest at the trachea and drops rapidly towards the bronchial tree. The total airway resistance is 600 Pa · s/l, and is surprisingly low. The time constant for inspiration/expiration is 0.75 s.

9.8 Lung ventilators and oxygen generators

The lungs are susceptible to many types of acute and chronic diseases. That's because the airways are quite open not only to fresh air, but also to any kind of pollution, dust, smoke, viruses, and bacteria. Many lung diseases, such as emphysema and fibrosis, can be detected by measuring tidal volume and lung compliance. Another simple but effective diagnostic tool is the stethoscope, which can be used to identify various sounds due to airflow when breathing. X-rays and CT scans are routinely used for further diagnosis. MRI is becoming routine for lung scans to avoid radiation exposure. XRT and MRI procedures are described in Chapters 3 and 8 in Vol. 2. Several lung diseases require the support of noninvasive or invasive ventilators. A well-known example is the COVID-19 virus infection. We first present different ventilator options and then discuss technical realizations for oxygen generation.

9.8.1 Invasive and noninvasive ventilators

Four types of ventilators are used in practice to treat lung diseases, ranging from noninvasive low-care treatment to invasive high-care treatment. These treatment methods are briefly outlined below.

Nasal cannula: The simplest type of ventilators are small cannulas that are placed under the nose, see Fig. 9.18. Oxygen is supplied by a portable gas bottle for short term use or by a portable oxygen concentrator for longer-term applications. The working principle of oxygen generators and concentrators are discussed later in Section 9.8.2.

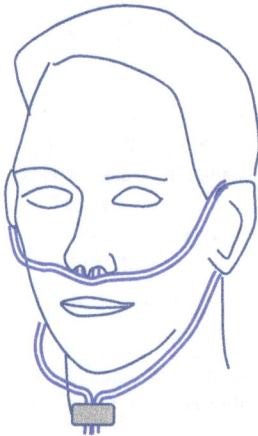

Fig. 9.18: Nasal cannula for ventilation with oxygen enriched air.

Noninvasive ventilator: If the need for oxygen is higher, noninvasive ventilators are used [11]. An airtight mask is placed on the face. A hose connects the mask with an oxygen supply and control system. Pure oxygen and air are mixed to preselected partial pressures in the control machine. A gas volume at a defined pressure is then delivered through the hose to the patient. The expiration gas is usually disposed of through an opening in the mask or close to the mask. Several options can be selected at the control console: continuous mandatory ventilation with either controlled air volume ventilation or controlled air pressure ventilation; assisted spontaneous breathing, where the patient's inhalation triggers the air/oxygen supply; or a so-called synchronized intermittent mandatory ventilation, which is a combination of the other two methods. The arrangement is shown schematically in Fig. 9.19.

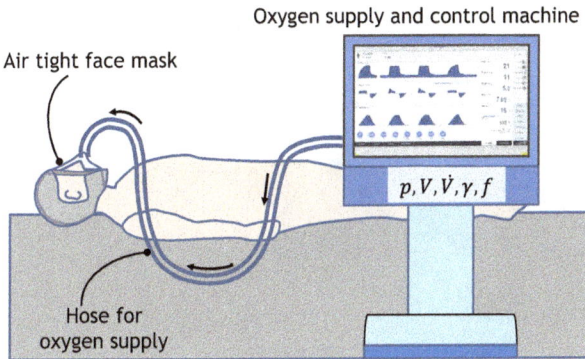

Fig. 9.19: Noninvasive ventilator with control over pressure, volume, flow, partial gas pressures and breathing frequency.

Invasive ventilator: The next level of high care ventilation is to use a tube that is inserted into the patient's airway [12] (see Fig. 9.20 for a schematic illustration). The supply tube is passed either through the mouth or nose or through an incision in the larynx directly into the trachea. The patient is placed in an artificial coma for intubation. The tube is connected to two hoses, one for the oxygen-enriched inspiratory air and the other for the exhaled air. Air-oxygen mixer, air compressor and humidifier are integrated in the rack. Valves control the ventilation process and monitor the pressure, gas flow, respiratory rate, tide or minute volume; determine the patient's status; and respond to the patient's needs via a feedback system. During inspiration, the inspiration valve opens and the expiration valve is closed. This allows pressure to build up in the lungs. At the end of inspiration by reaching a preselected pressure or gas volume, the inspiration valve is closed and the expiration valve is opened. Then the exhaust air can leave the system after monitoring. The expiration runs passively in contrast to the forced and pressure-controlled inspiration. All these procedures work well for the patient if gas exchange through membranes in the lungs is still functional. When the lung tissue is damaged to a degree that gas exchange is no longer guaranteed, the ultimate method must be used, which is discussed next.

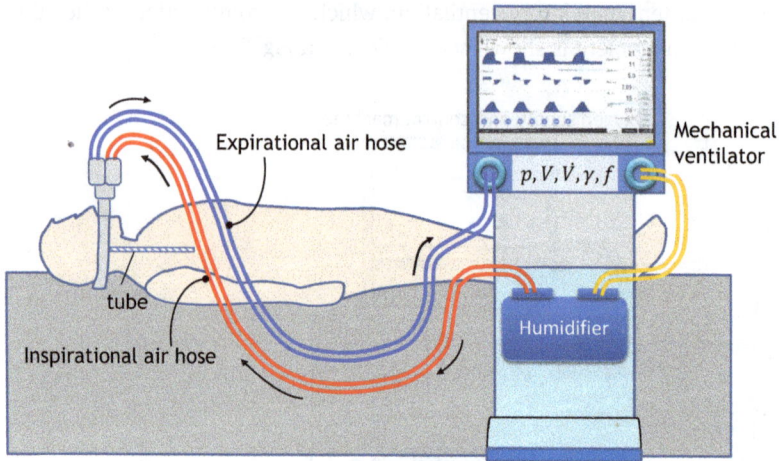

Fig. 9.20: Invasive ventilator with intubation and control of ventilation.

ECMO: The extracorporeal membrane oxidation (ECMO) [13] is the fourth and final method, a simplified scheme is shown in Fig. 9.21. Here the gas exchange takes place from the outside and the lungs are completely relieved. ECMO works similarly to a heart-lung machine (HLM), presented in Chapter 8. There are several types of ECMOs, one of the most common is veno-venous (VV-ECMO), which draws deoxygenated blood from large veins (such as the femoral vein or internal jugular vein). The blood is then enriched by an oxygenator and the oxygenated blood is returned to a large vein. There are two main differences compared to the HLM: With ECMO, in contrast to HLM, the heart is still actively involved in the circulatory system. Second, ECMO

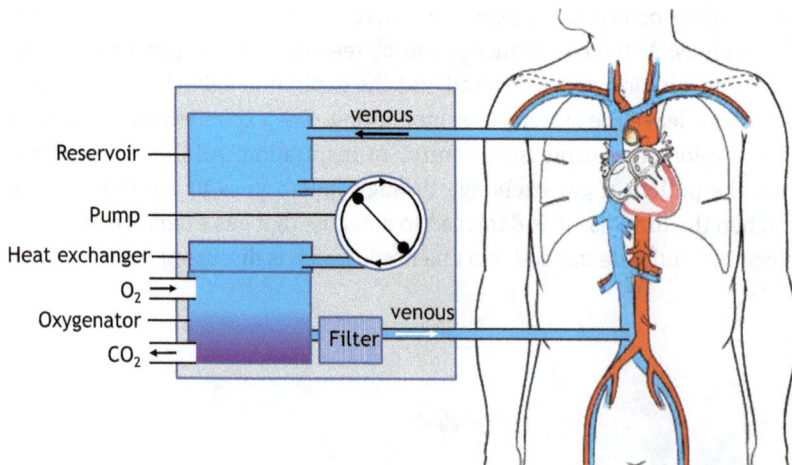

Fig. 9.21: Simplified scheme of a VV-type ECMO machine.

devices can be active for a few weeks while HLM only works for a few hours during surgery, giving the lungs time to heal without aggressive ventilation. Nevertheless, the ECMO is viewed as the last resort due to the high technical and personnel requirements, the costs and the risk of complications.

> Four types of ventilators are used for the treatment of lung diseases. In order of severity of the disease they range from nasal cannula, noninvasive ventilator, invasive ventilator, to extracorporeal membrane oxidation (ECMO). !

9.8.2 Oxgen generators

Ventilators use oxygen gas at a concentration higher than ambient. Machines that deliver high concentrations of oxygen are called oxygen generators or oxygen concentrators. There are two main types available in the market, and another one is in the developing stage. The working principles, as well as the advantages and disadvantages, are briefly explained below.

Cryogenic distillation: Oxygen generators via distillation [14] use the fact that the boiling point of oxygen (90 K) is higher than the boiling point of nitrogen (77.2 K). Therefore, when cooling air to a temperature below 90 K but above 77.2 K, oxygen will become liquid, while nitrogen remains in the gas phase. The schematic of a cryogenic destillator is shown in Fig. 9.22.

The ambient air is drawn in and compressed by a multi-stage air compressor to 5–8 bar. The compressed air is cooled, and water vapor in the incoming air is condensed and removed, as the air passes through a series of interstage coolers. Liquid nitrogen is used to cool the gaseous air in the heat exchanger. The cooled air is then transferred to the distillation column. Nitrogen gas is extracted from the top of the column due to its relatively lower boiling point, and oxygen is removed from the bottom of the column. The excess feed gas in the column is returned to the distillation column for further purification for several stages until the desired oxygen concentration is reached. The oxygen produced can be used either in liquid form or in compressed gas cylinders. The liquidation of gases via distillation is known as the Linde[7] process, which is based on the Joule[8]-Thomson[9] effect. The Joule-Thomson effect describes the cooling of gases upon decompression.

---

7 Carl von Linde (1842–1934), German engineer.
8 James P. Joule (1818–1889), English beer brewer and physicist.
9 William Thomson later Lord Kelvin (1824–1907), English physicist.

Cryogenic distillation has the advantages of high daily throughput (>100 tones per day) and an excellent oxygen purity (>99%). The disadvantage is the expensive and large infrastructure that extends normal clinic operation. Therefore, liquid oxygen or compressed gas bottles are delivered to the hospitals while the production remains off-site.

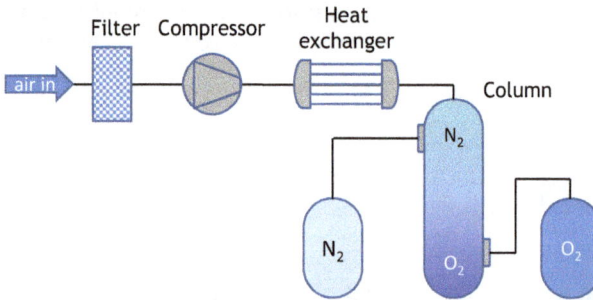

Fig. 9.22: Simplified scheme of a gas liquefier according to the Linde process.

Pressure swing adsorption (PSA): The most common method of generating oxygen is the pressure swing adsorption method (PSA) [15]. This process uses special selective adsorption properties of molecular sieves, such as zeolites. The two-step process is schematically shown in Fig. 9.23.

Air is first compressed and dried to remove water vapor. The compressed air is then passed through one of two cylinders filled with a molecular sieve, mostly zeolites. Mainly nitrogen is adsorbed in the sieve, while purified oxygen flows through the sieve and can exit through an outlet valve. Some of the oxygen is used in the reverse flow to regenerate the second cylinder. The pressure in the second cylinder is lowered before the oxygen can pass through. This removes nitrogen from the zeolite surface by desorption. In the second part of the cycle, the compressed and dried air is now passed through the second cylinder, which before has been cleaned. Some of the oxygen is diverted again to clean the first cylinder for the next lap. With this alternating adsorption-desorption cycle combined with alternating high-pressure and low pressure, a continuous flow of oxygen is available at the outlet.

Zeolites are predominantly used as molecular sieves [16]. These particles give the best results in terms of gas separation and purity. Most common are Li and Na zeolites. The Li-zeolites are more expensive but more effective. In general, zeolites belong to a class of materials that is characterized by an open pore structure and a very large internal surface area, which is particularly favorable for the adsorption of gases.

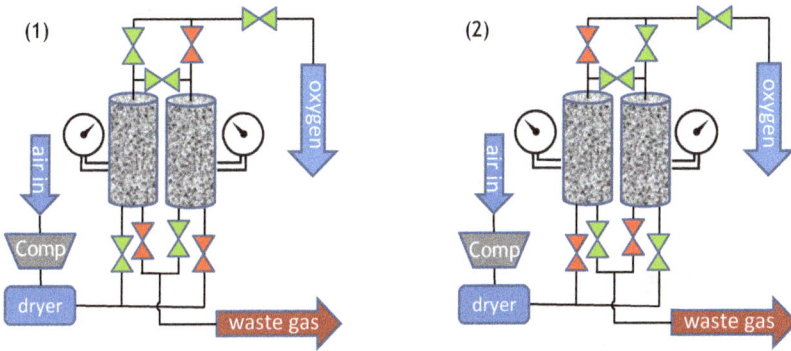

Fig. 9.23: Oxygen generator via the pressure swing adsorption method. Shown are two cycles: (1) adsorption at high pressures in the left cylinder, desorption at low pressures in the right cylinder. (2) reverse flow, adsorption in the right cylinder, and desorption in the left cylinder. Green: open valve; red: closed valve. Comp = compressor. Pressure gauges indicate high and low pressures during airflow.

Membrane diffusion: The two methods of oxygen production described above are technically mature and deliver large amounts of very pure oxygen (see Tab. 9.7). However, both are energy-consuming and therefore expensive. For this reason, another method of oxygen production based on diffusion through membranes has been developed for many years [17]. Membrane technologies are seen as inexpensive, environmentally friendly, and energy-saving alternative gas separators and oxygen generators. The schematics of membrane gas separator is shown in Fig. 9.24.

The physical basis of the membrane method is the diffusion of gases through polymers in the form of threads or films. If a gas mixture flows through the membranes and the permeability of one gas is greater than that of the other, separation occurs. In analogy to Fick's first law combined with the law of Hagen–Poiseuille, the gas flow can be described by:

$$I_V = \dot{V} = P_i \frac{A}{L} \Delta p. \tag{9.37}$$

Here P_i is the permeability of gas component i in a gas mixture like air, A is the cross-sectional area of the membrane and L is its length. Δp is the pressure difference between inlet and outlet. The permeability is therefore defined as:

$$P = \frac{\dot{V}L}{A\Delta p}. \tag{9.38}$$

Under ideal circumstances, the selectivity of two gas components A and B may be determined by the ratio:

$$S = \frac{P_A}{P_B}. \qquad (9.39)$$

In general, oxygen diffuses faster than nitrogen, and therefore $S = P_{O_2}/P_{N_2} > 1$. So far, selectivity of better than $S > 7$ and oxygen purity of about 40% have been reached, which is rather promising [17].

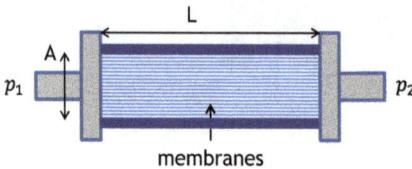

Fig. 9.24: Polymer membranes are stacked together for diffusional separation of gases.

Tab 9.7 summarizes the various oxygen production processes including production capacity and purity of the oxygen produced. The oxygen generators are used on the one hand for large-scale industrial oxygen supply and, on the other hand, for inpatient and outpatient care. The oxygen generators require different infrastructures and must therefore be considered individually. The liquefaction of oxygen is very complex and is therefore carried out in specialized factories. Tanks of different sizes as needed are then delivered to hospitals. Small tanks of 1–2 l are also distributed directly to patients for ambulatory use. PSA systems are available in every size, from large-scale high throughput industrial applications to complete hospital care and to portable devices for mobile outpatient use. PSA devices have the advantage that they continuously deliver oxygen without changing tanks. There are now available small portable devices powered by high energy density batteries to ensure an oxygen supply for several hours. Another alternative is a high-pressure oxygen cylinder. Oxygen cylinders are easy to handle, but as the supply time is relatively short, they are only practical for outpatient and mobile patient care. Since the membrane technology is not yet fully developed, it will not be discussed further. In summary, and as can be seen from the table, for hospitals as well as for outpatient care, PSA is the most practical and convenient oxygen generator. Further information is given in the reviews [18, 19].

Tab. 9.7: Comparison of different production technologies for oxygen gas.

	Cryogenic distillation	PSA	Membrane
Portability	no	yes	yes
Scalability	no	yes	yes
Production volume	>100 tones/day	20–100 tones/day	10–25 tones/day
Gas purity	>99%	95%	25–40%

Three methods for oxygen production are available: cryogenic distillation, pressure swing adsorption (PSA), and membrane diffusion. PSA is the most practical and convenient oxygen generator for hospitals and outpatients. !

9.9 Summary

S9.1   The respiratory tract consists of the pharynx, larynx, trachea, bronchi, bronchioles, and terminal alveoli.

S9.2   The task of the lung is O_2-CO_2 gas exchange.

S9.3   Gas exchange is controlled by perfusion of the lung with blood, ventilation of the lung with air, and diffusion of gases across the blood-air barrier.

S9.4   The gas exchange occurs in the alveolar space, which provides a surface of about 80 m^2.

S9.5 CO_2 transport in the blood is very different from O_2 transport.

S9.6   The gas exchange occurs rapidly across the blood-air barrier and follows the concentration gradient via diffusion.

S9.7   Gas exchange rate is about 250 ml O_2/min or 360 l/day.

S9.8   Alveolar pressure difference is responsible for gas transport in and out of the lung through the airways.

S9.9   Intrapleural pressure difference is responsible for maintaining an underpressure in the intrapleural space.

S9.10   The lungs work rhythmically via under-pressure and overpressure for inhaling and exhaling air, respectively.

S9.11   The mechanical power required by the lung is very small because of the elastic recoil properties of the thorax.

S9.12   The tidal volume is 0.5 l, the breathing frequency is about 15 per minute.

S9.13   Surfactants in the alveolar membrane reduce the tension by a factor of 9. Without surfactants, breathing would not be possible.

S9.14   Lungs and thorax form a joint elastic system. At rest and in equilibrium, the lung is expanded and would contract without coupling to the thorax, while the thorax is compressed and would expand without coupling to the lung.

S9.15   The compliance of the lung is high but decreases with increasing lung volume.

S9.16   The total compliance of thorax and lung is evaluated by adding their reciprocal values.

S9.17   The airway resistance in the bronchial tree is lower than in the trachea in spite of lower individual cross sections because of laminar flow and an increasing number of parallel airways.

S9.18   The airflow in the trachea is partially turbulent but becomes laminar in the bronchus.

S9.19 There are four types of lung ventilation and oxygen supply ranging from low care to high care: cannula, noninvasive ventilation, invasive ventilation (intubation), and ECMO.

S9.20 The two main methods of oxygen production are pressure swing adsorption and cryogenic distillation.

Questions

Q9.1 What are the three main organs involved in respiration?

Q9.2 Where does the gas exchange take place?

Q9.3 Which gases are exchanged?

Q9.4 What is the function of the pleural cavity?

Q9.5 What happens if the pleural cavity is ruptured?

Q9.6 What is more important for oxygen transport in blood, the physical solution or the chemical binding?

Q9.7 How is CO_2 transported in blood?

Q9.8 The vital capacity of the lung is composed of which lung volumes?

Q9.9 How big is the tidal volume?

Q9.10 What is the breathing frequency at rest?

Q9.11 The pressure in the alveoli varies sinusoidally during inhalation and expiration. What is the pressure amplitude?

Q9.12 Why is the volume work of the lung during breathing negligible?

Q9.13 How is the compliance of the lung/thorax affected by diseases?

Q9.14 Why are surfactants of the alveoli important?

Q9.15 What is the main contribution to the airway flow resistance during respiration?

Q9.16 Which types of ventilators are used in practice for the treatment of lung diseases?

Q9.17 Which methods for producing oxygen are technically available?

Attained competence checker + 0 −

	+	0	−
I can name the organs which belong to the respiratory system			
I know what the three main steps are enabling gas exchange in the lung			
I know which physical parameters determine the speed of gas exchange in the lung			
I can distinguish between tidal volume and vital capacity			
I know how the air pressures in the alveoli can be change to allow inspiration and expiration			
I know the concept of compliance and I can apply this concept to the lungs			
It is clear to me that surfactants are essential to lower the surface tension of the alveoli			
I can describe how a spirometer works			
I appreciate that the respiratory resistance is highest in the trachea, and I know why			
I realize that the airflow in the trachea may be turbulent			
The airway resistance of the lung is surprisingly low, and I know why			

I know what surfactants are and how they affect the alveoli of the lung

I know how lung diseases like emphysema and fibrosis can be distinguished

I know under which circumstances different types of ventilators are used

I know how oxygen gas for ventilators are produced and what the advantages and disadvantages are

Suggestions for home experiments

HE9.1 **Tidal volume**

Test your own individual tidal volume, vital capacity, and breathing frequency by exhaling into an empty balloon.

HE9.2 **Lung model**

With a simple arrangement, you can explore the working principle of the lungs. Get an empty and clear plastic water bottle. Cut the bottom of the bottle with a knife or a pair of scissors without hurting yourself. Seal the screw opening with a small balloon (red in the picture) and pull the balloon inwards. Strech a second balloon (blue in the picture) over the lower end. Now pull the blue balloon and you will see the red balloon inflate.

Exercises

E9.1 **Oxygen concentration:** Please verify the following numbers cited in the text: "Each erythrocyte carries about 10^9 O_2 – molecules in saturation. This makes about 5×10^{21} oxygen molecules per liter blood, or roughly 0.01 mole O_2/l, much more than physical solution would deliver".

E9.2 **Flow resistance and critical radius:** When breathing, there is a risk of turbulent flow in the trachea. With a tidal volume of 0.5 l/s, calculate the critical radius of the trachea at which turbulence can be expected. The density of air is 1.2 kg/m^3; the viscosity is 17×10^{-6} Pa · s.

E9.3 **Smokers lung:** The flow resistance is greatly increased by smoking. Assume that the flow resistance in a patient is 4 times that of a healthy person. How does this affect inspiration time and oxygen supply?

E9.4 **Alveoli volume and surface:** The lungs contain 300 million alveoli. These can be filled with the vital volume of the air. Use this number to estimate the volume of an alveolus and the total surface area of all alveoli together.

E9.5 **Diffusion in the alveoli:** During inspiration, a healthy, resting person absorbs 0.5 liters of air in 1 s. 30% of the inhaled oxygen is exchanged via the air-blood barrier with a wall thickness of 1 µm. Calculate:

  a. the number of O_2 molecules that enter the blood during inspiration;
  b. the diffusion resistance R_D, which follows from the particle flow:

$$I = \frac{\Delta N}{\Delta t} = \frac{\Delta n}{R_D},$$

  where ΔN is the total number of particles flowing per unit of time Δt, and $\Delta n = \Delta N/V$ is the number density;
  c. the diffusion constant for a total surface area of the alveoli of approx. 80 m^2;
  d. What is the estimated permeability P based on 1 µm wall thickness?

E9.6 **Time constant:** Derive a differential equation such that the expirational volume as function of time is $V(t) = V_0 exp(-t/\tau)$, where $\tau = R_L C_L$ is the time constant, R_L is the flow resistance and C_L the compliance of the lung (not to be mixed with the respiratory coefficient RC). Consider a circuit with R_L and C_L arranged in a series.

E9.7 **Oxygen supply:** A patient with COPD (chronic obstructive pulmonary disease) will be supplied with oxygen from a storage bottle. The oxygen bottle has an internal volume of 2 l and contains pure oxygen with a pressure of 200 bar. The oxygen flows continuously and without any change in temperature via a pressure regulator with 1 bar and 1.5 l/min. How long will the oxygen supply last?

E9.8 **Cryogenic oxygen supply:** An outpatient will be supplied with a liquid oxygen bottle. The bottle contains 2 l of liquid oxygen at 90 K. The oxygen flows continuously via a valve at ambient pressures of 1 bar and a rate of 1.5 l/min. How long will the oxygen supply last?

References

[1] Boron WF, Boulpaep EL. Medical physiology. 2nd edition, Philadelphia, London, New York: Saunders W.B. Elsevier; 2012.
[2] Guyton AC, Hall JE. Textbook of medical physiology. 11th edition, Philadelphia, Pennsylvania, USA: Elsevier Saunders; 2006.
[3] Martini FH, Nath J, Bartholomew EF. Essentials of anatomy and physiology. 7th edition, New York: Pearson; 2017.
[4] Chaudhry R, Bordoni B. Anatomy, thorax, lungs. In: StatPearls [Internet]. Treasure Island (FL): StatPearls Publishing; 2021, Available from. https://www.ncbi.nlm.nih.gov/books/NBK470197/.
[5] Faller A, Schünke M, Schünke G. The human body. An introduction to structure and function. Stuttgart, New York: Thieme Verlag; 2004.

[6]   Sander R. Compilation of Henry's law constants (version 4.0) for water as solvent. Atmos Chem Phys. 2015; 15: 4399–4981.

[7]   Scheid P. Chap. 10. p. 255–324. In: Pape H-C, Kurtz A, Silbernagel S, eds. Physiologie. 7th edition, Stuttgart, New York: Thieme Verlag; 2014.

[8]   Evans P. Properties of Air at atmospheric pressure. The Engineering Mindset.com, 2015.

[9]   Rudersdorf D. Entwicklung und Anwendung eines Beatmungssystems zur Untersuchung der Lungenfeinstruktur mittels He-3-Magnetresonanztomographie. Dissertation, JG University Mainz, 2010.

[10]   Irvin CG, Bates JHT. Measuring the lung function in the mouse: The challenge of size. Respir Res. 2003; 4: 4.

[11]   Osadnik CR, Tee VS, Carson-Chahhoud KV, Picot J, Wedzicha JA, Smith BJ. Noninvasive ventilation for the management of acute hypercapnic respiratory failure due to exacerbation of chronic obstructive pulmonary disease (Review). Cochrane Database Syst Rev. 2017; (7).

[12]   Stefan MS, Nathanson BH, Higgins TL, Steingrub JS, Lagu T, Rothberg MB, Lindenauer PK. Comparative effectiveness of noninvasive and invasive ventilation in critically ill patients with acute exacerbation of COPD. Crit Care Med. 2015; 43: 1386–1394.

[13]   Makdisi G, Wang I. Extra corporeal membrane oxygenation (ECMO) review of a lifesaving technology. J Thorac Dis. 2015; 7: E166–E176.

[14]   Hegemann KR, Guder R. Linde-Fränkl-Verfahren der Sauerstoff-Gewinnung. In: Stahlerzeugung. Heidelberg New York: Springer Verlag; 2020, 47–49.

[15]   Ruthven DM, Farooq S, Knaebel KS. Pressure swing adsorption. Weinheim, Germany: Wiley-VCH; 1993.

[16]   Breck DW. Zeolite molecular sieves: Structure, chemistry, and use. New York, London, Sydney, Toronto: Wiley; 1973.

[17]   Chong KC, Lai SO, Thiam HS, Teoh HC, Heng SL. Recent progress of oxygen/nitrogen separation using membrane technology. J Eng Sci Technol. 2016; 11: 1016–1030.

[18]   Magnussen H, Kirsten AM, Köhler D, Morr H, Sitter H, Worth H. Guidelines for long-term oxygen therapy. German society for pneumology and respiratory medicine. Pneumologie. 2008; 62: 748–756.

[19]   Hardavella G, Karampinis I, Frille A, Sreter K, Rousalova I. Oxygen devices and delivery systems. Breathe. 2019; 15: e108–e116.

Further reading

Cameron JR, Skofronick JG, Grant RM. Physics of the body. 2nd Edition, Madison, Wisconsin: Medical Physics Publishing; 1999.

10 Physical aspects of the urinary system

Physical properties of the urinary system

Number of nephrons per kidney	1×10^6
Renal plasma flow	600 ml/min
Weight of kidneys	160 g each kidney
Renal blood flow	0.5 l/min for each kidney
Glomerular filtration rate	120 ml/min
Filtration fraction	20%
Urine production	1.5–2 l/day
Urine flow	1–1.5 ml/min
Fractional excretion of plasma	1%
Fractional excretion of inulin	100%
Plasma creatinine density	0.01 mg/ml
Bladder volume capacity	300–600 ml
Bladder filling rate per kidney	0.5–1 ml/s
Urethra flow rate	10–30 ml/s
Flow resistance	1–3 hPa·s/ml
Bladder compliance	20–60 ml/hPa

10.1 Introduction

A widespread misconception is that whatever we drink will finally end up in the bladder and be disposed of as urine. However, this notion is much too simple. What we drink goes through the normal digestive tract, i.e., from the esophagus to the stomach into the intestine. From there, part of the fluid is cleared through the intestinal wall into the blood circulation and to the liver for processing. The rest continues through the intestines to the fecal exit. The kidneys continuously filter the blood. Water in the plasma that is no longer needed will finally be disposed into the urinary tract. Therefore we can conclude that anything cleared by urinary excretion has been in the blood circulation before. But the reverse conclusion does not hold because water and other waste products also go out through the intestines and the lung. In fact, the kidneys are only one of three emunctories the body maintains for waste disposal. The other two are the lungs for the disposal of CO_2 and water, and the intestines/colon/anus for the disposal of feces.

The kidneys are a multitasking organ for maintaining an array of body functions [1, 2]:

1. Regulation of water level in the body;
2. Control of electrolytes and pH balance;
3. Removal of metabolites from blood and excretion of urine;
4. Synthesis of hormones to support the endocrine system (erythropoietin, renin, vitamin D3).

https://doi.org/10.1515/9783110756951-010

Metabolites are metabolic end products. In particular, nitrogen-containing molecules must be excreted by the kidneys such as ammonia, urea, urea acid, and creatinine. Some functions of the kidneys are treated in the next paragraphs related to filtration, diffusion, and perfusion, including discussions of malfunctions and remedies such as dialysis. Our primary focus in this chapter is on renal clearance. In addition, we discuss some properties of bladder filling, pressure build-up, and voiding. For all other aspects, we refer to standard textbooks on physiology.

10.2 Global properties of kidneys

The kidneys are located left and right of the spinal column, below the diaphragm. The kidneys are slightly asymmetrically positioned between the 11th thoracic vertebra and 2nd lumbar vertebra for the left kidney, but one vertebra lower for the right kidney. The kidneys have the shape of a large bean, and they weigh between 80 and 160 g each, the left kidney being slightly heavier than the right kidney (Fig. 10.1).

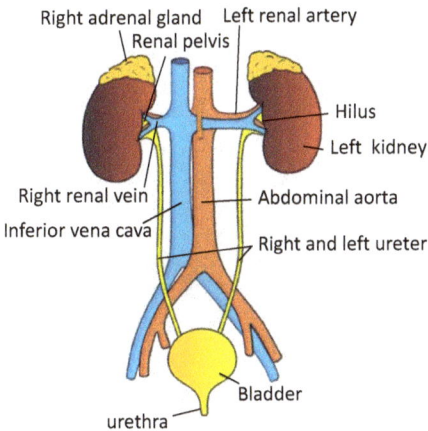

Fig. 10.1: Location of the kidneys and the ureter tract in the body (adapted from Wikimedia, © Creative Commons).

The blood volume rate, i.e., the blood volume flowing through the kidneys per time unit, is 1 l/min of both kidneys, called *renal blood flow* (*RBF*). This is about 20% of the cardiac output (CO) (about 5 l/min). Since only plasma is filtered but not hematocrit, only 0.5–0.6 l/min of plasma runs through both kidneys. This volume is called the *renal plasma flow* (*RPF*). From this 0.6 l/min a mere 20% or just 120 ml/min = 172 l/day is filtered, which is called the *glomerular filtration rate* (*GFR*). The ratio:

$$FF = \frac{GFR}{RPF} \tag{10.1}$$

is the *filtration fraction* (*FF*), which is 120 ml/600 ml = 0.2 or 20%. The production of *primary urine* is about 170–180 l/day. However, most of this is reabsorbed back into the venous bloodstream. Only 1–2 l/day of *secondary urine* is disposed of through the ureter to the bladder. Carl Ludwig[1] was the first to realize that primary urine is produced by filtration in the glomerulus and that the final urine arises from reabsorption in the tubule.

The final urine leaves the kidneys through the left and right ureter and is collected in the bladder. The production rate of the final urine and the filling capacity of the bladder is such that voiding through the urethra is necessary every 3–4 h.

To understand the filtration process's complexity, it is helpful to consider a simplified block diagram according to Fig. 10.2. Part of the arterial blood influx goes into the kidneys via the renal artery. The blood is then processed in *nephrons*. There are about a million nephrons in each kidney, and all work in parallel. Each nephron consists of two main subsystems: *glomerulus* (light blue) for filtration and *tubule* (beige) together with *loop of Henle* for reabsorption and secretion. The filtered and cleaned plasma is fed back into the renal vein and from there to the vena cava. The end urine leaves the kidneys through the ureter tubes. The nephron processing of the plasma is highly redundant. With a loss of 50% nephrons, full functionality is still guaranteed. The cleaning of the blood plasma works according to the house cleaning principle: everything goes out of the house for cleaning and only useful items are allowed to be taken back in again, the rest is disposed.

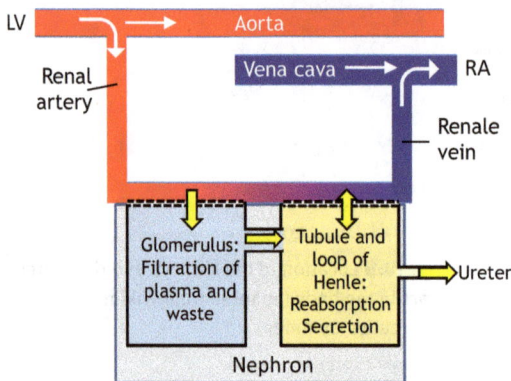

Fig. 10.2: Block diagram of the kidney parts and functions. LV = blood flow from left ventricle; RA = blood flow to right atria.

---

1 Carl F.W. Ludwig (1816–1895), German physiologist.

10.3 Structure of kidneys

Figure 10.3 shows an anatomic cross section through one of the kidneys. This graph gives a first impression of the kidneys' enormous complexity. The main components are the influx of blood through the renal artery, the distribution through capillaries in the renal cortex, and the medulla for the filtration and reabsorption. The outflow of filtered plasma occurs via the renal vein on the one hand and urine through the renal papillae into the ureter, on the other hand, leaving the kidneys at the renal pelvis. The most prominent features are the pyramidal-shaped medullas which reach from the renal pelvis to the cortex. There are seven of those in the human kidney on either side.

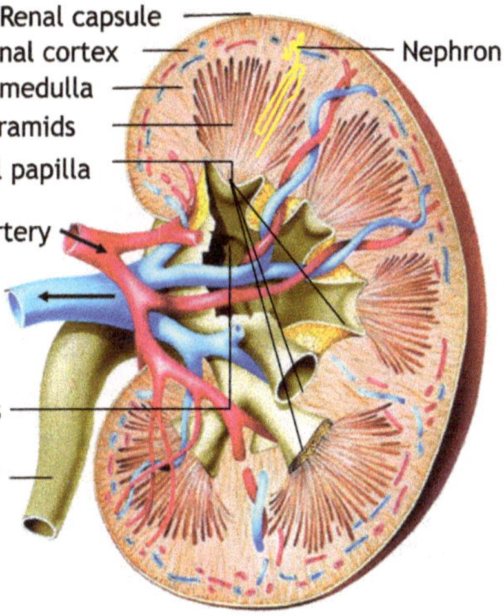

Fig. 10.3: Cross section of the kidney. Most prominent are seven pyramidal-shaped medullas containing nephrons for filtration of the blood. Only one nephron out of a million is sketched (reproduced from OpenStax Anatomy and Physiology, © Creative Commons [3]).

The most crucial functional subunit in the renal medulla is the nephron. Nephron is the Greek word for kidney. However, these days, the term "nephron" is used only for the filtration subunit. The nephron extends from the cortex, where the filtration unit is located, to the lower part of the medulla, where reabsorption and secretion occurs. The main components of the nephron are shown schematically in Fig. 10.4.

The nephron consists of a glomerular capsule known as *Bowman's capsule*,[2] of the proximal tubule, the *loop of Henle*,[3] a distal tubule, and a collecting duct for the urine. The plasma is "pressed" out of the afferent arterioles in the glomerular capsule.

---

2 William Paget Bowman (1816–1892), British physiologist and ophthalmologist.
3 Friedrich Gustav Henle (1809–1885), German anatomist.

Subsequently, the plasma is filtered as primary urine and reabsorbed in the proximal and distal tubules. The remaining filtrate is collected in the duct leading to the ureter and finally to the bladder. In the following, we discuss these different parts in more detail.

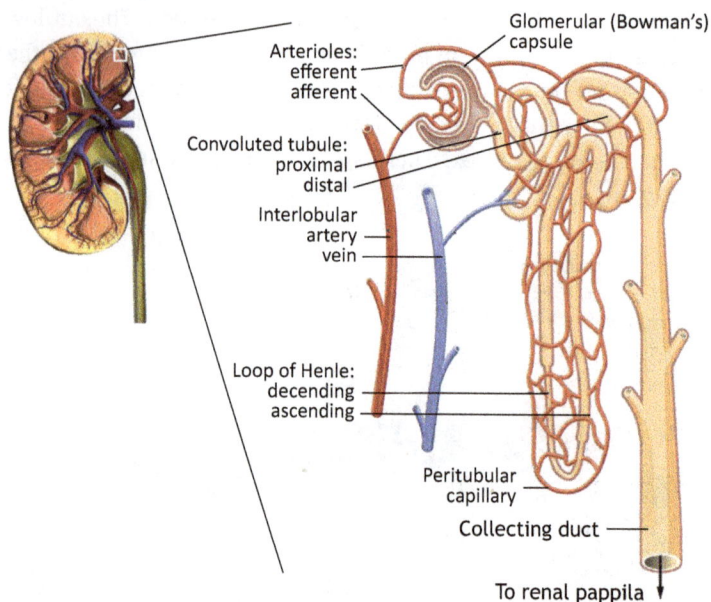

Fig. 10.4: Main parts of a nephron (adapted from Anatomy & Physiology, Connexions Web site. http://cnx.org/content/, © creative commons).

10.4 Filtration

Filtration takes place in the glomerular capsule (Bowman's capsule), sketched in Fig. 10.5. The glomerulus located inside the capsules is a tiny ball-shaped structure composed of a dense network of arterioles. The arterioles have pores on the inside and outside for passing the plasma into the proximal tubule. A pressure difference between the renal afferent arteriole and the proximal tubule is required for effective filtration. If the *effective filtration pressure* p_{eff} is reached, the plasma is pressed out of the blood through narrow holes acting as a sieve.

Filtration is achieved by a three-layer sieve system shown in Fig. 10.6. Starting from the blood side, we first encounter the *endothelial fenestration* (opening) of the glomerulus. This filter stops the erythrocytes but allows all other components of the plasma to pass. The second filter is the *glomerular basement membrane (GBM)* which hinders the filtration of larger proteins. And finally, there is a third membrane filter consisting of *podocyte pedicels* (little feet), preventing filtration of medium-size proteins.

Fig. 10.5: Schematics of the Bowman's capsule containing arterioles and filters.

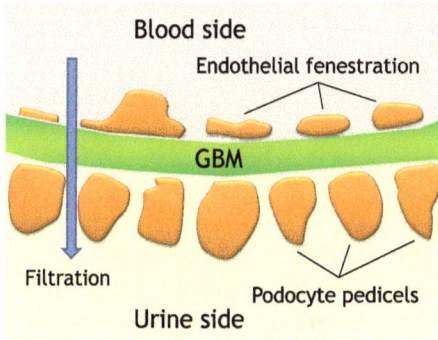

Fig. 10.6: Filter in the glomerulus consisting of three sieves with different pore holes. GBM = glomerular basement membrane.

Unlike ion-channels in cell membranes, these three filters act as a mechanical sieve separating particles only by size but not by function or charge. In Fig. 10.7 is plotted the relative filtrate versus the molecular weight in atomic mass units (amu). The *relative filtrate* is defined as the ratio of particle concentration in the plasma after filtration versus concentration in the plasma before filtration. Anything above 20 000 amu is not filtered out. In particular, the erythrocytes and hemoglobin do not pass the filter but smaller proteins can pass. For nanomedical applications discussed in Chap. 17/vol. 3, it is important to know what the renal clearance of nanoparticles is with respect to size. Investigations have shown that nanoparticles (NP) with a hydration diameter (HD) of 5 nm or less are filtered out, whereas NPs with HD > 8 nm are not capable to glomerular filtration. In the intermediate size range, filtration depends on the specific shape and surface composition of NPs [4]. In any case, the glomerular filtrate, also called primary urine, contains mainly water, glucose, salts, urea ($CO(NH_2)_2$) as the end product of protein metabolism, and urea acid ($C_5H_4N_4O_3$) as the end product of purine metabolism.

> The kidneys task is the cleaning (clearance) of the blood plasma from metabolic end products and to keep the water balance of the body. Clearance is achieved by filtration and reabsorption.

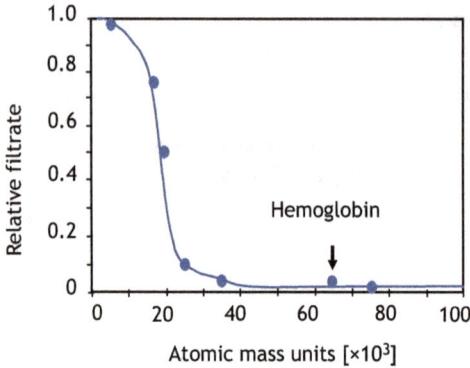

Fig. 10.7: Relative filtrate as a function of atomic weight. [5].

About 180 l of glomerular filtrate is produced daily, but only 1–2 l is disposed as urine. As discussed in the next paragraph, the other 99% of the filtrate is being resorbed. Tab. 10.1 gives an overview of the components of the filtrate in relation to the plasma.

Tab. 10.1: Components in the plasma in order of molecular weight, concentration, and filterability [5].

Component	Molecular weight [amu]	Concentration in plasma [mmol/l]	Filter-ability
Water	18	55,5	1
Sodium	23	135	1
Urea	60	3–7	1
Glucose	180	4–5	1
Inulin	5200		1
Myoglobin	17,000		0.75
Hemoglobin	68,000		0.005

The *effective filtration pressure* p_{eff} is given by the blood pressure in the glomerular capillaries p_G, minus the hydrostatic pressure p_H in the Bowman capsule, and minus the osmotic pressure Π_{Os} from colloids in the plasma:

$$p_{eff} = p_G - p_H - \Pi_{Os} \tag{10.2}$$

Using typical values [6] for $p_G \approx 100$ hPa, $p_H \approx 20$ hPa, and $\Pi_{Os} \approx 40$ hPa, the effective filtration pressure p_{eff} is about 40 hPa. If p_{eff} would drop to zero, active filtration would cease and only diffusion through the filter will continue, which, however, is by far too slow. Both the *renal plasma flow* (RPF) and the *glomerular filtration rate* (GFR) depend critically on the blood pressure p_G in the glomerulus. If the pressure is too low, both rates will go down. If the pressure is too high, RPF and GFR will increase. In the intermediate plateau region from 100 to 230 hPa, RPF (~600 ml/min) and GFR (~120 ml/min) do not critically depend on p_G. RPF and GFR are plotted as a function of average blood pressure in Fig. 10.8. This relative constancy of GFR and RPF is referred to as *autoregulation*. The autoregulation operates by changing the arterial vessel tension with a negative feedback system: increasing blood pressure will increase the tension, thereby decreasing the vessel radius and the flow rate; with decreasing pressure, the opposite reaction occurs. Furthermore, the kidneys are located at about the *hydrostatic indifference level*, where the blood pressure is independent of the posture, standing up, or lying down (see Chap. 8.3.1).

> Filtration takes place in the Bowman capsule acting as a molecular sieve. All molecules with a weight less than 10 000 amu are filtered out of the blood plasma. !

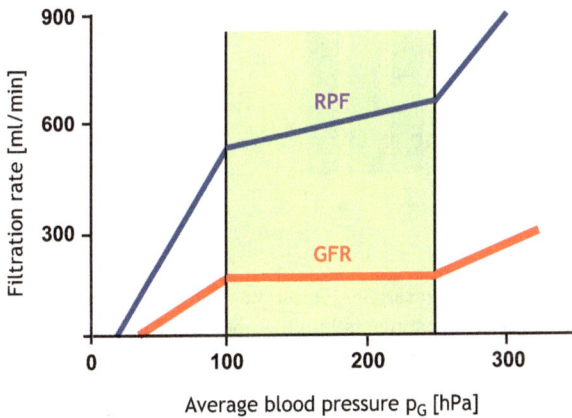

Fig. 10.8: Renal plasma flow (*RPF*) and glomerular filtration rate (*GFR*) as a function of blood pressure p_G in the glomerulus. Autoregulation occurs in the green area.

Infobox I: Mass transport in the body

Mass transport in the body is an omnipresent and permanently active process. On the macroscopic level, we recognize the mass transport of blood, respiratory gases, interstitial fluids, urine, and feces. Mass transport also takes place on the molecular level as seen, for instance, in the nephrons of the kidneys. Four different mass transport mechanisms can be distinguished: diffusion, osmosis, filtration, active secretion. and reabsorption, shown schematically in the lower panel:

(a) Diffusion follows a concentration gradient and is described by Fick's first and second laws. Exchange of molecules across a barrier increases the entropy but does not cost energy.

(b) Osmosis occurs if there is a concentration gradient of the solute across a barrier, but particle diffusion for equilibration is not allowed. Then water volume is shifted across the barrier, equilibrating the osmotic pressure on the expense of volume work.

(c) Filtration is controlled by the pore size of the sieve and the pressure difference on both sides. The pore size determines the maximum particle size of the filtrate, and the pressure difference controls the filtration rate.

(d) Active secretion and reabsorption are both particle transport processes against the concentration gradients. Therefore, these processes are actively supported by a pumping mechanism, in general by the ATP pump.

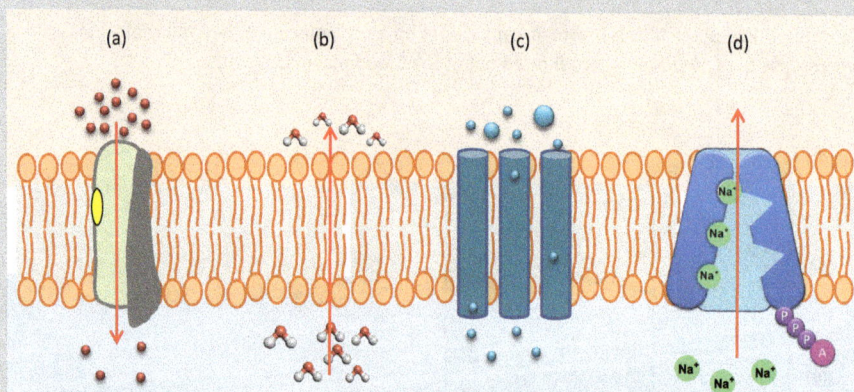

What is the difference between reabsorption and resorption? Resorption implies the elimination of cells or tissues and the absorption of its constituents into the body's circulation. For instance, bone resorption is performed by osteoclasts via break down of bone tissue and release of minerals to the blood. In contrast, reabsorption is the process by which nephrons remove water and solutes from the tubular fluid (primary urine) and return them to the circulating blood. In the context of this chapter, the term "reabsorption" describes the return of water and solutes that have been filtered out of plasma back to the renale vein.

10.5 Reabsorption

To disentangle the complexity of the distal convoluted tubule and the peritubular capillaries depicted in Fig. 10.4, the graph in Fig. 10.9 shows simplified schematics, focusing on the functional parts of the nephron. As already discussed, the blood reabsorbs much of the filtrate in the proximal convoluted tubule within the proximal and distal tubular system. Figure 10.9 depicts only the tubule for clarity, which is strongly intertwined with arterioles for water and ion exchange, as indicated in Fig. 10.4.

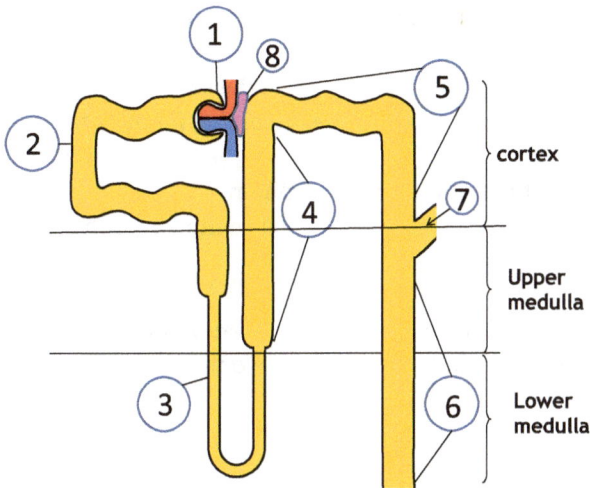

Fig. 10.9: Different parts of the nephron: 1. Bowman's capsule; 2. Proximal convolute; 3. Loop of Henle, descending part; 4. Ascending part of the loop of Henle; 5. Distal convolute; 6. Collecting duct with connection to the ureter; 7. Influx from neighboring duct; 8. Macula densa. For more details are given in the text.

The different parts of the nephron have the following main functions, where numbers in the list refer to the numbers in Fig. 10.9:

1. *Bowman's capsule* contains the glomerulus and the arterioles that filter blood plasma into the proximal tubule, as shown in Fig. 10.5 and discussed in the previous section.
2. *Proximal convolute tubule* is responsible for the reabsorption of most of the water and solvents into the efferent arterioles for delivering them back to the circulation. In fact, the proximal tubule reabsorbs already 2/3 of the water in the filtrate or 80 ml/min. The rest is reabsorbed later, and only 1% of the total filtration is finally excreted into the urine. Furthermore, the capillary system reabsorbs glucose completely and a large fraction of salt ions (Na^+, K^+, Ca^{2+}). The distal tubule processes the remaining filtrate.
3. The *loop of Henle* consists of a thin water-permeable part and a thicker water-impermeable part. The loop extends like a needle from the cortex into the medulla and back again. The task of the thin part is to maintain a hypertonic environment in the medulla via opposite diffusion currents. This passive process helps stabilize an osmotic gradient in the medulla, which is necessary for concentrating urine.

4. In the thicker part of Henle's loop, reabsorption of the cations Na^+, K^+ and the anion Cl^- into the arterioles continues, as well as reabsorption of divalent ions, such as Mg^{2+}. As ions leave the tubule while water cannot penetrate the wall, the osmotic pressure decreases leading to hypotonic conditions in the loop.
5. In the distal tubule, reabsorption of ions into the arterioles continuous. But with the absorption of K^+ ion into the tubule, the conditions become less hypotonic and more isotonic.
6. In the collecting tube, a final control of the water level, the urea concentration, and the acid-base balance takes place before the remaining filtrate goes into the ureter.
7. Influx into the collecting duct from a neighboring nephron.
8. Macula densa, determines and controls the NaCl – concentration in the distal tubule.

Figure 10.10 shows the production of urine schematically starting from the initial renal blood flow. After processing in the kidney, a fraction of only 0.1% of the initial blood volume is finally excreted in the urine.

This is a very brief overview of the parts and functions of the kidneys. Further details can be found standard biology or medical textbooks.

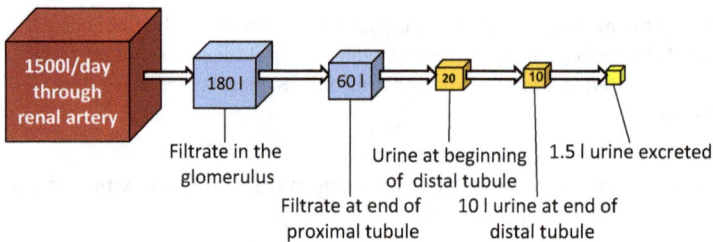

Fig. 10.10: Production of 1.5 l/day urine from a renal blood flow of 1500 l/day.

> From 180 l filtered blood plasma, only 1.5 l urine is excreted daily.

Some additional remarks on the proximal tubule are obligatory, as various processes take place there that are of physical interest. In Infobox I, respective terms are explained. Figure 10.11 schematically shows the proximal tubule. The gray shaded area lists all mass transport processes that occur in the nephron. First, there is the active filtration in the Bowman's capsule based on a pressure gradient between the glomerular capillary and the capsule, as already discussed. Next, diffusion occurs from the proximal tube to the peritubular capillary and vice versa due to a concentration difference for one specific substance. The diffusion process is followed, in general, by a shift of water to keep the osmotic pressure constant. Furthermore, ATP ion pumps actively assist ion transport in the proximal tubule. In summary, ion and water transport in the nephron can either be active with the help of pressure differences or assisted by ATP pumps, or diffusive following concentration gradients. All three processes work together for the most effective filtering and reabsorption of the plasma that cleans the blood, regulates the water level, and controls the pH balance.

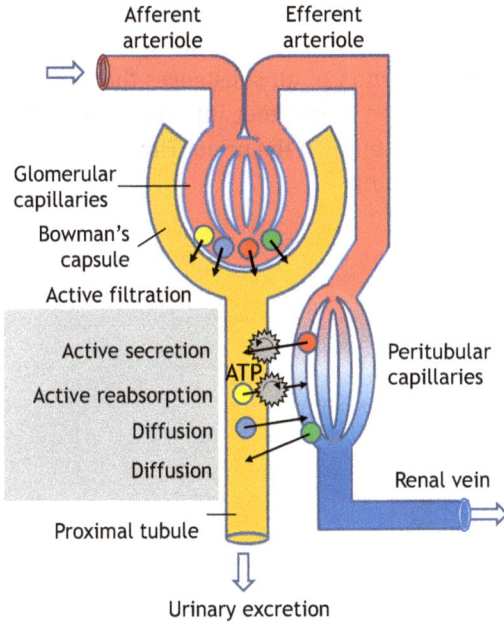

Fig. 10.11: Diffusion and reabsorption in the proximal tubule. Yellow, blue, red, and green circles symbolize different molecules passing the glomerular filter into the proximal tubule. From there the molecules diffuse back and forth or may actively be transported between proximal tubule and peritubular capillaries. Gear wheels indicate active transport.

10.6 Renal clearance

10.6.1 Renal clearance fundamentals

Renal clearance is an important concept, not just for a basic understanding of how the kidneys work. Renal clearance is the volume of blood that is cleared per unit of time. Therefore, the clearance has the unit of a volume rate [volume/time]. Clearance can tell us if the kidneys work properly or if there are any signs of malfunction and disease. Clearance measures the overall function of nephrons: glomerular filtration, tubular reabsorption, and tubular secretion related to arterial input and venous or urinal output, as shown in Fig. 10.12. Clearance is an integral measure of all

individual processes active one after the other at different points along the nephron. In addition, a total of one million nephrons per kidney work in parallel with a
high degree of redundancy for renal clearance. Therefore, clearance measurements
cannot localize individual damage to nephrons but give general information about
the condition of the kidneys. For more detailed microscopic examinations, high-
resolution imaging techniques are required, as explained in Vol. 2. For example,
Fig. 9.13/Vol. 2 shows the time-resolved perfusion of the radioisotope ^{99m}Tc-MAG3
by means of scintigraphy to examine renal clearance.

Renal clearance is based on mass conservation of all substances that are neither metabolized in the kidneys nor synthesized. What goes into the renal artery
must come out either through the renal vein or through the urine. Thus the arterial
input must equal the sum of venous and urine output.

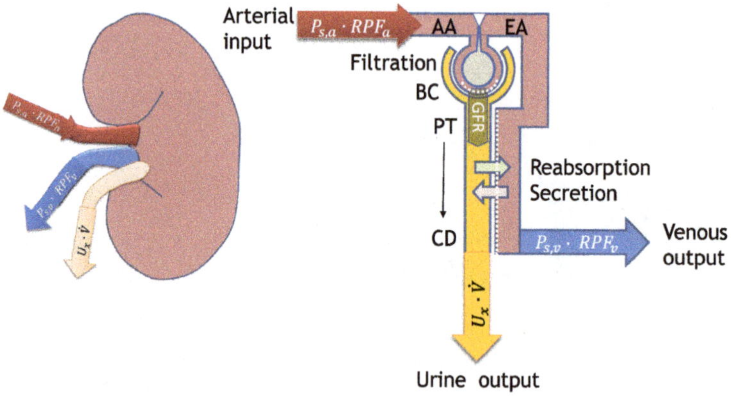

Fig. 10.12: Left panel shows one input and two outputs of the kidney. In equilibrium, the sum of
venous and urine output must equal the arterial input. Right panel shows a simplified schematics
and overview on the main functions of nephrons in kidneys: filtration, reabsorption, and secretion.
The other acronyms stand for: AA = afferent arteriole; EA = efferent arteriole; BC = Bowman's
capsule; PT = proximal tube; CD = collecting duct; RV = renal vein. Arrows indicate flows:
RPV = renal plasma flow, GFR = glomerular filtration rate, $\dot{V}$ = urine flow.

The mass flow rate $\dot{m}$ is the product of mass density and volume flow rate:

$$\dot{m} = \rho \dot{V}. \tag{10.3}$$

Because of mass conservation, the following equation holds for any substance S in
the blood, where ρ_S is the mass density of S in units of [g/l] and $\dot{V}_S$ is the volume
rate or volume flow in units of [l/min]:

$$\rho_S \dot{V}_{Sa} = \rho_S \dot{V}_{Sv} + \rho_S \dot{V}_{Su} \tag{10.4}$$

Here the subscripts a, v, and u refer to arterial, venous, and urine.

In physiology, a different nomenclature is used for the same expression, which we want to adopt for the remainder of this chapter [6]:

$$P_{Sa} \times RPF_a = PS_v \times RPF_v + U_S \times \dot{V} \tag{10.5}$$

In this expression, P_{Sa} and P_{Sv} are the densities of substance S in the renal artery and in the renal vein, respectively. Although the letter P usually stands for pressure, here it is used for the mass density of substance S in the arteries and veins. RPF_a is the renal plasma flow in the renal artery and RPF_v Is the corresponding flow in the renal vein. U_S is the density of substance S in the urine, and $\dot{V}$ is the urine flow. The product $U_S\dot{V}$ is also known as the *urinary excretion rate* of substance S. The plasma carries all substances, and therefore the plasma flow is also the flow for the substances dissolved in the plasma. Therefore, the subscript S can be omitted for RPF_a and RPF_v. It is useful to remember that the RPF is 600 ml/min, GFR is about 120 ml/min, and the urine flow $\dot{V}$ is about 1 ml/min. If a substance S is dissolved in the plasma, the flow of S is the same as the flow of the renal plasma.

Since we are only interested in the arterial input versus urine output, we rephrase the last eq. (10.5):

$$P_{Sa} \times \left(RPF_a - \frac{P_{Sv}}{P_{Sa}} RPF_v \right) = U_S \times \dot{V} \tag{10.6}$$

or

$$P_{Sa} \times C_S = U_S \times \dot{V} \tag{10.7}$$

With

$$C_S = RPF_a - \frac{P_{Sv}}{P_{Sa}} \times RPF_v \tag{10.8}$$

C_S is called *renal clearance* of substance S. C_S has the units of flow: $[C_S] =$ ml/min. According to eqs. (10.7) and (10.8), we have two possibilities to determine the clearance. One is according to eq. (10.8) to take the difference between arterial plasma flow into the kidneys and the venous plasma flow coming out of the kidneys, weighted by the density ratio P_{Sv}/P_{Sa}. The other option is to determine the clearance from the urinary flow:

$$C_S = \frac{U_S}{P_{Sa}} \times \dot{V} \tag{10.9}$$

Here we need to determine the density U_S of the substance S in the urine and the density of the same substance P_{Sa} in the arterial blood plasma. The urine flow $\dot{V}$ is more or less constant at 1 to 1.5 ml/min. Note that renal clearance requires urine flow. A substance can only be partly or fully cleared from the blood plasma if some fraction is disposed through the ureter. Table 10.2 provides an overview on the terms used in this Chapter.

Tab. 10.2: Some important terms frequently used in the text.

RPF	Renal plasma flow	600 ml/min
GFR	Glomerular filtration rate	120 ml/min
FE	Fractional excretion	0–5
FF	Filtration fraction	0.2
$\dot{V}$	Urine flow	1–1.5 ml/min
P_S	Density of substance S in the capillaries	
U_S	Density of substance S in urine	
C_S	Clearance of substance S	
$U_S \times \dot{V}$	Urinary excretion rate	
PAH	para-aminohippuric acid	

Now we can rephrase the term *renal clearance* once more in light of the previous definitions:

> *Renal clearance C_S of a substance S is defined as the volume of plasma completely or partially cleared of that substance by the kidneys per unit time.*

Clearance compares a plasma volume per time (=flow) in the afferent arterioles before filtering (RPF_a) and the same plasma volume per time after filtering in the venous output (RPF_v). If the same amount of substance S is still present, the clearance is zero. If the substance is completely removed, the clearance is identical with the RPF_a: $C_S = RPF_a$. For instance, the solute para-aminohippuric acid (PAH) has the property that it is completely removed from the plasma by filtering and secretion, and therefore PAH can be used for determining the RPF_a, as described in more detail later. In the following, we will consider different filtering and clearance cases to illustrate the kidneys' function.

> **!** Renal clearance of a substance is the plasma volume cleared of that substance per unit time.

10.6.2 Case studies of renal clearance

Some illustrative examples for renal clearance are sketched in Fig. 10.13 for different cases of filtration, reabsorption, and secretion.

In panel (a), the substance S_1 marked in red and dissolved in the plasma has the same density in the afferent arterial plasma and the renal vein: $P_{S1a} = P_{S1v}$; and the plasma flows are identical: $RPF_a = RPF_v$. Hence, the clearance is $C_{S1} = RPF_a - \frac{P_{S1v}}{P_{S1a}} \times RPF_v = 0$.

In panel (b), the substance S_2 with density P_{S2} is partially filtered out in the Bowman's capsule. As no reabsorption from the tubule into the efferent veins or secretion from the veins into the tubule takes place, the glomerular filtration flow of that substance $(P_{S2} \times GFR)$ must equal the urinary excretion $(U_{S2} \times \dot{V})$:

$$P_{S2} \times GFR = U_{S2} \times \dot{V} \tag{10.10}$$

Therefore, the glomerular filtration rate is:

$$GFR = \frac{U_{S2}}{P_{S2}} \times \dot{V}. \tag{10.11}$$

Here, eq. (10.11) is identical with eq. (10.9) and therefore, the glomerular filtration rate must be identical with the clearance: $C_{S2} = GFR$.

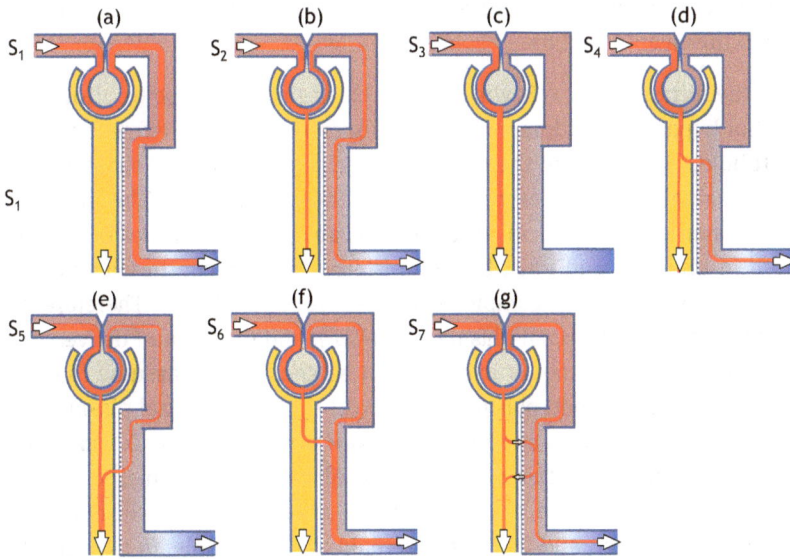

Fig. 10.13: Seven different possibilities for clearance of substance S_i marked in red. (a) no clearance; (b) partial filtering and no reabsorption; (c) complete filtering and no reabsorption; (d) complete filtering and partial reabsorption; (e) partial filtering and secretion; (f) partial filtering and complete reabsorption; (g) partial filtering, partial reabsorption and secretion.

In panel (c), we discuss, as an example, filtration of the polysaccharide inulin S_3, not to be confused with the hormone insulin. The Bowman's capsule completely filters inulin, and neither reabsorption nor secretion in the tubule takes place. Therefore, inulin serves as a *"gold standard"* for renal clearance. We assume that the administered density of inulin in the plasma is $P_{inu} = 1$ mg/ml. As we know the values for GFR (=120 ml/min) and the urinary flow $\dot{V}$ (=1 ml/min), the density of inulin in the urine is: $U_{inu} = GFR \times P_{inu}/\dot{V}= 120$ mg/ml and the density ratio: $U_{inu}/P_{inu} = 120$. Due to the glomerular filtration, densification of inulin in the tubule occurs by a factor of 120, while the flow is reduced by a factor of 120 to keep the balance. We may also argue that a density measurement of inulin in the urine yields $U_{inu} = 120$ mg/ml. Therefore, the excretion rate must be the same as the filtration rate: $P_{inu} \times GFR = U_{inu} \times \dot{V}$. Furthermore, the glomerular filtration rate of inulin must be identical with the clearance: $C_{inu} = GFR = 120$ ml/min. This shows that by measuring the clearance of inulin, one can determine the GFR of the kidney. Higher or lower values than the standard may indicate malfunctions in the nephrons.

The clearance of a substance S_x is often normalized by the inulin clearance; the ratio is called *fractional excretion (FE)*. The FE for inulin is 1 [1, 6, 7].

In panel (d), another substance S_4 is also completely filtered out in the Bowman's capsule. However, in contrast to the previous case, 50% is reabsorbed in the peritubular capillaries and 50% remains in the tubule. Therefore, the ratio must be $U_{S3}/P_{S3} = 60$, the clearance is: $C_{S3} = GFR \times 0.5 = 60$ ml/min, and the $FE = 0.5$. This example refers to the clearance of urea.

In panel (e), the filtration in the Bowman's capsule is similar to case (b). However, the portion that bypasses the glomerular filter is secreted from the efferent arterioles into the tubule. In the end, in the renal vein, the plasma is free of substance S_5, and the total original density is excreted through the urinary tract. This situation occurs for para-amino hippuric acid (PAH). PAH is an organic acid that is not normally present in the body and must be administered through continuous intravenous infusion for testing. PAH is completely cleared from the plasma. Therefore the clearance is equal to the total renal plasma flow (RPF), and the amount delivered to the kidneys is equal to the amount excreted. To be specific, we assume that the PAH density in the arterial blood plasma is $P_{PAH} = 0.1$ mg/ml. The renal plasma flow RPF is 600 ml/min. Thus the amount $RPF \times P_{PAH} = 600$ ml/min $\times$ 0.1 mg/ml = 60 mg/min will be loaded into the glomerulus for filtration, from which only 20% or 12 mg/min will be filtered and excreted into the tubule. The rest will remain in the plasma, bypass the filter, and finally be secreted into the tubule. In the end, the total amount of PAH in the urinary tract will be again = 60 mg/min, which is the excretion rate:

$$U_{PAH} \times \dot{V} = RPF \times P_{PAH}. \tag{10.12}$$

The clearance of PAH is therefore:

$$C_{PAH} = \frac{U_{PAH} \times \dot{V}}{P_{PAH}} = RPF. \tag{10.13}$$

Substituting numbers, we find:

$$C_{PAH} = \frac{60\,\text{mg/min}}{0.1\,\text{mg/ml}} = 600\,\frac{\text{ml}}{\text{min}}.$$

Therefore, using the PAH test, the RPF can be determined. The fractional excretion is $FE = C_{PAH}/C_{inulin} = 600/120$ ml/min $= 5$. This excretion is the highest FE known.

In panel (f), the substance S_6 is partially or entirely filtered and subsequently completely reabsorbed from the proximal tubule to the efferent arterioles. Therefore, after filtration, the density in the tubule $U_{S6} = 0$ and the urine flow of substance S_6 is $\dot{V}_{S6} = 0$. Hence, the renal clearance is $C_{S6} = 0$ ml/min, although glomerular filtration has taken place. S_6 will not appear in the urine; everything is back in the circulation. This clearance is, in fact, the case for glucose, which is reabsorbed entirely from the proximal tubule such that $FE = 0$.

Finally, the discussion of panel (g) is left to exercise E10.4.

10.6.3 Creatinine clearance

Although inulin serves as a standard, a simpler test can actually be performed with creatinine [8]. Creatinine is always present in the body as a metabolic waste product of creatine. Creatine is produced by the body to supply energy mainly to muscles for anaerobic activity. Creatinine is removed from the body entirely via glomerular filtering in the kidneys, similar to inulin. Thus the equation for creatinine clearance holds:

$$C_{Cr} = \frac{U_{Cr}}{P_{Cr}} \times \dot{V} = GFR. \tag{10.14}$$

This equation predicts that in a steady-state situation, the metabolic production in muscles equals the urinary excretion rate $U_{Cr} \times \dot{V}$ of creatinine. Assuming that creation and excretion rates remain fairly constant, the product $P_{Cr} \times C_{Cr} = $ const. Therefore, P_{Cr} should be inversely proportional to C_{Cr}. In Fig. 10.14, the plasma creatinine concentration P_{Cr} is plotted versus the glomerular filtration rate GFR. According to this figure, the prediction appears to hold. For instance, for a healthy kidney with a GFR of 100 ml/min, the plasma creatinine density is approximately 0.01 mg/ml. The product $GFR \times P_{Cr} = 100$ ml/min $\times$ 0.01 mg/ml $= 1$ mg/min corresponds to the production rate of creatinine. Thus creatinine production rate and creatinine excretion rate are equal. If by some reason the GFR drops to 50 ml/min, the creatinine level in the plasma will rise to 0.02 ml/mg, while the product $GFR \times P_{Cr}$ remains constant. Creatinine tests are easy to perform. One just needs a sample of venous blood and a similar sample of urine and analyze them with respect to the creatinine density, which yields the desired ratio. If the kidney function is not normal, the creatinine level in the blood will increase since the glomerulus filters less creatinine and therefore the urine excretes less creatinine.

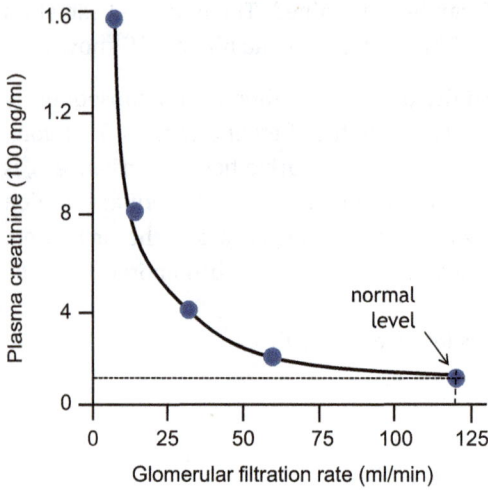

Fig. 10.14: Density of creatinine in the plasma is plotted versus the glomerular filtration rate (adopted from [6]).

> **!** The creatinine concentration in the blood is a reliable indicator for the glomerular filtration and consequentially for the functioning of nephrons.

10.6.4 Urinary excretion

Aside from very big proteins and erythrocytes, almost anything is being filtered in the glomerulus. However, what will finally be excreted through the ureter duct is a question of reabsorption and secretion in the tubules and collecting duct. The amount of substance excreted in the ureter can be calculated as follows. The filtered part (glomerular filtration rate) is

$$GFR \times P_S.$$

From this, we subtract the reabsorbed part with the rate R_S and add the secreted part with the rate S_S, yielding:

$$(GFR \times P_S) - (R_S - S_S)U_S. \tag{10.15}$$

This sum equals the urinary excretion rate defined before:

$$\dot{V} \times U_S = (GFR \times P_S) - (R_S - S_S)U_S. \tag{10.16}$$

In case that $R_S = S_S = 0$, then $\dot{V} \times U_S = GFR \times P_S$. In fact, this is our case in panel (c) for inulin. Therefore inulin can be used to determine the GFR, as noted before. The reabsorption rate R_S can also be determined in case the secretion rate $S_S = 0$:

$$R_S = GFR \times \frac{P_S}{U_S} - \dot{V} \tag{10.17}$$

Similarly, if the reabsorption rate R_S is zero, we find for the secretion rate S_S:

$$S_S = \dot{V} - GFR \times \frac{P_S}{U_S} \tag{10.18}$$

Now we come back again to panel (d) in Fig. 10.13, which also represents the situation for the plasma flow itself. At the artery input, plasma arrives with a $PRF=$ 600 ml/min, but only 20% are filtered; the remaining 80% bypass the glomerular filter. So the GFR is 120 ml/min. Most of the plasma filtered into the tubule is reabsorbed into the efferent arteriole. Only 1% of filter plasma volume remains in the tubule. Without secretion, we have for the urinary excretion of plasma:

$$\dot{V} \times U_{Plas} = (GFR_{Plas} \times P_{Plas}) - R_{Plas} \times U_{Plas}. \tag{10.19}$$

For the plasma fluids, the densities in the arterial capillaries and in the tubule are the same and correspond to water: $P_{Plas} = U_{Plas} = 1$ g/ml. Then we have for the plasma clearance:

$$C_{Plas} = \frac{U_{Plas}}{P_{Plas}} \times \dot{V} = GFR_{Plas} - R_{Plas} \times \frac{U_{Plas}}{P_{Plas}}, \tag{10.20}$$

or

$$C_{Plas} = \dot{V} = GFR_{Plas} - R_{Plas}, \tag{10.21}$$

substituting numbers:

$$C_{Plas} = 120 \text{ ml/min} - 119 \text{ ml/min} = 1 \text{ ml/min}$$

The fractional excretion of plasma is:

$$FE_{Plas} = C_{Plas}/C_{inulin} = 1 \text{ ml min}^{-1}/120 \text{ ml min}^{-1} \approx 0.01 = 1\%$$

This corresponds to the daily urinary volume of about 1.5 l.

Tab. 10.3 lists different possible relations between glomerular filtration (F), renal clearance C, and urinary excretion E and conclusions drawn from these (in-) equalities. Tab. 10.4 is an overview of clearance and fractional excretion for a few standard solvents, and Tab. 10.5 summarizes different blood and urine tests discussed in the text. Finally, kidney functions can also be tested by radioactive markers, as will be discussed in Chap. 9/Vol. 2.

Tab. 10.3: Overview of filtration, excretion, and clearance of substances in blood plasma. F_S = Filtration of substance S; E_S = excretion of substance S; C_S = clearance of substance S.

Molecule S filtered at glomerulus	Renal processing
$F_S > E_S$	Net reabsorption of S
$F_S < E_S$	Net secretion of S
$F_S = E_S$	No exchange of S
$C_S < C_{inulin}$	Net reabsorption of S
$C_S > C_{inulin}$	Net secretion of S
$C_S = C_{inulin}$	No exchange of S

Tab. 10.4: Clearance and fractional excretion of different representative substances in the kidney.

Substance	Clearance [ml/min]	Fractional excretion
Glucose	0	0
Plasma	1	0.01
Urea	60	0.5
Inulin	120	1
Creatinine	120	1
PAH	600	5

Tab. 10.5: Overview of different renal tests. Samples may be taken from the blood and/or from the urine.

Test	Blood	Urine
Inulin	GFR	
PAH	RPF	
Creatine	GFR	GFR
Urea	Rest volume	
Glucose	Rest volume	

10.7 Artificial filtering: Hemodialysis

If the creatinine level in the blood is too high compared to standard values of 0.007–0.012 mg/ml, this may indicate a reduced *GFR*, resulting from a kidney malfunction.

One of the reasons for kidney malfunction is *diabetes mellitus* [9], see Infobox II. Usually, glucose in the plasma is completely removed from the plasma by glomerular filtration, followed by complete reabsorption from the tubule into the peritubular capillaries. Under normal conditions, the urine is free of glucose.

If the blood contains a fairly high concentration of glucose (>1.80 mg/ml), this overloads the ability of the tubing system to reabsorb glucose. Then more glucose will remain in the urine. This increases the osmotic pressure in the tubule and hinders the absorption of water into the blood. Then more urine is excreted that contains more glucose than normal. This process leads to a thickening of the glomerular basement membrane (GBM) (see Fig. 10.6) and swelling of the mesangial cells, located between the capillaries and the glomerulus supporting the capillary walls. The glomerulus begins to leak proteins into the tubule that opens up the filter, resulting in decreased *GFR*. In the end, normal kidney function stops working, including disposing of waste products and controlling water levels and electrolytes in the body. In this final stage, the patient can only be helped with artificial filtering and cleaning, known as hemodialysis. Kidney disease is noticed late. Serious clinical symptoms often do not appear until the number of functional nephrons is 70%–75% below normal. Indeed, with the remaining 25% to 30%, relatively normal blood levels of most electrolytes and normal body fluid volumes can still be maintained. However, if for some reason the kidneys completely fail and nothing is done to restore their function, the remaining life is only 10 days [11].

Dialysis machines are designed to mimic the function of kidneys by removing waste products and maintaining water levels, electrolytes, and minerals within acceptable margins. Two dialysis methods are commonly used [12]. *Hemodialysis* machines circulate blood outside the body. P*eritoneal dialysis* performs filtration within the body by using the peritoneum as a natural filter for wastes and water. Peritoneum is a membrane covering most of the abdominal organs. In either case, dialysis is an incomplete replacement of lost kidney function, as only filtration of blood is maintained. All other tasks of a normal kidney are not fulfilled, such as the control of electrolytes and the synthesis of hormones.

Infobox II: Diabetes mellitus

We differentiate between type 1 and type 2 diabetes mellitus. Most common is type 2 diabetes mellitus (T2DM), worldwide (90%). In type 1, insulin production is completely stopped; in type 2, the efficiency of insulin is reduced despite an increased production rate.

After a meal, the glucose level in the blood increases within 30 min. The pancreas senses the raise of glucose concentration and releases insulin to speed up the metabolism of glucose in the cells.

Insulin is a hormone (should not be confused with inulin) that is produced by the β cells of the pancreas and released into the blood when demanded. Insulin is needed for the cellular metabolism of glucose. Glucose that circulates in the blood can only get into the cells with the help of insulin. Insulin is a gate opener, so to speak. In the cell, the cellular metabolism of glucose generates energy, as described in Chap. 4. Glucose is absolutely necessary for the synthesis of ATP. Without glucose we cannot survive. Therefore, the glucose level is controlled by a negative feedback system.

In the case of T2DM, the number of binding sites for insulin on the cell surface decreases. The body tries to compensate for this growing insulin resistance by increasing insulin production. Nevertheless, there is a relative lack of insulin and an excessively high concentration of glucose in the blood.

In the late stages of T2DM disease, the constant need for insulin can deplete the pancreas to the point that insulin production decreases. Then an absolute insulin deficiency may develop, which can only be compensated for by insulin injections. At this late stage, T2DM evolves into T1DM.

The reasons for insulin resistance are not entirely clear, but apart from genetic effects, environmental conditions, diet and obesity appear to have a major influence. In any case, T2DM is viewed by the WHO as a global epidemic, which now affects around 10% of the world's population [10].

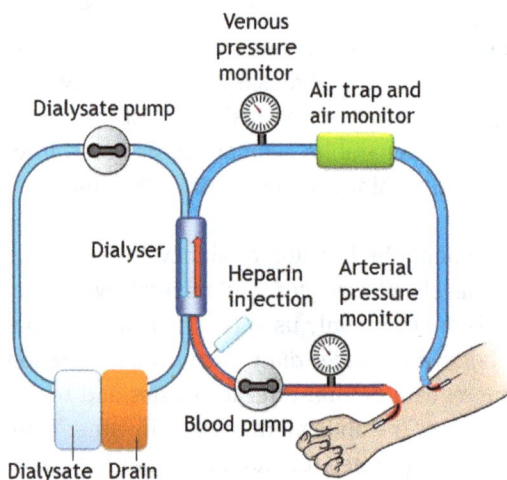

Fig. 10.15: Dialysis set up for cleaning of blood.

We will only and briefly discuss the first version: hemodialysis. Figure 10.15 shows a schematic representation of a hemodialysis setup. It consists of two loops: a primary and a secondary loop. In the primary loop, the blood is drained from an artery in the arm and, after being filtered, returned to a vein. Pumps, pressure regulation, and air pockets control the proper blood flow in the primary circuit. Heparin, a blood thinner, is injected to prevent blood from clotting. The central part of the dialysis machine is a semipermeable membrane that removes excess water and waste that is drained in the secondary circuit. During this process, various solutions from the dialysate are mixed with the blood to remove specific contaminants or correct certain conditions according to the patient's needs.

The semi-permeable filter membrane between blood and dialysis contains properly sized pores through which water, electrolytes, urea, and other waste products can pass, but not larger proteins and red blood cells. Thus, the semipermeable membrane has properties similar to the glomerular filter in the nephrons. The principle of the filter function is shown in Fig. 10.16(a). Fig. 10.16(b) schematically shows a possible technical realization. The filter consists of a bundle of hollow fiber membranes through which the blood permeates. Each fiber is embedded in dialysate fluid that flows countercurrently to the bloodstream. Tiny pores in the fibers on the one hand, filter the blood and, on the other hand, allow electrolytes from the dialysate to penetrate. It is essential for all designs that blood and dialysate flow in countercurrent to avoid contamination.

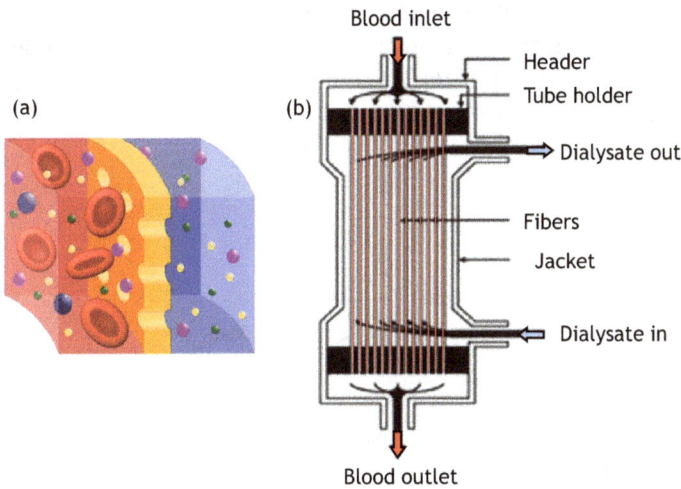

Fig. 10.16: (a) Schematics of a membrane filter in the dialyzer acting as a molecular sieve. Water and small molecules can pass, but not larger proteins and red blood cells. (reproduced from https://en.wikipedia.org/wiki/Semipermeable_membrane, © creative commons); (b) sketch of fiber membrane filter. Blood flows from top to bottom through the hollow fibers while dialysate flows in a countercurrent fashion from bottom to top.

10.8 Bladder, urethra, and urination

10.8.1 Structure and function

The bladder is a urine reservoir passed on from the kidneys via the ureter (see schematic cross section in Fig. 10.17). The bladder wall consists of a smooth muscle called the detrusor muscle. The detrusor muscle is thickened in a triangular area called the trigons containing the ureter's inlets. The outlet is the urethra at the bottom of the bladder. The urethra of males is much longer than that of females. The urethra of males runs from the bladder over the prostate and the pelvic floor to the penis with a total length of about 16 to 20 cm. In contrast, female's urethra is much shorter (~5 cm). The bladder is closed by two muscles at the bottom, the internal urethral sphincter and further down in the pelvic floor by the external urethral sphincter, which is part of a skeletal muscle. These two muscles open to urinate, while the trigone muscle closes the ureteral inlets to prevent reflux into the kidneys.

10.8.2 Neural control

Bladder filling and urination is a rather complex process controlled by a neural circuitry of the central nervous system regulating normal function. For a detailed description of the circuitry, we refer to [13, 14]. A simplified and schematic version is shown in Fig. 10.17 and explained as follows.

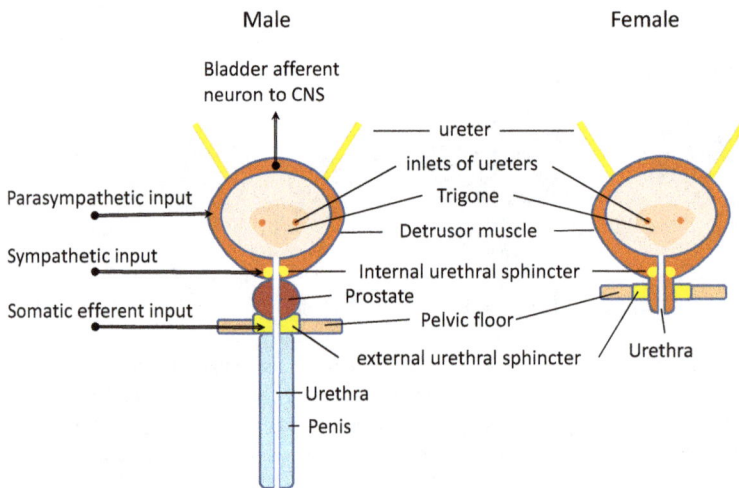

Fig. 10.17: Schematic cross section of the urinary system for males and females. In this scheme, the ureters arrive from the back side. The urethra is indicated as vertical straight tubes at the bottom of the bladder. The urethra in males is much longer than in females due to the passage through the prostate and penis. The opening of the urethra is controlled by two muscles: the inner and outer urethral sphincter, which are marked in yellow.

The detrusor muscle stretches as the bladder expands upon filling. Dendrites of afferent receptor neurons, located in the detrusor muscle, sense the tension in the wall. These sensory neurons project to the spinal cord and from the spinal cord to various regions in the brain for coordination and decision making. When the individual makes a conscious decision, efferent output signals to the detrusor muscle and its inner and outer urethral sphincters initiate urination.

Three main nerve innervations control the bladder function. Two belong to the autonomic or vegetative nervous system; the third one belongs to the somatic system. Their functions are as follows: parasympathetic input stimulates contraction of the detrusor muscle, whereas sympathetic input stimulates contraction of the internal urethral sphincter. Somatic efferent neurons stimulate contraction of the external urethral sphincter.

There are two operational modes of the bladder: urine storage and urination (voiding). During filling time, the parasympathetic input to the detrusor muscle is inhibited, allowing muscle expansion. Simultaneously, the sympathetic and somatic efferent pathways are activated, resulting in a contraction of the internal and external urethral sphincters. With increasing wall tension, the urge to urinate increases. However, the urethral sphincters do not relax, unless the brain makes a conscious decision to urinate (*micturition*) at an appropriate time and place. Then the parasympathetic efferent neurons become activated, causing contraction of the detrusor muscle. At the same time, there is a coordinated inhibition of the sympathetic efferent neuron and the somatic efferent neuron that relaxes the internal and external urethral sphincters. The various stimulations are summarized in Tab. 10.6. For more details of this complex control and feedback system, we refer again to [13, 14]. As this neural control system requires development during growth, it does not function until an age of 2–3 years. Vice versa, it can easily suffer from neural diseases or injuries. A similar neural control system also exists for the excretion of feces through the rectum and anus.

Tab. 10.6: Overview of bladder control by the autonomous and somatic nervous systems.

Nervous system	Somatic afferent	Somatic efferent	Sympathetic efferent	Parasympathetic efferent
Muscle control	pressure sensor in detrusor muscle	external urethral sphincters	internal urethral sphincters	detrusor muscle
Urine filling		activated	activated	inhibited
Urination		inhibited	inhibited	activated

10.8.3 Urostatics and urodynamics

The bladder filling capacity of adults depends on age, sex, and daily conditions but ranges on average from 0.3 to 0.5 l for females and 0.4 to 0.8 l for males. The filling rate is about 0.5 to 1 ml/min from each kidney. At a 50% level, the normal (healthy) bladder signals pressure. At a 75% level, people feel an urgency to release. A filling level of 100% causes pain and an urgent desire for voiding. The maximal breech pressure of the urethra is about 20–30 hPa. Figure 10.18 depicts an idealized time evolution of bladder filling, detrusor pressure, and voiding [15, 16]. At the beginning (point A), after emptying the bladder, the urine volume steadily increases. At the same time, the hydrostatic pressure in the bladder and the pressure exerted onto the detrusor muscle remain nearly zero. Upon reaching a urine volume of 150–200 ml, the pressure in the bladder increases, but remains below breech pressure (B). At the breech pressure (C) and after neural signaling to open the urethral sphincter valve, the detrusor pressure rapidly increases (D) voluntarily. The volume outflow rate reaches a maximum, and the urine volume in the bladder decreases down to some rest volume within a short period. In the depicted case, the total urine volume before voiding is 600 ml, the top volume rate is about 30 ml/s, the average volume rate is about 20 ml/s, and the voiding time is roughly 30 s. The total volume can easily be determined by the weight or volume of the excreted urine, a catheter can measure the pressure on the detrusor wall through the urethra with a pressure sensor, and the flow rate follows from urinating on a spinning plate, which records pressure variations as a function of time. Males may determine the breech pressure still by another method, discussed in exercise E10.7.

From the plots in Fig. 10.18, we can determine the elastic compliance of the bladder wall, i.e., the detrusor muscle. The compliance is defined as the ratio:

$$C_{bladder} = \frac{\Delta V_{bladder}}{\Delta p_{bladder}},$$

(10.22)

and measures the elastic volume response (expansion) upon pressure variations. As a reference, we chose the points A and C indicated in Fig. 10.18 (b): point A indicates the pressure at the start of the filling phase and point C refers to the pressure at the opening of the urethra sphincter valve (breech pressure). ΔV is the urine volume difference at these points. Typical values are 25 ml/hPa, but the values may range between 20 and 60 ml/hPa.

Measurement of the compliance is important for evaluating the healthy state of the bladder and, in particular, of the detrusor muscle. Similarly, we can determine the flow resistance of the urethra R_u, which is defined as

$$R_u = \frac{\Delta p_{bladder}}{\dot{V}_{bladder}}.$$

(10.23)

Here we take the maximum pressure during voiding divided by the maximum flow rate. This yields a typical resistance value of (2 ± 1) hPa $\cdot$ s/ml $= (2 \pm 1) \times 10^5$ Pa $\cdot$ s/l. This flow resistance is in the same order of magnitude as that of the blood circulation (compare section 8.5.3). Typical physical parameters of the bladder are listed in Tab. 10.7.

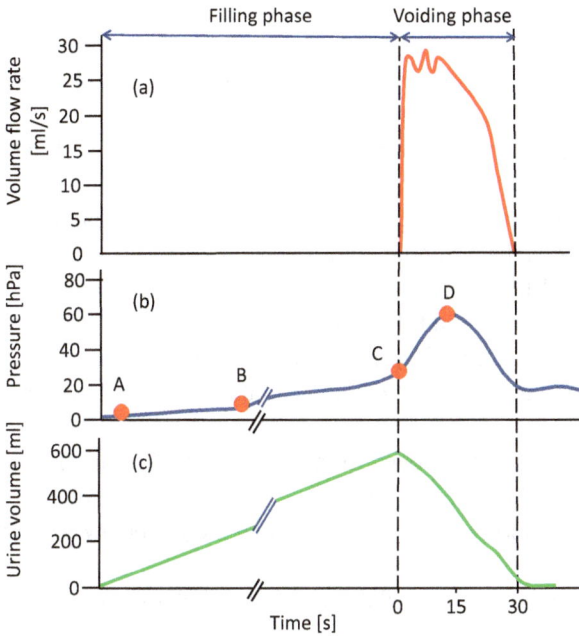

Fig. 10.18: Idealized time dependence of bladder filling, bladder voiding, and detrusor pressure variation. (a) volume rate of bladder during voiding, starting at time t = 0; (b) pressure variation over time during filling and voiding phase; (c) urine total volume as a function of time during filling and voiding phases. The points A-D indicate the start of the filling phase, midway of filling phase, breech pressure, and detrusor pressure during voiding.

Tab. 10.7: List of typical physical parameters of the bladder [15].

	Units	Values
Volume capacity	ml	300–600
Filling rate per kidney	ml/s	0.5–1
Detrusor pressure during filling	hPa	10–20
Detrusor pressure during voiding	hPa	60
Flow rate	ml/s	10–30
Flow resistance	hPa·s/ml	1–3
Compliance	ml/hPa	20–60

10.9 Summary

S10.1 The three main tasks of the kidneys are: 1. Regulation of blood level; 2. Control of acid-base balance; 3. Removal of metabolites.

S10.2 The nephron is the most important structural unit in the kidney that acts as a microscopic filtration unit.

S10.3 The nephron consists of Bowman's capsule, including glomerulus, proximal tubule, loop of Henle, distal tubule, and collecting duct.

S10.4 There are three types of filtration/exchange of the nephrons: glomerular filtration, tubule reabsorption, and tubule secretion.

S10.5 Filtration and exchange is achieved by active transport and diffusion.

S10.6 Glomerulus is a dense mass of very fine blood capillaries in the Bowman's capsule that act as a filter.

S10.7 Glomerular filtrate is the liquid removed from the blood by filtration in the Bowman capsules of the kidneys.

S10.8 Filtration occurs only for molecules with a molecular weight below 10 000 amu.

S10.9 Renal clearance of a substance is defined as the volume of plasma completely cleared of that substance by the kidneys per unit time.

S10.10 Fractional excretion of inulin is 1, fractional excretion of glucose is 0.

S10.11 Creatinine production rate and creatinine excretion rate are equal.

S10.12 1500 l of blood run through the kidney daily. The production of primary urine is about 170–180 l/day. The final secondary urine is 1–2 l/day.

S10.13 Dialysis machines are designed to mimic the function of kidneys by removing waste products and maintaining water levels, electrolytes, and minerals within acceptable margins.

S10.14 The bladder function is controlled by three main nerve innervations: two of the vegetative nervous system; one of the somatic system.

S10.15 The bladder capacity is about 300 ml.

? Questions

Q10.1 What are the main tasks of the kidney?

Q10.2 What is the most important unit in kidneys?

Q10.3 Which part of the nephrons is responsible for filtration?

Q10.4 Which part of the blood is filtered, plasma or hematocrit?

Q10.5 What is the difference between the renal plasma flow rate and the glomerulus filtration rate?

Q10.6 Describe the different parts of a nephron.

Q10.7 Which feature of the glomerulus helps the process of filtration?

Q10.8 Name the five components of unfiltered blood which appear in the glomerular filtrate.

Q10.9 Why do blood cells and protein molecules not appear in the glomerular filtrate?

Q10.10 Which three components of the glomerular filtrate are reabsorbed?

Q10.11 Why is it important for these three components to be reabsorbed?
Q10.12 Which substances are present in the final urine?
Q10.13 What is the fractional excretion of plasma, and what is the total excretion of plasma per day?
Q10.14 Which physical processes take place for secretion and reabsorption in nephrons?
Q10.15 What is the reason for diabetes mellitus type 2?
Q10.16 When is hemodialysis indicated?
Q10.17 Which nerve has to be activated for voiding the bladder?
Q10.18 Which substance should not be found in the finale urine of a healthy person?

Attained competence checker	+	0	–	
I can name the main parts of the kidneys				
I know what the main functions of the kidneys are				
I know how primary urine is produced				
I know why secondary urine volume is much less than primary urine				
I know what the glomerular filtration rate is in numbers				
It is clear to me why the RPF is higher than the GFR				
I know that inulin filtration is a gold standard for testing				
I know why the FE for glucose is zero				
I realize that creatinine test is simpler than inulin test				
I know why the FE for PAH is highest among all filtration tests				
I know under what circumstances hemodialysis is indicated				
I can name the main parts of a hemodialysis set-up				
I can name the main muscles that control bladder function				
I know what diabetes mellitus is and I can distinguish between type 1 and type 2				

Suggestion for home experiments

HE 10.1 **Home experiment: Filling and voiding the bladder**
You may drink a liter of water within a short time and then watch how fast the bladder fills up. You may even sense the filling of the bladder by its expansion and bowing of the abdominal wall. Once the breech point is reached, measure the total volume and voiding time, from which you can determine the urethra volume rate.

ℹ **Exercises**

E10.1 **Water permeability:** The water permeability through the epithelium is different in the sections of the renal tubules. At which point is the transepithelial water permeability the lowest? Numbers refer to Figure 10.9.
   (A) the proximal tubule (2)
   (B) the thin descending part of Henle's loop (3)
   (C) the thick ascending part of the Henle loop (4)
   (D) the collecting tubulus (6)
   (E) the connecting tubule (7)

E10.2 **Clearance and filtration:** Discuss clearance and filtration according to the panel (g) in Fig. 10.13. Assume that for a particular substance S7 the glomerular filtration is 20% of RPF, reabsorption is 20% of GFR and secretion is 30% of RPF. What is the clearance and what is the fractional excretion?

E10.3 **Renal processing:** In Tab. 10.3 six different renal processes are listed. Give examples for each one of them.

E10.4 **Hemodialysis:** Name three clinical pictures that lead to a loss of kidney function and make hemodialysis necessary.

E10.5 **Voiding the bladder:** Let's assume that the voiding of the male bladder is solely determined by the gravitational pressure of the fluid in the bladder and the volume flow of the urethra. Determine the time it takes for complete voiding. Assume a filling of the bladder by 0.5 l in a cylinder of height 0.25 m. The urethra length is about 0.15 m and the diameter is about 0.0025 m.

E10.6 **Bladder pressure:** Assuming that the bladder is completely full. Discuss the pressure at the detrusor point of an upright standing person. What portion of the pressure is due to hydrostatic pressure, what is due to muscle tension?

E10.7 **Bladder pressure at detrusor point:** Determine experimentally (if possible) the internal pressure of your urinary bladder using the range of the urine stream in an upright position, similar to the Manneken Pis sculpture in Brussels. If you can not do the experiment, just estimate the range.

References

[1] Guyton AC, Hall JE. Textbook of medical physiology. 11th edition, Philadelphia, Pennsylvania, USA: Elsevier Saunders; 2006.

[2] Faller A, Schünke M, Schünke G. The human body. An introduction to structure and function. Stuttgart, New York: Thieme Verlag; 2004.

[3] OpenStax Anatomy and Physiology. https://openstax.org/details/books/anatomy-and-physiology

[4] Longmire M, Choyke PL, Kobayashi H. Clearance properties of nano-sized particles and molecules as imaging agents: Considerations and caveats. Nanomedicine. 2008; 3: 703–717.

[5] Aguilar-Roca N. Essential Physiology, Connexions Web site. http://cnx.org/content/, © creative commons).

[6] Boron WF, Boulpaep EL. Medical physiology. 2nd edition, Philadelphia, London, New York: Saunders W.B. Elsevier; 2012.

[7] Stevens LA, Coresh J, Greene T, Levey AS. Assessing kidney function – Measured and estimated glomerular filtration rate. N Engl J Med. 2006; 354: 2473–2483.

[8]   Shahbar H, Gupta M. Creatinine clearance. Statpearls [Internet]. Treasure Island (FL): StatPearls Publishing; 2021, Jan. Bookshelf ID: NBK544228.

[9]   Kharroubi AT, Darwish HM. Diabetes mellitus: The epidemic of the century. World J Diabetes. 2015; 6: 850–867.

[10]   Roglic G. WHO global report on diabetes: A summary. Int J Non-Commun Dis. 2016; 1: 3–8

[11]   Held PJ, Port FK, Wolfe RA, Stannard DC, Carroll CE, Daugirdas JT, Bloembergen WE, Greer JW, Hakim RM. The dose of hemodialysis and patient mortality. Kidney Int. 1996; 50: 550–556.

[12]   Murdeshwar HN, Anjum F. Hemodialysis. In: StatPearls [Internet]. Treasure Island (FL): StatPearls Publishing; 2021 Jan-.

[13]   Fowler CJ, Griffiths D, de Groat WC. The neural control of micturition. Nat Rev Neurosci. 2008; 9: 453–466.

[14]   de Groat WC, Griffiths D, Yoshimura N. Neural control of the lower urinary tract. Compr Physiol. 2015; 5: 327–396.

[15]   Abrams P. Urodynamics. London: Springer Verlag; 2006.

[16]   Yao M, Simoes A. Urodynamic testing and interpretation. In: StatPearls [Internet]. Treasure Island (FL): StatPearls Publishing; 2021.

Further Reading

Layton AT, Edwards A. Mathematical modeling in renal physiology. Berlin, Heidelberg: Springer Verlag; 2014.

11 Physical aspects of vision

Physical properties of the visual system

Refractive power of the cornea	43 dpt
Refractive power of the lens	15 dpt
Combined optical power of the eye (relaxed state for infinite object distance)	58 dpt
Average refractive index	1.4
Accommodation width	Young: 12 dpt; old: 2 dp
Resolving power	3 μm
Visual acuity	1'
Intraocular pressure	10–25 hPa
Visual field	Horizontally: 200°, vertically: 130°
Total number of cones	7×10^6
Total number of rods	120×10^6
Density of cones and rods in the fovea	Cones: $120 \times 10^3/mm^2$; rods: $50 \times 10^3/mm^2$
Density of cones and rods outside the fovea	Cones: $50 \times 10^3/mm^2$; rods: $120 \times 10^3/mm^2$
Spectral range of the eye	400–700 nm
Dynamic range of the eye	$10^6 – 10^{-6}$ cd/m^2
Number of muscles controlling the eye ball	9
Number of pigments in a rod/cone	10^9
Membrane potential of cone/rod in dark state	−30 mV
Number of layers in the retina	5
Cell types in the retina	6
Thickness of the retina	100–300 μm

11.1 Introduction

It has been estimated that 80% of our sensory information from the environment is perceived through our eyes. The processing of visual perception takes up one-fourth of our brain. Infants' visual intelligence develops earlier and completes before verbal skills improve. All these facts show how important the visual system is [1]. Loss through blindness is a serious handicap.

This is the first of two chapters on our sensory system. The next chapter deals with the auditory and vestibular systems. Of course, there are many more senses in our bodies that we constantly draw information from such as pressure, taste, smell, and temperature. The visual and auditory sense are, however, excellent examples of how biology converts physical properties, such as light and sound, into action potentials (APs). The APs are stored and processed in our brain to enable decision-making using memory and experience.

https://doi.org/10.1515/9783110756951-011

As we will notice, visual perception is an extremely complex and intriguing signal transduction. Roughly speaking, one can distinguish three main parts contributing to visual perception: (1) optics of the eye; (2) photon detection and initial image processing within the retina; and (3) signal transmission and further processing in the visual cortex of the brain. This chapter gives a brief overview of mainly the first two parts with some remarks on the third part.

The eyeball as an optical instrument is the least impressive part of our visual system. Every compact camera delivers sharper images than the human eye. On the other hand, photon detection in the retina is much more efficient than photon detection with a CCD chip. The sensitivity of the retina spans ten orders of magnitude from weak to bright sunshine. Although the range of photon wavelengths that we can recognize is rather limited from around 400 nm (violet) to about 700 nm (red), we can distinguish between millions of different hues within this wavelength range. Most notable is the signal processing that begins in the retina and continues in the visual cortex. We perceive more with our brain than we can recognize with our eyes [2, 3].

As before, the main focus of this chapter is on the physical aspects of the eye. For all other physiological and medical facets, we refer to specialized literature and textbooks listed at the end of this chapter.

11.2 Anatomical aspects of the eye

Figure 11.1 shows a cross section of the eyeball labeling the most important parts: *cornea* together with lens for refraction, zonula fibers for accommodation, iris for aperture action (pupil), retina for detection of optical signals, and optic nerve for transmission of APs to the brain. The fovea is the area of the highest optical resolution on the retina and the highest density of receptors. The center of the fovea and the center of the lens defines the location of the optical axis. The optical disk is optically "dead" as here the optic nerve bundle punches through the retina on the way to the optical cortex in the brain.

The *choroid* is penetrated by blood vessels for oxygen supply and for cooling the back of the retina, which is warmed up by focused light. The eyeball is filled with a transparent glassy material called *vitreous humor*. The anterior chamber is filled with a watery fluid called *aqueous humor*. Both vitreous and aqueous humor have the same refractive index. The lens, in contrast, has a higher refractive index produced by an arrangement of fibers accumulating proteins known as *crystallins*. Most important is the cornea with a slightly lower refractive index than the lens but a much higher curvature that accounts mainly for the eye's refractive power.

The iris has the function of an aperture controlling the light intensity, which is focused on the retina by increasing or decreasing the pupil's diameter. The color of the eye is the color of the iris. Brown eyes contain melanin pigment in the fibers (stroma) of the iris, which absorbs the blue part of the spectrum, leaving a brown impression upon light reflection. However, in case of blue eyes there is no equivalent blue pigment. The color blue is the result of light scattering in the tissue of the iris, until blue is left over, an effect known as Tyndall[1] effect or Rayleigh[2] scattering, which is also responsible for the blue color of the sky [4]. The lack of melanin in the iris results from a mutation that is believed to have occurred some 10 000 years ago. The diameter of the pupil ranges from a minimum $d_{min} = 1.5$ mm to a maximum $d_{max} = 7.5$ mm.[3] This corresponds to an area ratio of 25 or roughly one decade with respect to intensity. However, the *eye's dynamic range* covers a much larger range of about 10 decades from 10^{-6} to 10^4 cd/m^2. Therefore, there must be an additional mechanism in the eye to adapt to the light intensity aside from the iris/pupil size, which we will learn about in later sections of this chapter.

The response time of the iris on intensity changes is very rapid, taking only 200–500 ms for contraction, which is important for protecting the retina against too high intensity. The iris *contraction muscle* controlling the pupil (aperture) and the *ciliary muscle* controlling the curvature of the eye lens (accommodation) form one unit in the ciliary body.

A total of six extraocular muscles control the eyeball movement. Four of them are straight muscles: top (rectus superior), bottom (rectus inferior), left (rectus lateralis), and right (rectus medialis), controlling tilting movement, and two more oblique muscles, inferior and superior, controlling rotational movement.

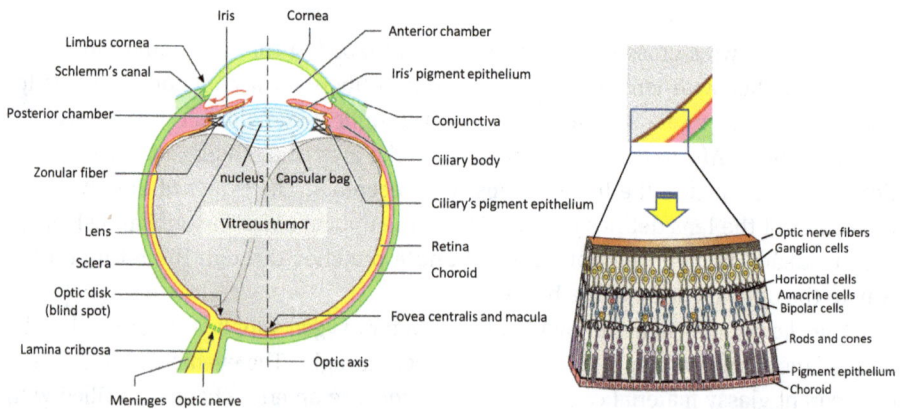

Fig. 11.1: (a) Cross section of the eyeball; (b) enlargement of the lens system; (c) cross section of the retina-containing photoreceptors and various neurons for signal processing.

1 John Tyndall (1820–1893), Irish physicist.
2 John William Rayleigh (1842–1919), English physicist and mathematician.
3 In camera language, this corresponds to *f*-values of 16 and 3.2, respectively.

> Nine muscles control the eye. Six extraocular muscles move the eyeball. Two intraocular muscu-les adjust the iris and one more muscle adapts the lens accommodation.

In total, there are nine muscles per eye: six for controlling the eyeball movement, two muscles for adjusting the iris (pupil dilator muscle and pupil constrictor muscle), and one ciliary muscle for accommodation. The ciliary muscle is a ring-like sphincter mus-cle. The muscles for left and right eye movement are synchronized. If they move paral-lel in the same direction, the eye movement is called conjugated. If the eye movement is tilted inward, the movement is convergent. The eye movement allows a biaxial fixa-tion on single objects that is necessary for obtaining distance information. All eye muscles are at constant tension, enabling fast reaction, much faster than any other body movement [5]. The eye balls are also connected to the vestibular organ to release the *vestibular–ocular reflex*, as we will see in Section 12.6 of the next chapter.

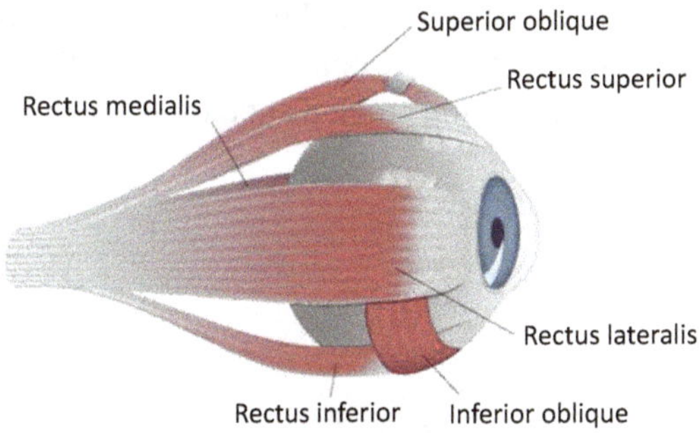

Fig. 11.2: Six exterior muscles of the eye control eye movement. They are also known as extraocular muscles. Three more muscles are inside for control of aperture and accommodation.

11.3 The dioptric eye

11.3.1 The refractive power of the eye

Dioptrics is the branch of optics that studies refractive systems. The name was in-troduced by Kepler[4] to distinguish it from ray tracing optics via reflection and trans-mission. Gullstrand[5] received the 1911 Nobel Prize in Physiology "for his work on the dioptrics of the eye." The term is still used in physiology textbooks, but can rarely be found in physics books.

---

4 Johannes Kepler (1571–1630).
5 Allvar Gullstrand (1862–1930), Swedish ophthalmologist and Nobel laureate in physiology 1911.

In the first approximation, the focal length f of the eye must match the inner diameter of the eyeball. Only then will objects located at an infinite distance be focused on the retina. We can estimate this distance to be 24 mm. We usually observe objects at distances greater than twice the focal length or at distances greater than 50 mm. The near point at which we can still get a sharp picture from objects is 100–120 mm at a young age. Thus, all objects that we see in focus are at distances much larger than twice the focal length. At this distance, all sharp images on the retina are reduced in size and inverted compared to the real object. If we look at an object at a distance of approx. 25 cm, we have the subjective impression that we perceive the correct size of the object. At shorter distances, objects appear enlarged; at larger distances, objects appear smaller. In all cases, the image size on the retina is always reduced.

Next we realize that the eye is completely filled with transparent material and all light rays entering through the cornea stay inside the glassy and jelly-like body of the eye called vitreous humor. Therefore we have to treat the eye as a thick homogeneous lens with a focal length f_2 on the image side within the eye and a different focal length f_1 outside in front of the eye. Using the lens equation for thick lenses, we find for the focal lengths [6, 7]:

$$\frac{1}{o} + \frac{n_2/n_1}{i} = \frac{1}{f_1} \tag{11.1}$$

and

$$\frac{n_2/n_1}{o} + \frac{1}{i} = \frac{1}{f_2}. \tag{11.2}$$

Here the refractive index is $n_1 = 1$ for air and $n_2 = 1.33$ for a watery substance, o is the object distance to the cornea, and i is the image distance between cornea and retina. For an object at infinite distance these equations simplify to:

$$\frac{n_1}{f_1} = \frac{n_2}{f_2}. \tag{11.3}$$

Since $f_2 = 24$ mm, we find for $f_1 = 18$ mm. The *refractive power* is defined as $1/f_1$ or n_2/f_2 and is in our case 55.4 m^{-1} = 55 dpt. The unit is diopter (dpt), where 1 m^{-1} = 1 dpt. Without taking any measurement, we arrive at a pretty good estimate for the focal lengths and the refractive power of the standard eye, which comes close to the real value of about 58 dpt.

Refractive index

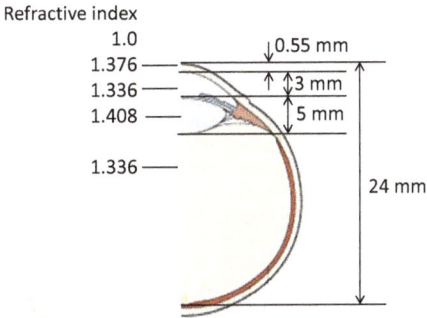

1.0
1.376
1.336
1.408

0.55 mm
3 mm
5 mm

1.336

24 mm

Fig. 11.3: Refractive indices and distances of different parts of the eye for a wavelength of 590 nm.

The eye is in fact not a homogeneous spherical body but consists of several parts with different refractive indices, as shown in Fig. 11.3. The cornea at the air/eye interface has the largest curvature and therefore causes the major refractive effect. The thickness of the cornea is less at the center (550 µm) than at the edge (650 µm) and would therefore act as a diverging lens unless it is internally supported by a refractive index matching fluid (vitreous humor). The biconvex lens has the highest refractive index and the curvature can be adjusted to accommodate the image on the retina. The average refractive index of the lens is 1.408. Closer inspection reveals that there is a gradient in the refractive index perpendicular to the lens. The refractive index is highest in the middle of the lens and decreases on both sides toward the outside [8, 9]. As already mentioned, aqueous and vitreous humor have identical refractive indices similar to water.

Now we are prepared to reconsider the optics of the eye by starting without the lens. For the standard eye, the main refraction occurs at the air/cornea interface. From the curvature of the cornea ($R = 7.7$ mm) and using the lens maker equation:

$$f_1 = \frac{n_1}{n_2 - n_1} R, \tag{11.4}$$

we find $f_1 = 23$ mm assuming an average refractive index of $n_2 = 1.336$. Then $f_2 = f_1 \times n_2 = 31$ mm, and the refractive power is therefore 43 dpt.

With these values, the image of objects at infinite distance lies at 31 mm, which is 7 mm beyond the retina (see Fig. 11.4(a)). The lens adds another 15 dpt to the refractive power to reach the focal point on the retina at a distance of 24 mm from the cornea in a relaxed state (Fig. 11.4(b)). This corresponds to an average refractive index of 1.4, a refractive power of 58 dpt total, and a focal length of 17 mm in front of the eye.

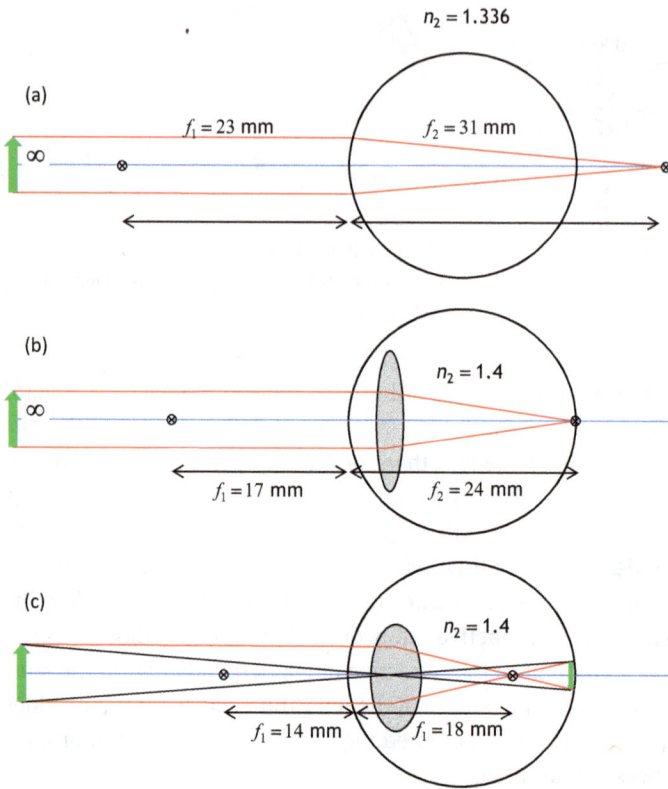

Fig. 11.4: Imaging with the eye as a thick lens. Note that the focal lengths before and in the eye are different: (a) refraction only due to the cornea and assuming a homogeneous refractive index within the eye; (b) inserting a lens with a higher refractive index places the focus for infinite objects on the retina; and (c) for objects, which are closer to the eye, the lens has to accommodate the focal length by increasing its curvature.

For accommodation, the lens has to increase its curvature to increase the total refractive power. Then the focal length f_2 becomes shorter and f_1 also decreases. Objects at the near point (green arrow) than appear sharp on the retina (Fig. 11.4(c)). Table 11.1 lists the average optical parameters of the standard eye.

Tab. 11.1: Average optical parameters for the standard eye. The focal lengths given are for the far point.

Average reflractive index n_2	1.4
Refractive power	58 dpt
Focal length in front of the eye	17 mm
Focal length inside the eye	24 mm

11.3.2 Accommodation

The eye's ability to produce sharp images of objects located at far distance and upto the near point is called *accommodation*. Accommodation width ΔA is defined as the difference of the refractive power for objects at the near point and the far point:

$$\Delta A = \frac{1}{L_{near}} - \frac{1}{L_{far}}.$$ (11.5)

The distance L between the near/far point and the eye is measured in meters. For normal-sighted (emmetrope) young individuals, the far end is at infinite distance and the near point is about 0.08 m; the accommodation width is therefore is

$$\frac{1}{0.08 \text{ m}} - \frac{1}{\infty} = 12.5 \text{ dpt}.$$

For short-sighted elder individuals, we assume a far distance of 1 m and a near point of 0.33 m; then the accommodation width is

$$\frac{1}{0.33 \text{ m}} - \frac{1}{1 \text{ m}} = 2 \text{ dpt}.$$

In Fig. 11.5 we show the age dependence of the accommodation width [10]. The ability to accommodate firmly depends on age and starts to decrease already at age 10, much before reaching adulthood. Accommodation is usually completely lost at the age of 50. This is one of the strongest age-dependent properties of the body. Reading glasses become a necessary tool already from the age of 40.

The refractive power of the eye (58 dpt) is mainly owing to the curvature of the cornea (43 dpt) with the assistance of the lens (15 dpt) for accommodation.

Accommodation is the cooperative result of the ciliary muscle and zonula fibers. In the relaxed state of the eye, the ciliary muscle is relaxed and the zonula fibers are pulled straight, which flattens the lens (Fig. 11.6). In this state, the lens contributes 15 dpt to the refractive power of the eye, as we have already seen. However, if we want

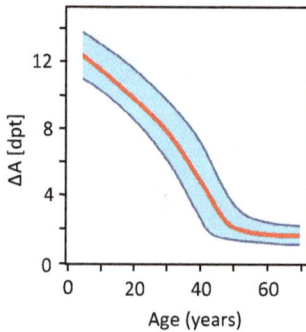

Fig. 11.5: Accommodation width as a function of age. The blue band indicates a one-sigma standard deviation from the mean.

to see objects at near distance, we need to focus the image on the retina. This is achieved by contracting the ciliary muscle, which releases the zonula fibers causing a rounding of the lens. The higher curvature of the lens increases the effective refractive power, bringing the image onto the retina. The description of the accommodation mechanism goes back to Helmholtz[6] and still holds today. With aging, the lens becomes increasingly stiffer and remains flat, reducing the ciliary contraction effect. Different options for restoring the accommodation power by surgical means are currently under investigation; for a review, we refer to Ref. [11].

Presbyopia, the gradual age-related loss of accommodation, occurs mainly as a result of a steady stiffening of the lens fiber material. There is no evidence that the loss of ciliary muscle contractility would decrease with age and contribute to presbyopia. The stiffening of the lens fibers is surprising as they are continuously renewed throughout life. As new fibers are added from the outside and move toward the center, there is an age-related gradient in the fibers, with the oldest ones sitting in the lens's core, making the lens thicker over time. The age gradient is accompanied by a refractive index gradient that is higher in the central part than in the shell. The refractive index gradient is due to a higher protein concentration in the core. Although the stiffening of the lens has been confirmed in numerous studies, the origin remains unsolved [12].

11.3.3 Resolving power

According to the Rayleigh criterion for the resolution of optical instruments, two objects a and b can be recognized as separate if the zeroth-order intensity maximum of the diffraction pattern from object b falls into the first minimum of the diffraction pattern of object a. For circular apertures of radius r, the diffraction pattern

---

6 Hermann von Helmholtz (1821–1894), German physicist and physiologist.

relaxed   contracted

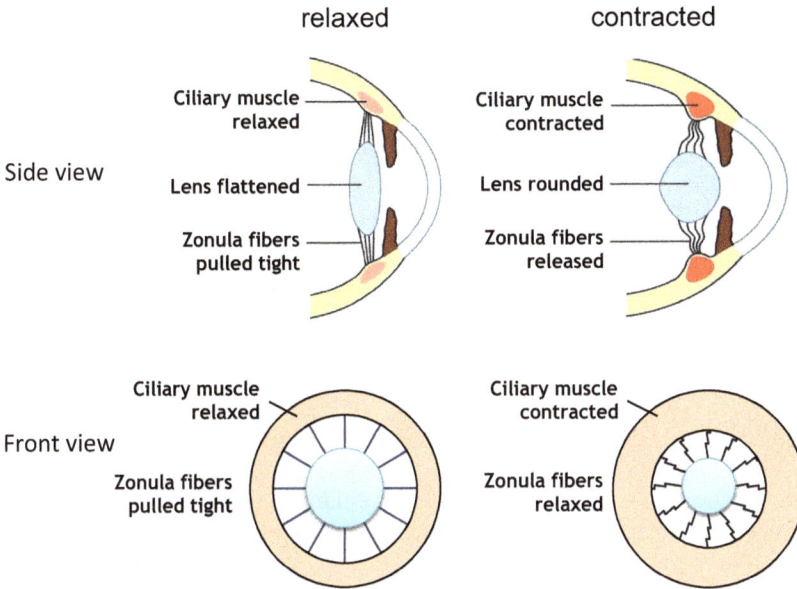

Fig. 11.6: Ciliary muscle shown in relaxed and contracted position. Top: side views; bottom: front view.

consists of airy discs. In this case, the Rayleigh criterion is expressed in terms of a minimum angle α_{min} at which two objects can be separated:

$$\alpha_{min} = 1.22 \frac{\lambda/n}{2r}. \tag{11.6}$$

For the eye, we use average numbers: $2r = 5$ mm for the pupil, $\lambda = 550$ nm, and $n_{eye} = 1.4$. The minimum angle is then related to the minimum resolving distance d_{min} of two objects via (Fig. 11.7):

$$d_{min} = \alpha_{min} f = 1.22 \frac{\lambda/n}{2r} f, \tag{11.7}$$

where $f = 24$ mm. Substituting numbers, we find a minimum distance on the retina of $d_{min} = 2.3$ µm for two objects that are considered separate. How does this fit the average spacing of the cones within the fovea of the retina? In the foveal area with a diameter of 500 µm, the cones are rather small and their density is highest with an average distance of about 2.5–3 µm (Fig. 11.19). Therefore, the optical resolution of the eye and the density of the light receptors in the foveal area match very well. In fact, the physical resolution is slightly better than the physiological limit: we can distinguish two objects a and b on the retina as different if the images of a and b fall on two receptors that are separated by at least one receptor in between, which results in a physiological spatial resolution of approx. 5 µm.

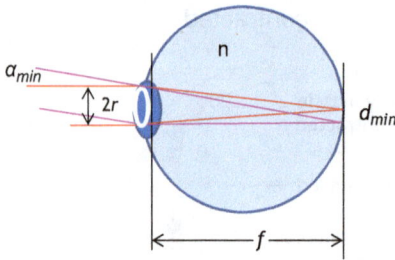

Fig. 11.7: Resolving power of the eye.

11.3.4 Visual acuity

At a distance g_0 of 25 cm we see objects with a psychophysical magnification of 1. Our personal judgment tells us that at this distance, objects appear to us by their proper size. At shorter distances, objects appear enlarged; at larger distances, they appear reduced. The distance g_0 is called the *distance of most distinct vision*. If G is the object height, we observe objects, in general, at a visual angle of

$$\varepsilon = \frac{G}{g},\qquad(11.8)$$

and $\varepsilon_0 = G/g_0$ is the most distinct visual angle. $\varepsilon = 1°$ corresponds to an image height of 300 μm on the retina. The *visual acuity* is defined as the reciprocal value of the visual angle:

$$\alpha = \frac{1}{\varepsilon}.\qquad(11.9)$$

If a gap in the Landolt ring, shown in Fig. 11.8(a), can be recognized under a visual angle of 1' (1' = 1°/60), then the visual acuity is 1. This corresponds to a separation on the retina by 5 μm. The visual acuity depends on the retinal location of the image projection and on the brightness. The visual acuity is highest in the area of the fovea with the highest density of cones and drops off outside with increasing eccentricity (see Fig. 11.19). The fovea is at the center of the optical axis (see Fig. 11.1(a)). In comparison, we observe the sun and the moon under the same visual angle of 30'.

Visual acuity is often expressed in terms of a fraction, such as 20/40. The first number refers to the distance (in meters or feet) a person with impaired eyesight can read the line with the smallest letters on a chart correctly. The second number indicates the distance at which a person with normal eyesight could read the same line of letters correctly. The smaller the fraction, the worse is the eyesight of the patient.

The *vernier acuity* is the ability to recognize a shift in the contour of an object, as shown in Fig. 11.8(b). The vernier acuity is about six times higher than the visual acuity and better than what could be explained physically by the receptor density. Therefore, this high vernier acuity can be seen as a first hint to a physiological effect of enhancement and inhibition within receptor fields discussed further in Section 11.6.3.

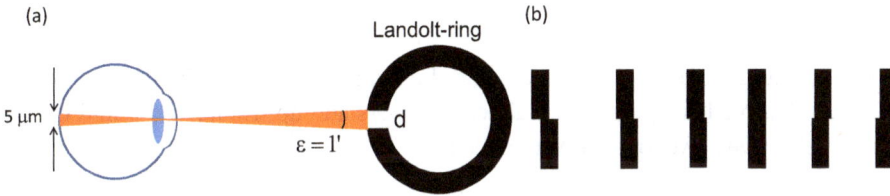

Fig. 11.8: (a) Visual acuity is defined by recognizing a 1′ gap in the Landolt ring; (b) vernier acuity is higher than visual acuity.

> The resolving power of the eye is adapted to the density of light-sensitive receptors in the retina. The resolution of the eye is about 5 μm, and the acuity is about 1′.

11.3.5 Lens aberrations

The most frequent aberrations of the human lens are *myopia* and *hyperopia*. Eyeglasses can easily correct for both.

Myopia or near-sightedness refers to the fact that the focal length of an object at infinite distance lies before the retina. This is shown by red lines in Fig. 11.9(a).

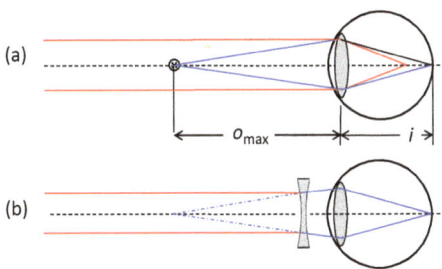

Fig. 11.9: Diverging biconcave correcting lens for myopia.

At the object distance o_{max}, a sharp image occurs on the retina, shown in Fig. 11.9(a) by blue lines. Then the lens, eq. (11.1), becomes

$$\frac{1}{o_{max}} + \frac{1}{i} = \frac{1}{f_e} = D_e, \tag{11.10}$$

where D_e is the refractive power of the myoppic eye. The goal is a sharp image on the retina of objects at an infinite distance, as indicated by the red and blue lines in Fig. 11.9(b). This is achieved with the help of a diverging lens. Then the lens equation changes to:

$$\frac{1}{\infty} + \frac{1}{i} = \frac{1}{f_e} + \frac{1}{f_{dl}} = D_e + D_{dl}, \tag{11.11}$$

where f_{dl} and D_{dl} are the focal length and the refractive power of the diverging lens, respectively. In a lens system, the refractive powers add up. Therefore we find the required refractive power of the diverging lens:

$$D_{dl} = - \frac{1}{o_{max}} \tag{11.12}$$

Similar arguments hold for imaging with hyperopia or far-sightedness (Fig. 11.10(a)). At a minimum distance o_{min}, a sharp image is projected on the retina (blue line), but for closer object distances, the image lies behind the eye (red dashed line). Therefore a converging correcting lens is required to bring the image back onto the retina (red and blue lines in Fig. 11.10(b)).

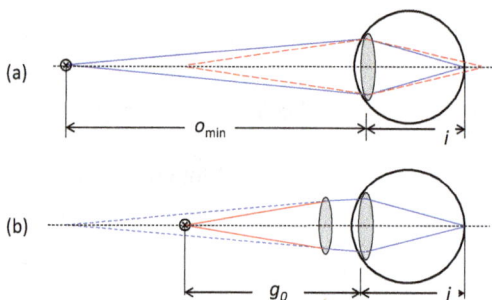

Fig. 11.10: Converging biconvex correcting lens in case of hyperopia.

The lens equation without correcting lens for the minimum distance is:

$$\frac{1}{o_{min}} + \frac{1}{i} = \frac{1}{f_e} = D_e, \tag{11.13}$$

With correcting lens it is

$$\frac{1}{g_0} + \frac{1}{i} = \frac{1}{f_e} + \frac{1}{f_{cl}} = D_e + D_{cl}, \tag{11.14}$$

For the refractive power D_{cl} of the converging correcting lens we therefore find:

$$D_{cl} = \frac{1}{g_0} - \frac{1}{o_{min}} = 4 - \frac{1}{o_{min}}. \qquad (11.15)$$

In the last equation we have set $g_0=0.25$ m for the distance of most distinct vision.

Three more aberrations are usually discussed in physics and optics textbooks: spherical aberration, chromatic aberration, and astigmatism.

Spherical aberration refers to widening the focal spot for rays far from the optical axis compared to those close to the optical axis. Spherical aberration is due to the spherical curvature of lenses deviating from the ideal parabolic shape. When we reduce the pupil, i.e., the aperture of the eye, the spherical aberration becomes negligible.

Chromatic aberration refers to the widening of the focal spot due to dispersion. The refractive index depends on the wavelength. Assuming a normal dispersion, the red light is refracted less than blue light. Therefore, the focal length for red light is longer than for blue light. The difference is, however, small because of the short distances in the eye and therefore insignificant for normal vision. The chromatic aberration in the eye is calculated and discussed in Exercise E11.5.

The first two aberrations play no role in the optics of the eye. But astigmatism frequently is observed because the cornea often exhibits a nonspherical shape. Usually, astigmatism goes hand in hand with myopia or hyperopia and is corrected simultaneously by properly crafting the correcting lenses. A critical review of all aspects of the eye's optics can be found in [13].

Corrections of myopia and hyperopia by laser applications, in particular, photorefractive keratectomy (PRK) and laser-assisted interstitial keratomileusis (LASIK), are presented in Chapter 6/Vol. 3.

Cataract is a disease of the visual system that causes the lens first to become opaque and later block the light from transmission toward the retina. It is one of the several possibilities causing blindness. Fortunately, it is the one that can nowadays be restored by replacing the lens with an artificial one. Cataract surgeries are the most frequent and successful eye surgeries that are performed. A short outline of the procedures taken is presented in the Infobox on cataract.

> Eye lens aberrations can be corrected by appropriate correcting lenses: diverging lens for correcting myopia, converging lens for correcting hyperopia. Further aberrations (spherical, chromatic) can be neglected, while astigmatism is compensated for together with the correcting glasses for myopia and hyperopia. **!**

> **Infobox I: Cataract symptons and remedies** **i**
> Some cataracts are related to inherited genetic disorders. Others can be caused by medical conditions such as diabetes, trauma, past eye surgery, or eye injuries. Cataract is not necessarily age-related, although the lenses' fibers are responsible for both accommodation loss and opacity.

Once cataract symptoms have started, they are progressive. In the opaque stage, light becomes strongly scattered such that the image appears blurred, dimmed, and grayish. This type of scattering is known as Mie scattering. It is, for instance, responsible for the scattering of light in clouds and requires objects with a size on the order of the wavelength of light.

Recent research has shed light on the possible cause of cataracts [14]. The lens requires a higher refractive index than the surrounding vitreous and aqueous humor, while remaining transparent for visible light. This is achieved by proteins called crystallins. Two proteins are mainly involved; we call them for simplicity A and B. The crystallins A and B tend to clump together unless kept separate by another "chaperone"-type protein C. There is only a finite amount of proteins C in the lens. When depleted by some reason or other, proteins A and B may aggregate, causing the lens to become cloudy. Knowing the cause, there is hope to find a biochemical/pharmaceutical cure of cataracts. Until then, the only treatment is a surgical replacement of the lens.

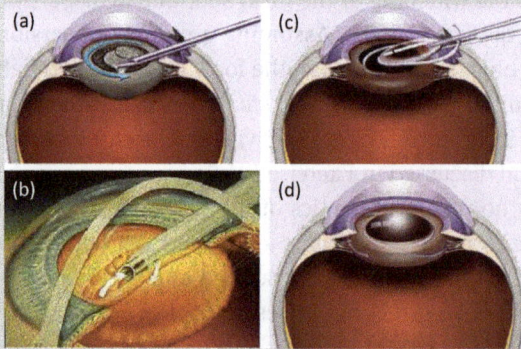

Operational steps of clinical cataract treatment.

Cataract can be treated surgically in four steps as shown schematically in the figure above. The outpatient procedure, which takes less than 10 min, begins with a small 3 mm incision through the cornea and into a pouch containing the vitreous material around the lens called the capsular bag. The sac covering the lens is ruptured in a circular motion known as continuous curvilinear capsulorhexis (CCC). A sharp stainless steel knife is used for the cut and the CCC procedure is performed with tweezers. Today, a femtosecond laser (see Chapter 6/Vol. 3) is often used for this procedure.

In the second step (b), the lens material (nucleus) is fragmented using either ultrasonic techniques, or more recently, nanosecond or femtosecond laser. This process is called phacoemulsification of the nucleus. The residues are sucked out by a tiny pump.

In the third step (c), the original lens is replaced with an artificial lens. The refractive power of the artificial lens is preset and cannot be changed once it has been inserted. Plastic springs on the rim hold the lens in place.

In the fourth and last step (d), the cornea is closed again in a self-sealing manner and without stitches. After a short time, the tunnel through the corneal sclera closes and heals, and the procedure is over. Vision is restored by this procedure, but accommodation is lost. Of course, it is extremely important to keep the eye free from contamination and especially bacteria during the operation. Therefore, cleanliness is the key to success. The whole process is well-documented and explained in videos [15] published on the internet.

The cataract treatments are to be distinguished from the LASEK/LASIK procedures for myopia and hyperopia correction, which are described in Chapter 6.5/Vol. 3.

11.3.6 Intraocular pressure

The intraocular pressure (IOP) of eyes is on the order of 27 hPa. The static pressure is maintained by a balance of inflow and outflow of aqueous humor. A proper IOP is important for the spherical shape of the eye ball and for keeping all inner parts of the eye in place (Fig. 11.11).

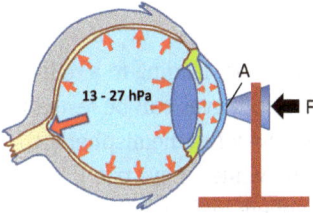

Fig. 11.11: The hydrostatic pressure of the eye is dynamically maintained by inflow and outflow of aqueous humor. If the pressure is too high, the retina and the optic nerve may be damaged. A tonometer can be used for measuring the intraocular pressure.

Aqueous humor is a transparent fluid that consists mainly of water (98%) similar to blood plasma, providing nutrition and oxygen to the cornea, lens, retina, and vitreous humor, i.e., to all those parts of the eye which do not have access to the blood circulation. Aqueous humor is filtered and secreted from blood in the ciliary body behind the iris. The drainage goes through the trabecular meshwork into the *canal of Schlemm*[7] and back into the circulatory system. The balance of in and outflow is critical and usually controlled by stimulation of β-receptors. If the pressure is too low, the cornea will swell; if the pressure is too high, there is a potential danger of damaging the retina and the optic nerve. Therefore it is important to measure the IOP during an eye examination. Normal IOPs are between 13 and 27 hPa. Ocular h*ypertony* is defined as pressures beyond 27 hPa; ocular *hypotony* is a condition for pressures below 6.5 hPa.

The IOP can be measured with a *tonometer* as illustrated in Fig. 11.11. The cornea is touched with a flat conical-shaped prism after a topical anesthetic has been administered to the patient. Then the force F is measured that is required for flattening the cornea such that the area of the conical dip equals the indented area A on the cornea. A standard area A of 3.06 mm diameter is used. Under this condition, the inside pressure equals the outside pressure $p = F/A$. However, the measured values can be affected by the thickness of the cornea and its rigidity. The standard and average thickness of the cornea is 555 μm. If the corneal tissue is thicker than average, the tonometry reading is falsely high. Vice versa, if the cornea is thinner than average, the tonometry reading is falsely low. The tonometer reading can be corrected by using an additional ultrasonic determination of the corneal thickness.

---

7 Friedrich Schlemm (1795–1858), German anatomist and surgeon.

Another and contact-free method measures the deflection of the cornea upon a pulsed air pressure (air-puff). This method is quick but not error-free, as the pressurized air pulse may be scattered at eyelids or damped by eyelashes. There are several other methods for IOP determination, but ophthalmologists most frequently use the two methods mentioned.

11.3.7 The visual field

Hypertonia can damage the optic nerve and/or parts of the retina, leading to a reduced visual field. With an optical perimeter, the visual fields for both eyes can be tested independently (see Fig. 11.12). A perimeter consists of a hollow white hemisphere. One eye is in the center, fixing a light at the center of the sphere, while an eye patch blocks the other eye. A tiny spot of light flashes across the hemisphere in different places and with different intensities. While maintaining fixation, the patient presses a button as soon as he sees the point of light within the hemisphere. Several sweeps are performed to correct for errors. When the procedure is complete, a visual field map highlights areas of high and low light sensitivity.

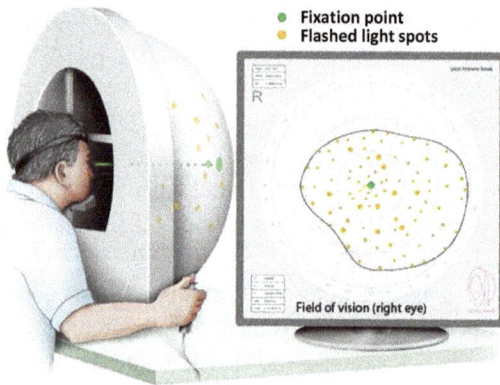

Fig. 11.12: Optical perimetry for examining the visual field. The screeen shows the result of the test for the right eye (adapted from information health at https://www.informed health.org/what-kinds-of-eye-examina tions-are-there.html).

In a normal eye fixed to look forward, the *visual field* extends by 130° in the vertical direction and 180–200° in the horizontal direction (Fig. 11.13). This extremely large visual field exeeds the values reached by super-wide angle cameras and is due to the very short focal length of only 24 mm. There is only one *blind spot* on the visual field where the optic nerve and blood vessels enter the retina called the *optic disk*.

The visual field of a normal person mapped onto a circular disk and color-coded is reproduced in Fig. 11.14(a). The blind spot appears on the right side of the visual field, implying that the map is from the right eye with the blind spot on the left side of the vidual field (compare Fig. 11.34). The visual field of another person with *glaucoma* is demonstrated in Fig. 11.14(b). The brown-colored area indicates regions with

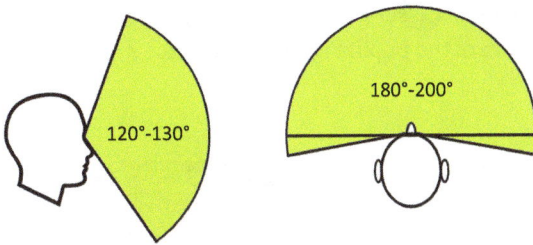

Fig. 11.13: Visual field of a normal eye in the vertical and horizontal directions.

reduced light sensitivity. With progressing glaucoma, those areas may expand and turn black like in the central part of the blind spot unless measures are taken to reduce the IOP. It should be mentioned that glaucoma may also develop at normal IOP, called *normal-tension glaucoma* (NTG). NTG damages the optic nerve rather than the retina. The reason for NTG is not clear presently. Perimeters are also used to distinguish between glaucoma and any damage to the optic nerve due to stroke. More information on this topic follows in Section 11.4.6.

> The visual field of the human eye covers a large angular range in the horizontal and vertical direction. Possible losses of the visual field may be due to glaucoma or stroke and can be examined by a perimeter. With an ophthalmoscope the fundus of the eye can be scrutinized. !

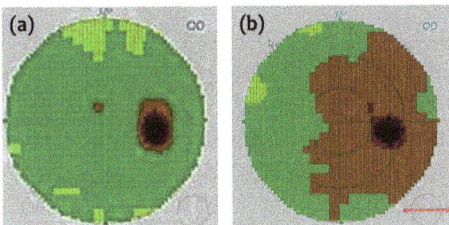

Fig. 11.14: Visual field mapped onto a circular disk-shaped area (a); visual field of an average person and a person with glaucoma of the right eye (b). The location of the optic disk appears black.

Deterioration of the retina can also be examined with an *ophthalmoscope* invented by Helmholtz. In the early version, a hemispherical or parabolic mirror was used to focus light through the patient's eye lens onto the retina. The examiner watches the light reflected back through a hole in the center of the mirror. Figure 11.15(a) shows the scheme of a more modern version, where a mirror or prism replaces the hole. In order to examine a large area of the retina, the pupil must be dilated. There are currently three versions used to image the fundus, i.e., the inner back of the eye: (1) direct ophthalmoscope provides a magnified, virtual, and upright image of the retina with a magnification of about 15; (2) indirect ophthalmoscope provides a real, inverted image with a magnification of about 8; (3) the scanning laser ophthalmoscope provides good images of the fundus even without dilation of the pupil. Figure 11.15(b) shows an example of a normal, undamaged fundus.

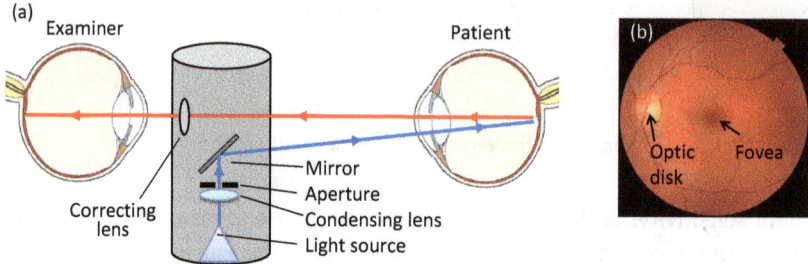

Fig. 11.15: (a) Schematics of a direct ophthalmoscope for the examination of the fundus; (b) image of a normal fundus of the left eye. The optic disk and the macula are visible (from *Medical gallery of Mikael Häggström 2014. Wiki Journal of Medicine* 1(2). DOI:10.15347/wjm/2014.008. Public Domain).

In recent years, optical coherence tomography (OCT) has greatly enhanced the examination capabilities of the fundus. Not only the thickness of the retina can be determined. Spectroscopic OCT allows in addition to examinate the functioning of the different layers. OCT is a topic presented in Section 2.5.6/Vol. 2.

In most cases, damage to the retina or the optic nerve due to glaucoma is irreversible. This is because neither ganglions in the retina nor nerve fibers regenerate even if the IOP has been lowered by surgical means. Therefore, glaucoma surgery serves the purpose of preventing further damage rather than restoring the visual field that has already been lost. Medication may reduce the IOP, and surgeries may increase the drainage of aqueous humor either by extra filtration or an artificial implant [16]. Presently biocompatible *microstents* are being tested for increasing the outflow of aqueous humor, as displayed in Fig. 11.16.

Nevertheless, glaucoma remains one of the main reasons for developing blindness. Unfortunately, pharmaceutical treatments and/or various types of surgeries are still by far less successful than treatments of cataracts. Completely blind patients may gain some limited vision by implantation of epiretinal or subretinal microelectrodes discussed in more detail in Chapter 13.

11.4 Photoreception

11.4.1 Structure of the retina

This section analyzes the light path from absorption in the retina to the firing of an AP that travels via the optic nerve to the brain's visual cortex. Figure 11.17 shows a

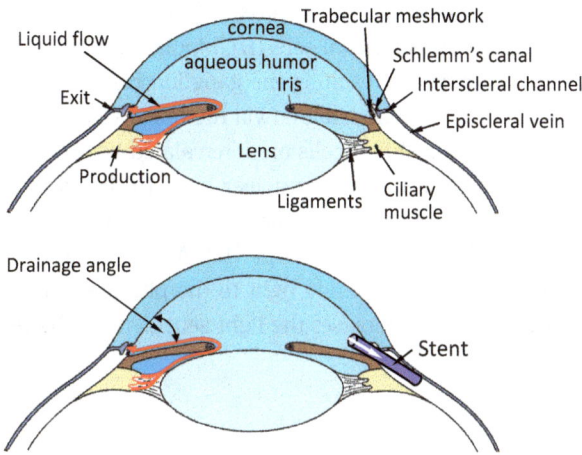

Fig. 11.16: Aqueous humor that fills the anterior chamber is produced by the ciliary body (muscle) and flows between the iris and lens through the pupil and to the drainage angle at the junction of the iris and the cornea. Aqueous fluid exits the eye through the trabecular meshwork into the Schlemm's canal, from where it goes back into the episcleral vein. Stent implants increase the outflow rate of aqueous humor to lower the intraocular pressure.

cross section of the retina, which consists of several layers. The light first traverses a nerve fiber bundle from the proximal side to the distal layers and then passes through a layer filled with *ganglion* cells (G). From there, light travels from the inner plexiform layer, a network of axons and dendrites, to the outer plexiform layer, passing through three distinct cell types: *amacrine* cells (A), *bipolar* cells (B), and *horizontal* cells (H) in the inner core layer. The outer plexiform layer contains nerve endings from bipolar cells, horizontal cells, and photoreceptor cells. Ultimately, light is absorbed in two types of photoreceptors: rods and cones.

The retinal pigment epithelium is a supply layer for retinal molecules that are used in photoreceptors for light absorption. Due to its black color, caused by high melanin levels, the epithelial layer absorbs most of the light that is not already captured by the receptors. This way, the epithelial layer prevents the light from reflecting back to the retina. Otherwise, it would interfere with the image recognition in the retina. Thus, the pigment epithelial layer protects the photoreceptors from harmful levels of light intensity. But the absorption of the epithelial layer is not perfect. In pictures taken with flash light, red eyes can be seen from back reflection. The choroid layer contains blood vessels and capillaries for oxygen and nutrients supply, collecting waste products, and cooling the back of the retina, which is exposed to the high light intensity at the focal point.

The retina is a highly complex arrangement of receptors cells described further below. In short, light is absorbed in cones with trichromatic color sensitivity and in rods with light/dark sensitivity. Light causes electrostatic hyperpolarization in rods and cones. The polarization moves from the distal to the proximal side of the retina and is processed by three different cell types: horizontal cells (H), bipolar cells (B), and amacrine cells (A) in the inner core layer before reaching the ganglion cells (G). The ganglion cells fire APs when stimulated that are transmitted via the optic nerve to the visual cortex in the brain for final processing. Müller cells are intercalated between all of the aforementioned cells and extend from the outer nuclear layer to the ganglia layer. Müller cells support the retina's metabolism by recycling neurotransmitters necessary for the synaptic activity of the cells in the inner retina [17]. At the same time, they structurally support the retinal network and guide light to the photoreceptors, which reduces the signal noise (N/S) ratio and increases the light sensitivity of the receptors [18].

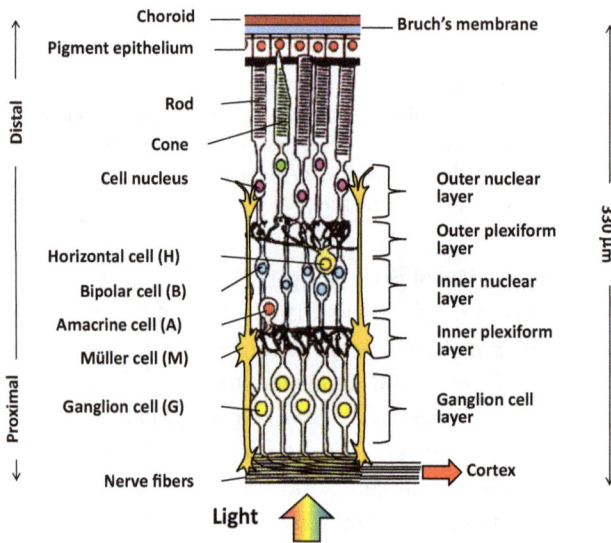

Fig. 11.17: Cross section of the retina showing different layers together with cell arrangement. Light comes from the bottom up while the cell's membrane polarization travels from the distal end at the rods and cones to the ganglion cells at the proximal end, and from there via the nerve fibers to the visual cortex.

There are approximately 120×10^6 rods on the retina, but only 7×10^6 cones; the ratio is roughly 20:1. Rods and cones are unevenly distributed over the retina. Cones have a very high density in the foveal area (Fig. 11.18) which extends over an area of about 1.5 mm in diameter. In the middle of the fovea, with a diameter of about 0.2–0.3 mm, the cones are even slimmer than on the outside for an increased density of about 150 000–200 000 mm^{-2}. Within the foveal area, green and red-sensitive cones dominate, and the density of blue-sensitive cones increases in the parafoveal region. Outside the fovea, the cone density drops rapidly (see Fig. 11.19). In contrast, there are no rods in the inner part of the fovea. The rod density increases in the par-afoveal area up to a maximum and then decreases again outside the macula as can be seen in Fig. 11.18. The fovea is the region with the highest visual acuity and lies on the optical axis of the eye [19].

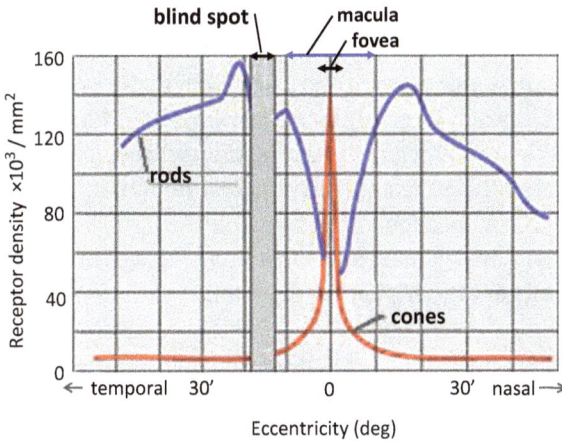

Fig. 11.18: Receptor density as a function of eccentricity. Note that the distribution of cones and rods is highly uneven in the retina.

Figure 11.19 again shows a cross-sectional view of the retina in the region of the fovea and macula. For increased sensitivity to light, the fovea is not obstructed by nerve fibers and blood vessels and can be recognized by a depression (fovea centralis) on the retina. The macula centered around the fovea has a diameter of about 5.5 mm. It is the area of the highest cone and rod density and therefore the area with the highest optical resolution. The macula is also called "yellow spot" because of its yellowish appearance.

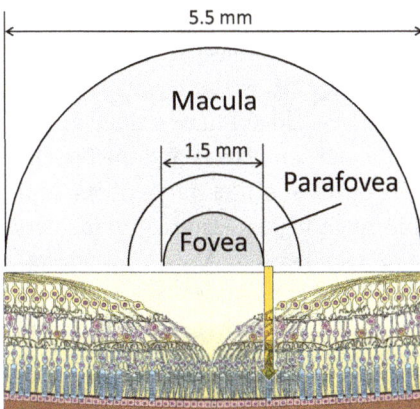

Fig. 11.19: Cross section of the retina in the macula area. The fovea is the region of maximum sensitivity where the density of cones is highest. The central rod-free region of the fovea has a diameter of about 1.5 mm and a thickness of about 80 μm. The yellow arrow indicates the direction of the incident light.

Finally, rods and cones converge to form about 10^6 ganglion cells. Each ganglion cell is connected to one nerve fiber that is bundled in the optic nerve leading to the optical cortex in the brain.

Closer inspection of the cells in the retina shows that they feature different dendritic field sizes [20]. We distinguish the following cells (see Fig. 11.20):

- P cells (parvocellular cells) are small ganglion cells with small dendritic fields. They occur mainly in the area of the fovea. Because of their small size, they are responsible for collecting information on shape and colors with high resolution;
- M cells (magnocellular cells) have a large cell body and large dendritic field. They are located outside of the foveal region. Because of their large size, they are more sensitive to light intensity but feature lower spatial resolution. In general, M cells collect information on location and movement;
- H-cells (heterogeneous cells) have a small nucleus and a sparsely branched dendtritic field. They sit mainly at the periphery of the retina.

We will notice later that the separate perception of shape and color in the P cells and movement in the M cells is retained when projecting the nerve fibers into the visual cortex.

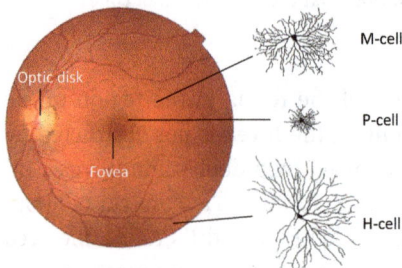

Fig. 11.20: Three different cell types are present in the retina. P, M, and H cells with their characteristic dendritic fields sketched on the right side.

A final note: it seems awkward that the nerve fibers lie on the proximal side of the retina instead of the distal side, requiring a blind spot on the retina. Compound eyes do not have this constructive problem. Each facet is connected to a nerve fiber that goes directly to the insect's "brain." And even squids, which have a retina like us, have the fibers on the distal side and consequently do not have a blind spot. In the human eye, however, the contact between the rods and cones and their supplying epithelial layer is more important than the position of the nerve fibers. Since the light absorption of the retinal epithelium is much higher than that of the nerve fibers, the nerve fibers must be located on the proximal side.

! In the retina, rods and cones are the light-sensitive receptors. The distribution of receptor cells in the retina is not homogeneous. Within the area of the fovea only cones are located, sensing color and shape. Outside the fovea, cells respond to location and movement.

11.4.2 Sensitivity and adaptation

11.4.2.1 Intensity range

Rods and cones form two independent visual systems: *scotopic* (darkness) and *photopic* (brightness) systems. The monochromatic vision of rods is adapted to work at dim light. In contrast, cones are adapted and optimized for trichromatic vision at daylight. The sensitivity regions for scotopic and photopic vision as a function of luminance are shown in Fig. 11.21(a). The units are candela per square meter (cd/m^2). The total sensitivity range covers 10–11 orders of magnitude. This is a unique dynamic range of light sensitivity, only matched by the auditory sensitivity, which also spans some 10–12 decades (Chapter 12). The lower limit of light sensitivity corresponds to single-photon detection, whereas the retina is exposed to millions of photons per second in the upper limit.

Figure 11.21(b) presents the dark adaptation of rods and cones as a function of time. Plotted is the threshold light intensity that a subject can recognize after switching off a bright light. The adaptation is fast in the beginning, with increasing sensitivity by a factor up to 500 after 5–8 min, followed by a slower adaption for the next 20–30 min. During this extended adaptation time, the sensitivity increases by another factor of 2000. The first fast drop is due to cones. However, their sensitivity

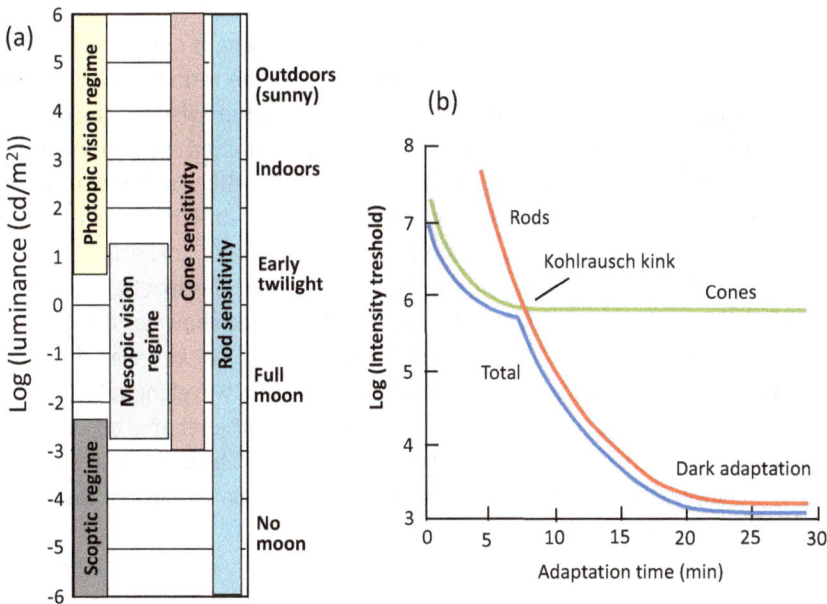

Fig. 11.21: (a) Dynamic range of photosensitivity of rods and cones. Marked are also the regions for scoptic vision, photopic vision, and mesopic vision. (b) Dark adaptation of rods and cones as a function of time.

levels off, and the additional sensitivity increase is due to rods. The combined curve shows a kink at the cross point, referred to as the *Kohlrausch*[8] *kink*. This kink was an early indication of the existence of two visual systems in the eye, the scotopic and photopic system. Three strategies are followed to enhance the sensitivity even further: increased pupil, regeneration of visual pigment in the rods, and neural enhancement; the latter is discussed in Section 11.6. Table 11.2 compares and summarizes the main features of the scotopic and photopic visual systems.

Tab. 11.2: Main characteristics of the photopic and scotopic visual systems.

Property	Photopic	Scotopic
Receptor	Cones	Rods
Color	Tricromatic	Monochromatic
Pigment	Rhodopsin	Rhodopsin
Sensitivity	Low	High
Usage	Daylight	Twilight
Location	Fovea	Outside fovea
Acuity	High in fovea	Low overall

11.4.2.2 Spectral range

The sensitivity spectrum of cones and rods covers the wavelength range from about 400 to 700 nm. Sensitivity means that the absorption of the retinal receptors for wavelengths in this range is high and zero outside. Within the "visible" range, the sensitivity of rods and cones is different. Cones have a combined maximum sensitivity at 555 nm, while the rod sensitivity is blue-shifted to 498 nm, known as the *Purkinje*[9] *shift* (Fig. 11.22). Blue objects are easier to see in the dark than red objects, although both appear gray. The absorption maximum of blue cones is 420 nm, green cones 534 nm, and red cones 564 nm [21, 22]. However, the spectral response of the three cone types is extended, i.e., their sensitivity curves show large overlaps. Therefore, the designations "red," "green," and "blue" should not be taken literally. More adequate is the labeling "long wavelength" (L), "middle wavelength" (M), and "short-wavelength" (S), adapted in Fig. 11.22. Also, note that the relative absorbance of the green and blue cones is less than that of the red cones. The color perceived by our brain is a complex mixture of the cones' spectral distributions, the discussion of which is continued in Section 11.8.

---

8 Friedrich Kohlrausch (1840–1910), German physicist.
9 Jan Evangelista Purkyně (1787–1869), Czech physiologist.

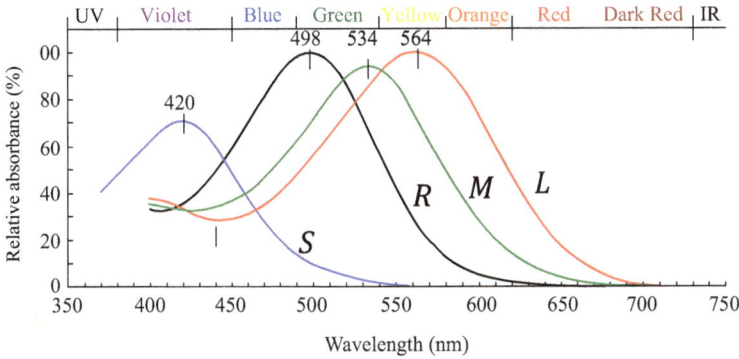

Fig. 11.22: Relative absorbance of rods (black line, R) and the three types of cones (colored lines, S, short wavelength; M, medium wavelength; L, long wavelength). The absorbances of the L cones and rods are normalized to 100% (adapted from: https://www.unm.edu/ ~ toolson/human_cone_response.htm).

As already mentioned and also shown in Fig. 11.22, the bandwidth of the visual sensitivity covers the wavelength range from at most 400 nm to 700 nm. This bandwidth is due to the combined sensitivity of rods and cones. If we could invent a method to increase the visual bandwidth of the receptors, would the cornea, aqueous humor, lens, and vitreous humor system be transparent to this increased bandwidth? Rephrasing this question: How large is the spectral transmission range of the eye's optical system? Measurements have shown that the optical transparency extends from 400 nm to around 1400 nm [23] (see Fig. 11.23), but from 700 nm to 1400 nm we are blind. So another receptor in this "blind" region would be useful. Night vision devices have sensors with sensitivity in the infrared range, which are then wavelength shifted by photocathodes and phosphor screens in the visible range. Blindness in the infrared range can be dangerous when working with infrared lasers of high

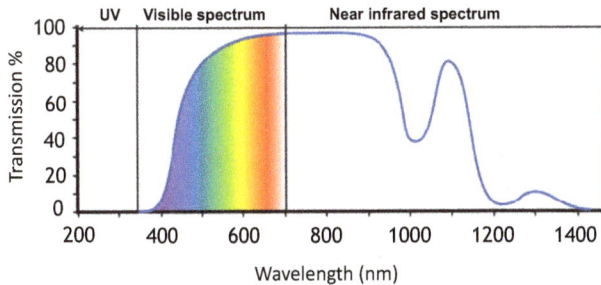

Fig. 11.23: Spectral transmission of the eye's optical system. The light sensitivity of rods and cones is shown in colors. The sensitivity band of the receptors matches the optical transmission band at small wavelengths in the UV region. In the infrared region the transparency of the optical system is much more extended than the sensitivity of the receptive field (adapted from [23]).

intensity because there is no immediate sensory reflex to withdrawal. Corneal burns and cataract development may result from laser exposure. Similarly, workers exposed to high-intensity infrared light can suffer from cataracts, such as glass blowers and steel workers. Potential hazard also occur in the UV range and beyond. In the UV range, however, the lens of the eye is no longer transparent.

11.5 Phototransduction

11.5.1 Visual pigment rhodopsin

Now we study how rods and cones convert light stimuli into membrane potentials. The conversion of light into an electrical signal is known as *phototransduction*. Solar cells also convert light into electrical voltage. However, comparing the complexity of the conversion process in solar cells and in the retina, the phototransduction in the retina turns out to be by far more intricate. As a result, the quantum efficiency is also much higher than that of the solar cell. More on quantum efficiency is posted in the Infobox III.

Figure 11.24 shows the important cellular parts of rods and cones in the retina. We distinguish between an outer segment and an inner segment. The outer segment contains a stack of membrane discs that hold the visual pigment rhodopsin. The disk-shaped membrane consists of a double lipid layer inside of the receptor's cytosol. The inner segment contains the essential cell components: mitochondria, nucleus, and dentrides. The synaptic terminal is connected to bipolar cells via synaptic gaps (Fig. 11.17). Light travels from the inner to the outer segment and is eventually absorbed by visual pigments in the disk membranes.

Each rod contains about 1000 disks [24]. The disks are permanently refreshed: those next to the pigment epithelium are replaced by fresh ones grown at the connecting cilium. The membrane of each disk holds about 10^6 visual pigments (see the inset in Fig. 11.24). Therefore, each rod encloses about 10^9 pigments. Each pigment consists of the protein opsin and the light-sensitive retinal; both are called rhodopsin. The protein opsin is composed of seven transmembrane helices connected by loops [25]. It belongs to the class of so-called *G-protein-coupled receptors* that are also responsible for taste and smell. In rhodopsin, the light-sensitive receptor is an 11-cis retinal, an aldehyde of vitamin A_1, i.e., an $H-C=O$ group attached to vitamin A [26]. The rhodopsin protein attached to a disk membrane is sketched in Fig. 11.25. The membrane is, as usual, a dense package of lipid molecules arranged in a double layer with hydrophobic tails inside and hydrophilic heads sticking out into the cytoplasm. The structural analysis of rhodopsin was a

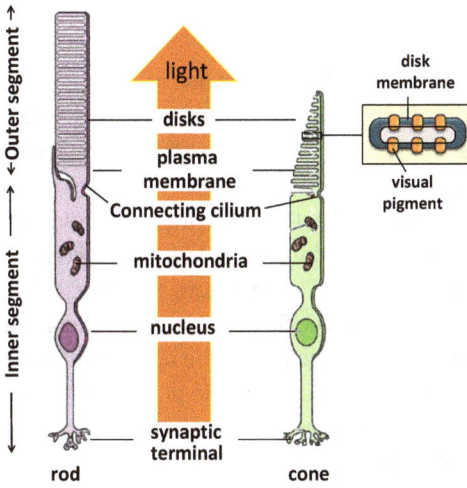

Fig. 11.24: Inner and outer segments of rods and cones. The outer segment holds a stack of disks, which contain visual pigments.

milestone in the understanding and description of the phototransduction process. The importance of rhodopsin for the visual process was discovered by Wald.[10]

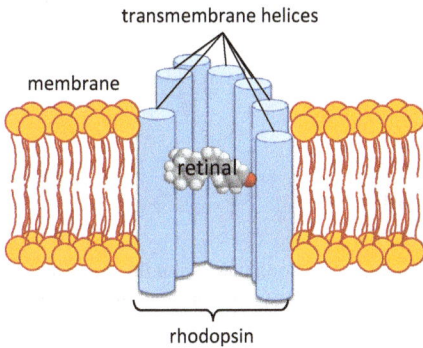

Fig. 11.25: Rhodopsin, composed of the protein opsin and the retinal, is embedded in the disk membrane. The protein opsin is composed of seven transmembrane helices connected by loops (here omitted for clarity). Each disk in rods and cones contains about 10^6 rhodopsin proteins.

Each receptor contains about a billion pigments; each pigment is composed of the protein opsin and the light-sensitive retinal; both together called rhodopsin.

10 George Wald (1906–1997), American biochemist, Nobel Prize 1967.

11.5.2 Photoisomerization

When a photon is absorbed by a photoreceptor, a *photoisomerization* of the retinal occurs, changing the bent 11-cis configuration to the straight all-trans form (Fig. 11.26). This structural change is also known as *configurational isomerization* of the retinal chromophore in rhodopsin. It requires a rotation of the molecule about one of the double bonds by 180°, indicated by a red arrow in Fig. 11.26. The trans-configuration is an excited state of the retinal R^*. The cis–trans transition takes less than 200 fs [27]. This ultrafast photochemical reaction initiates vision. The processes that then follow are sketched in Fig. 11.27, and the numbers refer to specific states in the cycle of rhodopsin as detailed below. The excited trans-form R^* remains in the opsin until transducin molecules are activated (state (2)). Then the trans-retinal R^* no longer fits into opsin; the weak covalent bonds between retinal and opsin are broken, and all-trans retinal molecules are released from the opsin protein, a process which is known as "bleaching out" (state (3)).

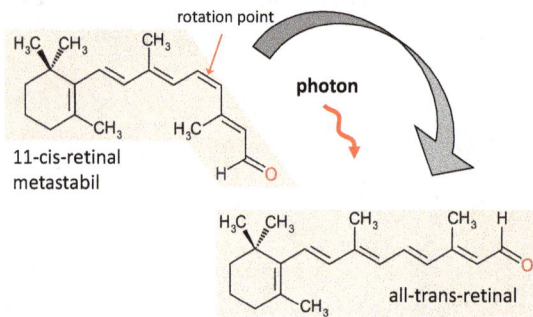

Fig. 11.26: Photoisomerization of the retinal from the kinked 11-cis configuration to the all-trans form. The cis–trans configurational change occurs in less than 200 fs (adapted from: https://commons.wikimedia.org/w/index.php?curid=18461428).

Opsin is now in an inactivated state, and all parts must be reassembled for the next photon absorption. Once opsin and retinal have split up, the retinal drops into the stretched ground state (state (3)). Restoring the kinked state requires energy. The energy is delivered by an ATP–ADP enzymatic reaction, where ATP is supplied by mitochondria that are stored in the inner segment. Opsin and retinal rejoin and are ready for the next photon absorption (state (4)). The potential energy landscape for the retinal is shown in Fig. 11.28 for a complete cycle; the numbers in parantheses refer to the same states as in Fig. 11.27.

> **!** Light absorption by the retinal causes a cis–trans isomerization of the retinal that initiates vision.

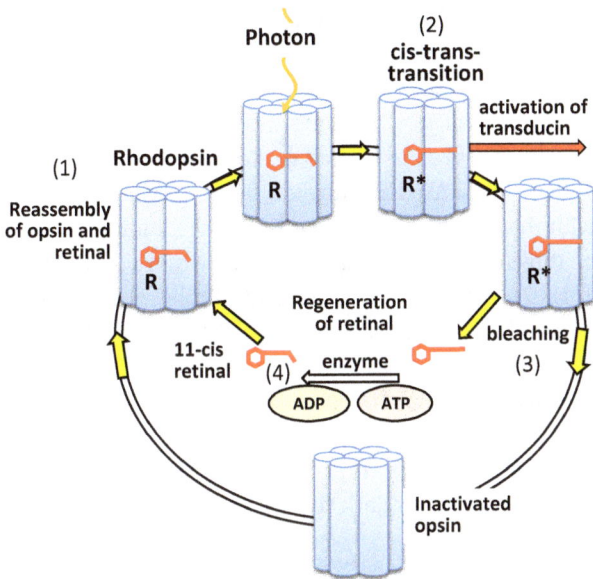

Fig. 11.27: Cycle of rhodopsin after photon absorption. The retinal goes from the cis to the trans form (2), activates transducin, and releases from the opsion (3). The retinal is regenerated by the help of the ATP–ADP encymatic process (4) and then restored as rhodopsin (1).

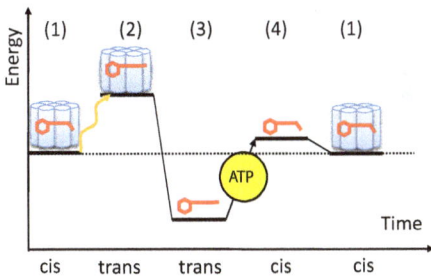

Fig. 11.28: Potential energy landscape for a complete retinal cycle. The ground state is an isolated and stretched retinal (3). Bending requires energy delivered by ATP (4). The energy is first lowered by the opsin-retinal reaction (1), followed by a strong increase during photoisomerization in a confined environment (2), causing the retinal molecule to stretch and the opsin molecule to strain. After release of the retinal, the retinal goes back into the ground state (3).

11.5.3 Three cycles to control receptor potential

The task of the configurational isomerization is, ultimately, the change of the receptor potential for light "on" compared to light "off." Usually, the K^+ channels are open in cells, and the Na^+ channels are closed, yielding a resting potential of about −70 mV (see Chapter 5). However, this is not so in the photoreceptors. Here the Na^+/Ca^{2+} channels in the outer segments are gated by cyclic guanosine monophosphate (cGMP) molecules. cGMP molecules attach to the Na^+/Ca^{2+} channels and keep them open [28]. Because of open Na^+/Ca^{2+} channels, the transmembrane potential is only −30 mV in the resting (dark) state instead of the usual −70 mV.

Photoisomerisation sets off three interlinked biochemical reaction cycles in the cytoplasm of the receptor cells with the goal to reduce the cGMP concentration for closing the Na^+/Ca^{2+} channels. The three cycles are schematically shown in Fig. 11.29: (1) rhodopsin cycle shown in Fig. 11.27; (2) transducin cycle; and (3) phosphodiesterase (PDE) cycle. Together, they reduce the cGMP concentration by hydrolizing cGMP to GMP (see Infobox II). Then the cGMP-gated Na^+/Ca^{2+} channels close, the K^+ channels in the inner segment of the receptor cell remain open, and the cell potential hyperpolarizes to −70 mV. The PDE cycle is completed when the cGMP concentration is restored to its dark value. Conversely, in the dark, cGMP binds back to Na^+/Ca^{2+} channels and opens them again, accounting for dark current in rods and cones.

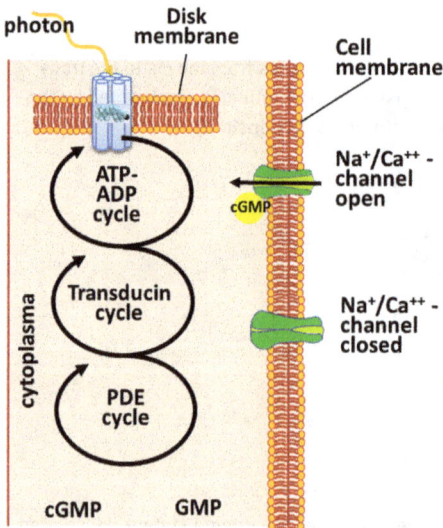

Fig. 11.29: The photoisomerization of the retinal triggers three reaction cycles. The ATP–ADP cycle restores the retinal. The transducin cycle and the PDE cycle are responsible for reducing the cGMP concentration in the cytoplasm so that the cGMP-gated Na^+/Ca^{2+} channels close and the transmembrane potential drops from −30 to −70 mV, as seen in Fig. 11.30.

Infobox II: Transducin

Transducin is a protein consisting of three subunits. It binds to rhodopsin in the inactivated state R and is released in the activated state R*. The transducin's task is to activate PDE. PDE is an enzyme that converts cGMP into GMP. cGMP regulates the ion channel conductance such as the Na^+/Ca^{2+} channel. When converted to GMP, cGMP levels drop. Then cGMP-gated Na^+/Ca^{2+} cation channels close, resulting in a hyperpolarization of the receptor potential. Guanosine monophosphate (GMP) has only one phosphor group and is to be distinguished from guanosine diphosphate (GDP) and triphosphate (GTP). The chemical structure of GMP and cGMP is shown below (source: https://commons.wikimedia.org).

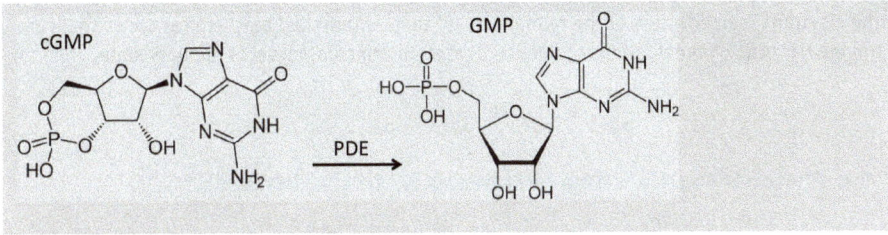

Figure 11.30 shows the membrane potential in the dark state and in the bright state. In the dark state (a) the inactivated rhodopsin is embedded in the disk membrane. Transducin is attached to rhodospin and a high concentration of cGMP keeps the Na^+/Ca^{2+} channels open. As the K^+ channels in the plasma membrane stay open, the plasma membrane potential remains at an intermediate level of -30 mV. In the bright state (b), light absorption causes isomerization of the retinal, and rhodopsin goes into the excited state R* and releases transducin. Transducin binds to PDE and activated PDE converts cGMP (green) to GMP (red). The cGMP levels drop and Na^+/Ca^{2+} channels close, allowing the transmembrane potential to hyperpolarize to -70 mV.

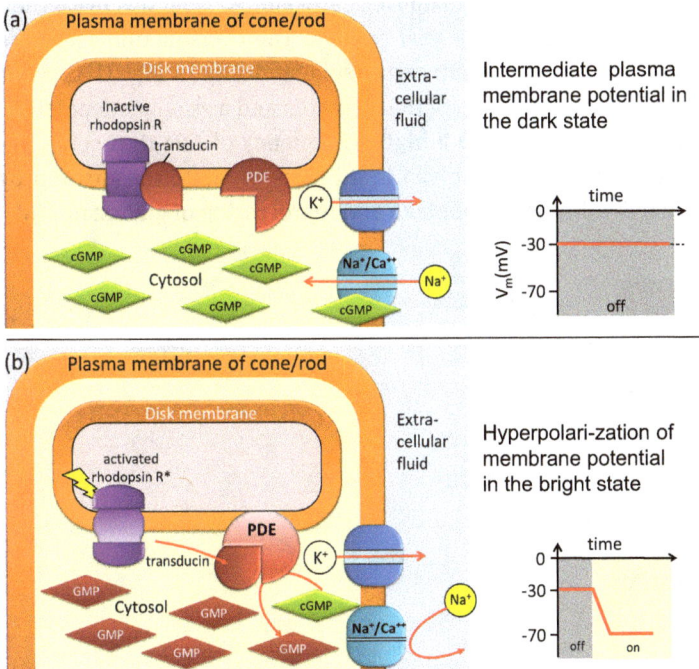

Fig. 11.30: Potential change from the dark state (-30 mV) to the bright state (-70 mV), after closing the cGMP-gated Na^+/Ca^{2+} channels.

> ❗ The cis–trans isomerization of the retinal sets off three interlinked biochemical cycles that control the Na^+/Ca^{2+} channel potential, keeping it at an intermediate level in the dark state.

11.5.4 Phototransduction: from photoreceptors to ganglion cells

The phototransduction from the photoreceptors to the ganglion cells now continues as follows [10, 24]. In the dark state, the rods are depolarized to −30 mV, and the Na^+/Ca^{2+} channels are open. At the same time, the neurotransmitter glutamate is emitted at the synaptic terminals between the rods and the bipolar cells, which prevents depolarization of one class of subsequent bipolar cells (the "ON" cells, see below). When light is absorbed, the receptor potential drops to −70 mV, which reduces glutamate secretion at the synapse. The bipolar cells can now depolarize and transfer the depolarization to the ganglion cells. In response, the ganglion cell triggers an action potential (AP).

In contrast to A, B, and H cells, ganglion cells fire APs in response to depolarization when the degree of depolarization exceeds the threshold potential. Thus, the hyperpolarization in the receptor cells is finally converted into an AP of ganglion cells. For the sake of simplicity, intermediate steps are omitted from this discussion and are described in more detail in the next section. The transduction sequence from hyperpolarization to the AP is summarized schematically in Fig. 11.31. More light causes a stronger hyperpolarization of the receptors and a stronger depolarization of the bipolar cells, which leads to a higher frequency of APs in G cells. The ubiquitous Müller cells are actively involved in all parts of phototransduction by regulating synaptic activity through the uptake and reprocessing of glutamate [18].

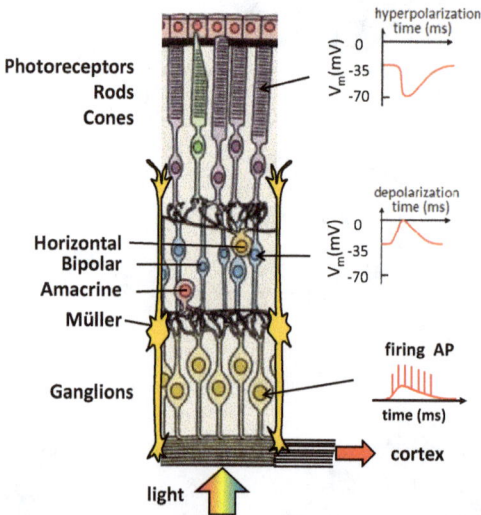

Fig. 11.31: From hyperpolarization to depolarization and action potential in the ganglion cells.

Two distinctive and notable features in the phototransduction process differ from all other receptor cells. First, the stimulation of photoreceptors causes hyperpolarization instead of depolarization. And, second, in the dark, the inactivated Na^+/Ca^{2+} channels are open and generate a dark current. Thus the inactivated state is actually the most active state.

While the sensitivity of the retina is extremely high, the data processing is rather slow, although the first step is on the femtosecond timescale [26]. A certain slowdown is due to the three reaction cycles shown in Fig. 11.29 (ATP, transducin, PDE). The reaction cycles are chemically or more precisely diffusion-controlled and therefore much slower than the light absorption process. On the other hand, the transfer of graded potentials from receptors to bipolar cells and from bipolar cells to ganglion cells is very rapid. The ultimate time-limiting process triggers APs in the ganglion cells with a maximum frequency of about 50 Hz.

Light absorption leads to a hyperpolarization of receptor cells causing a depolarization of bipolar cells, which finally releases an action potential in the ganglion cells that propagates to the visual cortex. !

The signal transduction in cones is similar to the one in rods. There are three cones named according to the color of light they absorb: blue, green, and red; the absorption spectra are shown in Fig. 11.22. In all three color cones, the retinal is identical, but the binding sites in the opsin molecule are slightly different due to slightly different amino-acid sequences in opsin. The response time of cones is faster than that of rods, but their overall sensitivity is lower. The density of red, green, and blue cones in the fovea is not equal and not homogeneously distributed, as we have already seen. Red cones are most frequent, followed by green cones, blue cones have the lowest density in the foveal area [32]. The generation of intermediate hues is due to the activation of more than one type of cone simultaneously, as already mentioned.

Infobox III: Quantum efficiency of photon detection

ℹ

What is the quantum efficiency of photon detection that our eyes can achieve? Quantum efficiency is defined as follows: when a molecule is hit by 100 photons, one after the next, a fraction $x = N/100$ of them are being absorbed; x is the quantum efficiency. In the case of the photon-induced isomerization of the retinal chromophore we have to distinguish between an isolated retinal molecule in watery solution specifically prepared for spectroscopic experiments and in vivo experiments with the molecule in the native environment. In the first case using time-resolved femtosecond spectroscopy, a very fast absorption process on the timescale of about 200 fs with a quantum efficiency of 67% was discerned [27]. In the in-vivo case, the absorption of only one photon activates one retinal molecule R to the excited state R^*. The excited state activates about 1000 transducin molecules within 100 ms. Each transducin molecule intracts with one PDE, and each PDE deactivates about 1000 cGMP molecules. This results in an amplification factor of at least 10^6. The cGMP concentration in the cytoplasm then drops by about 8%, closing about 250 Na^+ – channels out of a total of 10^4 channels, i.e., closing some 2.5% of all channels. This is sufficient to fire an AP and to recognize a single photon in the visual cortex [29, 30]. The cones and rods of the retina have indeed been considered as one of the best-known quantum detectors [30]. However, only 8 out of 100 photons which cross the cornea are absorbed in the retinal, and the others are lost by scattering, reflection, or absorption in the pigment epithelium. A recent re-evaluation of the quantum efficiency of photons in the human eye shows that one has to distinguish between absorptive quantum efficiency of the receptors and light perception by a proband [31]. Any photon that is absorbed causes a receptor response. However, the lowest number of photons crossing the cornea that a proband perceives as light flash is on average about 70 photons [31]. Assuming that more than 90% of the photons are lost on the path to the photoreceptor, the perceived quantum efficiency is still about 20%.

11.6 Retinal signal processing

11.6.1 On–off bipolar cells

The retina is not just a photoreceptor. It is also an essential part of signal processing, which starts in the inner nuclear layer of the retina and is completed in the brain's visual cortex [24, 33, 34]. Figure 11.32 symbolically shows six different cells that take part in the preliminary information processing within the retina.

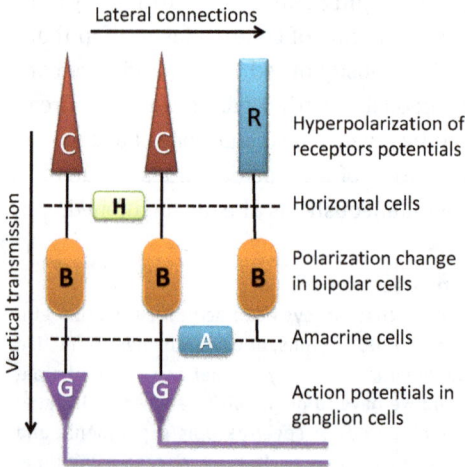

Fig. 11.32: Schematics of the six different cells in the retina with lateral and vertical connections.

Signal processing in the retina has its origin in the hyperpolarization of the receptor potentials as we have seen in the previous section. Receptor cells can only hyperpolarize from an intermediate negative transmembrane potential to a lower potential from −30 to −70 mV. The hyperpolarization is transmitted via synaptic connections to bipolar cells in the retinal network. The bipolar cells may either hyperpolarize or depolarize depending on the input but not stay at an intermediate level. Furthermore, bipolar cells receive input from either rods or cones but not from both. Therefore, bipolar cells are either of rod or cone type. Synaptic connections to the ganglion cells communicate the response of bipolar cells. The horizontal cells are responsible for lateral connections and inhibition between bipolar cells. The amacrine cells connect bipolar cells of rods with ganglion cells. Like the horizontal cells, amacrine cells make lateral connections, and in most cases, their action is inhibitory. The ganglion cells fire APs in response to the depolarization of bipolar cells. All other cells in the retina respond by graded potential variations. Finally, one nerve fiber is attached to each ganglion cell that connects to the visual cortex.

Each receptor in the fovea attaches to a single bipolar cell and a single ganglion cell (Fig. 11.33). Within the fovea, there are exclusively cones. Each cone in the fovea directly connects to the brain, allowing an exact registration of the input location. Several receptors, mainly rods, converge to one ganglion cell outside the fovea. With increasing eccentricity, the number of rods connecting to one ganglion cell increases and reaches up to 150.

Summation of rod signals increases the sensitivity for bright/dark contrast, but the spatial resolution (acuity) suffers outside of the macula region. Vice versa, in the central rod-free area of the fovea with a diameter of about 500 µm, the density of cones is highest, and the 1:1 ratio between cones and ganglions supports this high spatial resolution. The foveal region provides the highest visual acuity. Areas on the retina with high spatial resolution require more space for image processing in the brain.

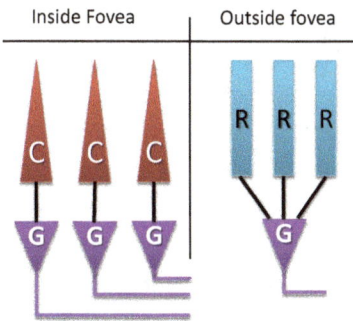

Fig. 11.33: Receptor fields inside and outside of the fovea in a simplified version without showing bipolar cells. Inside the fovea, there is a 1:1 relation between cones and ganglions. Outside, many rods converge to one ganglion.

The signal pathway and graded potential variations for cones are tabulated in Fig. 11.34. As already mentioned earlier, upon light absorption in rhodopsin, Na^+/Ca^{2+} channels close and cause a hyperpolarization. The hyperpolarization, in turn, reduces the *glutamate* concentration in the synaptic space between the cone and bipolar cell. Glutamate is a neurotransmitter that controls the conductivity of ion channels in the synaptic space (see Chapter 6).

In fact, each cone connects to two antagonistic bipolar cells: *metabotropic ON-bipolar cells* and *ionotropic OFF-bipolar cells* (Fig. 11.35). They react oppositely with respect to glutamate concentration: ON-bipolar cells close their Na^+/Ca^{2+} channels in the presence of high glutamate concentration, causing hyperpolarization. OFF-bipolar cells open their Na^+/Ca^{2+} channels in the presence of high glutamate concentration, causing depolarization. Hyperpolarization of cones activates ON-bipolar cells, depolarization of cones inactivates ON-bipolar cells. OFF-bipolar cells react oppositely.

The ON (OFF)-bipolar cells connect to ON (OFF) ganglion cells, which react in the same manner as the bipolar cells. The various dependencies are summarized in Fig. 11.34 for future reference. We note that ON bipolar cells always reverse the potential with respect to the receptor potential, while OFF-bipolar cells keep the same potential change as the receptor.

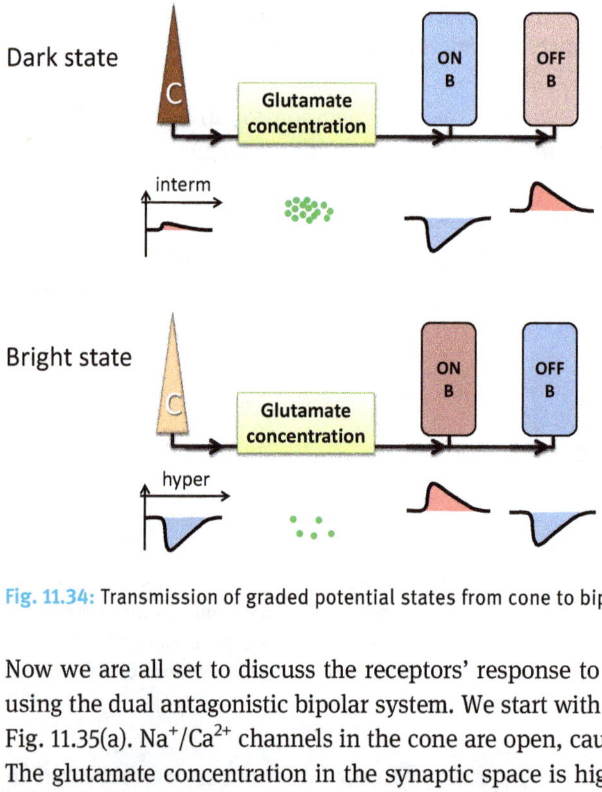

Fig. 11.34: Transmission of graded potential states from cone to bipolar cells.

Now we are all set to discuss the receptors' response to light absorption once more using the dual antagonistic bipolar system. We start with the dark response shown in Fig. 11.35(a). Na^+/Ca^{2+} channels in the cone are open, causing the cone to depolarize. The glutamate concentration in the synaptic space is high, causing Na^+ channels in ON-center cells to close and to turn off ON-center ganglion cells. Vice versa, the same receptor state causes the OFF-bipolar cell and the OFF ganglion to be activated. The OFF ganglion cells fire a high-frequency AP in the activated state. Simultaneously, the hyperpolarized ON ganglion cells are inactivated. When light is turned on, the situation is reversed (Fig. 11.35(b)): the receptor cell hyperpolarizes, ON cells become depolarized, and ON ganglion cells fire APs, while OFF cells become inactivated.

Why do we have two antagonistically operating bipolar cells that tell us about the same state? It should be possible to differentiate between bright and dark states with a single light-sensitive detector, including all intermediate shades. However, the eye uses two detectors, one responsible for brightness and the other for detecting darkness. The dual system gives us a much more precise and weighted impression of the actual state of bright versus dark. The two systems are compared in Fig. 11.36. A single detector measures the intensity on an absolute scale, here normalized from 0 to 100 in arbitrary units. The visual system with two antagonistic detectors, on the other hand, determines brightness and darkness independently of one another and processes the difference Δ. This is a comparative and weighted measurement instead of a single and absolute measurement. It enables us to assess on which brightness level contrast is perceived.

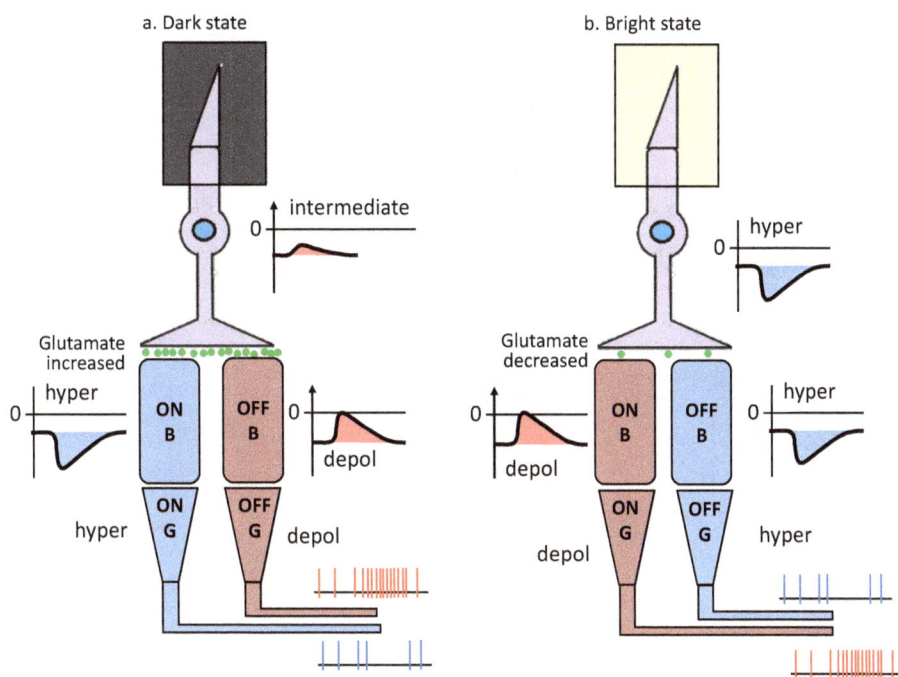

Fig. 11.35: Signal pathway from cones to ganglion cells (G) via two antagonistic ON and OFF-bipolar cells (B): (a) bipolar potentials for darkness and action potentials of the ganglion cells; (b) potential changes upon light absorption. Note that ON-ganglion cells fire high-frequency action potentials (red thin lines) in the bright state. In the dark state, ON-ganglion cells fire low-frequency action potentials (thin blue lines). The response of the OFF ganglion cells is reversed.

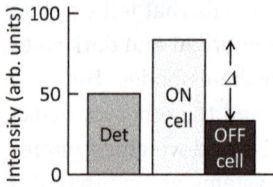

Fig. 11.36: Intensity measurement with a detector system (left) and with the antagonistic visual system (right).

The signal pathway for rods is shown in Fig. 11.37. In rod bipolar cells (RB), reduced glutamate concentration in response to light leads to an opening of Na^+/Ca^{2+} channels and its depolarization. The depolarization excites an amacrine cell, which connects to ON and OFF synaptic connections at ganglion cells. With this detour, rods trigger the same on–off response to light as cones do. The detour is necessary as rods do not connect directly to ganglion cells.

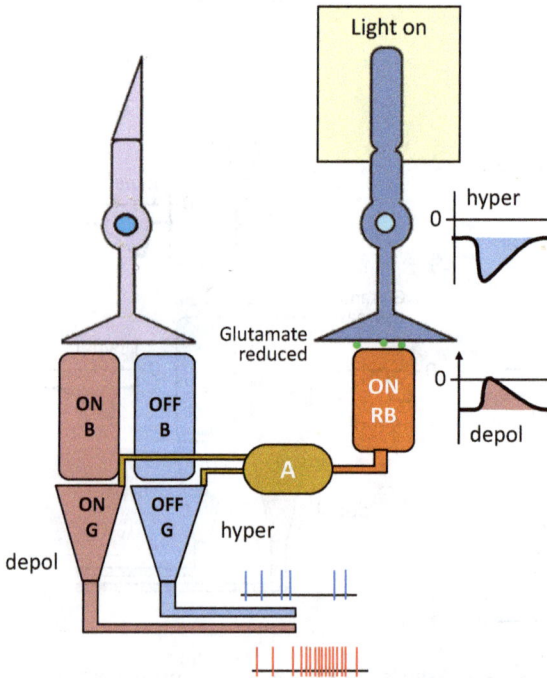

Fig. 11.37: Signal pathway in rods: hyperpolarization of rods causes depolarization of bipolar cells, initiating action potentials in ganglion cells. The signal detour involves amacrine cells.

> **!** Cones connect to two antagonistic ON/OFF-bipolar cells, which determine the difference between brightness and darkness. Rods connect to ON and OFF ganglion cells, yielding the same antagonistic light response as cones.

11.6.2 Receptive field

Each of the five neurons (photoreceptor, bipolar, amacrine, horizontal, ganglion) covers an area of vision in the retina. This area, where an appropriate stimulus (light) modifies a particular neuron's activity, is called this neuron's receptive field [35, 36].

The receptive field of a single photoreceptor cell may be limited to a tiny spot of light that corresponds to the precise location of this receptor on the retina. However, with successive layers of the retina, the receptive field increases in space and becomes increasingly complex because of numerous lateral and vertical interconnects. Furthermore, the size of the receptive field depends on the location in the retina. The receptive field is small In the foveal area but increases with increasing eccentricity according to the increasing dendritic extension from the P to the M-cells (see Fig. 11.21).

As an example, we consider the receptive field of bipolar cells. The receptive field of each bipolar cell is approximately circular. A light ray striking the center of the field has the opposite effect of one striking the surrounding area, called *"surround"*. The difference is due to the ON and OFF-bipolar cells, a discovery made by Keffer Hartline.[11]

A light stimulus applied to the center of a receptive field will cause the ON-center cell to be activated (see Fig. 11.35(b)). The same light hitting the surround has the opposite effect on such an ON-center cell. An ON-center cell will be deactivated in the surround, while OFF-center cells are being activated. The different

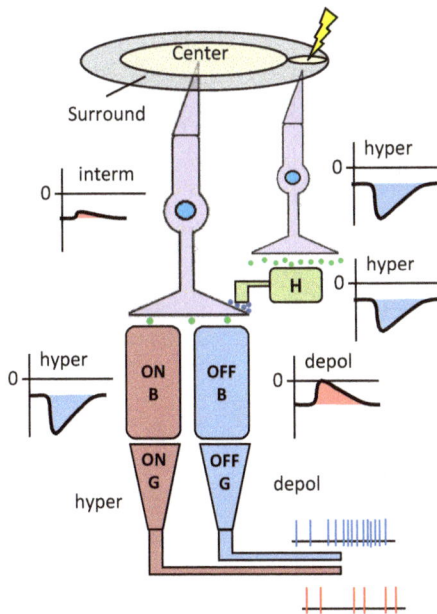

Fig. 11.38: Light in the surround of a receptive field has an inhibiting effect on ON-center bipolar cells by intervention of horizontal cells connecting receptors in the lateral direction.

11 Haldan Keffer Hartline (1903–1983), American physiologist, Nobel Prize 1967.

responses are due to horizontal cells, which connect the receptors in the lateral direction, as shown in Fig. 11.38. All signals become inverted by illumination of a spot in the surround: center receptors react by OFF response, surround receptors react by ON response. Conversely, light at the center turns off the OFF-center cells, while light on the surround turns on the OFF-center cells.

Just like bipolar cells, ganglion cells have concentric receptive fields with a center-surround antagonism. But contrary to the two types of bipolar cells, ON-center ganglion cells and OFF-center ganglion cells do not respond by depolarizing or hyperpolarizing. They respond by increasing or decreasing the frequency with which they discharge APs. In Fig. 11.39, we consider the response of ON ganglion cells to light in the center of a receptive field and its surround and compare them with the equivalent response of ON-bipolar cells.

In the dark state, ganglions in the center and in the surround react in the same way by releasing APs at low frequency. When light is turned on in the center, the AP frequency of the center ganglions increases. Simultaneously, the AP frequency of ganglions in the surround decreases in comparison to the dark state. The decrease of APs in the surround is due to the inhibiting impact of the horizontal cells, serving contrast enhancement. Turning on the light in the surround and leaving the center dark will cause the center ganglions to reduce the AP frequency and increase

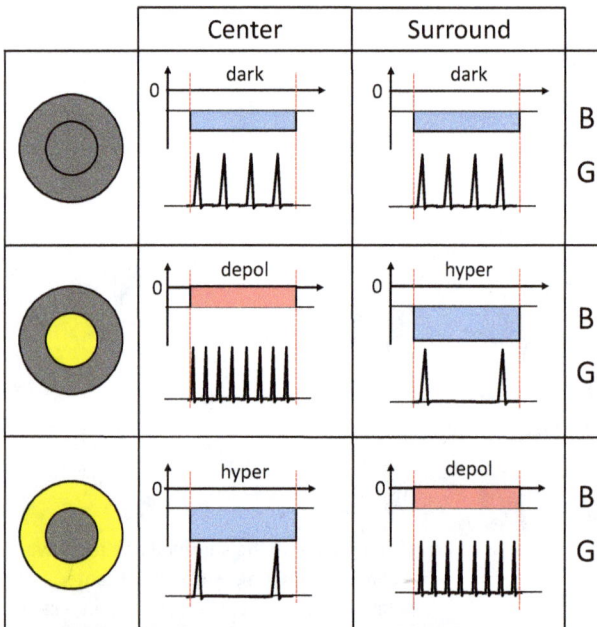

	Center	Surround	
	dark	dark	B / G
	depol	hyper	B / G
	hyper	depol	B / G

Fig. 11.39: Polarization of ON bipolar cells and action potentials of ON ganglion cells responding to light excitation in the center and surround. B, bipolar cell potentials; G, ganglion action potentials, both as function of time. Activation is during the time between the red-dashed lines.

the AP frequency of the surround. The response of OFF-bipolar and ganglion cells to light in the center and surround is opposite to the ON cells. This example shows the principle of the receptive fields' antagonistic working mechanism.

Bipolar and ganglion cells also react to different light intensities: bipolar cells vary the amplitude of their graded potential. In contrast, ganglion cells respond by changing the frequency of APs. Bipolar cells act as analog devices, ganglion cells act as digital devices. Center and surround are always connected by the intervening horizontal cells that gradually change the potentials of the center. The response to center stimulation of the receptive field is always slightly inhibited by simultaneous stimulation of the surround.

11.6.3 Lateral inhibition

The lateral inhibition serves to enhance the contrast. This is examplified in Fig. 11.40. Panel (b) shows two connected areas with different gray levels, i.e., with different reflected intensities, here 100 and 20 in arbitrary units. The receptor field "observes" these two fields with an overall 10% lateral inhibition. By applying the lateral inhibition, the gray field intensities registered in the bipolar cells are reduced to 80 and 16, as shown in panel (a). However, at the edge between the gray fields, the intensity is raised to 88 on one side and lowered to 8 on the other. At the edge, the perceived intensity ratio is not 100:20 = 5, but 88:8 = 11. This doubles the intensity contrast in the retina, which is not physically present. We could not prove this contrast enhancement with a light-sensitive detector. Contrast enhancement is solely the result of retinal processing.

An impressive example of lateral enhancement is shown in panels (c) and (d). The lower panel (d) shows a continuous lateral gradient of gray tones. In image (c), the linear gradient is divided into bands of constant shades of gray. We observe strong contrast enhancement at the edges between the bands due to lateral inhibition. The bands are known as Mach[12] bands.

Figure 11.41 shows another illustrative example of contrast enhancement: points in the receptive field at the crossing points between black squares flicker because of switching between center and surroundings. This is one version of the so-called Hermann[13] grid. Several other versions can be found in the literature. Most optical illusions are based on center-surround perception coupled with lateral inhibition. Lateral inhibition in the retina is an extremely effective process for modifying receptor potentials and also helps us read.

---

12 Ernst W. J. W. Mach (1838–1916), Austrian physicist and philosopher.
13 Ludimar Hermann (1838–1914), German physiologist.

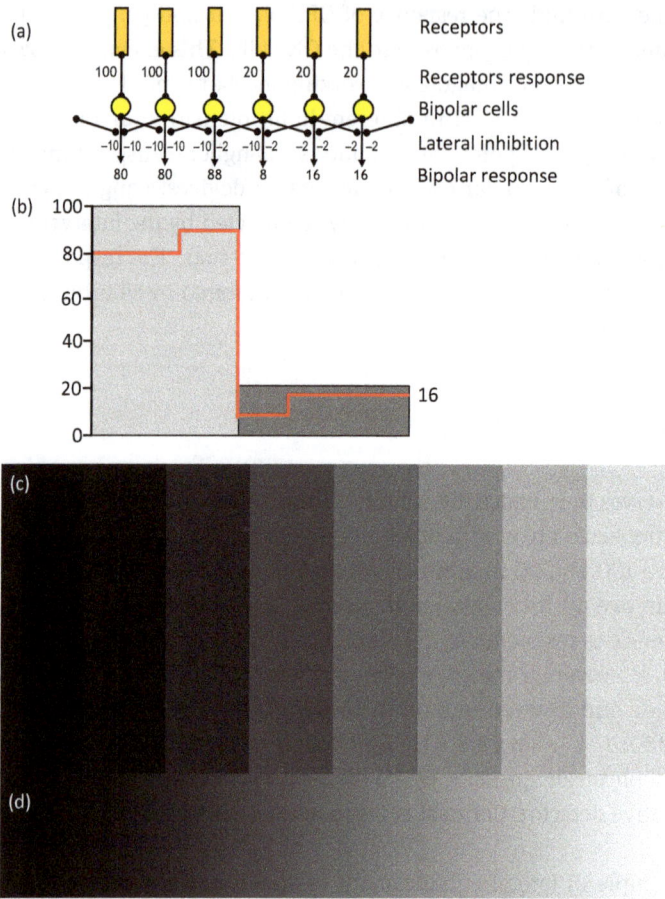

Fig. 11.40: (a) Lateral inhibition of 10% applied to all bipolar cells; (b) contrast enhancement occurs at the edge between two areas of different gray scale; (c) Mach bands of gray shades; (d) linear gradient of gray shade over the same range as in panel (c) (adapted from https://de.wikipedia.org/wiki/Machsche_Streifen, graph generated by Polini).

> ! The receptive field of bipolar and ganglion cells distinguishes between center and surround. Illumination of the center causes high-frequency action potentials of center ganglions and reduces the action potential frequency of ganglions in the surround. Lateral inhibition enhances contrast.

Fig. 11.41: Shown is the optical illusion of disks at the center of cross points between black squares due to center enhancement and lateral inhibition.

11.7 Optic pathway

11.7.1 The optic nerve

Up to here, we have considered the retina's signal processing of optical images. The final signal processing takes place in the visual cortex of the brain. The optic nerve transmits the information from the retina to the visual cortex, passing the optic chiasma and the lateral geniculate nucleus (LGN) (see Fig. 11.42). The optic chiasma, meaning crossing point, is known since antiquity [37], the signal processing in the visual cortex was described first in the 60s of the last century by Hubel[14] and Wiesel,[15] both Nobel Prize winners in medicine and physiology 1981 [38].

Now let us scrutinize the path of the optic nerve through the brain. The visual field of each eye is divided into two hemispheres: an outer (temporal) segment and an inner (nasal) segment. Light entering from the temporal side through the iris strikes the retina on the nasal side of the retina; light entering from the nasal side strikes the retina on the temporal side. The optic nerve begins in the retina and ends in the visual cortex of the cerebrum. Part of the optic nerve fibers cross at the optic chiasma and the other part does not. As shown in Fig. 11.42, the optic nerve fibers from the temporal area of the retina do not cross, while all nerves from the nasal part cross. The partial crossing of the optic nerve fibers entails light entering from the right side and detected on the left side of the visual field (temporal side for the left eye and nasal side for the right eye) to be both projected on the left side of the visual cortex. Vice versa, light entering the eye from the left side is detected on

---

14 David Hubel (1926–2013), Canadian neurophysiologist, Nobel Prize in Medicine 1981.
15 Torsten N. Wiesel (*1924), Swedisch-American physiologist, Nobel Prize in Medicine 1981.

the right side of the visual field and is projected onto the right side of the visual cortex. Objects which move in front of our eyes from right to left elicit signals which move from left to right in our visual cortex. The crossing point is known as optic chiasm (or chiasma). The convergence of crossing and noncrossing optic nerves is necessary to distinguish between left and right: the left visual field is projected onto the right visual cortex and the right visual field is projected onto the left visual cortex. This is the case in all mammals due to the body's bilateral symmetry and is also true of most vertebrates. Even the chameleons with independent left and right eye control show an optic chiasm.

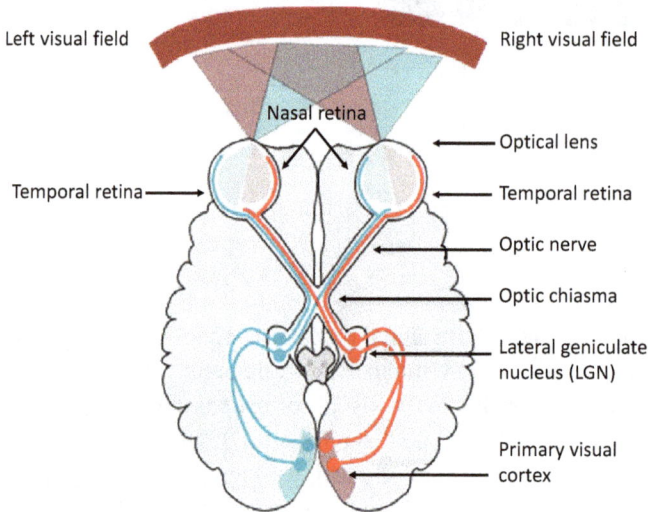

Fig. 11.42: Optic pathway from the visual field of the retina to the cerebral cortex (adapted from http://en.wikipedia.org/).

Lesions or damage of the visual pathway due to stroke or injuries can be judged by examining the visual field with a perimeter (Fig. 11.14) and can be distinguished from impairments that are due to glaucoma. Figure 11.43 shows some typical scenarios [41]. Losses due to glaucoma exhibit irregular shapes in the visual field, whereas damage of nerve fibers before or after the chiasm generate patterns with particular symmetries that hint at the location of the damage.

11.7.2 Lateral geniculate nucleus and primary visual cortex

Along the optic nerve, we can distinguish four parts: (1) photoreceptors (cones and rods); (2) bipolar cells; (3) ganglion nerve fibers; (4) optic radiations. The first three parts are located within the retina, as we already discussed in Sections 11.4–11.6. The

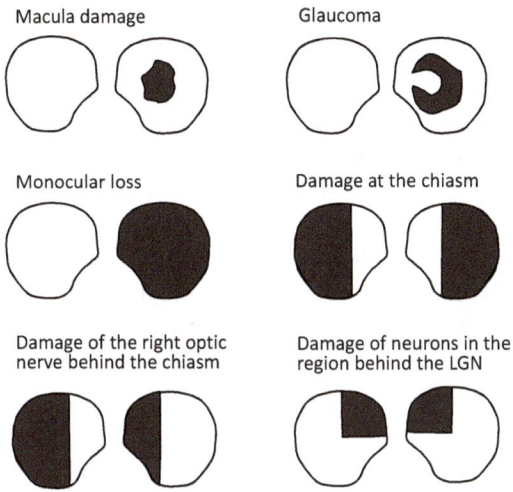

Fig. 11.43: Black indicates damage of the visual field, and white signifies normal functionality.

fourth synapse between ganglion nerve fiber and optic radiation fibers lies within the *lateral geniculate nucleus*, which is part of the thalamus. The LGN is also known as *corpus geniculatum laterale*. Because of the synaptic connection, the signals from the retina can be modulated in a manifold way on the path to the visual cortex. Similar to retinal signal processing, signals may be enhanced, inhibited, or modified in the LNG. The LGN has a laminated structure consisting of six layers, two magnocellular layers (M layers 1 and 2), and four parvocellular layers (P layers 4–6), schematically shown in Fig. 11.44. These layers project back to the corresponding P- and M-cells in the retina (Fig. 11.21). Signal input from the contralateral eye (the eye located on the opposite side to the visual cortex) goes into layers 2, 4, and 6 (blue and yellow lines in Fig. 11.44); signals from the ipsilateral eye (the eye located on the same side as the visual cortex) go into layers 1, 3, and 5 (green and red lines in Fig. 11.44). M-layers are sensitive to movement, P-layers are sensitive to shape and color, just like in the retina. The inputs from the M- and P-layers arrive in different sublayers within the primary visual cortex called V1. In the V1 layers, most of the optical information is processed. Within V1, the P-layers (4–6) project into the layer IV-Cβ, and the M-layers (1–2) project into layer IV-Cα. Thus, the organization and layering are transposed from the retina via the thalamus to the cerebellar cortex. Within the cortex, an array of columnar structures encodes the image of objects with respect to shapes, orientation, direction, length, color, and movement [10, 24].

> The optic nerve passes the partial crossing point in the chiasm and ends in synaptic junctions in the LGN within the thalamus. From there, the optic radiator projects the layered organization of the LGN into the visual cortex.

11.7.3 Binocular vision

Binocular vision enables us stereoscopic perception. Stereoscopic perception is necessary to judge the distance of objects in relation to each other and correctly assess movement directions. Examples are provided in Infobox IV. Additional layers V2–V6 can be distinguished in the visual cortex (not shown in Fig. 11.44), where further processing, computation, association, and memory take place. The details are the subject of current brain research [39]. A detailed description of these processes goes beyond the scope of this text. However, in the following, we want to provide some heuristic arguments.

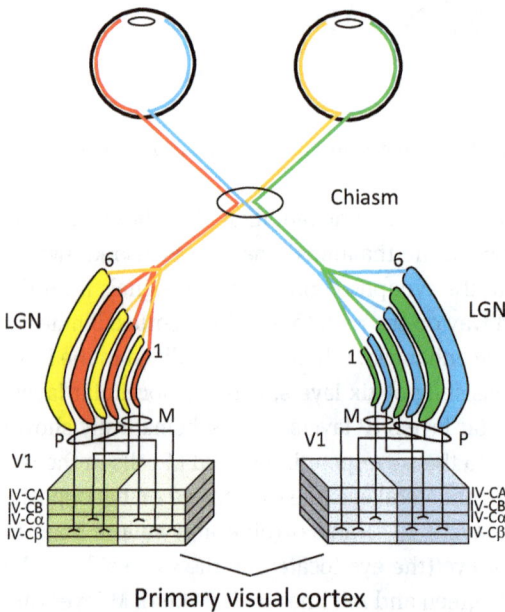

Fig. 11.44: Optic pathway similar to Fig. 11.42, showing the layered organization of the optic nerves in the lateral geniculate nucleus (LGN) originating from P- and M-ganglions and their projection into the primary visual cortex V1.

How do we perceive a single image of an object with two eyes? How do we perceive objects at different depths, and how can we follow the trajectory of objects? The answer to these questions lies in what is known as *binocular disparity*. First, we notice that points lying on a horopter circle in Fig. 11.45(a) project to corresponding points on the retina and are perceived as a single image of an object. The horopter circle runs at a distance d through the fixation point and the nodes of the eyes S_1 and S_2. The fixation point is rather a small fixation area where different objects inside this area are fused together and recognized as a single object. Outside this area, the same objects are seen as separate. If objects are shifted laterally along the dashed line as indicated in panel (b), the corresponding image points on the receptive field move in the same direction, leading to a disparity regarding the horopter projection. By changing the depth of an object as in panel (c), the corresponding image points on the retina move in opposite directions. The blue object point is perceived more closely if the corresponding points on the retina are directed temporally; The purple point is perceived further away when the corresponding points are directed nasally. The rest is left to the computation of our brain [40].

The limiting value for the lateral disparity is an angle change of at least 20 s. Stereoscopic vision is only possible up to a distance of 100 m. Foveal retinal images with a deviation of more than 1.6′ are perceived as separate objects. Squinting prevents the assignment of corresponding points on the receptive field. Depth sensitivity is nevertheless possible. Monocular cues of depth are provided by size, occlusion, brightness, shadowing, and movement of objects. Various binocular function tests are described in detail in [41].

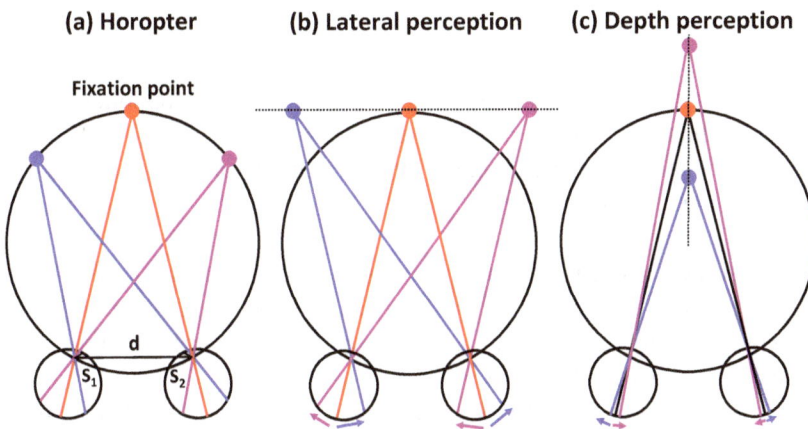

(a) Horopter   **(b) Lateral perception**   **(c) Depth perception**

Fig. 11.45: Horopter is a circle that passes through both eyes: (a) the diameter is chosen such that points lying on a horoptic circle project to corresponding points on the retina and are perceived as a single image of an object; (b) lateral movement of an object's causes leads to disparity of the image location on the retina; (c) depth reception occurs when the images on the retina of the left and right eye move in opposite directions.

11.8 Color perception

11.8.1 Tricolor and four-color systems

Color perception is the result of neuronal processing of light of different wavelengths in the visual cortex. Rhodopsin in the cones selectively absorbs light in three broad frequency bands (Fig. 11.23), which we interpret in the visual cortex as red, green, and blue light. These spectral bands cover the entire visible spectrum from 400 to about 750 nm.

Color perception can be differentiated according to hue, saturation, and brightness. We can distinguish between:

- 200 hues (shades)
- 20 levels of saturation
- 500 levels of brightness

These color qualities yield a total of 2×10^6 differentiation possibilities for color perception. In contrast, with monochromatic vision, only 500 levels of brightness are available.

Historically, two different color theories were proposed:

1. The trichromatic color theory was advocated by Young,[16] Helmholtz,[17] and Maxwell.[18] It states that any color can be created by additive color mixing of three monochromatic and complementary colors: red, green, and blue, the so-called RGB colors.

2. The color opponency theory was proposed by Hering.[19] This theory states that color pairs such as red and green, blue and yellow, and white and black inhibit each other and result in all color combinations. The opponent color theory corresponds to the processing mechanism of the opposite color neurons in the visual cortex. It explains why greenish-red and bluish-yellow do not occur in the receptive field.

Today, both theories are equally valid. The three-color theory explains the mechanisms on the level of the photoreceptors of the retina. The opponent color theory takes the color sensitivity of the receptors into account, and in particular, the color-antagonistic ganglion cells of the retina.

Figure 11.46(a) shows the additive color mixing by superimposing the three primary colors: red (r), green (g), and blue (b). Adding red and green yields yellow (y), adding blue and green yields cyan (c), and adding red and blue yields magenta (m). Red and blue are not neighbors in the light spectrum and therefore the mixed-color magenta does not occur in nature. Combination of all three colors yields white. Combining red with cyan, green with magenta, or blue with yellow also yields white because magenta, cyan, and yellow are already two-color combinations. Therefore, magenta, cyan, and yellow are called complementary to the fundamental colors: red, green, and blue.

---

16 Thomas Young (1773–1829), British physicist and ophthalmologist.
17 See footnote no. 2.
18 James Clerk Maxwell (1831–1879), Scottish mathematician and physicist.
19 Ewald Hering (1834–1918), German physiologist and brain scientist.

Figure 11.46(b) shows the four-color wheel according to Hering, in which the color yellow takes a prominent position as one of the basic colors. The wheel shows those colors, which are naturally perceived by the human eye. Opposite colors on the wheel (red and green, or blue and yellow) are not complementary, but antagonistic. The Hering's color system has been endorsed as the basis for the so-called natural color system, an internationally recognized system.

Now we want to analyze further the additive color mixing by superimposing the primary RGB colors. The percentage mix gives the hue. We define the sensitivity level for the color red with X, for green with Y and for blue with Z. Then the brightness of a particular color mix is defined as

$$H = X + Y + Z. \tag{11.16}$$

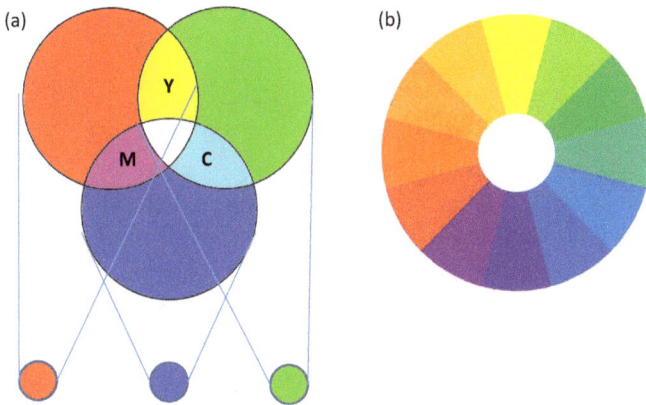

The relative brightness of the three colors is

$$r = \frac{X}{H}; \ g = \frac{Y}{H}; \ b = \frac{Z}{H}. \tag{11.17}$$

Now, the hue of a particular color c is then given by the sum:

$$c = r + g + b. \tag{11.18}$$

With the boundary condition:

$$r + g + b = 1. \tag{11.19}$$

This condition can be represented in a triangle of two components, where the boundary condition determines the third component. The color triangle is shown in Fig. 11.47(a). Here green is plotted on the y-axis and red on the x-axis. Blue is at the origin. Full saturation of the three colors is at the following values:

Red: $(r, g, b) = (1, 0, 0)$. Green: Green: $(r, g, b) = (0, 1, 0)$. Blue: $(r, g, b) = (0, 0, 1)$.

As there is no blue coordinate, blue is located in the corner at $r = 0; g = 0$. Mixed colors are found along the lines between the extremes. For instance, yellow is a 50:50 mixture of red and green and therefore has the value: yellow $= 0.5r + 0.5g$. The white point is at

$$r = g = b = 0.33. \tag{11.20}$$

The triangle would be perfect if 100% of the saturation of the three primary colors could be achieved. However, this is not the case, and therefore the color triangle also reflects the physiologically achievable saturation. The color triangle as printed in Fig. 11.47(a) represents the colors that can actually be seen by normal sighted and color-sensitive people. The number on the perimeter indicates the wavelength of the respective color in nanometers.

> **!** Two main color systems are known: three-color system RGB and four-color system RYGB. The most common color coding is the RGG system.

A disadvantage of the color triangle is that "black" is not displayed. A more logical representation of colors is given by using a color cube (Fig. 11.47(b)). The RGB colors are displayed along the x, y, and z-axes with values between 0 (black) and 1 (full saturation). Black has the coordinates $(0,0,0)$ and white is represented by $(1,1,1)$. The CMY colors are shown at the corners of the cube opposite to those of the primary colors and at the endpoints of the face diagonals: $C = (0,1,1)$, $M = (1,0,1)$, and $Y = (1,1,0)$. In computing, the range 0–1 is usually divided by 256 increments

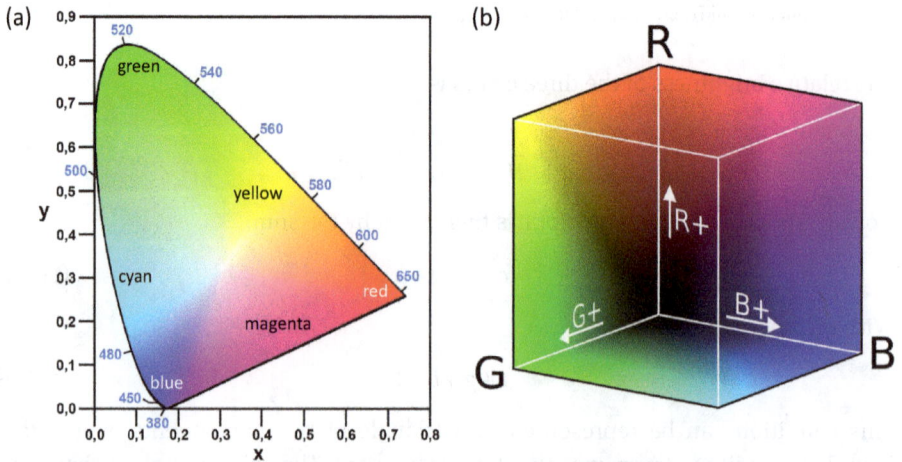

Fig. 11.47: RGB color triangle (a) and color cube (b). (reproduced from http://en.wikipedia.org/wiki/).

(0–255) for each of the R, G, and B primaries, making a total of 16 777 216 possible colot vomninszions. This is usually considered sufficient. The choice of 256 color codes is due to the eight-bit binary presentation, yielding $2^8 = 256$ values.

11.8.2 Additive and subtractive color mixture

Additive color mixtures are not that common in everyday life. Additive color mixing requires the projection of three-color sources onto a white screen such as in a cinema. In daily life, we are more often confronted with subtractive color mixtures. The color of any printing material, painting, wallpaper, etc. is due to subtractive color mixing. What's the difference? Usually we have a white source for lighting, either daylight or some kind of work lamp. When we look at a colored object in white light, the perceived color is that which is reflected from the object. For example, if we see a green object, the object is green because the colors red and blue have been absorbed by the object and only green is left over. We may also say that red and blue are subtracted from the white spectrum. An object appears black, when all colors are absorbed. An object is white, when none of the colors is absorbed, but all are reflected. Almost all colors that we observe in the environment are subtractive. There are few exceptions to colors produced by interference or dispersion, such as the colors of butterflies and the rainbow colors. Subtractive colors can also be seen on transparent materials such as colored window glasses when light shines through the window from the outside. Again, a green glass absorbs red and blue, while green light is transmitted. Both options for subtractive color perception by reflection and transmission are compared in Fig. 11.48.

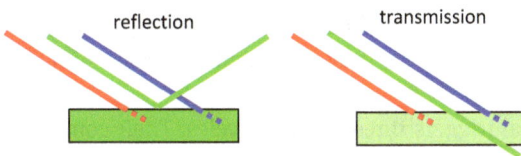

Fig. 11.48: Subtractive color mixing in reflection (a) and in transmission (b) mode.

Finally we should remember that colors represent sensations:
– which are neither carried along in the light;
– still arise in the eye;
– but only appear in the visual center of the brain.

The neuronal processing of light of different frequencies leads to color vision.

Colors of object that we observe are due to a subtractive color mixture. !

11.9 Summary

S11.1   For the eye, the image distance is fixed, and the focal length is variable.

S11.2   The iris controls the aperture of the eye and, thereby, the intensity.

S11.3   Seventy-five percent of the refractive power of the eye is provided by the refractive index and curvature of the cornea.

S11.4   The lens contributes another 15 dpt to the refractive power of the eye, which is in total 58 dpt.

S11.5   The lens has the ability to change the curvature for changing the accommodation.

S11.6   The age-related accommodation loss is due to an elastic stiffening of the lens.

S11.7   The optical resolution determined by the opening of the pupil matches well with the separation of rods and cones on the retina.

S11.8   The visus of the eye is about 1′, corresponding to a separation of images on the retina by 5 μm.

S11.9   Cataract is caused by increasing opacity of the eye and can be well treated by an artificial lens replacement.

S11.10 Glaucoma is due to an overpressure in the eye, which may lead to damage of the retina and optic nerve.

S11.11 The retina contains two visual systems: photopic supported by cones and scotopic supported by rods.

S11.12 Rods provide bright–dark sensitivity, and cones provide color sensitivity and brightness.

S11.13 The dynamical intensity range of the eye covers 10 orders of magnitude.

S11.14 The spectral bandwidth of the receptors ranges from 400 to 700 nm.

S11.15 In the foveal region, cones with green/yellow and red sensitivity dominate.

S11.16 The primary process of vision is the cis–trans conversion of the retinal by photoabsorption.

S11.17 Splitting of opsin and retinal triggers three reaction cycles. Their combined result is a hyperpolarization of the receptor potential and a recycling of the retinal.

S11.18 Cones are connected one by one to ganglion cells. In contrast, signals from many rods converge to one ganglion cell.

S11.19 Each cone is connected to two antagonistic bipolar cells, which determine the difference between brightness and darkness.

S11.20 Signals from bipolar cells to ganglion cells are modulated by horizontal and amacrine cells.

S11.21 Signals from the surround act inhibitive on receptor potentials in the center.

S11.22 Lateral connections are in most cases, inhibitive, and enhance the contrast.

S11.23 Only ganglion cells can fire APs.

S11.24 Any nerve fiber from a ganglion cell goes to the brain.

S11.25 Nerve fibers from nasal regions cross at the chiasm, nerve fibers from temporal regions do not cross.

S11.26 P and M cells in the retina project on P and M regions in the visual cortex.

S11.27 Two main color systems are known: three-color system RGB and four-color system RYGB.

S11.28 Colors can be mixed additively or subtractively.

Questions

Q11.1 What does the optical system of the eye consist of?
Q11.2 What is the reason for a gradual loss of accommodation capability?
Q11.3 How can intraocular pressure be determined?
Q11.4 How can myopia be corrected?
Q11.5 What is the difference between cataracts and glaucoma?
Q11.6 How many different kinds of cells are in the retina? Name them.
Q11.7 What is the function of Muller cells?
Q11.8 Why is the retinal epithelium layer located on the distal side of the retina?
Q11.9 What is the difference between fovea and macula?
Q11.10 How many visual systems are in the retina, and how are they supported?
Q11.11 What is the Kohlrausch kink, and what is it an indication for?
Q11.12 What is the dynamical width of the visual system from bright to dark?
Q11.13 What is the spectral band width of all receptors?
Q11.14 Where on the retina is visual acuity the highest?
Q11.15 Which protein is responsible for visual sensitivity?
Q11.16 Which primary process takes place when a photon is absorbed, and what is the time scale of this process?
Q11.17 Does light cause an AP in rods and cones?
Q11.18 Where is the AP generatated which goes to the visual cortex?
Q11.19 What is the difference between a CCD chip and the retina?
Q11.20 Which cells are activated in the dark state? Which ones in the bright ON state?
Q11.21 What is the main working principle of the receptive field?
Q11.22 How is contrast enhancement achieved?
Q11.23 Which are the light-sensing cells of the retina that mediate image formation?
Q11.24 What is the difference between P and M cells?
Q11.25 Why do some nerve fibers cross at the chiasm?
Q11.26 Which nerve fibeers cross, which do not?

Attained competence checker   + 0 −

	+	0	−
I know that the eye is a compounded optical system.			
I know which part of the eye has the highest refractive power.			
I know the origin of cataracts and of glaucoma, and I can distinguish between both conditions.			
I can determine the optical resolution of the eye.			
I can distinguish between optical resolution and acuity.			

I can name seven different cells in the retina.
I know which cells respond by graded potentials and which one trigger and AP.
I know which primary protein initiates vision.
I realize that visual perception is based on two systems: the photopic and the scotopic system.
I know what the optical bandwidth is, expressed in nanometers.
I know that the receptive field distinguishes between center and surround.
I realize that the visual system is a dual system represented by ON and OFF ganglion cells.
I know how contrast enhancement can be achieved.
I know that part of the optic nerve crosses at the chiasm.
I realize that different kinds of diseases of the eye can be distinguished by determining the left and right visual fields independently.
I realize that for the reproduction of colors two systems exist.

Suggestions for home experiments

HE 11.1 **Find your blind spot**

Watch the signs X and O on the screen or on a printout. Cover the right eye and fix the X with the left eye. If the screen is about three times the distance between the two letters, the O will vanish from the visual field. The triple distance is only a guideline. If the O can still be seen, then vary the distance to the screen until the O disappears. Conversely, for the blind spot of the right eye: cover the left eye, fix the O, and the X disappears.

O   **X**

HE 11.2 **Dark adaptation**

On a clear moonless night, watch the stars and notice the increasing number of stars you observe as you adapt more and more to darkness.

i Exercises

E11.1 **Image size on the retina**: At the distance of most distinct vision of 25 cm we see the objects with a magnification of 1. How big is the image actually on the retina? Calculate the image size for an object size of 1 cm. Is the image standing up or inverted?

E11.2 **Corrective glasses with two refractive powers**: You notice that you can only see objects clearly at a distance of between 0.75 and 2.5 m. You choose corrective glasses with two refractive powers. The upper part of the glasses allows you to see objects in the distance (object distance = infinite) sharply, and with the lower part you can see objects at a distance of 25 cm sharply. You may assume the glasses to be thin lenses. Calculate the refractive power: (a) for the upper part and (b) for the lower part.

E11.3   **Accommodation width:** In a myopic eye with a distance of 1 m from the far point, there is an age-related reduction in the range of accommodation to 2 dpt. What refractive index does a lens have to have so that objects can be seen clearly at a distance of 20 cm from the eye with maximum close-up accommodation? (The distance between the corrective lens and the eye is negligibly small.)

E11.4   **Complete hyperopia:** A shortsighted person needs for the far distance a correction lens with a refractive power of −2 dpt. The accommodation width is 0 dpt, i.e., complete hyperopia. For reading, the person uses reading glasses with a refractive power of +1 dpt. The distance between the corrective lens and the eye can be taken as negligibly small. The image distance is 25 mm. In what distance to the eye are objects, which the person can see clearly with the use of his reading glasses?

E11.5   **Chromatic abberation:** Chromatic aberration refers to the shift of focal lengths due to the dispersion of the lens; the refractive index depends on the wavelength. Assuming that the dispersion of the cornea and lens is similar to quartz glass, the difference of the refractive index Δn for blue and red light is about 0.015. The values are $n_{blue}(400\ nm) = 1.415$, $n_{red}(700\ nm) = 1.400$.
a.   Calculate the focal distance for both wavelengths and their difference.
b.   In the center of the fovea, there are no blue cones. For the green and red cones, use the wavelengths provided in Fig. 11.23. Reevaluate the difference of focal lengths. Discuss your result given the thickness of the retina is about 350 μm. Is chromatic aberration an important issue to be considered?

E11.6   **Airy disk of a microscope:** The objective lens of a microscope has a focal length of 3.3 mm and an aperture diameter of 2 mm. (a) Calculate in what distance the image occurs, if the object is located at 3.5 mm in front of the lens. (b) What diameter does the airy disk have at the calculated image distance and at a wavelength of 500 nm?

E11.7   **Lateral enhancement:** In a receptive field, the receptors record alternating potentials: 5, 10, 5, 10, etc. in arbitrary units. Now we invent horizontal cells that invert all lateral potentials by 100%. What is the contrast enhancement that can be achieved with such a system? Make a sketch of the receptor potentials before and after lateral inhibition. Use Fig. 11.40(a) as a guidance.

E11.8   **Optic nerve:** In pictures that show the eye with a horizontal optical axis, the optic nerve is often sketched as being bent upward and just as often as being bent downward. Which bend is right?

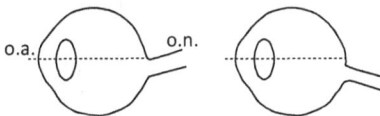

E11.9   **Hyperpolarization:** Describe why a receptor (rod or cone) can hyperpolarize and how this polarization is transmitted to the bipolar cells.

E11.10   **Visual field:** We discuss three different impairments of the optic nerve.
(a)   If the crossing of the nerve fibers is prevented in the optic chiasm, how does this affect the field of vision of the right and left eyes?
(b)   If the right optic nerve is blocked in front of the chiasm, what effect does this have on the field of vision of both eyes?
(c)   Can the field of vision distinguish a blockage of the right optic tract *after* the chiasm?

For all three cases, draw the visual field and the affected nerve tracts. Refer to Figs. 11.43 and 11.44 to answer these questions.

E11.11 **Color contrast:** The Mach bands shown in Fig. 11.40 are examples of lateral inhibition of rod receptors. Please test the lateral inhibition in cone cells by producing Mach bands in red, green, and blue. Can you observe colored Mach bands, and if yes, which color shows the strongest lateral inhibition. Use your graphics program and recall that in the RGB code red is (255, 0, 0), green (0, 255, 0), blue (0, 0, 255), and the separation of the bands is 25 units.

References

[1] Hoffman DD. Visual intelligence. New York: How to create what we see. New York: W.W. Norten&Company; 1982.

[2] Yong ED. Inside the eye: Nature's most exquisite creation. National Geographic; February 2016.

[3] Marr D. Vision. A computational investigation into the human representation and processing of visual information. New York: Freeman WH and Company; 1982.

[4] Wang H, Lin S, Liu X, Kang SB. Separating reflections in human iris images for illumination estimation. Tenth IEEE international conference on computer vision. 2005; 2: 1691–1698.

[5] McDougal DH, Gamlin PD. Autonomic control of the eye. In: Terjung R, ed. Comprehensive physiology. 2015; 5: 439–473.

[6] Tipler PA, Mosca G. Physics for scientists and engineers. 6th edition. New York, London: W. H. Freeman and Co.; 2006; 2007.

[7] Klein MV, Furtak TE. Optics. 2nd edition. New York, London, Sydney, Toronto: Wiley&Sons; 1986.

[8] Augusteyn RC. On the growth and internal structure of the human lens. Exp Eye Res. 2010; 90: 643–654.

[9] Uhlhorn SR, Borja D, Manns F, Parel JM. Refractive index measurement of the isolated crystalline lens using optical coherence tomography. Vision Res. 2008; 48: 2732–2738.

[10] Eysel U. In: Pape H-C, Kurtz A, Silbernagel S, eds. Physiologie. 7th edition. Stuttgart, New York: Thieme Publishe; 2014.

[11] Glasser A. Restoration of accommodation: Surgical options for correction of presbyopia. Clin Exp Optom. 2008; 91: 279–295.

[12] Michael R, Bron AJ. The aging lens and cataract: A model of normal and pathological aging. Phil Trans R Soc B. 2011; 366: 1278–1292.

[13] Navarro R. The optical design of the human eye: A critical review. J Optom. 2009; 2: 3–18.

[14] Kingsley CN, Brubaker WD, Markovic S, Diehl A, Brindley AJ, Oschkinat H, Martin RW. Preferential and specific binding of human alpha b-crystallin to a cataract-related variant of gamma s-crystallin. Structure. 2013; 3: 2221–2227.

[15] Cataract Surgery in 6 minutes. Narrated by Dr.Sibley, Florida Eye Center, https://www.you tube.com/watch?v=rUCoQzui704

[16] Richter GM, Coleman AL. Minimally invasive glaucoma surgery: Current status and future prospects. Clin Ophthalmol. 2016; 10: 189–206.

[17] Bringmann A, Grosche A, Pannicke T, Reichenbach A. GABA and glutamate uptake and metabolism in retinal glial (Müller) cells. Front Endocrinol. 2013; 4: 48.

[18] Franze K, Grosche J, Skatchkov SN, Schinkinger S, Foja C, Schild D, Uckermann O, Travis K, Reichenbach A, Guck J. Müller cells are living optical fibers in the vertebrate retina. Proc Natl Acad Sci. 2007; 104: 8287–8292.

[19]   Kolb H. Facts and figures concerning the human retina. 2005. In: Kolb H, Fernandez E, Nelson R, eds. Webvision: The organization of the retina and visual system [Internet]. Salt Lake City: University of Utah Health Sciences Center; 1995.

[20]   Dacey DM, Petersen MR. Dendritic field size and morphology morphology of midget and parasol ganglion cells of the human retina. Proc Natl Acad Sci. 1992; 89: 9666–9670.

[21]   Bowmaker JK, Dartnall HJ. Visual pigments of rods and cones in a human retina. J Physiol. 1980; 298: 501–511.

[22]   Cohen MA, Dennett DC, Kanwisher N. What is the bandwidth of perceptual experience? Trends Cogn Sci. 2016; 20: 324–335.

[23]   Boettner EA, Wolter JR. Transmission of the ocular media. Invest Ophthalmol. 1962; 1: 776–783.

[24]   Kandel ER, Schwartz JH, Jessell TM, Siegelbaum SA, Hudspeth AJ. Principles of neural science. 5th edition. New York, Chicago, San Francisco: McGraw Hill Medical; 2013.

[25]   Sakmar TP. Structure of rhodopsin and the superfamily of seven-helical receptors: The same and not the same. Curr Opin Cell Biol. 2002; 14: 189–195.

[26]   Palczewski K, Kumasaka T, Hori T, Behnke CA, Motoshima H, Fox BA, Le Trong I, Teller DC, Okada T, Stenkamp RE, Yamamoto M, Miyano M. Crystal structure of rhodopsin: A G protein-coupled receptor. Science. 2000; 289: 739–745.

[27]   Wang Q, Schoenlein RW, Peteanu LA, Mathies RA, Shank CV. Vibrationally coherent photochemistry in the femtosecond primary event of vision. Science. 1994; 266: 422–424.

[28]   Chabre M, Bigay J, Bruckert F, Bornancin F, Deterre P, Pfister C, Vuong TM. Visual signal transduction: The cycle of transducin shuttling between rhodopsin and cGMP phosphodiesterase. Cold Spring Harb Symp Quant Biol. 1988; 53(Pt 1): 313–324.

[29]   Hecht S, Schlaer S, Pirenne HP. Energy, quanta and vision. J Gen Physiol. 1942; 25: 819–840.

[30]   Rieke F, Baylor DA. Single-photon detection by rod cells of the retina. Rev Modern Phys. 1998; 70: 1027–1036.

[31]   Manasseh G, de Balthasar C, Sanguinetti B, Pomarico E, Gisin N, de Peralta RG, Gonzalez Andino SL. Retinal and post-retinal contributions to the quantum efficiency of the human eye revealed by electrical neuroimaging. Front Psychol. 2013; 845: 1–13.

[32]   Wikler KC, Rakic P. Distribution of photoreceptor subtypes in the retina of diurnal and nocturnal primates. J Neurosci. 1990; 10: 3390–3401.

[33]   Sagdullaev BT, Ichinose T, Eggers ED, Lukasiewicz PD. Visual signal processing in the inner retina. In: Tombran-Tink J, Colin J, eds. Visual transduction and non-visual light perception. Berlin, Heidelberg: Barnstable Springer Verlag; 2009, 287–304.

[34]   Kaneda M. Signal processing in the mammalian retina. J Nippon Med Sch. 2013; 80: 16–24.

[35]   van Wyk M, Wässle H, Taylor WR. Receptive field properties of ON- and OFF-ganglion cells in the mouse retina. Vis Neurosci. 2009; 26: 297–308.

[36]   Takeshita D, Smeds L, Ala-Laurila P. Processing of single-photon responses in the mammalian On and Off retinal pathways at the sensitivity limit of vision. Philos Trans R Soc Lond B Biol Sci. 2017; 372: 20160073.

[37]   Costea CF, Turliuc S, Buzdugă C, Cucu AI, Dumitrescu GF, Sava A, Turliuc MD. The history of optic chiasm from antiquity to the twentieth century. Childs Nerv Syst. 2017; 33: 1889–1898.

[38]   Hubel DH, Wiesel, TN. Brain mechanisms of vision. Sci Am. 1979; 241: 150–163.

[39]   Kwak Y, Curtis CE. Unveiling the abstract format of mnemonic representations. Neuron. 2022; 110: 1–7.

[40]   Kirschfeld K. How we perceive our own retina. Proc Biol Sci. 2017; 284: 20171904.

[41]   Jeon HS, Choi HY. Binocular function test. In: Lee JS, ed. Primary eye examination. Singapore: Springer; 2019.

Further reading

Kandel ER, Schwartz JH, Jessell TM, Siegelbaum SA, Hudspeth AJ. Principles of neural science. 5th edition. New York, Chicago, San Francisco: McGraw Hill; 2013.

Purves D, Augustine GJ, Fitzpatrick D, Katz LC, LaMantia AS, McNamara JO, Williams SM, eds. Neuroscience. 2nd edition. Sunderland (MA): Sinauer Associates; 2001. Online textbook can be accessed but not browsed. www.ncbi.nlm.nih.gov/books/NBK11059/

Hubel D. Eye, brain, and vision. Online textbook: https://epdf.pub/eye-brain-and-vision.html

Schwartz SH. Visual perception: A clinical orientation. 4th edition. New York, Chicago, San Francisco: McGraw-Hill Prof Med/Tech; 2009.

Kolb H, Nelson R, Fernandez E, Jones B, eds. Webvision. The organization of the retina and visual system. Free Webbook at http://webvision.med.utah.edu/

Pape H-C, Kurtz A, Silbernagel S, eds. Physiologie. 7th edition. Stuttgart, New York: Thieme Verlag; 2014.

Duane's ophthalmology: www.oculist.net/downaton502/prof/ebook/duanes/index.html

12 Sound, sound perception, and balance

Physical properties of sound and sound perception

Middle-ear amplification factor	625
Audible threshold pressure amplitude	20 µPa
Audible threshold intensity	10^{-12} W/m^2
Audible frequency range	20–20 000 Hz
Length of the cholea	35 mm
Number of inner hair cells	4000
Number of outer hair cells	12 000
Frequency discrimination	0.2%
Interaural time difference discrimination	10 ms
Angular accuracy of sound localization	2°
Expiration pressure for glottis opening	400–600 Pa
Sound velocity in air	330 m/s
Sound velocity in water	1500 m/s
Wavelength of sound in air at 1 kHz	0.33 m
Wavelength of sound in water at 1 kHz	1.5 m

12.1 Introduction

Hearing is the ability of the ear to detect sound waves, convert pressure waves of sound into receptor potentials (RPs), and process the signals in the brain's auditory cortex. This entire process is known as auditory pathway or auditory perception. It is one of the most important senses of humans and other Vertebrates next to visual perception and one of the most intriguing from a neuroscience point of view. Similar to smelling or seeing, hearing is a remote sense, i.e., the source of information is at a certain distance to the body and requires a medium to be detected by specialized receptors. In sound, a compressional pressure wave travels through the air as the transmitting medium between source and ear. The ear as sound wave detector consists of three main parts (Fig. 12.1): (1) the external ear with the pinna (auricle), an about 25 mm long ear canal (auditory meatus), and at the end, the eardrum (tympanum or tympanic membrane); (2) the middle ear with the ossicles hammer (malleus), anvil (incus), and stapes (stirrup); (3) the inner ear with a bony spiral (cochlea) that contains the actual sound-sensitive nerve receptors. These three parts of the ear have different tasks: the outer ear acts as a sound resonator. The middle ear takes care of impedance matching of sound waves between air and body. The inner ear holds receptors sensitive to pressure amplitudes and sound frequencies. These wave properties are then converted into electric signals and transmitted via the auditory nerve to the auditory cortex in the brain for analysis, storage, and comprehension. In addition, the pair of ears is also a sensor for sound location and movement.

https://doi.org/10.1515/9783110756951-012

Audibility implies coding frequencies and amplitudes of sound waves into electrical signals that the brain "understands." Nature does not use Faraday's laws of induction. A simple microphone would suffice to convert pressure waves into a corresponding electrical signal output if it did. Instead, nature works with receptors and receptor/action potentials. Therefore, another signal transduction has to be invented that converts minute mechanical motion into electrical signals that nerve fibers can guide to the brain.

The organ of equilibrium, also called the vestibular apparatus, is attached to the inner ear and forms one unit with the cochlea. It consists of three semicircular canals for sensing rotational acceleration and two sensors for horizontal and vertical acceleration. The cochlea is a sensor for sound waves from the outside, whereas the vestibular apparatus senses the body's movement without external contact. Nevertheless, the sensory device is similarly constructed in both cases: a fluid flow that tilts tiny hair cells that changes a receptor potential (RP).

Before discussing sound and balance perception, we will first derive some important relations of sound propagation in different media that are needed here and in Chapter 1/Vol. 2 on ultrasound imaging.

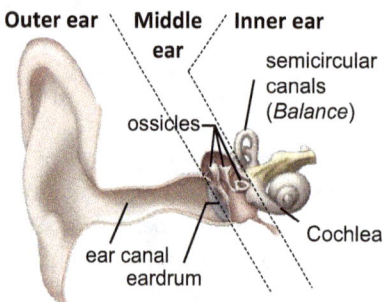

Fig. 12.1: Overview on the different parts of the ear (adapted from https://www.ncbi.nlm.nih.gov/pub medhealth/).

12.2 Sound waves

12.2.1 Acoustic impedance

The main task of the ear is to make sound waves audible. Sound waves are propagating pressure waves, which displace molecules from their "equilibrium" position in gases, liquids, or solids. We focus here on longitudinal compression waves since transverse waves have no restoring force in gases or liquids and therefore cannot propagate. For longitudinal waves, the propagation direction and the displacement amplitude are parallel. For the following, we assume that the propagation direction, the displacement amplitude, and the velocity all point along an arbitrary x-direction.

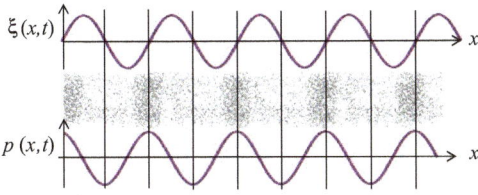

Fig. 12.2: Displacement and pressure of a propagating longitudinal sound wave. Both waves are phase shifted by 90°.

The displacement of molecules as a function of position x and time t can be described by a harmonic wave (Fig. 12.2):

$$\xi(x,t) = \xi_0 \sin(kx - \omega t), \tag{12.1}$$

where $k = 2\pi/\lambda$ is the wave number and ω is the angular frequency of the sound wave propagating along the x-direction. The harmonic sound wave is a solution of the one-dimensional wave equation (see the mathbox for a derivation):

$$\frac{B}{\rho}\frac{\partial^2 \xi}{\partial x^2} = \frac{\partial^2 \xi}{\partial t^2}. \tag{12.2}$$

Here B is the *compression modulus* of gases or liquids and ρ is the respective density. The *phase velocity* of sound waves is

$$v_{sound} = \frac{\omega}{k} = \sqrt{\frac{B}{\rho}}. \tag{12.3}$$

For ideal gases, the compression modulus follows from the ideal gas equation and is $B = p$ for isothermic processes and $B = \gamma p$ for adiabatic processes. Here γ is the ratio of the specific heat at constant pressure C_p to the specific heat at constant volume C_V, and p is the ideal gas pressure:

$$\gamma = \frac{C_p}{C_V}. \tag{12.4}$$

For sound waves propagating in the air, the expression for adiabatic processes has to be used since for any frequency above 20 Hz, thermal equilibrium by heat exchange can never be reached. Thus the sound velocity in liquids and gases is

$$v_{sound}^{liquid} = \sqrt{\frac{B}{\rho}}; \quad v_{sound}^{gas} = \sqrt{\frac{\gamma p}{\rho}}. \tag{12.5}$$

In air at normal conditions of pressure and humidity, the *speed of sound* is about 330–340 m/s; in water it is 1500 m/s. Note that the velocities in air and water are different by a factor of 5, which leads to an acoustic mismatch, as we will see later. The construction of the ear has to deal with this big velocity difference!

Next, we determine the *particle velocity* in the sound wave from the first derivative of the displacement wave:

$$u_x = \frac{\partial \xi(x,t)}{\partial t} = -\omega \xi_0 \cos(kx - \omega t). \tag{12.6}$$

where u_x is sometimes called *amplitude velocity*. The particle or amplitude velocity should not be mixed with the phase or group velocity derived next.

The propagating pressure wave follows from Hooke's law via (see eq. (3.3) and Exercise E12.1):

$$p_x = -B \frac{\partial \xi(x,t)}{\partial x} = -B\xi_0 k \cos(kx - \omega t). \tag{12.7}$$

The respective displacement and pressure amplitudes are therefore

$$u_0 = -\omega \xi_0; \ p_0 = -B\xi_0 k. \tag{12.8}$$

The SI units are $[u] = $ m/s and $[p] = $ Pa. Both waves u_x and p_x are out of phase with the displacement wave ξ_x by 90° as indicated in Fig. 12.2.

To derive the time average power $\langle P \rangle$ of the wave and its intensity, we need to consider the time average of the product of force F_x and particle velocity u_x:

$$\langle P \rangle = \langle F_x u_x \rangle. \tag{12.9}$$

Dividing the power by the area A through which the sound wave propagates, gives the time-averaged intensity:

$$\langle I \rangle = \left\langle \frac{P}{A} \right\rangle = \left\langle \frac{F_x}{A} u_x \right\rangle = \langle p_x u_x \rangle = B\omega k \xi_0^2 \langle \cos^2(kx - \omega t) \rangle. \tag{12.10}$$

The last term integrated over one period yields a factor ½, such that the time-averaged sound intensity is

$$\langle I \rangle = \frac{1}{2} B\omega k \xi_0^2. \tag{12.11}$$

The unit of intensity is $[I] = $ W/m². The time average intensity can be rephrased by various expressions using the previously derived definitions:

$$\langle I \rangle = \frac{1}{2} Z u_0^2 = \frac{1}{2} \frac{p_0^2}{Z}. \tag{12.12}$$

where

$$Z = \rho v \tag{12.13}$$

is the *acoustic impedance* of the sound wave. The unit is $[Z] = $ kg/m² s.

The acoustic impedance is a materials property like the refractive index of transparent media. The acoustic impedance is essential for evaluating sound propagation, transmission, and reflection at interfaces between different media. Typical values for Z of some relevant materials in the present context are listed in Tab. 12.1.

> Acoustic impedance is the product of mass density and sound velocity. !

Mathbox:
Heuristic derivation of the wave equation for longitudinal compression waves

$$\xi_1 = \xi(x) \quad \xi_2 = \xi(x + \Delta x)$$

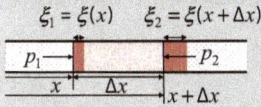

Consider a pipe filled with gas or liquid. A sound wave travels through the pipe and exerts a pressure p_1 at position x and a pressure p_2 at $x + \Delta x$. The pressures are different $p_1 \neq p_2$ and therefore the pressure difference Δp causes local compression (expansion) of the medium with amplitude $\xi(x, t)$ and acceleration of the gas particles inside. Thus the pressure difference is $\Delta p = F/A = m\ddot{\xi}/A = \rho V \ddot{\xi}/A = \rho \Delta x \ddot{\xi}$, where $\ddot{\xi}$ is the particle acceleration. As the cross-sectional area drops out of the relative volume change, we can set: $\Delta V/V = d\xi/dx$. Then using Hook's law: $\Delta p = B(dV/V) = B(\partial \xi/\partial x) = \rho \Delta x \ddot{\xi}$, we find $\frac{B}{\rho}\frac{(\partial \xi/\partial x)}{\Delta x} = \ddot{\xi}$, or $\frac{B}{\rho}\frac{\partial^2 \xi}{\partial x^2} = \frac{\partial^2 \xi}{\partial t^2}$. The last equation is the general wave equation $v^2 \frac{\partial^2 \xi}{\partial x^2} = \frac{\partial^2 \xi}{\partial t^2}$, with the propagation velocity $v^2 = \frac{B}{\rho}$.

Tab. 12.1: Sound velocity, density, and the acoustic impedance of different materials of the body.

	Sound velocity v (m/s)	Density (kg/m³)	Acoustic impedance Z (kg/m² s)
Air	330	1.3	430
Helium	1007	0.18	181
Water	1500	998	1.5×10^6
Fatty tissue	1470	970	1.38×10^6
Muscles	1570	1040	1.7×10^6
Bones	3600	1700	6×10^6

12.2.2 Crossing borders

Like optics, when sound waves cross boundaries between media characterized by different impedances, part of the intensity is transmitted and part is reflected as sketched in Fig. 12.3.

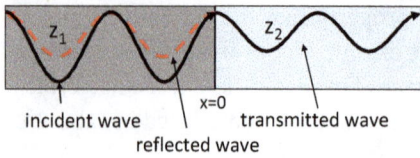

Fig. 12.3: Transmission and reflection of a sound wave at an interface between two materials characterized by different impedances Z_1 and Z_2 with $Z_1 > Z_2$.

For simplicity, we consider the normal incidence of sound waves toward a boundary. Energy conservation requires that the intensities of the incident (i), transmitted (t), and reflected (r) waves are related as

$$I_i = I_t + I_r, \tag{12.14}$$

$$Z_1 u_i^2 = Z_2 u_t^2 + Z_1 u_r^2.$$

Thus the waves in mediums 1 and 2 can be expressed as

$$\xi_1 = \xi_i \sin(k_1 x - \omega t) + \xi_r \sin(k_1 x + \omega t). \tag{12.15}$$

$$\xi_2 = \xi_t \sin(k_2 x - \omega t).$$

At the boundary $x = 0$, continuity of the slope of the waves is required. This entails the first derivative of the sound wave before and after the boundary to be equal. As a result, we find an additional condition for the particle velocities:

$$u_i + u_r = u_t. \tag{12.16}$$

Both eqs. (12.14) and (12.16) can be combined to yield for the particle velocities:

$$u_t = u_i \frac{2Z_1}{Z_1 + Z_2}; \quad u_r = u_i \frac{Z_1 - Z_2}{Z_1 + Z_2}. \tag{12.17}$$

Inserting these expressions in the equations for the intensities, we obtain

$$I_r = \frac{1}{2} Z_1 u_r^2 = I_i \left(\frac{Z_1 - Z_2}{Z_1 + Z_2} \right)^2. \tag{12.18}$$

$$I_t = \frac{1}{2} Z_2 u_t^2 = 4 I_i \frac{Z_1 Z_2}{(Z_1 + Z_2)^2}.$$

The reflected and transmitted sound intensities normalized by the incident intensity are then

$$R = \frac{I_r}{I_i} = \left(\frac{Z_1 - Z_2}{Z_1 + Z_2} \right)^2. \tag{12.19}$$

$$T = \frac{I_t}{I_i} = 4\frac{Z_1 Z_2}{(Z_1 + Z_2)^2}.$$

The intensity ratio R is known as *reflectivity* and T is called *transmissivity*. R and T fulfill the important relation confirming energy conservation:

$$R + T = 1. \tag{12.20}$$

In optics, these equations are known as Fresnel[1] equations for normal incidence if the refractive index n replaces the impedance Z.

In summary, the transmitted wave shows a different amplitude than the incident wave, a different wavelength, a different speed, and eventually a different direction of propagation. However, and most importantly, the frequencies of the incoming and transmitted waves are the same:

$$f_1 = \frac{v_1}{\lambda_1} = \frac{v_2}{\lambda_2} = f_2. \tag{12.21}$$

Now we discuss three limiting cases of the impedance change at the boundary, which we need later to understand the sound propagation in the ear:

a. For $Z_1 = Z_2$, $R = 0$, and $T = 1$, i.e., if the acoustic impedances are identical, although the materials may feature different densities and sound velocities, but the product ϱv is constant, there will be no reflection of sound waves.
b. For $Z_2 \rightarrow 0$ or $Z_2/Z_1 \rightarrow 0$, we find $R = 1$ and $T = 0$, i.e., the wave is completely reflected.
c. The same holds if $Z_1 \rightarrow 0$ and $Z_1/Z_2 \rightarrow 0$. Also here, we find $R = 1$ and $T = 0$, i.e., the wave again is completely reflected.

In case that reflection occurred at an interface to an acoustically less dense media characterized by $Z_1 > Z_2$, the amplitudes of the incident and reflected waves are in phase. Vice versa, for $Z_1 < Z_2$, the amplitudes of the incident and reflected waves are out-of-phase by π or 180°.

As an example, we consider the boundary between air ($Z_{air} = 430$ kg/m^2 s) and water ($Z_{water} = 1.5 \times 10^6$ kg/m^2 s). With these impedances, the reflectivity at the interface is $R = 0.9989\%$ or 99.89%. Therefore the transmitted intensity is $T = 0.001$ or 0.1%. Thus between air and water and vice versa, there is a severe *impedance mismatch*. Sound can propagate easily in the air or in water. But it cannot cross borders. Between air and water, there is an acoustic wall.

Sound detection requires acoustic waves to enter the body from the air. Assuming that the acoustic receptor is in a tissue with a water-like impedance, the transmitted wave has an intensity of 0.1% of the incident intensity. With such low

---

1 Augustin Jean Fresnel (1788–1827), French physicist and mathematician.

transmittance, we would be deaf. All three parts of the ear are designed to increase the transmission by a factor of approximately 625, so transmission is actually 62% instead of 0.1%. The mechanism leading to this amazing amplification is discussed in Section 12.5. The sonography presented in Chapter 1/Vol. 2 faces the same problem of coupling sound waves into the body for imaging. But the technical solution for ultrasound devices is different from the one that nature has chosen in the ear.

> **!** When sound waves cross borders, everything changes except the frequency.

12.2.3 Sound intensity and loudness

The physical definition of *sound intensity*, also known as *acoustic intensity*, can be expressed in several equivalent forms

$$I_{\text{sound}} = \frac{1}{2}\rho v_{\text{sound}} u_0^2 = \frac{1}{2}Z u_0^2 = \frac{1}{2}\frac{p_0^2}{Z}. \tag{12.22}$$

For simplicity, we have omitted here the bracket for the time average. The symbols are explained in Sections 12.2.1 and 12.2.2. The usual intensity range of auditory sound is

$$10^{-10}\ W/m^2 < I_{\text{sound}} < 10^{-2}\ W/m^2.$$

The audible threshold is even lower at 10^{-12} W/m², and the pain threshold is about 1 W/m². There are 12 orders of magnitude between threshold and pain!

Using the expression for the sound intensity and the acoustic impedance, we can calculate the pressure amplitude p_0 and the particle displacement ξ_0 for different intensities:

$$p_0 = \sqrt{2 Z_{\text{air}} I_{\text{sound}}}. \tag{12.23}$$

$$\xi_0 = \frac{p_0}{Z_{\text{air}} \cdot \omega}. \tag{12.24}$$

The pressure amplitude depends only on the intensity and the impedance but not on the frequency. In contrast, the displacement is, in addition, inversely proportional to the sound frequency. Typical values are given in Tab. 12.2. According to these values, the pressure for audible threshold in the frequency range of highest auditory sensitivity from 1 to 3 kHz is

$$2 \times 10^{-7}\ \text{Pa} = 2 \times 10^{-9}\ \text{hPa} = 2 \times 10^{-9}\ \text{mbar}.$$

This low-pressure amplitude is comparable to pressures experienced in an ultrahigh vacuum environment. The respective particle displacement of only 0.01 nm corresponds to a diameter less than the size of an atom. The actual displacement amplitude at the basilar membrane (see Section 12.5) is about 1 nm at the threshold pressure level. This is an incredible sensitivity that nature has achieved. In fact, the ear is the most sensitive receptor of the body.

Tab. 12.2: Typical values of pressure amplitudes and particle displacements from pain to audible threshold. Calculated values are for pressures in air and a frequency of 440 Hz.

I (W/m^2)	p_0 (hPa) in air	p_0 (kPa) in water	ξ_0 in air	ξ_0 in water
1	30 Pa	1700 Pa	160 μm	2.5 μm
10^{-2}	3 Pa	170 Pa	16 μm	250 nm
10^{-4}	300 mPa	17 Pa	1.6 μm	25 nm
10^{-10}	300 μPa	17 mPa	1.6 nm	0.025 nm
10^{-12}	30 μPa	1.7 mPa	0.16 nm	0.0025 nm

Sound intensity is a linear function of the squared pressure amplitude (eq. (12.22)). However, the sound reception is not a linear function of the sound intensity. Instead, the auditory perception depends logarithmically on the intensity, which is known as the *Weber–Fechner law*. Weber[2] and Fechner[3] noticed that sound intensity changes are not recognized as absolute increments of intensity ΔI, but rather as a relative increment $\Delta I/I$.

For example, we can easily distinguish between one sound source and two, say one or two violinists. In front of an ensemble of 10 players, it is difficult to distinguish the sound intensity from another player. However, we would notice the difference if another five violinists were added to the ones already there. So the relative change is the same in both cases. The audible sensitivity increment ΔS can thus be expressed as being proportional to the relative intensity increment $\Delta I/I$:

$$\Delta S = k\frac{\Delta I}{I} \tag{12.25}$$

where k is an appropriate constant. Taking the integral, we obtain

$$S = k\ln\left(\frac{I}{I_0}\right). \tag{12.26}$$

The ratio I/I_0 is called intensity level because it is a normalized intensity to a fixed reference value I_0, where $I_0 = 10^{-12}$ W/m^2 is the audible threshold value for the sensitivity. In the last equation, the integration constant is set to $-k\ln(I_0)$. The audible sensation level B is now defined as the logarithm to the base of 10 of the sound intensity level:

---

2 Ernst Heinrich Weber (1795–1878), German physiologist.
3 Gustav Theodor Fechner (1801–1887), German physiologist.

$$B = \log \frac{I}{10^{-12}}. \tag{12.27}$$

The sensation level B is expressed in Bel.[4] Although B is a pure number without units, Bel is treated as a unit. When the logarithm is multiplied by 10, the *sensation level* is measured in decibel or dB, which is one-tenth of a Bel:

$$dB = 10 \log \frac{I}{10^{-12}}. \tag{12.28}$$

> **!** The audible sensation level is a logarithmic function of the intensity level.

The threshold intensity has 0 dB, and the threshold of pain corresponds to an intensity level of 120 dB. Sound intensity and sound level are compared in Fig. 12.4.

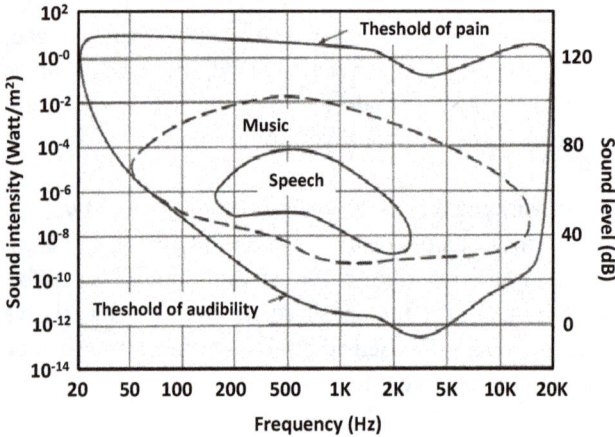

Fig. 12.4: Sound intensity–frequency map. The contours outline the threshold audibility and the threshold of pain. Typical intensities and frequencies for music and speech perception are encircled (adapted from [1]).

The sensitivity of the ear depends on the frequency. For the same intensity but different frequencies, the *loudness* of a tone judged by a person varies. The peak sensitivity is between 1000 and 3000 Hz and drops off for lower or higher frequencies. Young adults have an audible frequency range from 20 to 20 000 Hz. With age, the audible frequency range at the upper end drops to about 8000–10 000 Hz.

---

4 Alexander Graham Bell (1847–1922), British-American audiologist.

For quantifying the personal perception of sound intensity, a scale for loudness level is conveniently introduced by the definition:

$$L = 10 \log \frac{I}{10^{-12}}.$$ (12.29)

The unit of the loudness level is the phone. At 1000 Hz, the scale of the sound intensity level (dB scale) and the scale for perceived loudness level (phone scale) agree. At lower and higher frequencies, these two scales deviate.

Figure 12.5 is an isophone chart or equal loudness chart. It displays the sound pressure level in decibel, which is required for higher and lower frequencies to be perceived as the same loudness at the frequency of 1000 Hz. Since the sensitivity of the ear drops at lower and higher frequencies, this is compensated for by a higher pressure level. Only between 2 and 5 kHz in the region of the highest sensitivity of the ear, the pressure level can be reduced for the same loudness perception. Figure 12.5 also shows the disutility limit at which normal cognitive performance ceases. The other limits, pain threshold and hearing threshold, are self-explanatory.

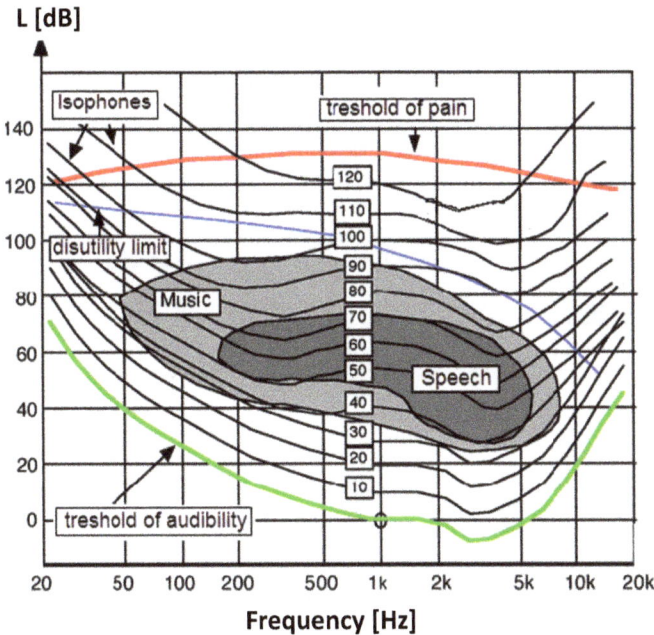

Fig. 12.5: Isophone chart comparing loudness (scale to the left) with sound intensity (dB values in square boxes). At 1 kHz both scales coincide (adapted from [1]).

12.3 Outer and middle ear

Despite the enormous acoustic impedance mismatch between air and body, we can nevertheless sense sound waves. How is this possible? More precisely: how can a sound wave with large displacement amplitude but small pressure amplitude in the air be converted into a sound wave with small displacement amplitude but high-pressure amplitude in fluids? Compare corresponding values in Tab. 12.2. To get an answer, we have to scrutinize the construction of the ear.

12.3.1 Outer ear

The outer ear has a length of about 25–30 mm and can be modeled as an *acoustic pipe* with one end closed by the eardrum and the other end open (Fig. 12.6). Resonance condition occurs for multiples of $\lambda/4$-waves with frequencies:

$$f_n = (2n+1)\frac{v_{\text{sound}}}{4L}; n = 1, 2, \ldots \tag{12.30}$$

With $v_{\text{sound}} = 330$ m/s we find the first resonance at 3300 Hz, the next at 9900 Hz, etc. In fact, the peak sensitivity of the ear is around 3000 Hz. What happens at other frequencies? The outer ear is not a perfect resonator, but one with a poor quality factor that allows all other frequencies to pass through. However, the sensitivity of the ear is highest between around 1000 and 8000 Hz. Sensitivity is also determined by other parts of the ear, which we will discuss later.

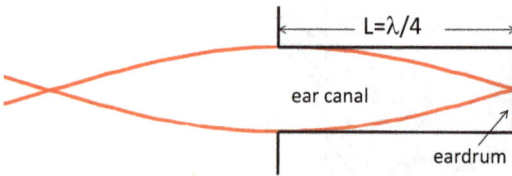

Fig. 12.6: Model of the outer ear canal in terms of an acoustic pipe with one end open. Indicated in red is the displacement wave $\xi(x)$, which is 90° out-of-phase with the pressure wave.

The middle ear is separated from the outer ear by the eardrum. The ambient pressure is the same on both sides of the eardrum because the middle ear is connected to the environment via the eustachian tube with the upper part of the pharynx. The middle ear contains the ossicles, malleus, anvil, and stirrup. Together they have the task of transmitting the sound pressure from the eardrum to the oval window of the cochlea (Fig. 12.7). The impedance mismatch between the outer and inner ear is overcome by the principles of hydraulics and levers, as we shall see next.

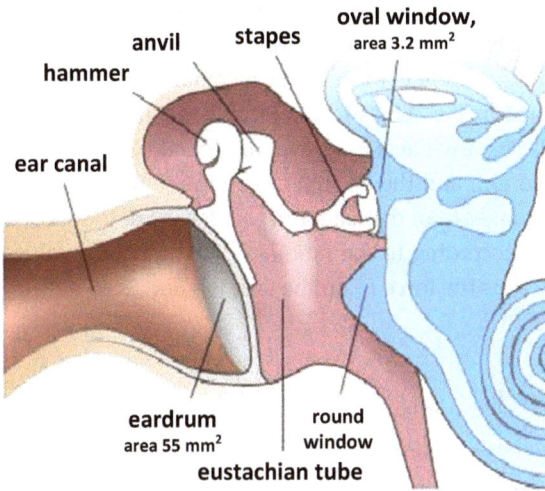

Fig. 12.7: Middle ear connecting the eardrum of the external ear with the oval window of the inner ear via the auditory ossicles hammer, anvil, and stapes.

We first consider the hydraulic action. The *hydraulic principle* implies that two cylinders of different diameters that are connected by an incompressible fluid in equilibrium experience the same pressure:

$$p_1 = \frac{F_1}{A_1} = \frac{F_2}{A_2} = p_2.$$ (12.31)

Here $F_{1,2}$ are the respective forces and $A_{1,2}$ are the respective base areas of the cylinders. Using the hydraulic principle, the forces are amplified when a small force is pressed on a small area, whereby a large force is obtained on a larger area. The hydraulic principle implies that the force per area remains constant when the area is increased. If the incompressible fluid is replaced with a rigid rod of constant diameter, the force remains constant, but the pressure can be increased. To illustrate this, let us assume that the hammer, anvil, and stapes are combined into a single bone that connects the eardrum to the oval window, as outlined in Fig. 12.8(a). The eardrum transmits the sound pressure on the connecting bone by a force F_1, which subsequently acts on the oval window. For a constant force, we have:

$$F_1 = p_{\text{drum}}A_{\text{drum}} = p_{\text{oval}}A_{\text{oval}}.$$ (12.32)

From this, we conclude that with constant force, the pressure is increased if A_{oval} is smaller than A_{drum}. The pressure amplification is proportional to the ratio of the areas:

$$p_{\text{oval}} = \frac{A_{\text{drum}}}{A_{\text{oval}}}p_{\text{drum}}.$$ (12.33)

Using typical numbers for the areas of the eardrum (55 mm^2) and oval window (3 mm^2), we gain a pressure enhancement by a factor of 18.

Next, we consider the lever in the middle ear formed by a combined action of all three ossicles. The lever is sketched in Fig. 12.8(b). The force F_1 acts from the ear drum via the hammer to the anvil. The anvil acts as a lever of second class (see Section 2.3.3). This means that the load arm and the lift arm are on the same side of the pivot point; and the lift arm (l_F) is longer than the load arm (l_L). Levers of the second class have a mechanical advantage according to the ratio l_F/l_L. Using typical values for the arms, the ratio is about 1.4. Thus the force F_2 on the oval window is by a factor of 1.4 or roughly 40% bigger than F_1.

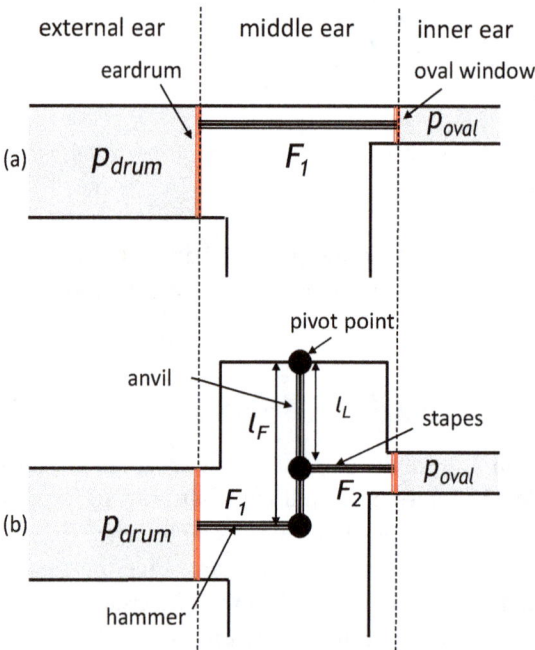

Fig. 12.8: (a) Mechanical analog of the hydraulic action of the middle ear; (b) schematics of the lever action, combining hammer, anvil, and stapes.

Combining the hydraulic principle and the lever action, we obtain a pressure amplification of 1.4 × 18 = 25. Since the transmitted sound intensity is proportional to the square of the pressure amplitude: $I_t \sim p_0^2$, the actual amplification is 625, corresponding to an intensity increase by 28 dB. Therefore, with the middle ear's action, 62.5% of the sound intensity is transmitted instead of 0.1% without impedance matching. This is an amazing amplification factor that our middle ear can achieve through mechanical principles alone!

The middle ear fulfills two more tasks aside from the sound pressure amplification. The transmitted frequency band in the audible region is very broad, stretching from 20 to about 20 000 Hz. The resonance frequency of the middle ear including the eardrum and oval window is at about 1400 Hz. In order to transmit a broad frequency band, the resonance curve should be broad and flat, which is indeed the case, similar to the low quality factor of the external ear (see Infobox I on quality factors). The other task is protection against too high intensities. This is achieved by an acoustic reflex triggering two sets of muscles, one tightening the eardrum, another one pulling the stapes away from the oval window to reduce the pressure transmittance to the fluid in the scala vestibuli. The latency is only 10 ms, much faster than the latency of the iris contraction, which is about 100 ms.

> The middle ear converts sound waves in air with large displacement amplitude but small-pressure amplitude into sound waves in fluids with small displacement amplitude but high-pressure amplitude.

Infobox I: Quality factor

We consider a harmonic oscillator of any kind, such as a classic pendulum that oscillates with its natural or "eigen"-frequency $\omega_0 = 2\pi f_0 = 2\pi/T$, where T is the oscillation period. From one turning point to the next, we measure the amplitude of the oscillation, for instance, as a function of the deflection angle $\alpha(t) = \omega_0 t$, described by $A(\alpha, t) = A_0 \sin(\omega_0 t)$. The oscillation amplitude A_0 of a real oscillator will decay exponentially over time because of damping, in contrast to an ideal one. Mathematically the damping can be taken into account by the expression: $A(\alpha, t) = A_0 e^{-\beta t} \sin(\omega_0 t)$, where β is the damping constant. Let us consider an oscillating system that is forced to oscillate with a frequency ω_e different from the natural frequency ω_0, where the subscript "e" stands for external. As an example of a powered oscillator, consider an old-fashioned pendulum clock. The damping of this system is overcome by feeding in energy from the outside and the oscillation amplitude remains constant over time. How does this system behave when we vary the forced frequency ω_e? For small frequencies, $\omega_e \ll \omega_0$, the system will just follow the forced frequency ω_e and the amplitude is comparable to the free oscillator. However, as the driving frequency approaches the natural frequency ω_0, the amplitude increases dramatically and is only limited by the damping constant β of the oscillating system. The plot in panel (a) shows the oscillation amplitude versus frequency for different damping constants $\beta_1 > \beta_2 > \beta_3$. The maximum of the oscillation amplitude is always at $\omega_e = \omega_0$. The width of this curve $\Delta\omega$ depends on the damping constant. The larger β, the larger is $\Delta\omega$. The ratio $Q = \omega_0/\Delta\omega$ is the quality factor of the oscillating system. When tuning into the frequency of a radio station, the amplitude of the resonance circuit should have a narrow frequency distribution and a high quality factor so as not to hear the transmission of another station at the same time. For the ear, however, the requirements are different. Both the outer ear and the middle ear should transmit a wide range of frequencies with as little loss as possible. Therefore the damping should be large, but not too large. The plot in panel (b) shows the sensitivity of the outer ear as a function of frequency. The sensitivity is highest between 2000 and 5000 Hz, but there is nowhere a sharp maximum, as it should be.

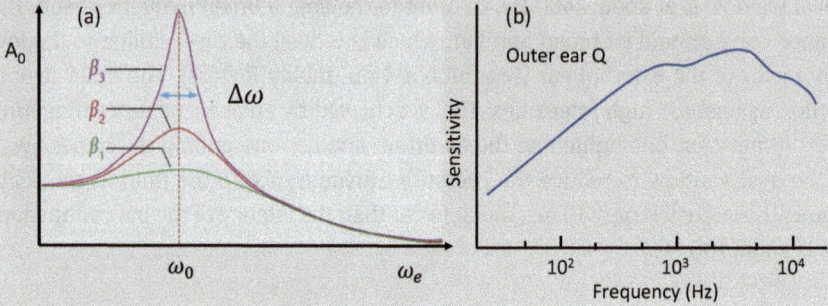

Panel (a), adapted from [2], shows the amplitude of a forced oscillator as a function of the driving frequency ω_0 and for different damping parameters beta. $\Delta\omega$ is the width of the resonance curve, which determines the quality factor of the resonator; panel (b), adapted from [3], shows the quality factor of the outer ear as a function of frequency.

12.4 Sound detection in the inner ear

Having solved the problem of impedance matching, we now turn to detection of sound in the inner ear. Sound detection should capture at least two properties of the sound wave: intensity (loudness) and frequency (tone). In fact, the ears are also sensitive to the location of noise and source movement, as we will see later. The inner ear's challenging task is to provide a range of sensitivity over ten orders of magnitude and to encode a frequency range of 20 000 different tones. Distinguishing sound frequencies can be viewed as the ability to recognize the "colors of sound" [4]. We have three color-sensitive cones in the retina to perceive different colors. How many receptors do we need to have in the cochlea to distinguish between different frequencies? And how can relatively slow nerve conduction pick up fast acoustic signals? In the following, we try to answer these questions, and we will notice that the inner ear is the most remarkable organ.

12.4.1 Structure of the cochlea

The answer to the posed questions lies in the complexity of the inner ear (*cochlea*) displayed in Fig. 12.9 [5]. The cochlea is a 35 mm long conical-shaped tube curled up in a snail-like fashion. When untwined, we recognize three parallel tubes: two outer tubes (*scala vestibuli* and *scala tympani*) and an inner cochlear duct (*scala media*) (see also Fig. 12.12). The inner duct separates the outer tubes with the basilar membrane at the bottom and the Reissner's membrane[5] at the top. The outer tubes contain a sodium (Na^+)-enriched perilymph liquid; the inner tube is filled with a potassium ion-rich (K^+) endolymph liquid. The vestibular tube is closed at the base by the *oval window* and the tympanic tube is closed by the round window. The fluids in the two outer tubes are in contact at the apex of the cochlea at a point called *helicotrema*. However, there is no contact between endolymph and perilymph because the basilar and the Reissner membrane keep the scala media tight. The relevance of these fluids will become clear later.

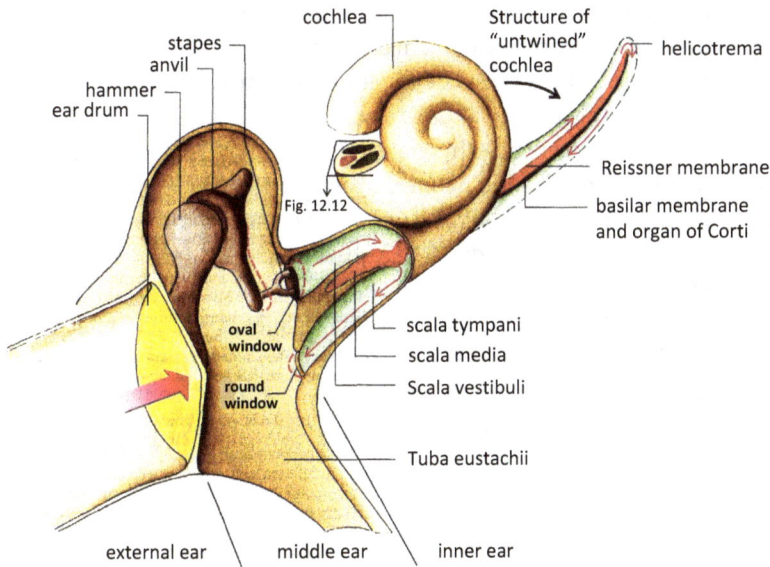

Fig. 12.9: Different parts of the ear in a cut-away view (adapted from [1] with permission).

The pressure wave of the sound, mediated by the stapes in the middle ear, forces the oval window to vibrate. The oval window, in turn, acts like a drum's membrane and creates a traveling wave in the scala vestibule, which moves back and forth in the rhythm of the stapes pounding. It is essential to recognize that the traveling wave in the scala vestibule involves a macroscopic fluid flow between the oval and round windows. This is in contrast to sound waves that entail just a local particle movement around an equilibrium position. The round window at the base of the cochlea serves as a pressure relief for the traveling and incompressible fluid wave. As the oval window moves inward, the round window moves outward and vice versa, as indicated by the red-dashed lines in Fig. 12.9. The fluid wave does not propagate through the cochlea to the apex. Instead, it stops at points along the

---

5 Ernst Reissner (1824–1878), Baltic physiologist.

basilar membrane specific to the wave's frequency (Fig. 12.10). At the stopping point, the basilar membrane bends like the buckling of a carpet when a runner suddenly stops. The curvature of the membrane is crucial for the detection of sound waves, as we will notice further on. For now, let us keep that the cochlea transforms pressure waves in fluid flow.

In Fig. 12.10, the cochlea is schematically shown as a thin uncoiled tube with a centerline, representing the 35 mm long basilar membrane. Sound arrives at the oval window via the vibration of the stapes. Low-frequency traveling waves propagate toward the apex and become strongly absorbed close to the apex by the membrane's elastic properties. High-frequency waves travel in the liquid only a short distance and become absorbed close to the base of the cochlea. Intermediate frequencies are detected somewhere in the middle between base and apex. Complex sound waves composed of different frequencies are absorbed at several positions along the basilar membrane.

The cochlea is a position-sensitive frequency detector similar to a harp when used as a resonator for different sound frequencies rather than as a musical instrument. High frequencies resonate with short strings at the knee, and low frequencies resonate with long strings at the shoulder. The spatial frequency analysis along the basilar membrane is referred to as *tonotopy*. The tonotopic frequency map is not linear from apex to basis but logarithmic. A tone with a frequency of 1000 is recorded at about the middle of the cochlea, 100 Hz close to the apex, and 10 000 Hz close to the base. The tonotopic map of the basilar membrane is a fundamental feature of auditory coding. It is preserved up to the auditory cortex in the brain [6].

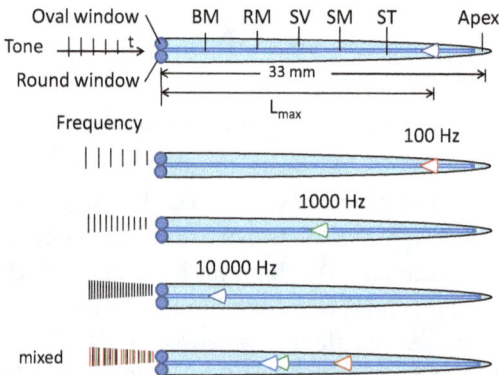

Fig. 12.10: Tonotopic absorption of sound waves in the basilar membrane (BM). The top panel designates schematically different parts of the cochlea, including oval window, round window, and apex (helicotrema). RM, Reissner's membrane; SV, scala vestibuli; SM, scala media; ST, scala tympani. L_{max} is the distance from the oval window to the position where the maximum absorption of the sound wave occurs. The frequency of different tones is indicated schematically by vertical bars on an arbitrary timescale. Sound consists of a mixture of tones, which are absorbed at different positions along the basilar membrane (adapted from [6]).

12.4.2 Elastic properties of the cochlea

The tonotopic mapping of sound waves is only possible if there is a gradient of the elastic properties along the basilar membrane. Indeed the elastic modulus of the basilar membrane continuously decreases from base (stiff) to apex (floppy) [7]. At the same time, the membrane becomes wider over the same distance from ~0.12 to ~0.5 mm (Fig. 12.11) in close analogy to the already quoted harp. The combined result of stiffness gradient and change of width yields a factor of about 100 in the variation of elastic modulus from base to apex.

Mathematically, the basilar membrane can be modeled as a damped harmonic oscillator driven by the pressure difference Δp that originates from the fluid flow in the scala vestibule and tympani, while the membrane's local elasticity constant acts as a restoring force. The equation of motion is then [2]:

$$m\ddot{h} + \beta\dot{h} + kh = -\Delta pA. \tag{12.34}$$

Here $h = h(x,t)$ is the displacement of the membrane about the equilibrium position, β is the damping constant, k is the restoring force constant, m is the mass, and A is the membrane's cross-sectional area. The derivatives of $h(x,t)$ with respect to the time are indicated by dots. The damping constant and the restoring force constant depend on the location x along the basilar membrane (basis: $x = 0$; apex: $x = 1$):

$$\beta(x) = \beta_0 e^{-\alpha x}, \tag{12.35}$$

$$k(x) = k_0 - x(k_1 - k_0) \tag{12.36}$$

This damped harmonic oscillator model has, however, an essential problem. The quality factor depends on the damping constant. High damping is necessary for fast response to changing input but results in a low quality factor. Low quality factor entails the loss of high-frequency selectivity. On the other hand, low damping implies high quality factor and high frequency-selectivity but insufficient response time. Obviously, this model is not appropriate to describe the complexity of the cochlea's oscillatory properties. The sharpness of the oscillator's frequency mapping hints at a nonlinear behavior close to resonance [8]. In fact, the cochlea has recently been modeled by a set of oscillators, and each one is kept at an instability point by a self-tuning mechanism with positive feedback. The concept of self-tuned critical oscillators combined with the traveling wave model can describe the main characteristics of hearing, particularly the sharp frequency-selectivity, the extreme sensitivity, and the large dynamic range [9]. This insight has now to be fed into a wider concept that accounts for the additional cellular amplification by the *outer hair cells* (OHCs) in the cochlea.

A recent study investigated the role of cochlear curvature [11]. This study shows that the increasing curvature redistributes wave energy density toward the outer

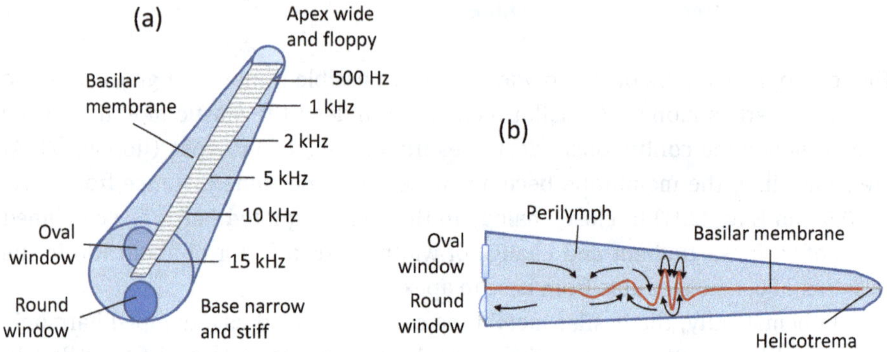

(a)

Apex wide
and floppy

Basilar
membrane

500 Hz
1 kHz

(b)

2 kHz

5 kHz

10 kHz

Oval
window

15 kHz

Round
window

Base narrow
and stiff

Perilymph

Oval
window

Basilar membrane

Round
window

Helicotrema

Fig. 12.11: (a) Mechanical model of the basilar membrane featuring a gradient of width and stiffness from the base to the apex. The basilar membrane is stiff and narrow at its base but floppy and wide at the apex. This causes the membrane to resonate at the base to high frequencies and at the apex to low frequencies. Therefore a tonotopic map of frequency location occurs along the basilar membrane. (b) Traveling waves in the fluid cause fluid flow and displacement of the basilar membrane that resonantly vibrate and damp at specific locations for different frequencies. The displacement amplitude of the basilar membrane is strongly exaggerated. Adapted from [10].

wall of the cochlea and affects the shape of the waves that propagate on the membrane, particularly in the region where low-frequency sound waves are processed. In this way, the shape also contributes to the amplification of the sound, corresponding to the phenomenon of the "whispering gallery" observed in St. Paul's Cathedral in London or by listening to an empty spiral-shaped shell.

The analysis of sound perception's physical and physiological aspects goes back to Helmholtz[6] in the nineteenth century, who described the cochlea as a sound receiver made of resonating strings. The analysis was continued by Békésy,[7] who discovered the tonotopic sound mapping in the cochlea via traveling waves. His traveling wave theory was later supplemented by taking into account the cellular amplification of hair cells in the cochlea.

12.4.3 Organ of Corti

Now we focus our attention on the fine structure of the inner duct, a cross-sectional view of the cochlea is shown in Fig. 12.12. Here we notice the basilar membrane and a tectorial membrane enveloping fine hairs (*stereocilia*) connected to hair cells. The

---

6 Hermann von Helmholtz (1821–1894), German physicist and physiologist.
7 Georg von Békésy (1899–1972), Hungarian physiologist and Nobel Prize laureate in physiology 1961.

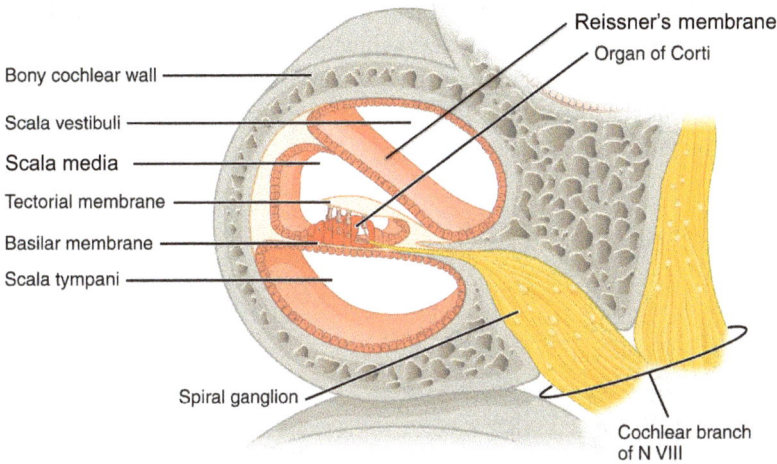

Fig. 12.12: Cross section through the cochlea showing the three main tubes: scala vestibule, scala tympani, and scala media. Inner and outer hair cells are arranged between the basilar membrane and the tectorial membrane (reproduced from OpenStax Anatomy and Physiology, © Creative Commons).

combination of hair cells and tectorial membrane is called the organ of Corti.[8] Wherever the traveling wave is resonantly absorbed along the cochlea, the basilar membrane bends, which inflects hair cells by moving against the tectorial membrane.

Figure 12.13 shows an enlarged cross-sectional cut through the cochlea, highlighting in more detail the organ of Corti [12]. In this cross-sectional view, we recognize three *OHCs* and *one inner hair cell* (IHC). OHC and IHC form together four rows on top of the basilar membrane from base to apex (Fig. 12.14). All hair cells are attached to a bundle of tiny hairs, so-called *stereocilia*. The outer hair cells and their stereocilia touch the tectorial membrane, while the IHCs sit in a pocket without touching the membrane. The tectorial membrane is a jelly-like substance that controls passively fluid flow in the scala media.

As the traveling fluid wave is absorbed in an area specific to its frequency, the basilar membrane becomes bent at the hinge point (Fig. 12.13(b)). The bending causes an inflection of the outer hair bundles against the tectorial membrane that depolarizes the outer hair cells. The depolarization activates motor proteins that contract the hair cells. This action is similar to the sarcomere contraction in muscle cells (see Section 2.4.2). The contraction amplifies the basilar membrane motion, acting as *positive feedback* on the original stimulus. Doing so enables fluid flow in the scala media passing by the inner hair cell bundle, bending them back and forth with the sound wave frequency. The positive feedback of the outer hair cells opposes the viscous

---

8 Alfonso Giacomo Gaspare Corti (1822–1876), Italian anatomist.

damping of the stereocilia vibration by the cochlear fluid motion. The total amplification factor gained by this mechanism has been estimated to exceed 4000 [8]! The ciliary vibration stimulates an RP, which, in turn, is perceived within the auditory cortex as the sound of a particular frequency. The details of the audible signal's mechano-electric transduction (MET) are discussed further.

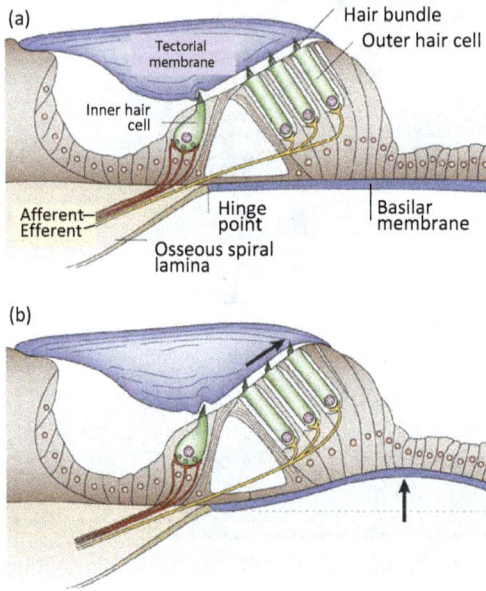

Fig. 12.13: (a) Cross section through the organ of Corti with basilar membrane and tectorial membrane eclipsing inner and outer sets of hair cells for sensing sound waves. (b) The arrow indicates the bending of the basilar membrane leading to the transverse motion of stereociliary bundles against the tectorial membrane (reproduced from Ref. [12] with permission of Nature Publishing Group).

12.4.4 Inner and outer hair cells

The frequency-dependent sensitivity of the inner ear is due to some ~16 000 hair cells lined up in four rows along the basilar membrane: one inner row of IHCs and three outer rows of OHCs. The inner ~4000 hair cells sense the sound intensity and frequency *tonotopically*, i.e., each hair cell is responsible for the detection of one narrow frequency band, such that frequencies from 16 to 20 000 Hz are logarithmically mapped onto these ~4000 inner hair cells. The outer ~3 × 4000 hair cells serve as amplifiers of the basilar membrane motion during resonance absorption. The highly ordered organization of both types of hair cells is shown impressively in the scanning electron micrograph of Fig. 12.14. Each V-shaped bundle of about 100 stereocilia is attached to one outer hair cell, and also each straight bundle of stereocilia is attached to one of the inner hair cells. The stereocilia can be visualized as elastically stiff cylinders tapered at the bottom, allowing them to bend in response to the external pressure wave. The length of the cylinders ranges from 4 μm for high frequencies at the basilar base up to 7 μm for low frequencies at the apex. The diameter is about 0.4 μm, and the tapered base has a diameter of about 0.1 μm. When the stereociliary cylinders move, they all move in the same direction. A rich structure of fiber connections ensures rigidity during movement. The upright position of the stereocilia and their arrangement in bundles is very delicate. Excessive sound intensity can irreversibly break them and disrupt proper alignment, leading to severe hearing loss [13].

The hair cells are supported by pillar cells on both sides and by the basilar membrane at the bottom. At the top, the stereocilia are bathed in the endolymph fluid that fills the scala media. Compared to other receptor cells, the hair cells are rather short. They have no dendrites and no axons. As typical receptor cells, they respond by graded potential variations.

So far, we have seen that the perception of sound in the inner ear requires a mechanical wave in the perilymph fluid that bends the basilar membrane. In regions where the traveling wave is resonantly absorbed, the outer hair cells are activated. Their active oscillation amplifies the wave amplitude and stimulates the inner hair cells. Now we come to the next important point: the inner hair cells respond by releasing a RP that activates an action potential in the spiral ganglion cells on their way to the cochlear nucleus. This last part of the signal transduction will be discussed next.

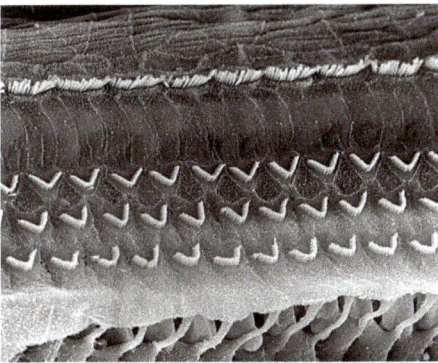

Fig. 12.14: Scanning electron micrograph showing the highly ordered organization of hair cells in the organ of Corti. A single row of inner hair cells is seen at the top, and three rows of outer hair cells are seen at the bottom. The scale bar corresponds to 15 μm (SEM image by Marc Lenoir, from the EDU website « Journey into the World of Hearing » www.cochlea.eu, with permission of R. Pujol et al., NeurOreille, Montpellier).

12.4.5 Mechanoelectric transduction

The cross section of IHCs and OHCs is schematically shown in Fig. 12.13. Their working principle is quite different from those of the other receptor cells. The difference is visualized in Fig. 12.15. As detailed in Chapter 5, cells are embedded in Na^+-rich extracellular fluid separated from the cytoplasm by a cell membrane. Two types of channels, one selective for K^+ ions and the other for Na^+ ions, connect inside and outside and control the cell's potential. During rest, the K^+ channels are open while the Na^+ channels are closed. As K^+ ions diffuse out, the resting potential drops to -70 mV. Upon stimulation, depolarization is achieved by opening the Na^+ ion channels. Repolarization occurs through the activation of ATP pumps. So much for the normal cells.

Hair cells have a K^+-rich cytoplasm like normal cells. However, the cell membrane contains only one type of channel, a mechanically gated K^+–channel. The hair cells are surrounded by an endolymph fluid, which is also K^+-rich and provides a potential of $+80$ mV with respect to the perilymph fluid. During rest, about 10% of the K^+ channels remain open, yielding a floating potential of about -40 mV. When stimulated, the gate opens, K^+ ions flow in, driven by the potential difference and cause a depolarization of the hair cell to about -20 mV. The K^+ ions needed for this process are recycled by a K^+/Cl^- – transport system on the side of the hair cell membrane that replaces the usual ATP-pump in other cells.

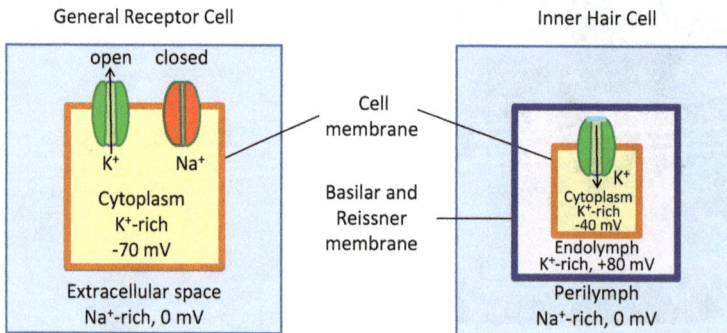

Fig. 12.15: Comparison between general cells and inner hair cells. In general, cells are embedded in extracellular fluid at zero potential. Extracellular fluid and cytoplasm are connected by two ion channels that control the cell's potential. Hair cells are embedded in a positively charged endolymph, and the cell potential is controlled by only one mechanically gated potassium channel. For further details, see text.

Within a single bundle, the stereocilia are graded in length [5, 6]. Tip-link fibers are attached at the top of the stereocilia. As the basilar membrane moves, the stereocilia flex back and forth, toward and away from their larger neighbor. Tilting it toward its taller neighbor bends the stereocilia at their tapered base and creates tension at the tip junctions that open the K^+ channels (Fig. 12.16). K^+ ions flow from the endolymph into the cell body, following the potential gradient, and depolarize the RP from −40 to −20 mV. When the stereocilia tilt to the opposite side, all K^+ channels close and the RP hyperpolarizes to −60 mV. Depolarization/hyperpolarization is proportional to the basilar membrane's bending amplitude and the stereocilia deflection angle. The maximum amplitude from hyperpolarization to depolarization is about 25–30 mV.

All stereocilia are interconnected by fibers not only at the top but also along the cilia body. This fiber network lends stiffness to the "floppy" cilia and ensures a motion in unison that produces just one particular potential change proportional to the deflection. Therefore, the mechanical control of the cell's potential is referred to as mechanoelectric transduction *MET* [14, 15].

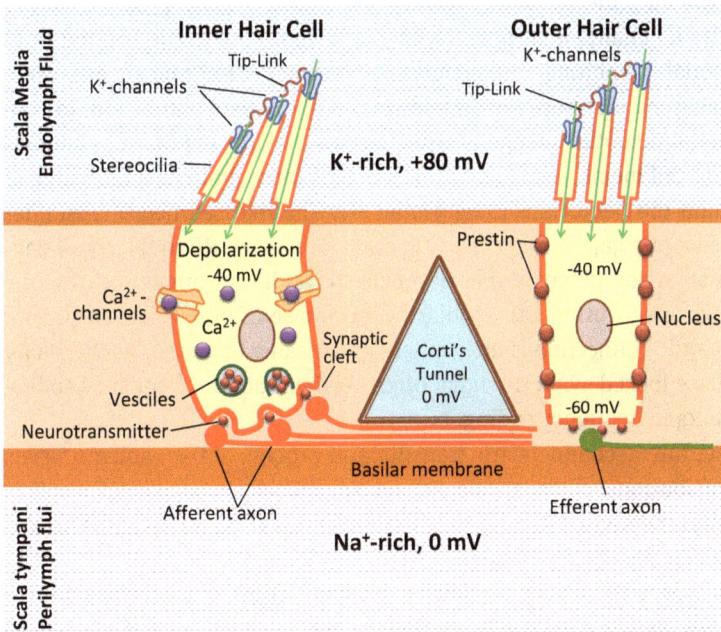

Fig. 12.16: Inner hair cell with stereocilia contains K^+ channels, which open on bending toward the taller neighbor, causing a depolarization of the hair cell. Simultaneously, Ca^{2+} channels open, stimulating the discharge of neurotransmitters. The outer hair cell acts similarly. In addition, depolarization contracts the cell body, whereas hyperpolarization stretches the cell, indicated by dashed lines and elongated prestin proteins.

The IHC and OHC receptors act similarly. However, opening K^+ channels in OHC does not just cause depolarization. It also creates a longitudinal vibration of the cell body, referred to as electromotility (EM) or OHC motility [16]. EM is caused by the protein prestin, which is integrated into the cell's membrane and reacts sensitively to the internal cell potential [17]. In the depolarized state, the prestin molecules in the cell membrane are contracted, in the hyperpolarized state, they are stretched. At a certain frequency, the entire cell body vibrates lengthways (Fig. 12.16). The oscillation modulates the fluid flow in the media scala, which bypasses the tectorial membrane. Then the tectorial membrane strengthens or inhibits the bending of the steriocilia attached to the IHC. The EM is comparable to piezoelectric transducers used for sound generation and ultrasound imaging. It has been estimated that the EM effect increases the bending of the basilar amplitude by a factor of 4000 [8], indicative of the inner ear's very high sensitivity.

> **!** Electromotility of the outer hair cells is responsible for an amplification of the basilar amplitude by a factor of 4000.

The IHCs detect the amplified basilar bending amplitude. The tip-links at the top open and close the K^+ channels in phase with the basilar movement and the fluid flow past the tectorial membrane. The membrane potential reacts quickly and oscillates with the same frequency between depolarization and hyperpolarization. In contrast to the OHCs, the potential change in IHCs triggers the opening of Ca^{2+} channels further down in the cell membrane (Fig. 12.16). The Ca^{2+} ions stimulate a discharge of vesicles that contain the neurotransmitter glutamate. The released neurotransmitter diffuses into the synaptic gap. In response, the axon directs the RP to the spiral ganglion, which fires an action potential that travels to the auditory cortex. Note that the Ca^{2+} channels only open during the depolarization phase of an oscillation period. Therefore, the synaptic connection is intermittent at the same frequency as the ciliary vibration. This phase-locked synaptic signal offers further auditory frequency coding in addition to tonotopic frequency coding.

To summarize, the mechanoelectric transduction process in the cholea consists of three main contributions:

1. First, the basilar membrane is set into motion by a traveling wave that is braked tonotopically between base and apex.
2. In the region where the traveling wave is stopped, OHCs are stimulated resonantly and their active vibration amplifies the basilar bending amplitude and tectorial fluid flow.
3. The amplified signal elicits a graded membrane potential in the IHCs that stimulates synaptic conductance.

Table 12.3 gives an overview of the properties of IHC and OHC. The mechanoelectric transduction is the auditory working principle in all mammals, never mind how small they are. Even in birds, the basic principle is identical. However, the avian middle ear appears to be somewhat simpler, featuring only one ossicle between tympanic membrane and oval window.

Tab. 12.3: Comparison of the physical and biophysical properties of inner and outer hair cells.

	IHC	OHC
Main task	Sound reception and signal transmission to the auditory cortex	Sound amplifier
Hair cells	One raw	Three raws
Hair cell organization	Straight	V-shaped
Innervation	Afferent	Efferent
Number of nerve fibers per cell	5–10	1
Prestin density	Zero	High
Motility	Passive motility	Active electromotility
Contact with tectorial membrane	None	Yes
Potential change	Graded	Graded
Response to potential change	Ca^{2+} release	Resonant oscillation

12.4.6 Frequency coding

There are about 30 000 afferent nerve fibers; most of these connect to the IHCs, 5–10 fibers to each inner hair cell, shown schematically in Fig. 12.17. Only a few efferent nerve fibers come from the superior olivary complex and connect to the OHCs one after the other. Their function is currently not well understood. However, it is known that the OHC has predominantly a modulating inhibitory influence on the sensitivity of the IHC. In fact, we not only perceive sound from the outside but we can also generate sound with our vocal cords. It is therefore sensible and necessary that there must be a feedback system between sound generation and sound perception, which most likely has its origin in the efferent connection of the OHC.

On the way up from the RP through the auditory nerve to the auditory cortex in the brain, the sound frequency is coded by two different methods: tonotopic localization and phase-locking.

As already described, the tonotopic localization of the sound frequency is achieved by connecting some 30 000 nerve fibers to 4000 inner hair cells. The one-by-one correspondence between frequency and location is preserved and mapped onto the auditory cortex (Fig. 12.18). The brain analyzes action potentials arriving in the auditory cortex versus localization and derives from this a characteristic frequency. One may consider such a process as a Fourier analysis of space.

For frequencies up to approximately 5 kHz, in addition to the tonotopic mapping, the sound frequency is also coded by phase-locking [6, 18]. Phase-locking can be viewed as a Fourier analysis of frequencies. By deflecting the stereocilia alone, the RP has the same frequency as the sound wave (Fig. 12.13). The postsynaptic action potential is phase-locked to the same sound wave frequency: with each deflection in the rhythm of the sound, an action potential is fired. However, intensity and frequency are encoded not only by the action potentials of a single fiber but also by

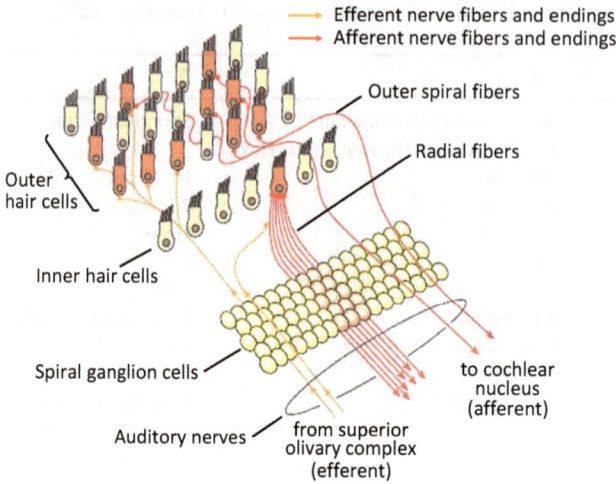

Fig. 12.17: Connection of afferent and efferent nerve fibers to inner and outer hair cells, respectively. Note that each inner hair cell is connected to many afferent nerve fibers (about 10) required for decoding frequency and intensity (adapted from http://www.open.edu/openlearn).

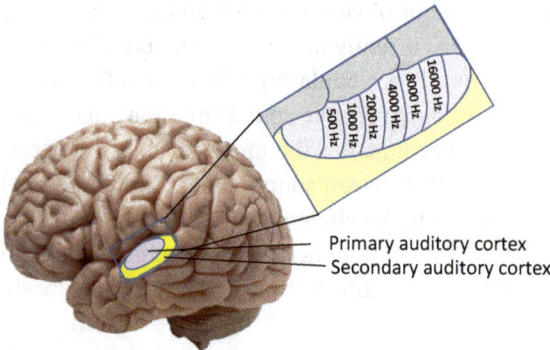

Fig. 12.18: Tonotopic mapping of sound frequencies in the cortex (adapted from http://www.open. edu/openlearn, © creative commons).

several afferent nerve fibers attached to the same hair cell, as shown in Figs. 12.17 and 12.19. While the first nerve fiber fires action potentials in each cycle in phase with the sound frequency, the second fiber fires – also phase-locked – but only in every fourth cycle, the third in every eighth cycle, etc. With increasing amplitude, the second fiber fires more often and possibly other neighboring fibers follow. The frequency is encoded by the number of action potentials per second carried by the first fiber. The number of fibers stimulated in addition to the first one encodes the intensity. From the RP to the action potential, we again observe the typical analog–digital conversion that is at work for all the senses in the body.

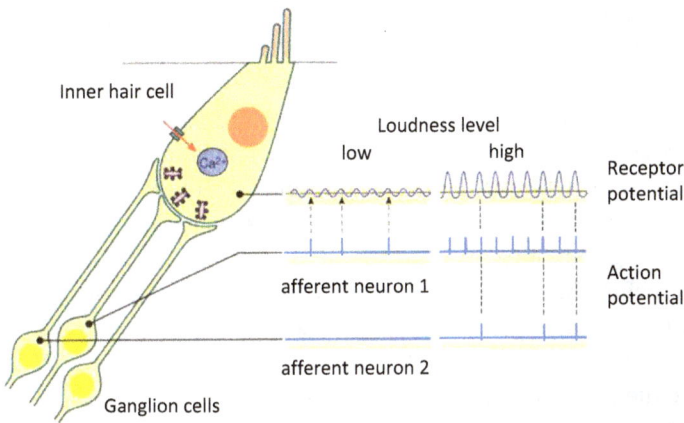

Fig. 12.19: Receptor potential and action potential of inner hair cells. The action potential is phase-locked to the frequency of the receptor potential. With increasing loudness, more fibers carry action potentials, all phase-locked to the frequency of the receptor potential (adapted from [19] with permission).

Phase-locking works up to a frequency of about 5 kHz, i.e., 0.2 ms between action potentials. Considering that action potentials require a refractory period of about 1 ms, phase-locking should stop beyond 1 kHz. However, for phase-locking it is not necessary that an action potential is fired on any period. As long as the period is in phase with the vibrational frequency, the brain can encode a sequence of action potentials belonging to that same frequency. The importance of phase-locking at high frequencies is still a matter of debate and subject of current research. An overview on the current state of debate is given in [18].

The inner ear has two methods to encode acoustic frequencies: tonotopic and phase-locking. Phase-locking works up to 5 kHz, higher frequencies are recorded tonotopically.

12.4.7 Selectivity, sensitivity, latency, and adaptation

As we have learned, the frequency selectivity of the inner ear is exceptionally high. Each hair cell reacts to a certain frequency band but has a characteristic frequency f_c at which the sensitivity is highest. The frequency selectivity, i.e., the just noticeable frequency difference Δf_c is approximately 0.2%. This difference is equivalent to the difference between the pure intonation and the equal-temperament scale. For example, it corresponds to the barely noticeable difference between C major and

D minor. In comparison, two adjacent keys on the piano are 8% apart. How this high selectivity of the inner ear is achieved is still the subject of current research. From a mechanical point of view, a large mismatch exists between the heavy basilar membrane with a high damping constant and the closely packed, light, and rapidly "dancing" stereocilia on top. Purely mechanical models may not be able to adequately describe such a system. However, the concept of tonotopically arranged oscillators that are self-tuned close to the critical frequency of the vibrating OHCs yields promising insight [8].

The depolarization of the hair cell is proportional to the deflection of the stereocilia in the direction toward the taller neighbor, whereas in the opposite direction, the cell becomes hyperpolarized, as indicated in Fig. 12.20. The change of the RP is extremely sensitive to the deflection direction and its amplitude. There is no threshold potential for complete depolarization. Instead, the depolarization is gradual and proportional to the deflection angle, which again is proportional to the wave amplitude or loudness. The pressure threshold amplitude for hearing is about 200 μPa, corresponding to a deflection of the stereocilia by about 0.1 nm, causing an RP of 100 μV. The threshold deflection sensitivity is, therefore, on the scale of atomic diameters and comparable to Brownian motion. The potential variation from hyperpolarization to depolarization has a sigmoidal shape with an amplitude of about 25–30 mV. At a deflection of 50 nm, about 80% saturation is already reached. At saturation, the deflection angle never exceeds 1° [6].

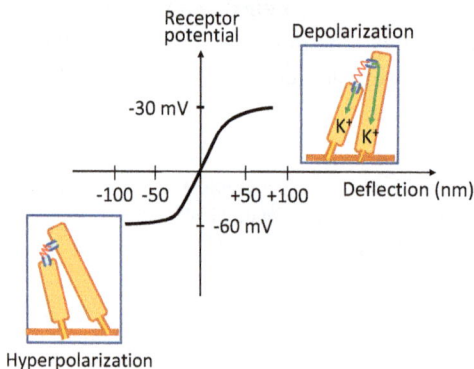

Fig. 12.20: Change of the receptor potential as function of stereocilia deflection.

The response time of the RP is very short. The latency period is only a few microseconds. This short time indicates that the mechanoelectric transduction is the only working principle of the auditory system. Because any intermediate chemical messenger would increase the latency period considerably up to milliseconds response time as is the case in the visual system.

Another distinctive feature of hair cells is adaptation. As mentioned earlier, the immediate response to hair cell potential changes is rapid and proportional to the stimulus. However, with more prolonged stimulation, the sensitivity decreases, the RP drops to the values for the resting phase, and the action potentials decrease. Figure 12.21 shows this graphically and schematically in a sequence from hyperpolarization to adaptation. Only a few action potentials are fired in response to a few open K^+ channels in the resting state. When the stereocilia bend to their shorter neighbors, the remaining K^+ channels close, the RP drops to the hyperpolarized value, and the number of action potentials decreases. When the deflection turns to the taller neighbors, K^+ channels open, Ca^{2+} channels follow, the RP depolarizes, and the number of action potentials increases. However, with continued stimulation, the RP drops again, and the number of action potentials decreases accordingly.

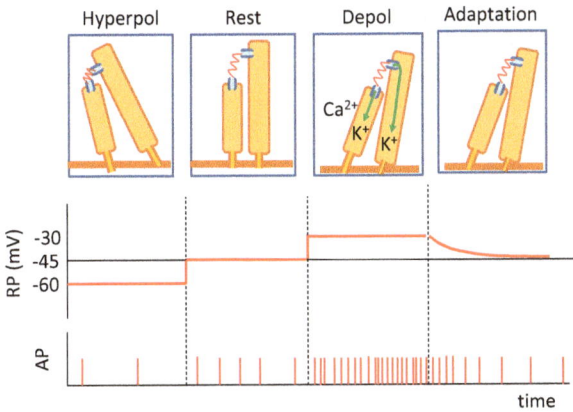

Fig. 12.21: Sequence of stereocilia tilts and respective response of the receptor potential (RP) and action potential (AP) as a function of time. The tilt of the cilia elicits hyperpolarization, resting polarization, depolarization, and adaptation.

12.4.8 Pathway to the auditory cortex

Next, we briefly sketch the path of the action potential to the brain's auditory cortex (Fig. 12.22). All 30 000 nerve fibers that emerge from the cochlea carrying information about sound frequency and intensity, and all nerve fibers that originate from the vestibular system conveying balance information are bundled in the sensory VIII cranial nerve as the main channel to the brain. The first station on the path to the brain is the cochlear nucleus, where the neuronal processing of digitized information from the inner ear occurs and cross-linking between fibers from the left and right cochlea occurs. The lateral superior olive (LSO) and the medial superior olive

(MSO) lie within the superior olive. LSO is important for detecting interaural sound level differences, while neurons in the MSO process microsecond time differences between the left and right ear, the major cue for localizing low-frequency sounds (see next section). The inferior colliculus is a relay station in the ascending part of the hearing system and is most likely used to incorporate information specific to locating the sound source from the superior olive before sending it to the medial knee joint body, which is part of the thalamus relay system. From there, the nerve fibers go on to the auditory cortex for final processing.

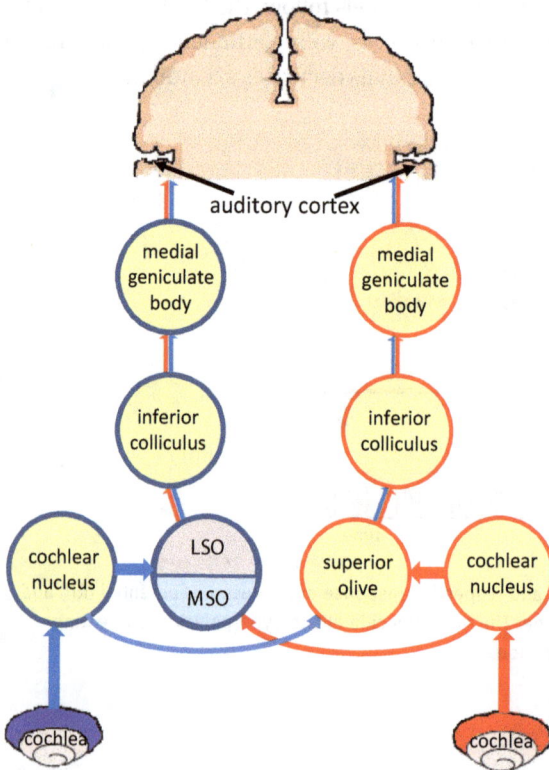

Fig. 12.22: Schematic outline of the auditory pathway from the left and right cochlea to the auditory cortex in the brain. Arrows indicate connections and cross linking of the auditory nerve on the way to the brain. The superior olive on both sides is subdivided in an LSO and an MSO (adapted from [20]).

12.4.9 Sound localization

Sound localization is part of the survival strategy and therefore essential for all vertebrates. Pairs of ears can localize sound whether it comes from the front or back, from the side or from sources above or below the azimuthal plane of the ear. Ears use different techniques to make these distinctions. For some techniques, monaural listening is sufficient; other techniques require a binaural ability. For frequencies f between 400 and about 1400 Hz, the source of sound is primarily localized by the *interaural arrival time difference* (ITD) Δt of the pressure amplitude. The associated path difference $\Delta s = v\Delta t$ is, according to Fig. 12.23, composed of $\Delta s = \Delta s_1 + \Delta s_2$, with $\Delta s_1 = (d/2)\sin \alpha$ and $\Delta s_2 = (d/2)\alpha$, yielding:

$$\Delta s = \frac{d}{2}(\alpha + \sin \alpha), \qquad (12.37)$$

where d is the interaural distance and α is the azimuthal angle of the sound source.

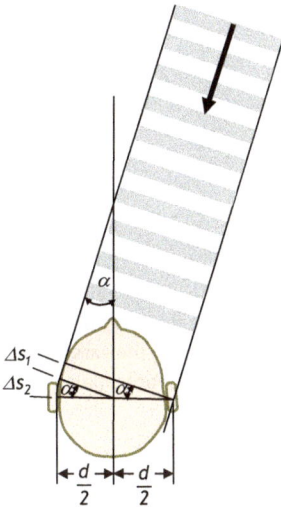

Fig. 12.23: In the far field, acoustic ranging is realized by interaural arrival time of sound waves.

As an example we take a sound source at an angle of $\alpha = 30°$, with a frequency of 1000 Hz, which propagates in air with a speed of sound of 330 m/s. For the interaural distance d we take a typical value of 0.20 m and assume that the sound source is far away compared to d, which justifies the use of plane waves. This results in an arrival time difference of $\Delta t = 0.31$ ms. The maximum time delay of 0.6 ms applies to sound arriving from the side at an angle of 90°. As the angle decreases, it becomes more and more difficult to differentiate between locations and requires a higher time resolution. A unique coincidence detector solves this problem in the *MSO* (Fig. 12.22), which compares the arrival time of action potentials from the left and right ears. This distinctive principle is known as *ITD* detection, first proposed by Jeffress[9] [21] and shown schematically in Fig. 12.24. Neurons arriving from the left and right fan out in the MSO into a horizontal arrangement of connected neurons that form a delay line. Only when excitatory inputs from both sides reach one of the joined neurons simultaneously does the sum potential stimulate an action potential. All other neurons remain quiet. From the location of the firing neuron, the brain can compute the delay time that yields the horizontal orientation of the sound. Only five joined neurons are shown in the schematics of Fig. 12.24. In reality, the coincidence detector in the MSO contains thousands of them. The time resolution of this coincidence detector is on the order of 10 μs, corresponding to an angular accuracy of about 2°. The neurobiology of ITD is reviewed in [22].

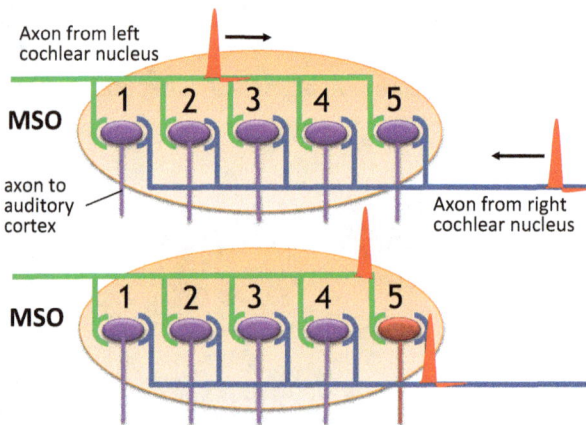

Fig. 12.24: Schematics of the coincidence detection of sound arriving at different times from the left and right cochlea for localization of the sound source in the azimuthal plane of the ears. An action potential is only fired in one of the joined neurons labeled 1–5 if the action potentials arriving from either side overlap, as indicated in the bottom panel at location 5.

The pair of ears is the only sensor in the body that measures time differences. ITD works well in the frequency range from 400 to 1400 Hz. Why ITD for humans fails at higher frequencies is not entirely clear since it was observed that other mammals have a higher ITD frequency range [23]. Alternatively, the location of sound sources is achieved at higher frequencies by *intensity level differences* (ILD) rather than by ITD. One ear is exposed to the sound wave intensity at high frequencies, while the opposite ear receives a lower intensity in the sound shadow. ILD is processed in the LSO (Fig. 12.22) by integrating ipsilateral excitatory inputs and contralateral inhibitory inputs [24].

---

9 Lloyd A. Jeffress (1900–1986), American audiologist and psychologist.

At low frequencies, in contrast, where the wavelength of sound exceeds the size of the head, diffraction effects distort the wave front, making it more and more difficult to localize sound sources. On the other hand, the asymmetry of the pinna assists in distinguishing between front and back and up or down (see Fig. 12.1). Sound becomes reflected from the rims that interfere with waves penetrating the outer ear canal without reflection. Interference and phase shift provide information that again allows acoustic ranging of the sound source.

Jeffress proposed ITD detection in a seminal publication of 1948 [21], and since then this intriguing mechanism was reproduced in all major physiology textbooks. However, recently it has been debated whether coincidence detection is the sole explanation for ITD. Inhibitory input and interaction among synaptic activities play a similar role for ITD as for ILD [24]. Coincidence detection, presented above as an instantaneous process, is oversimplified. Many more interactions causing internal delays appear to be operational [25].

To summarize, at low frequencies below 400 Hz, diffraction effects increasingly lead to confusion of sound localization. At intermediate frequencies between 400 and 1400 Hz sound localization in the horizontal plane is achieved by ITD in the MSO using contralateral excitatory inputs from both ears. At high frequencies, ILD is operational in the LSO, receiving ipsilateral excitatory inputs and contralateral inhibitory inputs. A concise review on the cell physiology of hearing is provided in [26].

> Sound localization at frequencies below 400 Hz can be confusing. Sound localization between 400 and 1400 Hz is achieved by interaural arrival time coding and above 1400 Hz by intensity level coding.
>
> !

The pair of ears is indeed a unique and highly sensitive sensor for detecting a wide range of sound frequencies, volume levels, and the spatial location of sound sources. In contrast to other sensory receptors in the body, auditory perception pursues different strategies to expand the sensitivity range: phase-locking for low frequencies and tonotopic mapping for higher frequencies; coincidence detection of the sound localization at low frequencies and interaural level difference detection at high frequencies. In comparison, visionary perception has a rather limited sensitivity range of perceived frequencies and no trick to extend it into the infrared or ultraviolet range.

12.5 From sound generation to hearing loss

12.5.1 Tone, sound, and noise

Generally, we distinguish between *tone, sound,* and *noise* (Fig. 12.25). The *tone* is a sound wave with a fixed frequency. Note that the frequency is responsible for the tone, not the wavelength, which changes from one medium to another. The frequency of sound waves is also sometimes called "pitch." For instance, the internationally accepted standard pitch for music instruments is 440 Hz. The ear has a very high-frequency (*pitch*) resolution: it is in the order of 0.2%, which means that one can distinguish between a tone of 1000 and 1002 Hz. Even though the left and right ear auditory nerves are connected, they keep a certain level of independence. If they are exposed to slightly different frequencies, they will recognize a beating effect despite the lack of any interference effect. The binaural beating effect appears to be of sole psychological origin. Only with increasing frequency difference will both ears recognize two distinct tones. You may test your pitch resolution and beating effects using an online tone generator [27].

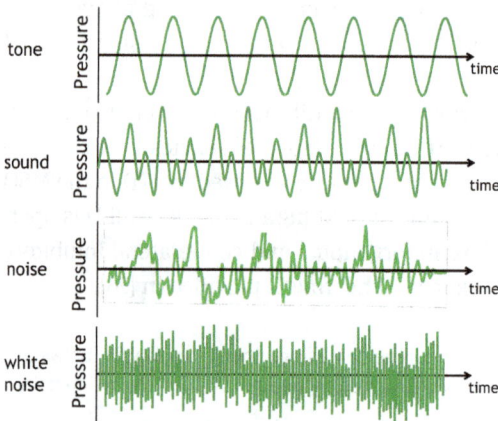

Fig. 12.25: Characteristic wave forms are shown for a single frequency sinewave tone, for the sound of a musical instrument, for some noise, and for a white noise source.

The term "sound" is generally understood to mean a superposition of tones with different frequencies and different pressure amplitudes, which, like the sound of a violin, are perceived as harmonic or consonant. Usually, these are overtones to a fundamental. In a Fourier analysis of a particular sound, we only find a few frequencies that are in a rational relationship to the fundamental frequency and are characteristic of the instrument producing the sound.

Noise, on the other hand, is an inharmonious mixture of sounds that is perceived as dissonant and often annoying. After all, we hear "*white noise*" from many different types of sound sources such as traffic and machines. The term "white noise" is reminiscent of "*white light sources.*" Similarly, white noise contains the entire frequency spectrum without any correlation or phase relationship between different frequencies.

12.5.2 Hearing loss

Loss of hearing can be tested by recording audiograms. The test measures the audible threshold as a function of frequency. If the hearing ability is reduced, for instance, by $\Delta L = 10 \times \log 10^{1.5} = 15$ dB, then the auditory sensitivity is the same as if the sound level were increased by a factor $I/I_0 = 10^{1.5} = 31.6$ in comparison to normal sensitivity. With age a decline of sound sensitivity is normal. A typical audiogram is shown in Fig. 12.26 for different frequencies and as a function of age. At high frequencies the loss is more severe than at low frequencies. Hearing loss with age is actually gender-specific and more acute for men than for women.

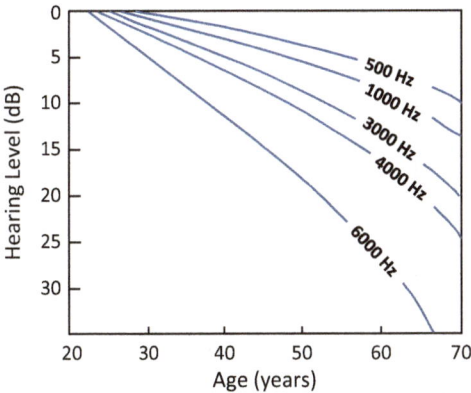

Fig. 12.26: Loss of hearing sensitivity with aging. Sensitivity loss is more severe at higher frequencies.

Loss of hearing ability is characterized by the following levels:
- 0–20 dB: Normal hearing
- 20–40 dB: Mild hearing loss
- 40–55 dB: Moderate hearing loss
- 55–70 dB: Moderately severe hearing loss
- 70–90 dB: Severe hearing loss
- >90 dB: Profound hearing loss to deafness

Most important for communication are sound frequencies in the range of 100–6000 Hz. If hearing loss is severe, a hearing aid may be able to compensate for the loss. But first the location of the loss should be identified. There are a number of potential reasons for dysfunctionalities causing hearing loss:
1. occlusion in outer ear canal;
2. eardrum stiffening (tympanic membrane);
3. middle ear (otosclerosis);
4. cochlea, damage of hair cells;
5. degeneration of nerve cells in the auditory cortex.

With an audiogram the cause for hearing loss cannot be identified precisely, but likely causes can be isolated (Fig. 12.27). In case of a conductive hearing loss of the middle ear, the hearing level drops homogenously for all frequencies. In case of damage of hair cells due to loud noise, a certain frequency band is usually affected, while other frequency regions behave normal. However, damage to hair cells may have additional adverse effects that reduce the ability to discriminate between sounds. This is often noticed as a reduced comprehension of speech, and simply amplifying the sound level as most conventional hearing aids do, is often insufficient to improve speech recognition.

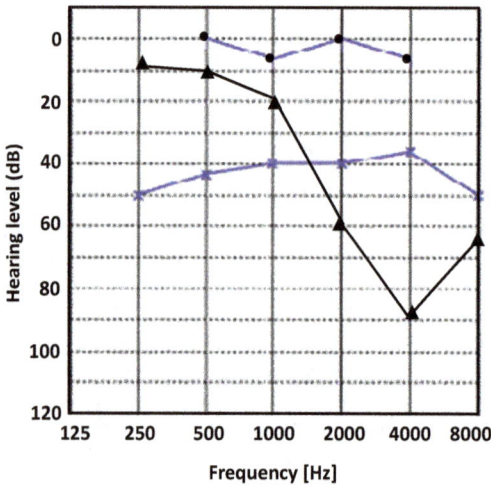

Fig. 12.27: Audiograms for different hearing losses: • normal hearing; × conductive hearing loss; ▲ damage in cochlea.

12.5.3 Hearing devices

Presently three types of hearing devices are available:
1. in-ear canal hearing devices;
2. middle-ear implants;
3. cochlea implants.

We have to distinguish between an outer part and an inner part of the hearing device in all three cases. The outer part contains the sound receiver, sound processor, and batteries. It can be hidden behind the pinna, attached to eye glasses, anchored on the bone, or made small enough to fit into the ear canal.

12.5.3.1 In-the-canal hearing aids
Providing that a good portion of hair cells in the inner ear are still functioning and that the middle ear is operational, an in-canal hearing device (aid) can improve

hearing loss. The inner part of hearing aids differs significantly depending on the application. The *in-ear* or *in-canal hearing* aids contain a loudspeaker that amplifies the sound and transmits it to the tympanic membrane (Fig. 12.28). This fairly simple device conceptually works like a loudspeaker and is the most common one in practice. But the technical realization of either analog or digital versions is quite demanding and uses the most complex electronics in miniaturized space. When active noise reduction is included, the audio signal is processed in the hearing aid by a spectral analyzer to filter out the "white" component. Modern electronics allow the amplification factor to be adapted to the individual audiogram. But feedback between the receiver and speakers may cause problems if adjustments and settings are incorrect. Nevertheless, the recognition of conversations in the midst of noise (conversation in a restaurant), speech from one side, and noise from the other side or the localization of a sound source in the horizontal plane at a certain distance are still challenging. In some cases, wearing bilateral hearing aids can help sound detection and localization [28].

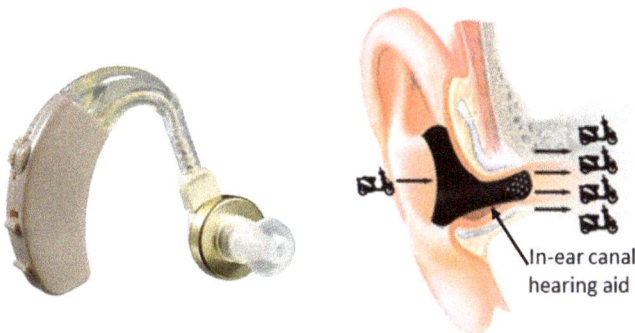

In-ear canal hearing aid

Fig. 12.28: In-ear canal hearing aid. The hearing aid amplifies external sound and the drum head is exposed to an increased pressure amplitude.

12.5.3.2 Middle-ear implants

The tympanic membrane (eardrum) is the interface between the outer ear and the middle ear. Myringoplasty is a reconstruction and enlargement of the eardrum via fascia or perichondria (firm tissue layer), which improves sound transmission to the middle ear [29]. Its brownish color indicates the improved eardrum in Fig. 12.29(a). If the middle ear is compromised, several solutions depend on the severity of the hearing loss and the cause of the loss. If the connection between incus and stapes does not work, special kinked middle-ear prostheses, as shown in Fig. 12.29(b), can already alleviate the problem. When the incus is no longer usable, a direct connection between the malleus and stapes can transmit the sound as indicated in panel (c). Even if the stapes is inoperative, a rigid connection between the incus and the oval window can restore the sensitivity to sound (d). In all cases, the implants are metal parts. 3D-printing capabilities are likely to revolutionize bone bridge implants in the future as they replicate the ossicles more naturally than any metallic implant. The main problem is the limited access to the tiny and sensitive ossicles. One solution can be minimally invasive interventions through the eustachian tube with the support of endoscopic methods. But the small image size of the endoscope, the difficult orientation, and the poor illumination have hindered further progress so far. As described in Chapter 2/Vol. 2, this can change with the availability of more powerful endoscopes. Endoscopy and middle-ear surgery are currently performed either by opening the eardrum or opening a new canal through the temporal bone behind the ear.

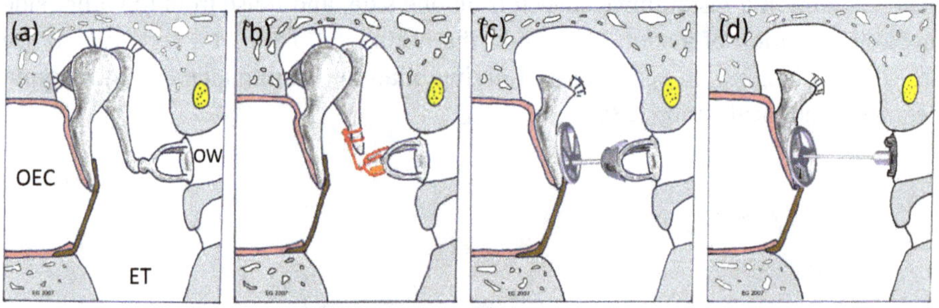

Fig. 12.29: Prostheses of the middle ear: (a) partial replacement of the outer ear drum; (b) kinked middle-ear prostheses; (c) direct connection between malleus and stapes; (d) rigid connection between the malleus and oval window. OEC, outer ear canal; ET, eustachian tube; OW, oval window (adapted from http://www.hno-arzt-augsburg.de/).

Alternatively, hearing loss in the middle ear can be revitalized with active middle-ear implants (MEI). In contrast to hearing aids, which only amplify the sound pressure in the external auditory canal, MEI converts sound from the environment into mechanical vibrations that stimulate bones and structures in the middle ear. One variant most commonly implanted in cases of hearing impairment is the so-called *vibrating sound bridge* MEI (Fig. 12.30) [30]. It consists of outer and inner parts as shown in Fig. 12.30(a). An audio receiver and processor is attached to the outside of the skin behind the ear. A pick-up coil (auditory ossicular prosthesis implant) is implanted under the skin. A small magnet keeps the outer and inner parts aligned. The signal from the pick-up coil is fed into the middle ear via a wire and activates a transducer that is attached to one of the ossicles or directly to the oval window. The transducer consists of a floating mass, which is a tiny but powerful SmCo magnet (Fig. 12.30(b)). The magnet sits in two antiwinding coils. When activated, the magnet bounces back and forth to the rhythm of the audio signal. The lateral movement of the magnet is cushioned like a spring on both sides by silica balls. The power supply battery is mounted outside in the audio receiver and can be replaced without surgical intervention. The vibrating sound bridge transmits a sound that is perceived as "natural" through the wide frequency range of up to 8 kHz. Lower frequencies can still be transmitted through the ear canal. In addition, the vibrating sound bridge can be combined with cochlear implants for a higher frequency range if required.

a. Soundbridge components

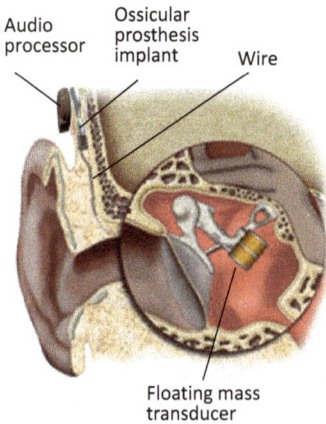

b. Floating mass transducer

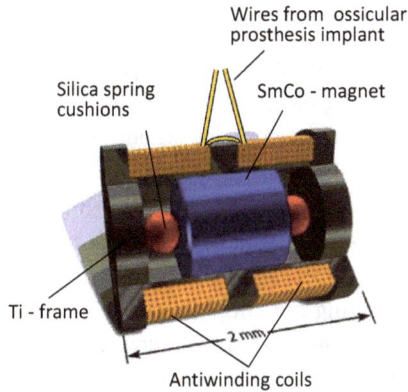

Fig. 12.30: Vibrating soundbridge middle-ear implant consists of an outer part and an inner part. Left panel: overview of the system composed of an audio processor and ossicular prosthesis implant behind the ear lap. A wire connects the pick-up coil in the implant with the floating mass transducer attached to one of the ossicles in the middle ear. Right panel: cross section of the floating mass transducer. A small SmCo magnet at the center moves back and forth when activated by the coils. Silica cushions on either end of the Ti-frame cushion the movement (adapted from Ref. [30] by permission of Taylor & Francis).

12.5.3.3 Cochlear implants

All implants discussed so far aim at restoring a partial loss of hearing ability, also known as conductive deafness. It presumes that the cochlea is still operational, at least partially, and that auditory signals can be sent via the cochlear nerve bundle to the brain's auditory cortex to decipher sound signals. Complete deafness or nerve deafness resulting from an infection, trauma, etc., is presently not curable. If cochlear hairs are damaged or lost, but the auditory nerve bundle is still functioning, cochlea implants may help gain a very rudimentary sound sensation.

Cochlear implants consist of a sound receiver and a wire. The output of the sound receiver is fed into a wire carrying electrode, which is implanted into the organ of Corti. The aim of the wire is to stimulate the auditory nerve cells (Fig. 12.31). However, there are two main problems: first, the implants reach only the high-frequency end of the cochlea close to the base and not the lower frequencies at the apex; second, the wire is immersed in a highly ionic and conducting fluid, causing shortening problems.

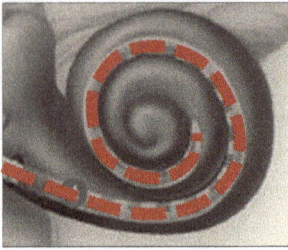

Fig. 12.31: The red-dashed line indicates the position of the implanted cochlear wire.

Because of this, the stimulation of frequency-dependent RPs is somewhat limited. Like in-the-canal hearing aids, cochlear implants are also applied on both ears to improve sound sensitivity. The electrodes on both sides must be positioned precisely and consistently to elicit the same tonotopic nerve fibers, which constitutes a considerable challenge. Alternatively, users of cochlear implants with residual hearing at the contralateral ear appear to have better speech recognition when using a contralateral in-canal hearing aid [31]. Then the high frequencies are represented by the implant, whereas the hearing aid supplements the lower frequencies. Cochlear implants are still under development. Presently they are only recommended in case of severe deafness due to damage of the sensory hairs in the organ of Corti. In addition, research groups are presently trying to restore lost cochlear hair cells by genetic means [32].

12.5.4 The making of sound

Language is a sequence of sounds that have a meaning. Sound reception is developed early in newborns while understanding its meaning takes more time. This example shows that there is a close connection between the auditory cortex and other parts of the brain. Hearing and speaking form a unit and are the most important instruments of communication between people. Without hearing, there is no speaking. Even deaf people without special training are mute; they are deaf-mute.

Two independent organs generate the human voice during expiration: the vocal cord and the oral cavity. The vocal cord is responsible for the pitch (frequency) and the volume; the oral cavity gives the sound a phonetic articulation so that one can distinguish between different vowels and consonants. You can check this by singing the vowel "a" out loud. Now you keep the pitch, but switch to a different vowel like "i." You can also change the pitch to a higher or lower frequency. As you do this exercise, you will find that you can change the pitch, the volume, and the articulation independently.

The vocal cords in the larynx can close the tube between the oral cavity and the trachea to increase the air pressure in the chest during talking (Fig. 12.32). If the pressure is high enough, the vocal cords will open again. Note that the terminology is a bit fuzzy. Sometimes only the gap between the vocal cords is called "glottis," and sometimes the vocal cords together with the gap is named "glottis."

The vocal cords approach each other due to muscle activity in the larynx. The expiration air is then forced to flow through the narrow gap left by the vocal cords. According to Bernoulli's equation:

$$p + 1/2\rho v^2 = \text{const.}, \tag{12.38}$$

the air flow decreases the pressure and the vocal cords close completely. At a threshold expiration pressure of 400–600 Pa the glottis open again. The intermittent air flow generates the pitch of the voice. The frequency depends on the length and the tension of the vocal cords. The frequency also depends on the sound velocity of the exhaled gas. This can be tested by inhaling first a tidal volume of helium gas. On expiration, you will notice that the pitch of your voice is much higher. The reason is the much higher sound velocity in He (1007 m/s) compared to air (343 m/s).

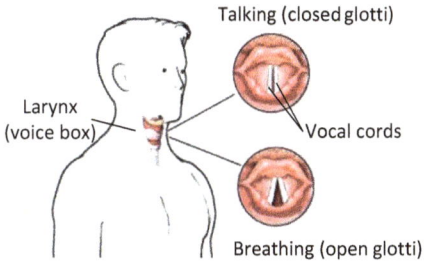

Fig. 12.32: Glottis are closed during talking but open during normal breathing.

Figure 12.33 shows the four main configurations of the vocal cords and the glottis controlled by muscles and cartilage in the larynx. The closed configuration is for speaking; the open configuration is for breathing. The extended configuration is for speaking with a modal voice and finally the triangular gap is for whispering. The open and whisper configurations are static, the closed and extended configurations are snapshots during speaking or singing. A pair of small, three-sided pyramidal cartilage called arytenoids are attached to the vocal cords. They support the movement of the vocal cords, in particular, to be tensed, relaxed, or approximated.

The pitch and volume generated by the glottis is further processed in the hollow space of the oral and nasal cavity to produce a sound (Fig. 12.34). This process is known as *phonetic articulation*. The oral and nasal cavities act as resonance bodies of various sounds. Each volume and shape have a different natural frequency, including harmonic overtones. Vowels (a, e, i, o, u) are formed in the oral space solely by changing the shape and volume of the cavity. Consonants are often frictive sounds that require the assistance of the tongue and lips to increase the friction, like "f, v, s, x." Explosive sounds like "p, t, k, q" are formed by buildup and sudden release of pressure in the cavity, and hissing sounds like "c, s, z" are formed by closing the teeth.

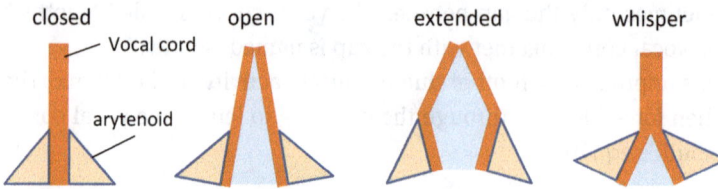

Fig. 12.33: Configurations of the vocal cords and the glottis with the help of the arytenoid cartilage for different speaking and breathing situations.

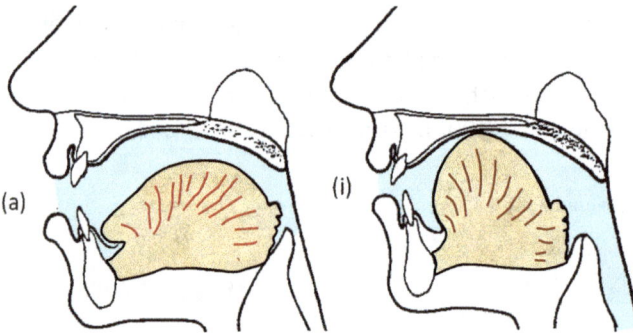

Fig. 12.34: Formation of the sounds "a" and "i" in the oral cavity with the help of the tongue.

Whispering does not require the vibration of the vocal cords. It is sufficient if the glottis opens a bit to supply air flow. The air flow can then be used to generate some articulation within the oral cavity. Further insight into the generation of the human voice can be found in [33].

12.6 Organ of equilibrium

The organ of equilibrium (vestibular organ) is a bilateral organ directly connected to the cochlea, as shown in Figs. 12.1 and 12.35(a). This small, labyrinthine organ hidden in the temporal bone was discovered rather late by Menière.[10] The vestibular importance for our body and the nervous system only became apparent over time; the milestones are described in [34, 35]. The cochlea recognizes and analyzes acoustic signals that come from the environment. In contrast, the vestibular system has no direct contact with the outside world. It perceives the coordinates, movements, and accelerations of the body from inside of the body. The organ of equilibrium is a navigation

---

10 Prosper Menière (1799–1862), French physician.

system, a natural gyroscope that is common to most animals. Although they perform very different roles, the cochlea and vestibular organ share the same endolymph and perilymph fluids. Both use the same principle of stereociliary deflection to trigger RPs.

The vestibular organ is constructed such that all six degrees of freedom are determined (Fig. 12.35(b)): three semicircular ducts arranged almost orthogonal to each other sense rotational acceleration about the x-, y-, and z-axes. In addition, one sensory area (macula utricle) is sensitive to lateral movement in the x,y–plane, and the other sensory area (macula saccule) oriented perpendicularly responds to acceleration in the z-direction. In total, all five sensors, three for rotation and two for translation, yield a complete vector analysis of the body's movement.

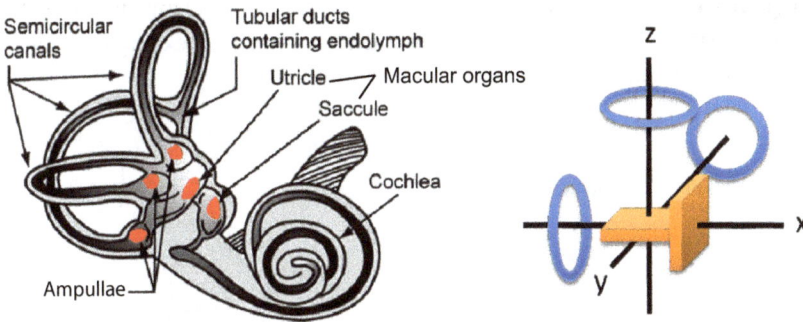

Fig. 12.35: (a) Vestibular system, semicircular canals with the sensory areas (ampullae), and otolith organs utricle and saccule; (b) three-dimensional coordinate system. Blue circles correspond to the semicircular canals, and orange areas symbolize utricle and saccule.

The macular organs and semicircular canals contain sensory cells in enlarged dome-shaped areas garnished with hair cells, supporting cells, and special membranes [3]. All hair cells in the vestibular organ carry a bundle of stereocilia (60–100). The stereocilia of each hair cell are connected by thin fibers (tip links) that can be stretched or compressed to open or close the inflow of K^+ ions from the endolymph (Fig. 12.36). The mechanism is the same as discussed in the section on IHC and OHC of the organ of Corti. Aside from these similarities, the sensory areas of the semicircular canals and macula are organized differently to serve their respective purposes [36].

The semicircular canals are almost circular-closed tubes filled with endolymph. The tube widens like a dome in the area of the ampulla. In this area, the canal wall contains the sensory epithelium with the hair cells, the so-called ampullary crista. The stereocilia of the hair cells protrude into a gelatinous membrane that covers the sensory epithelium. The gelatinous membrane, called the cupula, is firmly attached to the body of the hair cells and the top of the canal wall (Figs. 12.35 and 12.36). Cupula and endolymph fluid have exactly the same density. Therefore, translational acceleration or gravitational forces do not act on this organ since all forces acting on the semicircular canal are the same, left and right, and above or below. With rotational acceleration, however, the rigid walls of the semicircular canals move immediately, while the fluid drags behind due to the rotational inertia. This phenomenon can be observed when turning a coffee cup. The cup moves, but the coffee stays, creating an apparent flow of liquid against the wall of the coffee cup. This also happens in the semicircular canal. Due to the inertia of rotation, the cupula bends against the direction of acceleration. The bending deflects the stereocilia immersed in the cupula, and the direction of the tilt determines the frequency of the released action potentials (Fig. 12.36). If the stereocilia move in the direction of their taller neighbors, the RP depolarizes, and the frequency of action potentials increases. Conversely, if the stereocilia tilt toward their shorter neighbors, the RP hyperpolarizes and the frequency of the action potentials decreases. With this mechanism, the semicircular canals specialize in detecting rotational accelerations, for example, turning the head or any other rotational movement of the entire body.

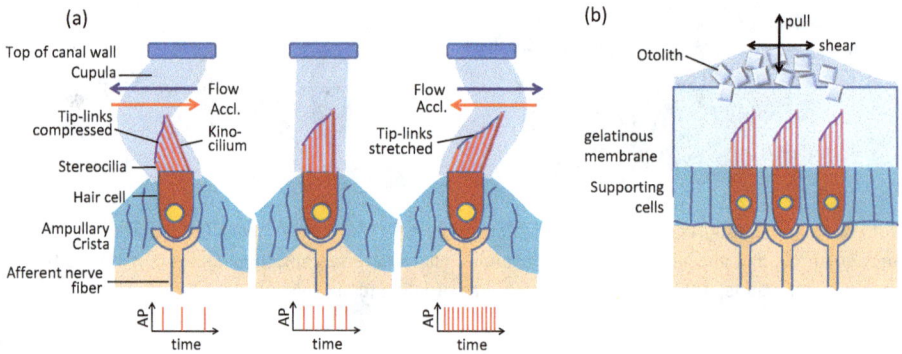

Fig. 12.36: (a) Sketch of the ampullary area of the semicircular canal. Bowing of the cupula tilts the stereocilia on top of the hair cells and elicits a receptor potential that generates an action potential (AP). Only one hair cell out of millions in the ampullary area is schematically shown; (b) the otolith organ has hair cells embedded in a two-component gelatinous membrane; a heavy membrane-containing otoliths tops a lighter membrane without otoliths. The heavier membrane pushes and squeezes on the more lightweight membrane during acceleration, causing tilting of the stereocilia.

The stereocilia of the macular hair cells are also embedded in a gelatinous membrane as shown in Fig. 12.36(b). In contrast to the cupula, the upper part of the macular membrane is filled with millions of tiny calcite crystallites ($CaCO_3$), the so-called otoliths ("ear stones"). These crystals increase the mass density of the membrane considerably from approx. 1 g/cm^3 for the endolymph and the membrane without otoliths to 2.2 g/cm^3 with otoliths. The greater the mass, the greater the force of attraction: $F = mg$, where $g = 9.8$ m/s^2 is the earth's constant of gravity. For example, in upward acceleration, the gravitational force on the heavier membrane is bigger than on the lighter membrane. Then the heavier upper membrane presses on the lower and lighter membrane, pushing it aside. This, in turn, deflects the stereocilia. When accelerating downward, the heavier membrane pulls on the lighter membrane dragging it up. This tilts the cilia again to the side. Similarly, when accelerating in the horizontal direction, the heavier membrane moves first, causing shear at the interface between the lighter and heavier membranes. In short, the otolith organs are gravity and linear accelerator sensors. Therefore, they also sense the tilt of the head, but not the rotation.

The vestibular organ has another important function, the so-called *vestibular-ocular reflex* (VOR). This reflex acts on the eyes when the head moves to the side. The reflex aims to stabilize the image on the retina while moving. Without this reflex, the visual perception during head movements would be confused. Thus, VOR detects the rotation of the head and commands the line of sight to remain fixed on the visual target via a complex neural network wiring. The compensatory movement of the eyes counteracts the movement of the head. For example, when the head moves to the left, the movement triggers an inhibitory signal to the extraocular muscles on the left side of the eye and an excitatory signal to the right side.

In summary, the vestibular system is important for sensing and judging the body's balance, equilibrium, and posture of head and body. The vestibular labyrinth consists of two subsystems: three semicircular canals for sensing head rotation and two otolith organs detecting gravitational forces, head tilts, and linear acceleration. All organs, semicircular canals and otolith organs, use hair cells as the primary sensor for movement, tilt, and acceleration. In addition, head movement triggers a VOR.

12.7 Summary

S12.1   Sound is a traveling longitudinal compression wave.

S12.2   Sound intensity is characterized by pressure amplitude and acoustic impedance.

S12.3   Acoustic impedance is determined by density of the media and sound velocity. It is the refractive index for sound propagation.

S12.4   Sound velocity depends on the density of the medium and its compressibility.

S12.5   At boundaries between two media characterized by different impedances, sound waves are partially reflected and partially transmitted.

S12.6   At boundaries to media with very low or very high impedance, sound waves become completely reflected. Such boundaries have an impedance mismatch.

S12.7   The boundary between air and water is acoustically mismatched.

S12.8   The task of the ear is to funnel sound into the body by compensating the impedance mismatch and to convert sound into action potentials.

S12.9   The ear consists of three parts: outer ear, middle ear, and inner ear.

S12.10 The middle ear is a mechanical pressure amplifier. It serves the purpose of overcoming the impedance mismatch between air and watery solution.

S12.11 The middle ear fulfills three main tasks: amplification of pressure amplitude, transmission of a large frequency band, and protection against destructive intensity levels.

S12.12 The cochlea of the inner ear acts as a mechanoelectric transducer.

S12.13 Hair cells in the basilar membrane of the cochlea together with the tectorial membrane form the *organ of Corti*.

S12.14 Hair cells along the basilar membrane provide a tonotopic mapping of sound frequencies.

S12.15 Inner hair cells are connected to afferent neurons. Their RP initiates action potentials traveling to the brain.

S12.16 The action potential of the inner hair cells is tonotopically frequency mapped and phase-locked to the frequency of the RP.

S12.17 Outer hair cells are connected to efferent neurons. Their RP causes longitudinal length changes of the cell, known as electromotility.

S12.18 The outer hair cells modulate the sound reception of inner hair cells.

S12.19 Sound localization is achieved by coincidence detection in the MSO that compares the arrival time of action potentials from the left and right ears.

S12.20 The sensitivity of the ear is extremely high and extends over 12 decades of sound intensity.

S12.21 Hearing loss is mainly due to loss of flexibility of the middle-ear bones, in severe cases due to irreversible breakage of the sensory hairs in the organ of Corti.

S12.22 Hearing aids can be placed in the external ear canal, or may involve a replacement of the middle-ear bones, or may require a cochlea implant.

S12.23 The frequency of the human voice is produced during expiration by an intermittent opening of the vocal cords.

S12.24 Implantable middle ear hearing aids can either passively restore bone conduction or actively bring the sound from the outside to the middle ear, for instance, by a vibrating sound bridge.

S12.25 The essential part of the vibrating sound bridge is a floating mass transducer that is activated by a receiver behind the auricle.

S12.26 Sounds are formed in the hollow space of the oral cavity acting as a resonance body.

S12.27 The vestibular system senses the body's balance and movement.

S12.28 The vestibular system consists of three semicircular canals and two otolith organs.

S12.29 The semicircular canals sense rotational acceleration, and the otolith organs sense linear acceleration.

Questions

?

Q12.1   Which three pieces of physical information do a sound wave carry?

Q12.2   Describe the main parts of the ear as a sound detector.

Q12.3   How is the acoustic impedance defined?

Q12.4   Which impedance is higher, water or air?

Q12.5   When do sound waves totally reflect at an interface to another medium?

Q12.6   How big is the acoustic mismatch between air and water?

Q12.7   What is the unit for the auditory level?

Q12.8   What is the difference between auditory level and loudness level?

Q12.9   What is the unit for the loudness level?

Q12.10   Isophones are equal loudness curves. Do isophones depend on the frequency?

Q12.11   What are the names of the three ossicles in the middle ear?

Q12.12   By what action is the acoustic mismatch overcome?

Q12.13   What is the amplification factor of the middle ear?

Q12.14   What are the three main tasks of the middle ear?

Q12.15   What is the task of the oval window?

Q12.16   What is the task of the round window?

Q12.17   The basilar membrane is a center line in the cochlea. What does it separate?

Q12.18   Along the basilar membrane, where are the high and low frequencies detected?

Q12.19   Frequency detection is achieved by a spatial separation. What is the technical term for such a separation?

Q12.20   The organ of Corti contains which two types of hair cells?

Q12.21   Are the Ca^{2+} channels of the inner hair cells potential gated or mechanically gated?

Q12.22   Are the K^+ channels of the inner hair cells potential gated or mechanically gated?

Q12.23   What is the task of the outer hair cells?

Q12.24   What is the task of the inner hair cells?

Q12.25   How many inner hair cells are responsible for detection a frequency range from 20 to 20 000 Hz?

Q12.26   The outer and inner hair cells are connected to afferent and efferent nerve fibers. Which ones are connected to which nerve fibers?

Q12.27   What are the unique properties of the outer hair cells, and how is this achieved?

Q12.28   When do the inner and outer hair cells depolarize or hyperpolarize by leaning to their longer or shorter neighbors?

Q12.29   How can the receptor in the organ of Corti be characterized?

Q12.30   How is the frequency and the sound intensity encoded on the way up to the auditory cortex?

Q12.31   How is sound in the equatorial plane spatially localized?

Q12.32   What is the difference between tone, sound, and noise?

Q12.33   Is age-dependent hearing loss more severe for high or for low frequencies?

Q12.34   Hearing aids can be distinguished according to their location. Name them.

Q12.35   Where does sound produced by the body originate from?

Q12.36   How many sensory areas does the vestibular organ have?

Q12.37   What is the purpose of the sensory areas in the vestibular organ?

Attained competence checker	+	0	-
I realize that sound waves are longitudinal compression waves propagating in a medium like gas, liquid or solid.			
I know that the sound intensity is proportional to the squared pressure amplitude.			
I know that the acoustic impedance characterizes the acoustic property of a medium.			
Knowing the impedance, I can calculate the transmittance of sound waves from one medium to another.			
I can distinguish between sound intensity and sensitivity level.			
I know in which units sound sensitivity is measured.			
When experiencing a hearing loss of 20 dB, I know how much higher the sound intensity must be to perceive the same loudness as before the loss.			
I can name the different parts of the ear and I know what their tasks are.			
I can roughly describe the inner ear and the organ of Corti.			
I can roughly describe the mechanoelectric transduction taking place in the cochlea.			
I realize how sound localization occurs with bilateral sound reception.			
I can recognize with the help of an audiogram the cause of a hearing loss.			
I know the advantages and disadvantages of different hearing aids.			
I know which parts of the body are involved in the generation of sound			
I know that the vestibular organ consists of five sensors.			
I realize how the vestibular sensors work and to which movement they response.			

Suggestions for home experiments

HE 12.1   Create your personal audiogram according to Fig. 12.27 using an online tone generator or an equivalent app.

HE 12.2   Test your interaural angular sensitivity. Place your tone generator about 5–10 m away. Take a seat in a swivel chair. Now slowly rotate your head and body from 0° to higher angles. In what angular range can you localize the sound source?

Exercises

E12.1 **Sound intensity:** Show that the eq. (12.12) $\langle I \rangle = \frac{1}{2} Z u_0^2$ holds.

E12.2 **Hearing loss:** A patient has a hearing loss of 30 dB, i.e., his hearing threshold is 30 dB higher than that of a person with normal hearing.
   a. By which factor is the threshold intensity I of the sound higher in the patient than in a person with normal hearing?
   b. By what factor does the pressure amplitude has to be increased in order to achieve the threshold intensity necessary for the patient?

E12.3 **Thunderstorm:** Four seconds after you see a lightening, you hear a thunder. Using a sound detector you determine a sensation level of 100 dB. Now take the origin of the thunder as a point source.
   a. What is the distance between you and the origin of the thunder?
   b. What is the sound intensity I (in W/m^2) at the point of detection?
   c. What is the power of the sound produced by the thunder?

E12.4 **Doubling distance, doubling the intensity:**
   a. A sound source treated as a point source at distance d_1 produces a sound with the sensation level I_1. Now you double the distance to the sound source to $d_2 = 2d_1$. What is the sensation level I_2 that you receive expressed in decibel?
   b. A sound source treated as a point source at a distance d_1 from an observer emits an intensity I_1. Now a second sound source at the same distance emits the same intensity $I_2 = I_1$. What is the intensity perceived by the observer in decibel?

E12.5 **Sensation threshold:** You hear a person speak at a distance of 1 m with a loudness level of 40 dB. Your threshold level is 20 dB. What is the distance up to which you can still hear this person speak?

E12.6 **Localization:** Consider a planar monochromatic soundwave that hits your head at an angle of 20°. What is the time resolution necessary to resolve such an angle if the interaural distance is $d = 15$ cm? Compare your answer with the temporal length of an action potential.

E12.7 **Pressure amplitude:** The middle ear converts sound waves in air with large displacement amplitude but small pressure amplitude into sound waves in fluids with small displacement amplitude but high pressure amplitude. Assuming a pressure amplitude of 3 Pa in air and a frequency of 440 Hz, what is the displacement amplitude of particles in air, the pressure amplitude in water with the help of the middle ear, and the displacement of particles in water?

References

[1] Klinke R. Hören und Sprechen; Kommunikation des Menschen. Chap. 21, p. 657–674 In: Pape H-C, Kurtz A, Silbernagel S, eds. Physiologie. 7th edition. Stuttgart, New York: Thieme Verlag; 2014.

[2] Marion JB. Classical dynamics. 2nd edition. New York: Academic Press; 1970.

[3] Mullin WJ, George WJ, Mestre JP, Velleman SL. Fundamentals of sound with applications to speech and hearing. Boston: Allyn and Bacon; 2003.

[4] Anikin A, Johansson N. Implicit associations between individual properties of color and sound. Atten Percept Psychophys. 2019; 81: 764–777.

[5] Hudspeth AJ. How the ear's works work. Nature. 1989; 341: 397–404.

[6]   Hudspeth AJ. The Inner Ear. Chap. 30, p.654–681. In: Kandel ER, Schwartz JH, Jessell TM, Siegelbaum SA, Hudspeth AJ, eds. Principles of neural science. 5th edition. McGraw Hill; 2013.

[7]   Zagadou BF, Barbone PE, Mountain DC. Elastic properties of organ of Corti tissues from point-stiffness measurement and inverse analysis. J Biomech. 2014; 47: 1270–1277.

[8]   Reichenbach T, Hudspeth AJ. Dual contribution to amplification in the mammalian inner ear. Phys Rev Lett. 2010; 105: 118102.

[9]   Duke T, Jülicher F. The Traveling Wave in the Cochlea. Phys Rev Lett. 2003; 90: 158101.

[10] Bear MF, Connors BW, Paradiso MA. Neuroscience: Exploring the brain. 5th edition. Philadelphia, New York, London: Wolter Kluwer; 2015.

[11] Manoussaki D, Dimitriadis EK, Chadwick RS. Cochlea's graded curvature effect on low frequency waves. Phys Rev Lett. 2006; 96: 088701.

[12] Fettiplace R, Hackney CM. The sensory and motor roles of auditory hair cells. Nat Rev Neurosci. 2006; 7: 19–29.

[13] Fitzakerley J. University of Minnesota Medical School Duluth. www.d.umn.edu/~jfitzake/Lectures/DMED/InnerEar/IEPathology/Fig.s/NormalHC.jpg

[14] Peng AW, Salles FT, Pan B, Ricci AJ. Integrating the biophysical and molecular mechanisms of auditory hair cell mechanotransduction. Nat Commun. 2011; 2: 523.

[15] Fettiplace R, Kim KX. The physiology of mechanoelectric transduction channels in hearing. Physiol Rev. 2014; 94: 951–986.

[16] Ashmore J. Cochlear outer hair cell motility. Physiol Rev. 2008; 88: 173–210.

[17] Takahashi S, Sun W, Zhou Y, Homma K, Kachar B, Cheatham MA, Zheng J. Prestin contributes to membrane compartmentalization and is required for normal innervation of outer hair cells. Front Cell Neurosci. 2018; 12(211): 1–11.

[18] Verschooten E, Shamma S, Oxenham AJ, et al. The upper frequency limit for the use of phase locking to code temporal fine structure in humans: A compilation of viewpoints. Hear Res. 2019; 377: 109–121.

[19] Speckmann EJ. Physiologie. 7. edition. Berlin, Frankfurt, München: Urban und Fischer in Elsevier Inc.; 2019.

[20] http://www.open.edu/openlearn, © creative commons.

[21] Jeffress L. A place theory of sound localization. J Comp Physiol Psychol. 1948; 41: 35–39.

[22] Ashida G, Carr CE. Sound localization: Jeffress and beyond. Curr Opin Neurobiol. 2011; 21: 745–751.

[23] Brughera A, Dunai L, Hartmann WM. Human interaural time difference thresholds for sine tones: The high-frequency limit. J Acoust Soc Am. 2013; 133: 2839.

[24] Grothe B, Pecka M, McAlpine D. Mechanisms of sound localization in mammals. Physiol Rev. 2010; 90: 983–1012.

[25] Franken TP, Roberts MT, Wei L, Golding NL, Joris PX. In vivo coincidence detection in mammalian sound localization generates phase delays. Nat Neurosci. 2015; 18: 444–452.

[26] Schwander M, Kachar B, Müller U. Review series: The cell biology of hearing. J Cell Biol. 2010; 190: 9–20.

[27] https://onlinetonegenerator.com/binauralbeats.html

[28] van Schoonhoven J, Schulte M, Boymans M, Wagener KC, Dreschler WA, Kollmeier B. Selecting appropriate tests to assess the benefits of bilateral amplification with hearing aids. Trends Hear. 2016; 20: 1–16.

[29] Wilson RM. Artificial eardrums get real. Phys Today. 2015; 68(7): 14.

[30] Marino R, Linton N, Eikelboom RH, Statham E, Rajan GP. A comparative study of hearing aids and round window application of the vibrant sound bridge (VSB) for patients with mixed or conductive hearing loss. Int J Audiol. 2013 Apr; 52: 209–218.

[31]   Zedan A, Jürgens T, Williges B, Kollmeier B, Wiebe K, Julio Galindo J, Wesarg T. Speech intelligibility and spatial release from masking improvements using spatial noise reduction algorithms in bimodal cochlear implant users. Trends Hear. 2021; 25: 1–14.

[32]   Du X, Cai Q, West MB, et al. Regeneration of cochlear hair cells and hearing recovery through Hes1 modulation with siRNA nanoparticles in adult guinea pigs. Mol Ther. 2018; 26: 1313–1326.

[33]   Tiwari M, Tiwari M. Voice – How humans communicate? J Nat Sci Biol Med. 2012; 3: 3–11.

[34]   Van de Water TR. Historical aspects of inner ear anatomy and biology that underlie the design of hearing and balance prosthetic devices. Anat Rec. 2012; 295: 1741–1759.

[35]   Wiest G. The origins of vestibular science. Ann NY Acad Sci. 2015; 1343: 1–9.

[36]   Goldberg ME, Walker F, Judspeth AJ. The Vestibular System. Chap. 40, p. 917–934. In: Kandel ER, Schwartz JH, Jessell TM, Siegelbaum SA, Hudspeth AJ, eds. Principles of neural science. 5th edition. McGraw Hill; 2013.

Further reading

Tipler PA, Mosca G. Physics for scientists and engineers. Vol. 1: Mechanics, oscillations and waves and thermodynamics. London: W. H. Freeman; 2003.

Kandel ER, Schwartz JH, Jessell TM, Siegelbaum SA, Hudspeth AJ. Principles of neural science. 5th edition. New York, Chicago, San Francisco: McGraw Hill; 2013.

Bear MF, Connors BW, Paradiso MA. Neuroscience: Exploring the brain. 5th edition. Philadelphia, Baltimore, New York: Wolter Kluwer; 2015.

Pape H-C, Kurtz A, Silbernagel S, eds. Physiologie. 7th edition. Stuttgart, New York: Thieme Verlag; 2014.

Kollmeier B, Klump G, Hohmann V, Langemann U, Mauermann M, Uppenkamp S, Verhey J. Hearing – From sensory processing to perception. Berlin, Heidelberg, New York: Springer Verlag; 2009.

13 Prosthetics

13.1 Introduction

When body parts and functions fail, there are two options for remedial action: regeneration or replacement. Regeneration is always the better option. However, if this option is not available, replacement with a donor or artificial parts is the next best choice. This chapter deals with the artificial replacement of organs, the so-called prosthesis. Prosthesis does not heal. It can only replace limbs or organs that are intended to take over the function of the healthy organ.

We distinguish between two types of prostheses:
– Exoprosthesis, prosthesis external to the body;
– Endoprosthesis, prosthesis within the body, also called implants.

In general, all prostheses have an internal and external connection to the body. In contrast, wheelchairs, walkers, or crutches are not considered prosthetics because they lack body connection.

Endoprostheses cover a wide range of specifications. Each implant has specific requirements, so a more general characterization would be inappropriate here. However, the overriding requirement for all implants is biocompatibility.

Orthopedics is a discipline that treats musculoskeletal disorders. In contrast to prosthetics, in orthopedics, lost organs are not replaced by artificial parts, but implants may be used to support existing organs.

Prosthetics and orthopedics are rapidly developing multidisciplinary fields of science and technology. But these fields are not modern inventions. The daughter of a priest in ancient Egypt, Tabaketenmut, who lived between 950 and 710 BC, had an artificial toe. The prosthesis was found as a burial object and showed signs of wear. Advanced materials science, microelectronics, and bioengineering offer new solutions for prosthetic applications. In this brief overview, some examples on limb and hip prostheses, stents, and retinal implants illustrate the challenges in this area.

13.2 Exoprostheses

In most cases, exoprostheses replace parts of the upper or lower limbs and are closely connected with the kinematics and elastomechanics of the body, as presented in Chapters 2 and 3. Paralympics are an excellent showplace for the development of new exoprostheses.

https://doi.org/10.1515/9783110756951-013

13.2.1 Lower limb prostheses

Lower limb prostheses mimic as well as possible the functions of legs and feet: standing, walking, running, turning, climbing, hunkering down, jumping, etc. Lower limb prosthetics aim to enable amputees to resume or improve their mobility level to some fraction of the original. The amputations are distinguished according to their location, as illustrated in Fig. 13.1. Walking, as discussed in more detail in Chapter 2, is the most basic mobility. At the same time, it is a highly complex movement involving the activation and coordination of many muscles, bones, and joints: rolling the toes, swinging the femur by lifting the leg, bending the knee, rolling the heels, stance of the foot, and repeating with the other leg in well-defined rhythmic sequence. While walking, many other features are taken care of, such as shock absorption when hitting the ground with one leg, supporting the entire body's weight when lifting the other leg, and balancing the upright posture. For obvious reasons, it is important for a lower limb prosthesis, whether the prosthesis starts above (transfemoral) or below the knee (transtibial). We first discuss prostheses that start below the knee.

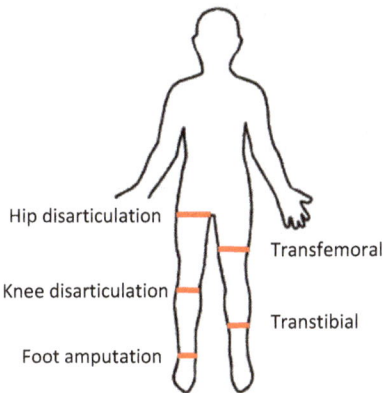

Hip disarticulation

Transfemoral

Knee disarticulation

Transtibial

Foot amputation

Fig. 13.1: Amputations of the lower limb.

13.2.1.1 Transtibial prostheses

The basic prosthesis design for transtibial amputations consists of three parts: a socket at the top fits the stump; a height-adjustable shin connects the socket at the top and the foot together with the ankle at the bottom (see Fig. 13.2(a)). The prosthetic foot and ankle serve a number of important functions during walking, such as shock absorption during loading, stability during weight-bearing, and smooth progression of the limb during walking. Furthermore, durability and structural stability are basic requirements for lower limb prostheses. Some variants are shown in Fig. 13.2. The basic design is a *solid ankle-cushioned heel* (SACH), shown in Fig. 13.2(a). The SACH foot includes a solid ankle and a rigid keel running the entire length of the prosthetic foot sole. The heel is composed of a foam wedge that provides cushioning in the heel section during heel strike. This foot is commonly used on pediatric or geriatric amputees as an early version after amputation and before fitting more advanced prostheses.

An advanced design is shown in Fig. 13.2(b), providing a good connection between stump and prosthesis. An inside liner at position (1) distributes pressures to create a precise fit and to increase the connection between stump and prosthesis. Furthermore, a vacuum system in position (2) draws the stump into the socket for a tight fit. Further down, a torsion and shock absorber at position (3) increase walking comfort. Carbon fiber springs at position (4) absorb energy by landing on the heels and release that energy to give a little push when leaving the ground. And finally, a split-toe design at position (5) adapts to uneven terrain, adding safety and stability when walking or standing. The toe design is shown in the inset without foot cover. Most of these prostheses are mechanical and made of advanced materials. Only the vacuum pump requires power during fitting.

The designs shown in Fig. 13.2(a) and (b) are cosmetically adapted to a real foot. For sports activities, however, functionality is more important than cosmetics. In this case, flexible carbon-reinforced blades are preferred by athletes (Fig. 13.2(c)). These blades are fabricated from carbon fiber-reinforced polymers (CFRP), and unlike all previously discussed foot prostheses, they transform kinetic energy from walking or jumping into potential energy, like a spring, allowing the wearer to run and jump. CFRP is a strong and light-weight composite material, usually made from epoxy mixed with carbon fibers, which is also used for molding car bodies and components of aircraft. In the CFRP blade manufacturing process, 30–90 sheets of impregnated materials are pressed onto a form to produce the final shape. The mold is then autoclaved to fuse the sheets into a solid plate. The carbon-reinforced blades have no motors, sensors, or microprocessors. The blades solely use elastic response returning 80% of the energy

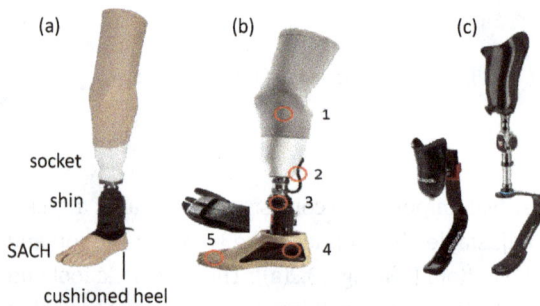

Fig. 13.2: Different prosthesis designs for lower limb transtibial amputation: (a) basic design consisting of socket, shin, and a solid ankle-cushioned heel (SACH) foot; (b) advanced design featuring inside liner at position (1), vacuum pump at position (2), torsion and shock absorber at position (3), carbon fiber springs at position (4), and split-toe at position (5); (c) carbon-reinforced blades for sports activities in two versions for transtibial and transfemoral amputations (reproduced from http://www.ottobockus.com/ by permission of Otto Bock).

stored during compression. A review of design issues and solutions found in active lower limb prostheses is given in [1].

13.2.1.2 Transfemoral prostheses

Amputations above the knee are called transfemoral amputations. They require considerably more complex prostheses because of the additional degrees of freedom that the knee provides. Present-day high prosthetics technology includes computer control of the entire course of motion during gait. Sensors measure the knee angle, the angular velocity of the knee joint, and the load of the knee and foot. The data acquisition is fed into a microprocessor that controls hydrologic valves for steering the knee joint and for determining the angular velocity of the lower leg, aiming at a regular gait pattern. The inbuilt microprocessor allows to mimic the actions of the knee and adapt to all walking speeds in real time.

Microprocessor-controlled prostheses are called C-legs; an example is shown in Fig. 13.3. They were introduced in the late 90s and are now in the fifth-generation development stage [2]. The C-leg is a single-axis knee joint system. It uses hydraulics with a two-way valve principle, in which the resistances in the flexion and extension directions can be continuously and independently adjusted to one another. The sampling frequency has been increased from 50 to 100 Hz in the current generation. According to [2], the major technological advance is the introduction of a new type of six-axis internal motion unit (IMU). The IMU uses its own microprocessor to calculate the position of the knee joint in space. Information about the solid angle, the rotation in space, and the acceleration of the knee in all three coordinate axes is available to the control algorithm. The knee extension moment is recorded to measure the force, which, together with the IMU and a knee angle sensor, provides reliable information about the movement and load status of the prosthesis.

People with C-leg perform significantly better on staircases than on mechanical knees during stair descent, but there is not much difference in stair ascent [3]. Although C-leg prostheses perform better than mechanical prostheses, C-legs come with certain disadvantages. Most microprocessor-controlled knee-joints are powered by a battery housed inside the prosthesis, which adds to the weight and needs to be recharged every one or two days, depending on the use. Nevertheless, a comparative study of people wearing C-legs and various nonmicroprocessor control (NMC) prosthetic knees could show that C-leg users have increased functional mobility and ease of performance. This improves their quality of life and lessens the impairment [4].

Myoelectric control, i.e., the control based on efferent neural signals from the CNS to peripheral muscles for movement, is a new development in lower limb prosthesis [5]. It has been used for upper limbs, as described in the next section, but so far not for the lower limbs. This technic may open new functionality and adaptability to different environments and tasks when fully developed.

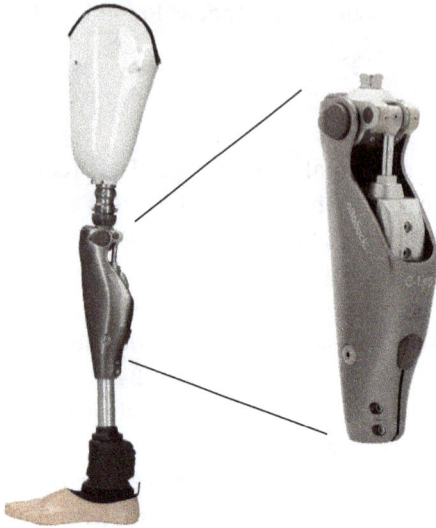

Fig. 13.3: Microprocessor-controlled prosthesis (C-leg) for transfemoral amputations. The enlargement shows the valve system inside the lower C-leg (reproduced from http://www. ottobockus.com/ by permission of Otto Bock).

13.2.2 Upper limb prostheses

In contrast to the lower limb system, the upper limbs have increased flexibility and functionality but no standard motion pattern. This makes the construction of prostheses very complex. We distinguish three types of upper limb prostheses:
1. Cosmetic prosthesis
2. Kinematic prosthesis
3. Myoelectric prosthesis

(1) Cosmetic prosthesis should look as natural as possible but have no function aside from holding or counteracting an object (Fig. 13.4(a)); (2) mechanic or kinematic prosthesis translates remaining muscle contraction in the remaining arm stump into the movement of artificial arm and fingers; (3) myoelectric prosthesis offers control of the arm and is the ultimate combination of function and natural appearance.

Sauerbruch[1] designed the first functional arm prosthesis. His design enabled simple movements to be executed with the remaining muscles (Fig. 13.4(b)). Kinematic prostheses are continuations of Sauerbruch's prosthesis, allowing the amputee to open and close a gripper hand through the motion of the shoulder. Thanks to continuous technological advances, the myoelectric-controlled prostheses are the most advanced ones for the upper limbs that feature astounding capabilities to control translational and rotational motion with the help of several tiny motors.

---

[1] Ferdinand Sauerbruch (1875–1951), German anatomist and surgeon.

Myoelectric control of the upper or lower limbs requires training of the patient and concentration during operation.

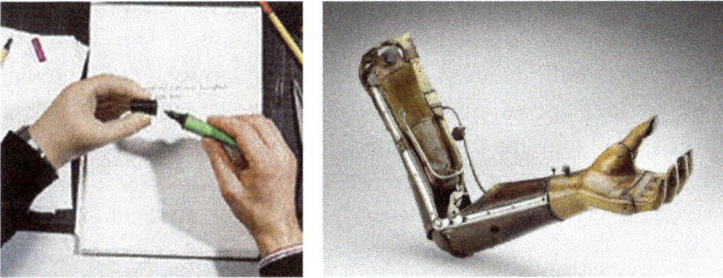

Fig. 13.4: Cosmetic hand (left) and original kinematic prosthesis designed by Sauerbruch (right).

Myoelectric prostheses use remaining efferent nerve fibers, which run from the brain to the arm. The surgical procedure aimed at improving the control of upper limb prostheses is referred to as targeted muscle reinnervation (TMR) [6]. After amputation of the upper arm, the residue nerve fibers are attached to muscle fibers on the chest, shown schematically in Fig. 13.5. Those nerve endings act as a phantom of the real arm. When deciding to lift the nonexisting arm, a corresponding muscle on the chest will contract. With the help of electromyography (EMG), explained in Section 6.9, the electrical potential generated by muscle cells can be detected with electrodes on the skin. This EMG signal is amplified and processed to drive tiny motors that activate arm or finger motion. But first the EMG signals on the chest must be mapped out and uniquely correlated with brain activities for specific motions: lifting the arm, turning a hand, grabbing an object, etc. This requires intensive training. Then the EMG signal can be used for myoelectrical control of motors that actually execute the intended motion. Myoelectric hand prostheses with many degrees of freedom and very good mechanical capabilities are now commercially available: elbows that flex and extend with muscle signals so the arm can be extended to reach for a beverage and bring it up to the lips; wrists that bend and rotate, allowing to position objects for convenient viewing and handling; hands that can carry a suitcase or hold an egg without cracking it; thumbs that can change orientation to multiple hand positions. If the upper arm still exists, the EMG signal can be taken directly from nerve endings in the upper arm. This dramatically facilitates the training process of the prosthesis.

EMG signals are weak and noisy, and the mapping is not very precise. Alternatively, one may record the transmural potential from nerve fibers directly. However, this would require an invasive attachment of electrodes under the skin with all adverse consequences that this procedure bears, such as inflammation and electrode corrosion. For these reasons, recording EMG signals on the skin remains the preferred procedure.

Despite impressive progress that myoelectric prostheses have seen in recent years, proportional and simultaneous control of many degrees of freedom is presently not available, resulting in swift and precise but unnatural arm movements. Although there is no standard pattern for hand and arm movement, unlike the normal gait, attempts are presently undertaken to recognize patterns and to predict movements of the upper limb that would eventually mimic a more natural course of motion. Commercially available prostheses and scientific/technological developments are changing quickly. The current status is reviewed in [6, 7], and the performance of different commercially available systems is compared and rated.

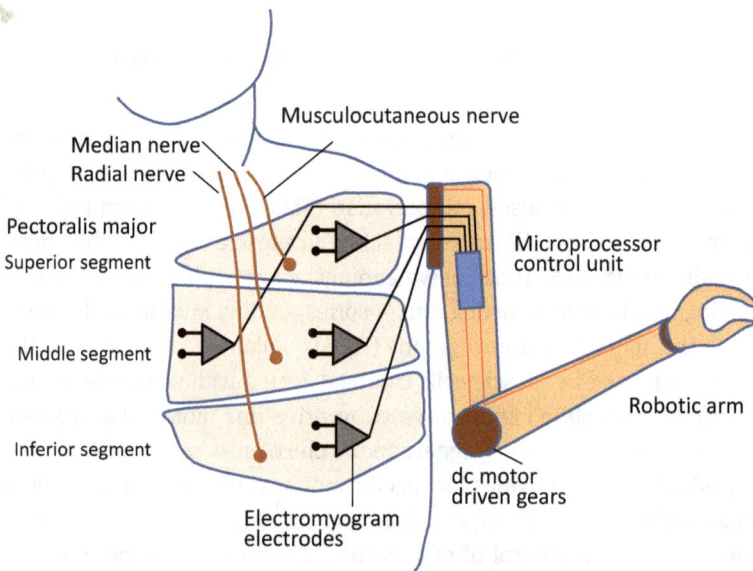

Fig. 13.5: Myoelectric control of upper limb prosthesis, also referred to as targeted muscle reinnervation (TMR). Nerve fibers from the brain, formerly ending in the arm, are connected to different locations of muscles on the chest. When nerve fibers from the brain activate a muscle, the corresponding EMG signal is detected with sensors on the skin. These signals are microprocessed and fed into a dc motor control for specific movements of the robotic arm.

The opposite neural direction, i.e., transferring signals from the prosthetic hand to the brain has also been implemented in some cases. Sensors for temperature, pressure, surface topography, vibrations, etc. have been tested and their electrical signal output was transferred to nerve fibers that are ultimately processed and interpreted by the brain. This reverse signal transfer of sensory information provides limited feedback, which is important for improving the dexterity of the hand and finger movement. There is still a long way to go. The main developments that can be foreseen for the future are incorporating pattern recognition and fast feedback systems. A concise review on these developments is given in [8].

Research and development of myoelectric control of upper limb prostheses have also been beneficial to paraplegia patients with spinal cord injuries (SCI). In the case of SCI, the limbs are still present, but the voluntary motor control below the level of spinal cord lesion is cut off. One recent study showed that intracortical-recorded signals from the motor cortex via a microelectrode array could be linked in real time to muscle activation to restore partial movement of a paralyzed patient [9]. Another recent study demonstrated that long-term training with a body-machine interface, spinal cord injured paraplegics could improve somatic sensation and regain voluntary motor control in muscles below the lesion [10]. In this unique study, gait training with a brain-controlled and EEG-monitored robotic treadmill could reactivate partial motor control. This neurological recovery could hint at a hitherto unknown cortical and spinal cord plasticity triggered by long-term training.

13.3 Hip replacement

As an example of bone fracture and repair, we discuss the femur fracture at the neck and hip replacement, one of the most common procedures.

13.3.1 Anatomical basics

The joint between the femur and pelvis is a ball and socket joint (see Fig. 2.10), providing three rotational degrees of freedom. Figure 13.6(a) shows a cross section of the femur and hip. The femur's hemispherical head (ball) fits into the socket in the pelvic bone, called the acetabulum. The bone's surfaces of both ball and socket are covered with articular cartilage, a smooth, slippery substance that protects and cushions the bones and enables them to move easily. In addition, the surface of the joint is covered by a thin lining called the synovium. In a healthy hip, the synovium produces a small amount of fluid, the synovial fluid that lubricates the cartilage and reduces friction. The cartilage and lubrication is not unique to the hip joint but is typical for all synovial body joints.

Several ligaments firmly connect the ball and socket joint of the hip. Not shown in the cross section of Fig. 13.6(a) is a small centered depression (pit) in the head called the fovea capitis. The fovea capitis contains a central ligament that internally connects the femoral head to the acetabulum. Three other ligaments (iliofemoral, ischiofemoral, and pubofemoral) externally reinforce the strong connection between the femur and pelvic bones. With a tensile force of about 3500 N, the iliofemoral ligament is the strongest ligament in the body [11]. With a conventional hip replacement, these ligaments have to be cut open. Most "hip fractures" are actually fractures of the femur at the neck position, indicated by a red line in Fig. 13.6(a). This is also the limit for shortening the femur before inserting a metal stem during hip surgery, as seen in image (b).

The replacement of the pivot head only is called *hemi endoprosthesis* (HEP); the replacement of both head and acetabulum is known as *total endoprosthesis* (TEP).

There are three main reasons which necessitate HEP or TEP: (1) osteoarthritis leading to a breakdown of the cartilage that covers both sides of the hip joint (60%); (2) fractures of the femoral neck due to osteoporosis (30%); (3) fractures due to various accidents (8%). Other reasons are bone tumors, rheumatoid arthritis, and osteonecrosis (2%). The percentages given are rough estimates. Osteoarthritis is by far the most prevailing reason for hip replacement [12].

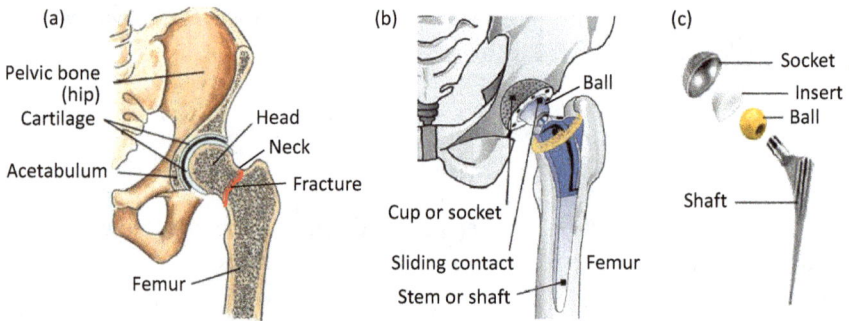

Fig. 13.6: (a) Cross section of the femur and pelvic bone. The head of the femur is anchored in the acetabulum, both forming a ball and socket joint for maximum mobility. The joint is kept together by strong ligaments, not shown in the cross section. The red line indicates the position where hip fracture usually occurs; (b) total endoprosthesis involves a replacement of the socket in the acetabulum and the head of the femur. Ball and shaft are drawn as inserted into the femur; (c) parts of a full prosthesis are shown, consisting of a socket, ball, and a shaft. The details depend on the supplier. The head and the socket may be fabricated of metal, ceramics, plastics, or material combinations. In some designs, an insert between ball and socket is used. The shaft usually is made of a sintered titanium alloy.

The first hip replacement was performed by Gluck[2] in 1891, who used ivory to substitute the ball. More hip replacements followed during the next years. But the number of replacements did not increase dramatically until the early 1960s. According to a recent study in the USA, the number of TEPs more than doubled over a period of 10 years from 2000 to 2010 [13]. It currently has a record high of 0.2–0.3% of the population undergoing HEP or TEP procedures annually. The prevalence of TEP increases with age and reaches around 10% in people over 80 years of age. What is more surprising is the fact that the TEP also increased by over 100% in the 45–64

---

2 Themistocles Gluck (1853–1942), German physician and surgeon.

age group over the same period. There are currently more than 4.5 million people in the United States with an artificial hip. This large number clarifies that, despite arthritis, a significant portion of the population can remain mobile with a hip replacement. This success is due to a multidisciplinary effort of physicists, materials scientists, engineers, and clinicians to develop and improve suitable procedures and materials.

> The three most frequent reasons for hip replacement are: (1) osteoarthritis at the hip joint; (2) fracture at femoral neck; and (3) accidents affecting the femoral joint.

13.3.2 Replacement procedures

According to Fig. 13.6(c), the implant for a TEP consists of a shaft, ball, and socket, and sometimes also an extra insert between the ball and socket. Hip replacement requires careful surgery as many muscles, vessels, and nerve fibers intercept the working space. The shaft is a long and bulky mechanical piece that requires careful insertion into the femur. Implants are not custom-made, but the best fitting one is chosen from a selection of sizes. With x-ray projection radiography, the required implant size is roughly estimated. Additional computer simulation aids the procedure by indicating, in particular, the position of the incision for access to the hip joint.

Infobox I: Procedures for hip replacement

After opening, the existing hip joint is completely removed. The natural socket is hollowed out and cleaned from the remaining cartilage. Then a socket is inserted into this hollow space, eventually cemented into the pelvis. The upper part of the femur, including the head, is removed with a swing saw and a hole is drilled and rasped into the femur for inserting the shaft of the artificial hip stem. One end of the stem is inserted into the trabecular bone of the femur; the other end is topped with a smooth ball to fit into the socket. Ball and shaft allow for a rough length adjustment via a conical fit inside the ball head. In the first test the required length is estimated, and later the ball is press-fitted onto the shaft without further adjustment or possibilities for correction. Once the ball and shaft are tightly attached and inserted into the femur, ball and socket are pressed into place and the incision is closed. There are standard and minimal invasive procedures offered for hip replacements. Minimal invasive procedures conserve all muscles and tendons, such that rehabilitation time of the patient is much shortened compared to the standard procedure.

13.3.3 Material considerations

Although hip replacement has become a standard and, in most cases, a very successful procedure, many issues call for attention. The success of the implant depends on the biocompatibility of all parts used, on a low wear rate between socket and ball, and on a biomechanical force distribution from the implant to the surrounding bone under dynamic load (walking, running, etc.), without causing permanent plastic deformation or strain on the femur. Some of the issues are discussed as follows.

13.3.3.1 Cemented versus uncemented parts

If cementation is used (Fig. 13.7), parts are secured to the healthy bone by means of acrylic cement. Uncemented parts are made from material that features a rough surface allowing bone to grow onto it and holding it in place. Often it is, *per se,* not obvious which procedure is the better option, and the decision may change during surgical procedures. In case of impaired bone quality, cementation is preferred. Some suppliers offer prostheses with intraoperative flexibility, changing from cementless to cemented implantation without having to change instrumentation. Cementless bone-implant interfaces have the chance to grow a cement line naturally and a bone precursor substance, ensuring bone-like mineralization at the interface [14].

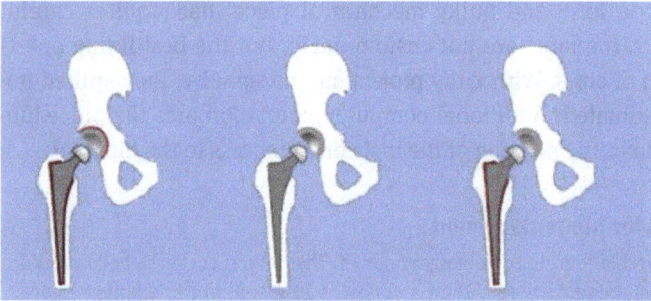

Fig. 13.7: Cemented, uncemented, and partial (hybrid) cemented implants. Cemented parts are outlined in red color.

13.3.3.2 Hip stem

The task of the hip stem is to absorb the load from the socket via the ball and to distribute the load homogeneously to the femur. By using finite element analysis, computer simulations help optimize the shaft's shape and stress distribution [15–17]. The implant may continue to settle after implantation and after receiving static and dynamic loads, which can alter both stress distribution and final leg length. The aim is to achieve as natural a stress distribution as possible in the femur, which corresponds to that before the implantation. Therefore, an optimal implant should provide a bone-loading pattern that closely replicates the preoperative physiological loading condition.

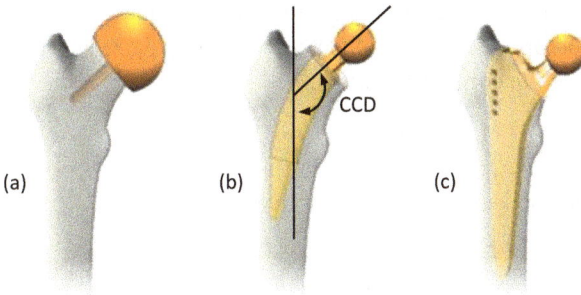

Fig. 13.8: Different shapes of hip stems: (a) ball replacement only;
(b) short shaft; (c) long shaft. The CCD angle is indicated.

13.3.3.3 Shafts

Shafts are distinguished by different neck inclination angles (CCD–angle, usually chosen between 125° and 135°), by different lengths (short or long shafts), by different shapes and surface coatings, and by different materials (Fig. 13.8). In the past, long shafts were used. Presently there is a tendency to insert shorter shafts. Shorter shafts allow replacement when the original one has worn out after some 15–20 years. Short shafts are therefore preferred for younger patients. An x-ray radiograph comparing short and long shaft implants in the same patient is shown in Fig. 13.9 [18]. Shafts are usually made of Co–Cr–Mo alloys, pure Ti, or Ti_6Al_4V alloys. All these metals and alloys fulfill requirements concerning corrosion resistance, durability, and biocompatibility. The tensile strength exceeds 1000 MPa, which is twice the maximum load ever expected in the femur. The alloy Ti_6Al_7Nb is mainly used for forging shafts that are used without cementation, as the surface is amenable to regrowth of bone tissue directly onto it, enhancing long-term stability and durability. The blades are often

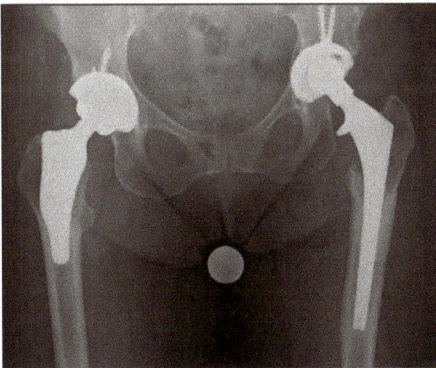

Fig. 13.9: X-ray radiograph of a short and long stem hip implant
(reproduced from Ref. [18] by permission of SLACK, Inc.).

covered by a plasma-spray coating of hydroxyapatite to further promote bone ingrowth, featuring a unique microporosity. This bioactive surfacing agent will ultimately provide strong bonding as the bone grows into it.

13.3.3.4 Material combinations

For the combination of ball and socket (cup) five alternatives are available:
a) metal on plastic (polyethylene or UHMWPE)
b) metal on ceramic
c) metal on metal (MoM)
d) ceramic on plastic (UHMWPE)
e) ceramic on ceramic (CoC)

Here UHMWPE stands for ultra-high molecular weight polyethylene. Metal-on-plastic has the longest tradition and is the most frequently and least expensive bearing [19]. The plastic consists of highly stable and reliable polymer material with greatly reduced signs of wear. Nevertheless, debris of wear eventually diffuses into the body, and as the body attacks them, osteolysis may develop, which necessitates a replacement of the implant. Therefore, lower wear rate is essential and requires smooth surfaces on either side of the bearing. The metal ball is usually Co–Cr–Mo alloy with a highly polished mirror finish that is traditionally used for articulating bearing surfaces in total hip joint replacements. However, a highly polished mirror finish with a less than 20 nm roughness does not necessarily guarantee a low wear rate. A surface finish with a roughness of more than 100 nm shows better wettability of lubricants resulting in a lower wear rate.

Metal-on-metal bearings were developed for lower wear rate and increased mobility because of larger size cups fabricated from metals. MoM bearings were therefore targeted to younger and sport active patients. However, the promise did not hold, and metal particles from wear distributed throughout the body, causing severe problems. Patients with MoM implants were called back to clinics for a survey and replacement. By now, MoM bearings are taken off the market.

Ceramic on plastic is a good combination of two very reliable materials. Ceramic heads are harder than metal heads and are the most scratch-resistant implant material. The hard and ultrasmooth surface can greatly reduce the wear rate on the polyethylene bearing. The potential wear rate for this type of implant is less than for metal on plastic.

Ceramic on ceramic is a combination of two very hard materials. There have been problems with catastrophic shattering and squeaking noises in the past. Both issues have been overcome. CoC is the hardest implant material that can be used in the body and has the lowest wear rate of all to almost immeasurable amounts (1000 times less than metal-on-polyethylene or about 0.0001 mm/year). Consequently, inflammation, bone loss, or systemic distribution of wear debris in the body is not an issue. New ceramic materials offer improved strength and more versatile sizing options. But it is also the most expensive option. The material combinations for hip replacement have been reviewed in [20].

13.3.3.5 Length adjustment
The standard hip prosthesis has only a rough length adjustment capability since a conical press fit joins the ball and shaft. Orthopedic shoe inserts compensate for any residual length difference between the left and right foot. However, it should be possible to improve the length adjustment by measuring the leg length before and after TEP. For this purpose, two LEDs are attached to the leg and by lifting the leg, the center of rotation can be determined, as shown in Fig. 13.10. This procedure is repeated after TEP. Suppose the circles indicate a shifted center of rotation due to a length change. In that case, the distance between ball and shaft can be individually adjusted by a screw connection replacing the traditional press fit [21]. Two LEDs are used instead of a single one to assure that the LEDs have not shifted along the leg during the intervention.

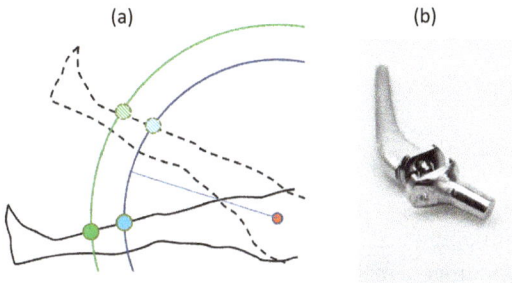

Fig. 13.10: (a) Length measurement of the leg with two LEDs attached to the leg describing a circle by lifting the leg. Measurements are taken before and after hip replacement. If there is a noticeable difference, the length can be changed with an adjustable shaft shown in (b) before closing the incision in the leg (adapted from https://www.fraunhofer.de/en.html).

13.3.3.6 Load sensor
Endoprostheses have been furnished with load sensors in the neck of the implant (Fig. 13.11). The sensors measure the load during different activities, walking, running, jumping, etc. [22]. Using six semiconductor strain gauges, the sensors are sensitive to three force and three moment components. Bluetooth techniques send the recorded data to a computer for analysis. The data may help to improve the design of

implants. Based on these in vivo sensors, it could be confirmed that during walking, the combined forces acting on the hip joint are about 2.5× BW. During slow jogging, the forces increase to about 5× BW. Standing on one leg causes a load nearly the same as for slow walking. When climbing a staircase, the forces in the joint are similar to those during walking, but the torque acting on the implants increases.

Fig. 13.11: Hip shaft with integrated strain sensors in the neck of the implant for measuring forces and torques during various activities (reproduced from [22] by permission of Elsevier Publishing Inc.).

13.3.4 Alternative approaches

Over the years, the materials used for artificial hip replacement have been continuously improved. The surgical instruments were also further developed. Furthermore, the surgical procedures themselves are optimized and assisted by computer simulations and x-ray monitoring. All these improvements follow the basic idea that the hip joint is to be replaced by hard, solid, durable, and biocompatible parts. This is indeed the only solution for femoral fractures. However, alternative approaches should be considered if the bone is still in good condition, but the cartilage is worn down. Researchers have now reported first success in the regrowth of cartilage from the patient's own stem cells. Once grown, the new cartilage has to be inserted into the joint. In the long run, this bioengineered cartilage may obsolete most solid hip replacements for curing osteoarthritis [23, 24].

13.4 Knee replacement

Knee replacement is a procedure similar to a hip replacement. The reasons for knee replacement are the same, i.e., predominantly osteoarthritis. The replacement rate is, however, twice as high. Also for knee replacements, we distinguish between partial and total knee prosthesis. Despite all similarities, there is one important difference. Knee joints are individual parts of the body, slightly different for each person. Standardized parts such as hip replacement do not work. Therefore, knee replacements require careful planning with the support of MRI and CT imaging, which is fed into a simulation program to find the optimal shape for a knee joint that is then individually fabricated. An uptodate overview of total knee arthroplasty techniques is given in [25].

The knee is the largest joint in the body (Fig. 13.12). It is made up of the lower end of the thigh bone (femur), the upper end of the shinbone (tibia), and the kneecap (patella). The point of contact of these three bones where they touch is covered with articular cartilage.

The knee joint is the so-called modified hinge joint (see Fig. 2.10) that allows a monoaxial extension and a limited rotation in the flexed position. Flexion occurs when the tibia swings into a bent position and moves toward the back of the thigh. Normal knee flexion is about 135° and is only limited physically by touching the calf and leg muscles in the back. Transfemoral prostheses are strictly monoaxial without much loss of mobility.

The menisci of the knee are located between the femur and tibia. These C-shaped menisci wedges act as "shock absorbers" cushioning the joint. Large ligaments hold the femur and tibia together and provide stability (Fig. 13.12). Those are, in particular, the lateral collateral ligaments (LCL) and the anterior and posterior cruciate ligaments (ACL and PCL). ACL and PCL prevent the tibia and femur from sliding forward, and LCLs control sideways movement. All remaining surfaces of the knee are covered by a thin lining, the synovial membrane. Similar to the hip joint, this membrane releases a fluid that lubricates the cartilage reducing friction to nearly zero in a normal knee.

The knee replacement, also called knee arthroplasty, is more accurately described as a knee "resurfacing" as only the bone surface is refinished rather than replaced. The different steps of a knee replacement surgery can be seen in a video, the link is provided under useful web pages.

During surgery, the ACL is removed. It is no longer needed because the design of the artificial implant provides the stability of the knee that the ACL guaranteed before. The PCL is likely removed during a total knee replacement, although it may sometime remain. The design of the artificial knee replacement prevents the femur from sliding. Thus, the PCL is not necessary. The LCL ligaments remain in the knee, providing further stability to the artificial knee replacement.

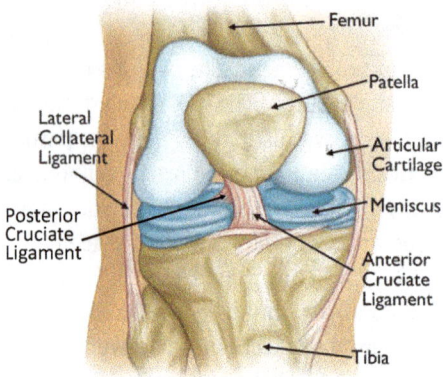

Fig. 13.12: Anatomy of the normal knee joining the lower part of the femur with the upper end of the tibia and the patella in front. The meniscus serves as shock absorber and the ends of the three bones, where they touch each other, are covered with articular cartilage that allows their relative movement. Strong ligaments, in particular, the cruciate ligament, keep the knee joint together (reproduced by permission of OrthoInfo. © American Academy of Orthopaedic Surgeons. http://orthoinfo.aaos.org).

Further details depend on whether a full or partial replacement is required; both versions are shown schematically in Fig. 13.13. In a partial replacement, only one knee side (one compartment) is replaced, preserving all ligaments. The material combination for knee implants is always metal on plastic. For the metal implant either Co–Cr–Mo or Ti_6Al_4V alloys are chosen, the plastic spacer is made of UHMWPE.

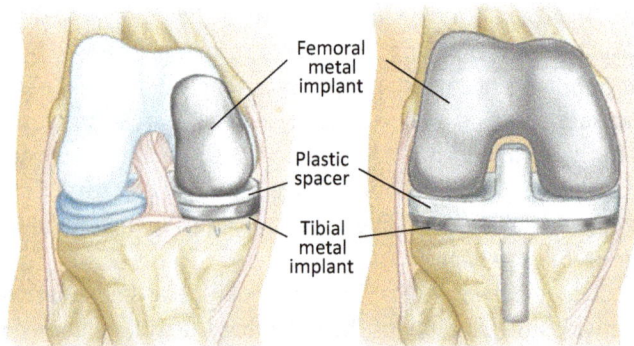

Fig. 13.13: Partial (left) and total (right) replacement of the knee. Between femoral and tibial components is a plastic spacer. The plastic spacer takes over the function of the meniscus (reproduced by permission of OrthoInfo. © American Academy of Orthopaedic Surgeons. http://orthoinfo.aaos.org).

13.5 Cardiovascular stents

Cardiovascular endoprostheses include artificial heart valves, cardiac pacemakers, and stents. In the following, we restrict the discussion to stents; pacemakers are discussed in Section 7.8.

Cardiovascular stents are used as vessel scaffolding to alleviate arterial narrowing, stenosis, and restoring blood flow to distal vasculatures. The procedure of placing a stent in an artery is termed *percutaneous coronary intervention* (PCI), also known as *coronary angioplasty*. PCI is a minimal invasive alternative to open-heart surgery for setting a bypass. The origin of the name "stent" is not clear. On the one hand, stent is an old English word for pillar or brace; on the other hand, the name may originate from Stent,[3] who tested materials for dental casts. Stents are small hollow and mashed tubes made from metals or synthetic fibers that can be slid into constricted arteries to restore the original blood flow. Stents were primarily used to treat coronary artery stenosis to prevent myocardial infarction. Currently, stents are also used to treat peripheral arteries, such as the carotid, renal, and femoral arteries. Furthermore, stents are inserted to widen the urinary tract, the biliary tract, lacrimal duct, and to increase the drainage of aqueous humor to lower the intraocular pressure (see Fig. 11.18). Reference [26] provides an overview on the present status.

Figure 13.14(a–h) shows the main steps of a PCI procedure. Catheterization can be used to locate the area of obstruction after a cardiac exam reveals reduced blood flow and insufficient oxygen supply to the myocardium. In this procedure, a catheter (plastic tube) is inserted into the patient's groin or wrist and guided up to the heart (a). Once the catheter has reached the heart, an iodine-containing dye is injected into the coronary arteries and the obstruction can be visualized with the aid of x-rays (b). Only obstructions of 70% and more are suitable for stent implantation. Once such a plaque has been identified, a metal guide-wire is inserted into the catheter (c) and threaded through the plaque (d). The plastic catheter is then pulled out again and replaced by another stent delivery catheter, wrapped at the proximal end with an inflatable balloon surrounded by a stent (e). The balloon is inflated once the stent is positioned at the plaque site using the guide wire (f). The stent expands plastically, keeping the blocked artery open. The balloon is then deflated again, and the catheter and guide wire is pulled out (g), while the stent remains expanded (h). The entire process is monitored by real-time x-ray fluoroscopy (see Section 8.4.2/Vol. 2). The stent delivery catheter has a radio-opaque tip for locating its position, while the guiding metal wire, although very thin, still has intrinsic radio-opacity. The stent delivery and especially the balloon expansion phase should be kept as short as possible as this phase completely blocks blood flow. On the other hand, insufficient pressure on the balloon will result in a premature collapse of the stent. Therefore a compromise has to be found.

---

3 Charles Stent (1807–1885), British dentist.

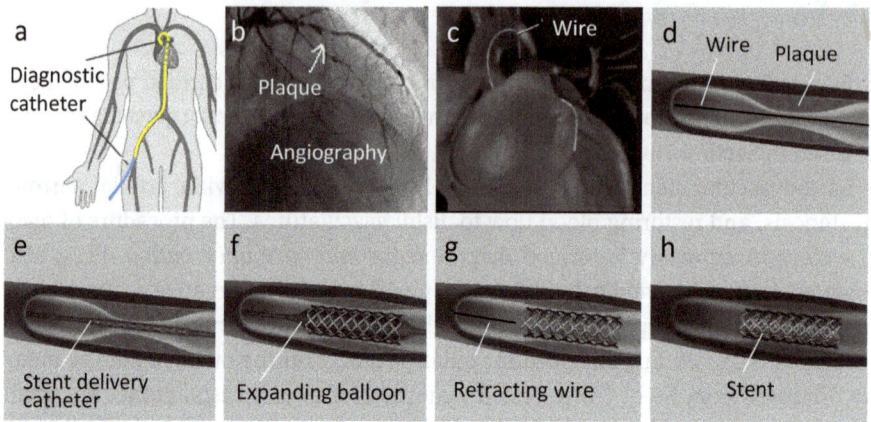

Fig. 13.14: Sequence of procedures for implanting a stent into a cardiovascular artery: (a) a catheter is inserted through the groin and pushed up to the heart; (b) after the catheter has reached the heart, an iodine-containing dye is spread through the catheter into the vasculature to identify obstructions by angiography; (c) once a plaque has been identified, a thin guiding wire is fed through the catheter; (d) the guiding wire is fed onward through the plaque; (e) while keeping the guiding wire in place, the original catheter is pulled back out and replaced by another catheter that is equipped with a balloon and stent on its proximal end; (f) once the stent is positioned at the location of the plaque, the balloon is inflated through the sheathed catheter. The stent expands, presses against the vessel wall, and opens the plaque; (g) the stent is now expanded and plastically stabilized. The balloon is deflated again; the catheter and guiding wire are retracted; (h) the expanded stent remains at the deployed position and keeps the plaque open.

While PCI appears straightforward, several aspects warrant closer scrutiny. These concern material choice, biocompatibility, mesh design of the scaffold, the use of drug-releasing stents, the plastic expansion behavior of the stent, the elastic recoil, and the change in local hemodynamics, just to name a few topics. There are many other physiological aspects to consider, such as restenosis and stent thrombosis, which are not considered here.

Four main types of stents are commonly used for PCI:
1. Balloon-expandable stents
2. Self-expandable stents
3. Shape memory alloy stents
4. Biodegradable polymer stents

Balloon-expandable stents deform plastically beyond the material's elastic limit and are relatively rigid at their expanded diameter. Elastic recoil occurs infrequently and is hindered by the shape of the mesh and large angle deformation of the struts during expansion. As the stent expands, it also shortens, which must be considered when determining the required length for the plaque. Bare metal stents (BMSs) are usually made of CoCr or PtCr alloys, which are considered biocompatible. The mesh is fabricated either by laser cutting thin hollow metal tubes or by weaving metal wires. Various manufacturers offer many mesh patterns, and it is not clear which pattern is the best. One example is shown in Fig. 13.15. The thinner the struts and the fewer there are, the less the hemodynamics of the blood flow is affected.

On the other hand, thinning the struts has a limitation in the plastic stability of the stents and minimal elastic recoil. Furthermore, x-ray opacity becomes a problem with decreasing thickness. The thinnest struts on the market are 74 μm thick and made of PtCr alloy. The hemodynamics is potentially affected by stents in a twofold way: first, erythrocytes in contact with foreign materials such as metals tend to cause clotting. And second, in the area of the stent, the blood vessel is not compliant to pulsatile pressure changes, which may cause turbulent flow. An upto-date review on BMS stents can be found in Ref. [27].

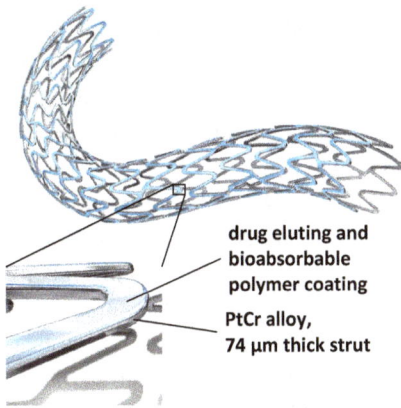

drug eluting and
bioabsorbable
polymer coating

PtCr alloy,
74 μm thick strut

Fig. 13.15: Stent made of a 74 μm thick PtCr–alloy strut providing high flexibility and coated with a 4 μm thick bioabsorbable polymer for drug delivery on the luminal side. This "high-tech" stent has recently been released for clinical use (image provided with courtesy of Boston Scientific. © 2017 Boston Scientific Corporation or its affiliates. All rights reserved).

Self-expandable stents store potential energy in a reduced spring-like fashion that opens up when released. Self-expandable stents tend to be more compliant to varying blood flow and blood pressure than balloon-expandable stents. Moreover, they are retrievable and repositionable, unlike expandable balloon stents. Otherwise, they are similar in material choice and mesh design.

Shape memory stents are made from NiTi alloys [28]. Their physical principle is based on a martensitic or structural phase transformation that changes the macroscopic shape of the metal without any internal atomic diffusion (Fig. 13.16). The alloy can be deformed to an arbitrary shape at low temperatures. However, it always flips back to the original shape when heated above the martensitic phase transition temperature [29]. This property is used in nitinol stents. The action is similar to elastically preloaded and self-expandable stents. But the release mechanism is different. The transformation temperature is set to about 30 °C. When delivered to the targeted area, premature expansion during delivery is constrained by a retractable sheath. Once released, the nitinol stent then conforms to the vessel wall at body temperature.

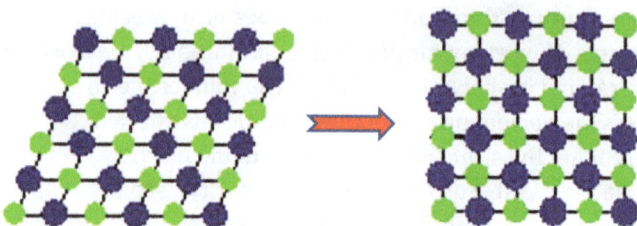

Fig. 13.16: Shape memory NiTi alloy. Left panel: deformed low-temperature crystal structure. Right panel: structural transformation to the intrinsic crystal structure after heating the alloy above the martensitic phase transition temperature.

Often BMS are covered with a biodegradable polymer coating containing drugs that prevent cell proliferation, called drug-eluting stents. The antiproliferative drugs are released slowly over time to help prevent tissue regrowth that can reclog the artery.

In addition to the aforementioned three types of metal stents, biodegradable or bioresorbable polymer stents (BPS) are also occasionally used. Bioresorbable polymer stents are being developed to improve metal stent's biocompatibility and drug reservoir capacity. In addition, they offer a temporary alternative to permanent metallic stent implants. In most cases, polyhydroxybutyrate is used for this. After about three months, the clogged artery has stabilized to the extent that the stent's scaffolding is no longer needed, but BMS remains in the body permanently while BPS dissolves. In this regard, BPS has an etch over BMS [30]. More uptodate information can be found in [31].

Stents are not only used to support the cardiovascular arteries. Other applications are in ophthalmology to reduce intraocular pressure (see Section 11.3.7), to overcome urethral strictures, and to open bronchi in the lungs.

13.6 Retinal implants

Retinal implants aim to restore some visibility that enhances blind people's visual recognition and unguided mobility, which significantly increases their quality of life. Retinal implants require full or partial functioning of the retina and in particular of the ganglion cells and the optic nerve fibers. We do not consider alternative strategies that directly target the optic nerve system, the dorsal lateral geniculate nucleus, or the visual cortex. Retinal implants are orders of magnitude more complex than hearing aids and auditory implants, and consequentially their present status is promising but not yet a breakthrough. Recent overviews on the state-of-the-art can be found in Refs. [32–35].

Retinal implants are classified according to their location as epiretinal or subretinal implants (Fig. 13.17). Epiretinal implants are located on the retina and directly interface with the ganglion cell layer beneath. Subretinal implants are positioned below the ganglion and bipolar cell layers in place of the photoreceptors (see Fig. 11.19). In both cases, the implants are pixelated solid-state device arrays that stimulate either ganglion cells or bipolar cells by changing electrical potentials in response to a light stimulus. Any single diode or electrode within an array is bigger than typical retinal cell size, and therefore stimulation proceeds via extracellular potential changes.

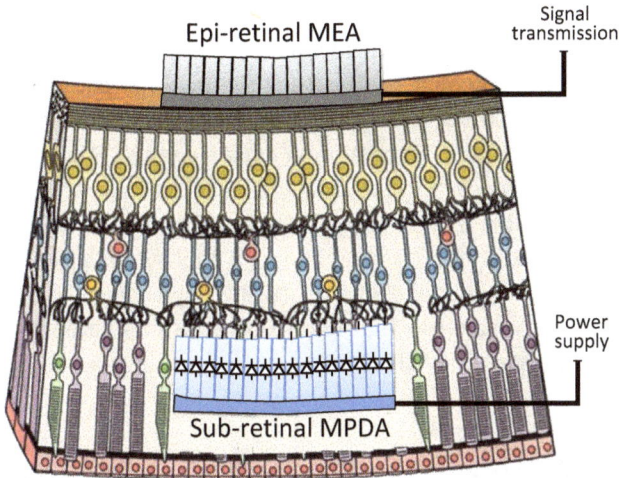

Fig. 13.17: Cross section through the retina showing the positions of retinal implants. Epiretinal implants are placed on top of the retina. Subretinal implants are at the location of the receptors. MEA, microelectrode array; MPDA, microphotodiode array.

In the following, we briefly discuss two distinct types of implants.

13.6.1 Epiretinal microelectrode array

In the case of epiretinal implants, the scattered light from the outside world that carries the information on surrounding objects is detected by a digital camera mounted into a pair of glasses. Hence, the camera lens bypasses the natural lens, which may still function. The optical pathway of the image and the parts required are shown in Fig. 13.18. The camera video images are processed and pixel-wise transferred wirelessly to a receiver, feeding an electronic case implanted on the outside of the eyeball. An epiretinal MEA is tacked onto the retina. Wires connect the electronic case on the eyeball and epi-MEA. The epi-MEA stimulates ganglion cells according to the processed images from the video processor. If the ganglion cells respond to electrode potential changes, they will fire an action potential that can be received in the visual cortex. This method can restore some visibility, albeit over a rather small visual field of about 20° combined with low spatial resolution and low dynamic range. The patient carries a power supply for the camera and wireless transmission in a bag. The person carrying the visual system has to turn the camera by moving the head instead of turning the eyes to focus on an object.

Although the optic pathway appears cumbersome, it is presently the only successful implant on the market, with about 500 implants worldwide. This general concept is followed and developed by three research-industrial consortia; the version Argus II of Second Sight Medical Products Inc. is shown in Fig. 11.50. It contains an MEA of 10×6 dots. Patients were screened with an assessment instrument that requires the fulfillment of a number of tasks. The conclusion from 26 participating patients was a "significantly improved completion of vision-related tasks with the device ON versus OFF" [36], i.e., with a binary on–off dynamic range. Similarly, the Argus II retinal prosthesis system was used to study whether patients can read a dot array normally used for tactile braille reading [37]. Instead of reading a 3×2 dot array through the camera, the dots directly stimulate a 3×2 subset of the 10×6 MEA for visual perception of individual braille letters. The high percentage of correctly read letters (more than 80%) suggests that the text can be stimulated and read as visual braille in retinal prosthesis by blind patients.

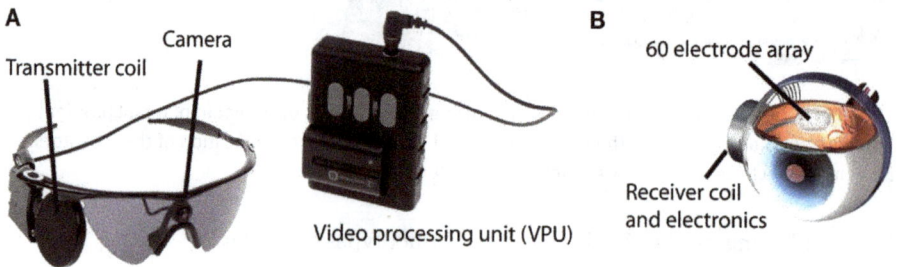

A
Camera
Transmitter coil
Video processing unit (VPU)

B
60 electrode array
Receiver coil and electronics

Fig. 13.18: Epiretinal implant of the Argus II type uses a camera incorporated in a pair of glasses. After processing the video images (VPU), they are sent wirelessly to a receiver implanted on the eyeball. From there, wires transfer the signal to a microelectrode array, tacked to the retina (reproduced from Ref. [33], © creative commons).

13.6.2 Subretinal micro-photodiode array

In contrast to epiretinal MEA, the subretinal implant uses the light that enters through the natural lens. The light is detected by a photodiode array on a chip that is implanted in the subretinal space. Each photodiode has attached a tiny electrode for stimulating ganglion cells proportional to the received signal amplitude. An intraocular wire connects the power supply with the microphotodiode array (MPDA). The advantage of this approach is a natural use of movement of the eyes so that an elaborate image processor becomes dispensable. The signals from the photodiodes are directly transferred to the inner nuclear layer expecting that the signal processing capabilities of the retinal network can at least partially be used for creating a signal output resembling the normal physiological one. The images observed by patients are still too crude to judge whether signal processing takes place in the retina.

The disadvantage of the subretinal integrated circuitry is a power supply for the photodiodes from the outside that generates heat. In one design, however, the power is solely delivered by light. Another drawback of MPDA is the sensitivity of the optic chip, which even in the best version, is much reduced in comparison to the sensitivity of a normal, healthy eye. Hence a powerful amplifier is mandatory for each photodiode, and higher light intensity is needed to achieve limited visibility. A hybrid system may overcome the mentioned problems: first, use of a camera for primary image recording; second, conversion of the visible image to the infrared regime and emission at much higher intensity. The infrared radiation goes through the eye and is then detected by a photodiode array in the subretinal space. Care must be taken that the high infrared intensity does not heat up the retina, causing damage. As in the previous case, the subretinal MPDA implant only works as long as ganglion cells and optic nerve fibers remain intact. This may not be the case in the late stage of glaucoma.

The realization of various design concepts may differ in pixel size, pixel density, and the total number of pixels. Furthermore, there are differences in the materials used. Chips have been fabricated using Si-technology, conjugated with light-sensitive polymers, or carbon nanotubes [38]. Also, microarrays may be patterned on flexible polymer sheets that conform better to the concave inner shape of the eyeball. Conformity becomes an issue when increasing the lateral array size for a larger visual field. In all cases, the biocompatibility of all components in contact with tissue is mandatory. This may require an encapsulation of the chip into a biocompatible envelope.

Whatever the design and choice of materials, retinal implants face a tremendous challenge concerning resolution, contrast, sensitivity, dynamic range, visual field, and visual acuity. Progress is slow, and the performance of retinal implants may not improve dramatically in the near future. Therefore, different strategies have been suggested, such as confocal image recording for better foreground–background recognition or enhanced image processing to simplify complex and cluttered images by emphasizing essential objects in the foreground [39]. An overview of the present status of microelectrode fabrication for retinal neural recording is presented in [40].

In this short chapter we have given an overview of the most important procedures in prostheses and implants. There are many other applications such as dental implants or full dentures, breast implants, heart valves, stents for the uterus, bile duct, bronchi, and blood vessels. Due to the rapid development of microelectronics, nanomedicine, and biotechnology, further areas of application for prostheses and implants can be expected in the future.

13.7 Summary

S13.1   Exoprostheses are prostheses external to the body; endoprostheses are prostheses within the body, also called implants.

S13.2   We distinguish between transtibial prostheses, i.e., protheses below the knee and transfemoral prostheses above the knee.

S13.3   Microprocessor-controlled transfemoral prostheses are called C-legs.

S13.4   Myoelectric prosthesis use remaining nerve fibers, which run from the brain to the arm. Myoelectrical control of upper limb prosthesis is referred to as TMR.

S13.5   There are three main reasons which necessitate hip replacement: (1) osteoarthritis leading to a breakdown of the cartilage that covers both sides of the hip joint; (2) fractures of the femoral neck due to osteoporosis; (3) fractures due to various accidents.

S13.6   Hip replacements require consideration of shaft, stem, ball, and socket. Important is the material combination, which determines wear and lifetime.

S13.7   Knee replacement is better described as knee "resurfacing" as only the surface of the bones is actually refinished rather than replaced.

S13.8   Cardiovascular stents are used as vessel scaffolding to alleviate an arterial narrowing, called stenosis, and to restore blood flow to distal vasculatures.

S13.9   The procedure of placing a stent in an artery is termed PCI.

S13.10   Four main versions of stents are commonly used for PCI: (1) balloon expandable stents; (2) self-expandable stents; (3) shape memory alloy stents; (4) biodegradable polymer stents.

S13.11   Retinal implants can be classified according to their location as epiretinal or subretinal implants.

S13.12   Retinal implants provide limited visibility to blind patients with a functional retina.

? Questions

Q13.1   What is the discipline called which provides support of the musculoskeletal system but not a replacement?

Q13.2   What are the main considerations for the construction of a computerized leg (C-leg)?

Q13.3   What is the challenge of the myoelectric control of the upper limbs?

Q13.4   What is the best material combination in the case of total hip replacement?

Q13.5   Stents are used for the biliary tract, lacrimal duct, and for increasing the drainage of aqueous humor to lower the intraocular pressure. Where are these tracts and ducts located in the body?

Q13.6   A stent design should fulfill which physiological and mechanical requirements?

Q13.7   What type of joint is the knee and which movement does it allow?

Q13.8   After replacement, how many degrees of freedom does the knee have?
Q13.9   Retinal implants are distinguished according to their location. Which ones are offered?
Q13.10   Which group of patients can be helped with retinal implants?

Attained competence	+	0	−	⚡
I know what prostheses are good for				
I can distinguish between transtibial and transfemoral prostheses				
I know what a C-leg is				
I know how to control the movement of a myoelectric prosthesis				
I can name the three most frequent reasons for hip replacement				
I know which material combinations are used for socket and ball				
I know what resurfacing implies for knee replacement				
I know what stents are good for and in which parts of the body they are used				
I have an idea of how stents are inserted into the cardiovascular artery				
I appreciate the potential problems that stents induce				
I know which materials are favored for stents				
I can distinguish between epiretinal and subretinal implants				

Suggestions for home experiments

HE13.1   Design your own stent via the open source software package. The SimVascular application enables users to model vessels, stents, and simulate blood flow. https://simtk.org/frs/?group_id=188

ℹ **Exercises**

E13.1 **Biocompatibility of alloys:** In angioplasty and endoprosthetis, materials choice and bio-compatibility are very important issues. Explain what
(a) biocompatibility implies,
(b) what binary and ternary alloys are, and
(c) why PtCr and CoCr alloys are considered biocompatible.

E13.2 **Stents:** Stents are short, flexible tubes that support cardiovascular pulsetile flow. Stents must be mechanically robust while reversibly tolerant of bending and compression.

We want to determine how a stent in the cardiovascular artery has to change in response to the arterial pulse wave amplitude of $\Delta p = 100$ hPa. The elasic modulus of the artery is $Y = 20$kPa.

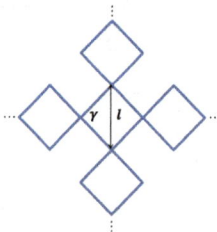

We assume the stent to be fabricated of a symmetric square mesh at rest and angle $\gamma = 90°$. The artery has an inside diameter of 3 mm. Ten squares close the inside circle.

What is the relative angular change $\Delta\gamma/\gamma$ and length change $\Delta l/l$ at the maximum amplitude of the pulse wave?

References

[1] Windrich M, Grimmer M, Christ O, Rinderknecht S, Beckerle P. Active lower limb prosthetics: A systematic review of design issues and solutions. Biomed Eng Online. 2016; 15(Suppl 3):140.

[2] Kampas P, Seifert D. Das neue C-Leg: Neue Funktionen und neue technologie. Orthopädie Technik. 2015; 66: 46–49.

[3] Wolf EJ, Everding VQ, Linberg AL, Schnall BL, Czerniecki JM, Gambel JM. Assessment of transfemoral amputees using C-leg and power knee for ascending and descending inclines and steps. J Rehabil Res Dev. 2012; 49: 831–842.

[4] Seymour R1, Engbretson B, Kott K, Ordway N, Brooks G, Crannell J, Hickernell E, Wheeler K. Comparison between the C-leg microprocessor-controlled prosthetic knee and non-microprocessor control prosthetic knees: A preliminary study of energy expenditure, obstacle course performance, and quality of life survey. Prosthet Orthot Int. 2007; 31: 51–61.

[5] Fleming A, Stafford N, Huang S, Hu X, Ferris DP, Huang HH. Myoelectric control of robotic lower limb prostheses: A review of electromyography interfaces, control paradigms, challenges and future directions. J Neural Eng. 2021; 18: 041004, p. 1–16.

[6] Atzori M, Müller H. Control capabilities of myoelectric robotic prostheses by hand amputees: A scientific research and market overview. Front Syst Neurosci. 2015; 9: 162.

[7]   Kabir R, Sunny MSH, Ahmed HU, Rahman MH. Hand rehabilitation devices: A comprehensive systematic review. Micromachines (Basel). 2022; 13: 1033.

[8]   Roche AD, Rehbaum H, Farina D, Aszmann OC. Prosthetic myoelectric control strategies: A clinical perspective. Curr Surg Rep. 2014; 2: 44

[9]   Bouton CE, Shaikhouni A, Annetta NV, Bockbrader MA, Friedenberg DA, Nielson DM, Sharma G, Sederberg PB, Glenn BC, Mysiw WJ, Morgan AG, Deogaonkar M, Rezai AR. Restoring cortical control of functional movement in a human with quadriplegia. Nature. 2016; 533: 247–250.

[10] Donati AR, Shokur S, Morya E, Campos DS, Moioli RC, Gitti CM, Augusto PB, Tripodi S, Pires CG, Pereira GA, Brasil FL, Gallo S, Lin AA, Takigami AK, Aratanha MA, Joshi S, Bleuler H, Cheng G, Rudolph A, Nicolelis MA. Long-term training with a brain-machine interface-based gait protocol induces partial neurological recovery in paraplegic patients. Sci Rep. 2016 ; 6: 30383.

[11] Hewitt JD, Glisson RR, Guilak F, Vail TP. The mechanical properties of the human hip capsule ligaments. J Arthroplasty. 2002; 17: 82–89.

[12] Maradit Kremers H, Larson DR, Crowson CS, Kremers WK, Washington RE, Steiner CA, Jiranek WA, Berry DJ. Prevalence of total hip and knee replacement in the United States. J Bone Joint Surg Am. 2015; 97: 1386–1397.

[13] Wolford ML, Palso K, Bercovitz A. NCHS Data Brief, No. 186, February 2015.

[14] Grandfield K. Bone, implants, and their interfaces, Phys Today. 2015; 68: 40–45.

[15] Srimongkol S, Rattanamongkonkul S, Pakapongpun A, Poltem D. Mathematical modeling for stress distribution in total hip arthroplasty. Int J Math Models Methods Appl Sci. 2012; 6: 885–892.

[16] Jasik A, Mroczek M. The influence of mechanical and material factors on the biological adaptation processes of the femoral bone implants. Arch Metall Mater. 2016; 61: 189–194.

[17] de Freitas Spinelli L, de Souza Macedo CA, Galia CR, Rosito R, Schnaid F, Corso LL, Iturrioz I. Femoral stem-bone interface analysis of logical uncemented stem. Brasilian J Biomed Eng. 2012; 28: 238–247.

[18] Patel RM, Lo WM, Cayo MA, Dolan MM, Stulberg SD. Stable, dependable fixation of short-stem femoral implants at 5 years. Orthopedics. 2013; 36: e301–e307.

[19] Fulín P, Šlouf M, Vlková H, Krejčíková S, Kredatusová J, Pokorný D. Comparison of the quality of the most frequently used new UHMWPE articulation inserts of the total hip replacement]. Acta Chir Orthop Traumatol Cech. 2019; 86: 101–109.

[20] Merola M, Affatato S. Materials for hip prostheses: A review of wear and loading considerations. Materials (MDPI Publication, Switzerland) 2019; 12: 495.

[21] BMWi – ZIM Kooperationsnetzwerk 027, www.zim-bmwi.de/.

[22] Damm P, Graichen F, Rohlmann A, Bender A, Bergmann G. Total hip joint prosthesis for in vivo measurement of forces and moments. Med Eng Phys. 2010; 32: 95–100.

[23] Pak J, Lee JH, Kartolo WA, Lee SH. Cartilage regeneration in human with adipose tissue-derived stem cells: Current status in clinical implications. Biomed Res Int. 2016; 2016: 4702674.

[24] Moutos FT, Glass KA, Compton SA, Ross AK, Gersbach CA, Guilak F, Estes BT. Anatomically shaped tissue-engineered cartilage with tunable and inducible anticytokine delivery for biological joint resurfacing. Proc Natl Acad Sci USA. 2016; 113(31): E4513–E4522.

[25] Varacallo M, Luo TD, Johanson NA. Total Knee Arthroplasty Techniques. In: StatPearls [Internet] Treasure Island (FL): StatPearls Publishing; 2022. Available from: https://www.ncbi.nlm.nih.gov/books/NBK499896/

[26] Borhani S, Hassanajili S, Ahmadi Tafti SH, Rabbani S. Cardiovascular stents: Overview, evolution, and next generation. Prog Biomater. 2018; 7: 175–205.

[27] Jorge C, Dubois C. Clinical utility of platinum chromium bare-metal stents in coronary heart disease. Med Devices (Auckl). 2015; 8: 359–367.

[28] Liu Y, Van Humbeeck J, Stalmans R, Delaey L. Some aspects of the properties of NiTi shape memory alloy. J Alloys and Compounds. 1997; 247: 115–121.

[29] Jani JM, Leary M, Subic A, Gibson MA. A review of shape memory alloy research, applications and opportunities. Mater Des. 2014; 56: 1078–1113.

[30] Eberhart RC, Su SH, Nguyen KT, Zilberman M, Tang L, Nelson KD, Frenkel P. Bioresorbable polymeric stents: Current status and future promise. Review J Biomater Sci Polym Ed. 2003; 14: 299–312.

[31] Schmidt T, Abbott JD. Coronary stents: History, design, and construction. J Clin Med. 2018; 7: 126.

[32] Ghezzi D. Retinal prostheses: Progress toward the next generation implants. Front Neurosci. 2015; 9: 290.

[33] Goetz GA, Palanker DV. Electronic approaches to restoration of sight. Rep Prog Phys. 2016; 79: 096701.

[34] Stingl K, Zrenner E. Electronic approaches to restitute vision in patients with neurodegenerative diseases of the retina. Ophthalmic Res. 2013; 50: 215–220.

[35] Ayton LN, Barnes N, Dagnelie G, et al. An update on retinal prostheses. Clin Neurophysiol. 2020; 131: 1383–1398.

[36] Geruschat DR, Richards TP, Arditi A, da Cruz L, Dagnelie G, Dorn JD, Duncan JL, Ho AC, Olmos de Koo LC, Sahel JA, Stanga PE, Thumann G, Wang V, Greenberg RJ. An analysis of observerrated functional vision in patients implanted with the Argus II retinal prosthesis system at three years. Clin Exp Optom. 2016; 99: 227–232.

[37] Lauritzen TZ, Harris J, Mohand-Said S, Sahel JA, Dorn JD, McClure K, Greenberg RJ. Reading visual braille with a retinal prosthesis. Front Neurosci. 2012; 6: 168.

[38] Bareket L, Waiskopf N, Rand D, Lubin G, David-Pur M, Ben-Dov J, Roy S, Eleftheriou C, Sernagor E, Cheshnovsky O, Banin U, Hanein Y. Semiconductor nanorod–carbon nanotube biomimetic films for wire-free photostimulation of blind retinas. Nano Lett. 2014; 14: 6685–6692.

[39] Jung JH, Aloni D, Yitzhaky Y, Peli E. Active confocal imaging for visual prostheses. Vision Res. 2015; 111(Pt B): 182–196.

[40] Ha Y, Yoo HJ, Shin S, Jun SB. Hemispherical microelectrode array for ex vivo retinal neural recording. Micromachines (Basel). 2020; 11: 538, p. 1–16.

Useful website

Video showing schematically a full knee replacement using a dry bone model. https://www.you tube.com/watch?v=lp4eRla1vFg

Appendix

14 Answers to questions

Chapter 1

A1.1   The primary function is to create a local environment that is different from the outside. Cell membranes contain ion channels that control the passage of ions and small molecules from the cytoplasm to the extracellular space and vice versa.

A1.2   Double lipid layers.

A1.3   About 30 trillion (30×10^{12})

A1.4   Cells organize themselves to form tissues, tissues form organs, and organs are assembled into systems.

A1.5   An organ has a specific function such as the lung. Several organs need to work together in order to fulfill a certain body task, like the respiratory system for the gas exchange.

A1.6   Negative feedback systems: Temperature control; filtration rate in the kidneys; water level and pH value. Positive feedback systems: Oxygen uptake in hemoglobin, contraction of outer hair cells in the cochlea.

A1.7   The heart acts as a pump for the circulatory system, activated by self-excitation.

A1.8   No. Heart and lung work independently and have different rhythms.

A1.9   The heart pumps through the right ventricle the exact same amount.

A1.10   Because in oxygen-rich blood the red color is less absorbed and more scattered than in oxygen-poor blood.

A1.11   The heart contains just one muscle, the myocardium.

A1.12   1. Filtering of blood and removal of toxic substances; 2. control of ion concentrations and the pH value; 3. control of water balance, i.e., the osmotic pressure.

A1.13   Deoxygenated blood is bluish, oxygenated blood is reddish.

A1.14   Blood has low absorption in the wavelength band from 600 to 700 nm, and the scattered light is perceived as red.

A1.15   The lung has the purpose of gas exchange, inhaling oxygen and exhaling carbon dioxide.

A1.16   Yes. The oxygen concentration during inspiration is not as high as it could be, if carbon dioxide were not present.

A1.17   Afferent nerve fibers transmit information from the periphery to the CNS, efferent nerve fibers are required to move body parts after processing in the brain or spinal cord.

A1.18   The digestive system encompasses the following organs: oral cavity including teeth and tongue, esophagus, stomach, liver, spleen, gallbladder, pancreas, intestines, colon, and rectum.

https://doi.org/10.1515/9783110756951-014

A1.19   The small intestine together with the colon removes solid waste, and the kidneys remove liquid waste.

A1.20   The body contains sensors for light, sound, temperature, pressure, balance, smell, and taste.

A1.21   Information processing takes place in the central nervous system, i.e., the brain and the spinal cord.

A1.22   The information of the receptors is transmitted to the CNS via nerve fibers.

A1.23   Each extremity has two antagonistic muscles, one for moving up, the other one for moving down, such as the biceps and triceps of the forearm.

A1.24   Melanin and blood (hemoglobin); melanin develops a brownish color upon sunshine, blood can turn the skin color more reddish or more bluish.

A1.25   The skin protects the organs, contains a number of sensors, controls the surface temperature, excretes waste products, and produces vitamins.

A1.26   Ejaculation is a parasympathetic reflex.

A1.27   Inside of the pelvic cavity.

Chapter 2

A2.1   Density of a body in relation to the density of water determines whether it will swim, float, or sink. From a physiological point of view, the proportion of fat can be estimated from the density.

A2.2   Because of higher calcium content.

A2.3   Anterior and posterior, or ventral and dorsal.

A2.4   Superior and inferior, or cranial and caudal.

A2.5   A parabolic trajectory, like throwing a ball.

A2.6   Because levers in the body are mainly constructed for mobility and agility but less so for lifting heavyweight.

A2.7   Only the foot is a lever with a mechanical advantage bigger than 1.

A2.8   For lifting, the biceps starts from a relaxed state. The force that can be developed in the relaxed state is much less than in the partially contracted state.

A2.9   Bones are not a rigid system. Osteoclast cells can under load eliminate old bone structures and build new ones, which are better adapted.

A2.10   Femur (hip) and shoulder (humerus).

A2.11   The knee is a modified hinge joint.

A2.12   Two main phases: stance and swing phase, and up to 16 subphases.

A2.13   Knee flexors: hamstrings, biceps femoris; knee extensor: quadriceps, rectus femoris.

A2.14   Bundles→fibers→myofibrils→myofilaments.

A2.15   On the sarcomere level.

A2.16   By movement of the myosin filament against the actin filament.

A2.17   On the cross section. The cross section increases with the number of fibers, each fiber contributing to the total force.

A2.18   Isometric tension is muscle activation without changing the muscle length; isotonic tension is a muscle contraction without a change of tension.

A2.19   Efferent somatic motor neurons activate muscles.

A2.20 Ca^{2+} is required for contraction, and ATP for release.

A2.21   A twitch is a short contraction of muscle fibers activated by just one motor unit in response to one action potential arriving at a somatic motor neuron.

A2.22   Low-frequency twitches, which do not overlap in time, cause unfused tetanus. High-frequency twitches, which overlap in time, cause fused tetanus.

A2.23   No, because the refractory time is rather long and does not allow a temporal overlap of two muscle contractions.

Chapter 3

A3.1   Linearity between strain and stress; no plastic deformation.

A3.2   For hydrostatic pressure, a medium is required to transmit the same pressure to all surfaces. The medium can be gas or a liquid like water. Therefore, the term "hydrostatic pressure."

A3.3   Elastic and plastic behavior can be distinguished by their hysteresis. In case of elastic response, there is no opening of a hysteresis. For plastic behavior, the hysteresis encloses an area that corresponds to the energy required to cause the plastic deformation.

A3.4   Ductile materials deform but do not break. Brittle materials break without deformation.

A3.5   Metals are usually ductile, and ceramic materials are usually brittle.

A3.6   By bending. Ductile materials deform, and brittle materials crack.

A3.7   Toughness implies material deformation beyond the yield point. A tough material accepts much plastic deformation until it breaks, a less tough material breaks already little beyond the yield point.

A3.8   Mainly shear type of fracture.

A3.9   Viscoelastic materials show elastic and viscous characteristics upon deformation. Part of the deformational energy is absorbed by viscous flow.

A3.10   The stress-strain curve depends on the rate of applied stress.

A3.11   In viscoelastic materials, the strain response on applied stress is governed by material flow and diffusion.

A3.12   Collagen and HA nanocrystals forming mineral reinforced fibrils.

A3.13   Collagen → fibrils → osteons → compact bones.

A3.14   Osteoblast and osteoclast cells. Osteoblasts build new cells, and osteoclasts eliminate old bone structures.

A3.15   At higher age, the osteoclast cells may dominate over osteoblast cells, diminishing the mineral content of bones, causing osteoporosis.

A3.16   The open spongy structure. In long bones also the medullary cavity.

A3.17   Osteons, Haver's canal, transverse Volkmann canals, and nerve fibers.

A3.18   Because bones are composite materials with hard nanocrystals embedded into softer fibrils, the latter ones can move against each other.

A3.19   The compact cortical bone is strong but not tough. In contrast, the trabecular part of the bone is less strong, but much tougher. The difference is due to the compact versus open meshwork of fibrils in the bone structure.

A3.20   Plastic deformation of bones is transformed into shear strain by cross-links in the interfibrillar matrix; glue filaments between fibrils help absorbing strain energy suppressing fracture.

Chapter 4

A4.1   Catabolic and anabolic reactions are both part of the cell's metabolism. Catabolic reactions degrade large organic molecules to smaller ones. Anabolic processes produce high-energy ATP.

A4.2   Within the mitochondria.

A4.3   ATP is used anywhere in the body where energy is required, for instance, for powering ion pumps and to activate myosin–actin pairs.

A4.4   35–38 ATP molecules are synthesized per molecule of glucose.

A4.5   Energy consumption per kilogram body weight per hour required for maintaining all functions of the inner organs without performing any physical work.

A4.6   BMR is the energy consumption required for maintaining all functions of the inner organs at rest. RMR is the resting metabolic rate and is defined similarly to BMR. MHR is the sum of BMR and heat produced by physical activity.

A4.7   One liter of oxygen is required for the production of 20 kJ of energy.

A4.8   8 MJ.

A4.9   Three conventional and one unconventional.

A4.10   For small temperature differences.

A4.11   Anaerobic: Energy is produced by splitting creatine phosphate (CP) into creatine and phosphate and converting glucose to lactate. Aerobic: Energy is produced by normal metabolism, i.e., combustion of food with oxygen into its constituents carbon dioxide (CO_2) and water (H_2O).

A4.12   Negative feedback system.

A4.13 Cardiorespiratory fitness corresponds to the maximum oxygen volume the body is capable of consuming. It can be determined by measuring the heart frequency, stroke volume, and the arteriovenous oxygen difference.

A4.14 The RER is defined as the ratio of CO_2 moles produced upon reaction to the number of O_2 moles consumed.

Chapter 5

A5.1 Double lipid layer.

A5.2 Na^+ channels.

A5.3 K^+ channels.

A5.4 In the cytoplasm: K^+; in the extracellular space: Na^+.

A5.5 The resting potential is determined by the K^+ ion concentration difference between the cytoplasm and extracellular space.

A5.6 −75 to −90 mV.

A5.7 Any kind of stimulus that surpasses the threshold potential.

A5.8 5 ms.

A5.9 $1\,\mu F/cm^2$.

A5.10 The strength of a stimulus is translated into the sequence and frequency of action potentials.

A5.11 Na^+–K^+ ion pump is a two-way ion channel for K^+ going into the cytoplasm and Na^+ going out.

A5.12 The ion pump is powered by the molecule ATP containing three phosphate groups. The energy is released by hydrolyzing ATP to ADP.

A5.13 Correct is: active ion transport.

A5.14 According to the reaction shown in Chapter 4: $C_6H_{12}O_6 + 6O_2 + 38ADP + 38\,P_i \rightarrow 6CO_2 + 6H_2O + 38ATP + 0.956MJ$, 1 molecule of glucose together with 38 ADP produces 38 ATP.

A5.15 ATP synthase takes place in the mitochondrion of cells.

Chapter 6

A6.1 Soma, dendrites, axon, axon terminal.

A6.2 Afferent neurons are sensory neurons, conducting receptor signals to the CNS, efferent neurons are motor neurons conducting action potentials from the CNS to the periphery.

A6.3 Myelinated axons support saltatory conduction of polarization current, speeding up the signal transmission in long nerve fibers. In the brain, the nerve fibers are not myelinated because of the short distances and limited space.

A6.4 They are distinguished by their response to external stimulus: temperature, pressure, or chemicals.

A6.5 Receptor potentials are graded, action potentials are all or none.

A6.6 The conversion takes place by a change from graded potentials to action potentials. Action potentials occur when the nerve fibers contain an increasing number of ion channels for fast depolarization.

A6.7 Yes, at all chemical synapsis, a DAC takes place via neurotransmitters.

A6.8 Automatic neuron reflexes take place by connecting afferent and efferent neurons via interneurons. In voluntary neuron activity, the afferent neuron is connected to the CNS (brain) for processing before an efferent neuron can cause any action.

A6.9 With EMG, muscle activity and conductivity is tested. With EEG, the brain activity is tested. With MEG also the brain activity is tested but with magnetic fields instead of electric fields.

A6.10 An action potential crosses a synaptic cleft by first converting the frequency of the action potential to a proportional amount of neurotransmitters. The neurotransmitter is converted back to an action potential by activating ligand-gated ion channels.

A6.11 The depolarized tail does not allow propagation into a zone that is already depolarized. Therefore, propagation of action potentials can only proceed in the direction that is not yet depolarized.

A6.12 Saltatory conduction is jump-like conduction of the polarization current along nerve fibers.

A6.13 Conduction in metal wires follows an electric potential gradient carried by a viscous flow of charge (electrons). Polarization current starts locally by an action potential across a cell membrane and propagates laterally due to electric dipole fields that opens neighboring-gated ion channels.

Chapter 7

A7.1 Absolute refractory is the time during which the heart is insensitive to a new action potential. The relative refractory period allows depolarization, but only with an enhanced threshold potential.

A7.2 Ca^{2+} ions prolong the plateau phase.

A7.3 The action potential of the SA node has a lower resting potential, lower threshold potential, no plateau phase, and repolarization that starts immediately after depolarization.

A7.4 The natural pacemaker is the sinoatrial node.

A7.5 The action potential is first distributed from the atrium to the AV node and from there along the highly conducting Purkinje fibers to the lower part of the myocardium.

A7.6   The Einthoven triangle has the base from right to left arm and the tip pointing down to the pelvic.

A7.7   The purpose of the Einthoven triangle is the measurement of potential differences between the extremities, which are representative for the cardiac action potential during a heart cycle.

A7.8   Einthoven leads provide potential differences between two extremities; Goldberger leads provide potential differences between one extremity and a neutral point.

A7.9   P-wave, QRS-complex, and T-wave.

A7.10   During the P-wave, the atria is depolarized. The QRS complex shows the depolarization of the ventricles. The T-wave reflects the repolarization of the ventricles.

A7.11   Left-oriented heart.

A7.12   ECG examination records 12 potential differences, 3 according to Einthoven, 3 according to Goldberger, and 6 according to Wilson.

A7.13   Single lead and wireless pacemakers.

A7.14   In case of a cardiac arrhythmia that causes cardiac fibrillation.

Chapter 8

A8.1   Pulmonary circuit and systemic circuit.

A8.2   Delivery of oxygen, and disposing carbon dioxide.

A8.3   They can be distinguished according to their blood pressure. Veins have low pressure; arteries have high pressure. Accordingly, the vessel walls are thin for veins and thick for arteries. Furthermore, all veins lead to the heart, all arteries lead away from the heart.

A8.4   No. Veins from the capillary bed to the heart have a low oxygen concentration; veins from the lung to the heart have a high oxygen concentration.

A8.5   Four, two for filling and two for ejection.

A8.6   (1) Inflow, (2) contraction, (3) ejection, and (4) relaxation.

A8.7   60%.

A8.8   5–6 l/min.

A8.9   5.6 W, efficiency of 25%.

A8.10   120 hPa.

A8.11   During the P-wave, the mitral valve is open. The P-wave causes a contraction of the atria that leads to a rapid filling of the ventricles. During the QRS complex, the ventricle becomes depolarized, causing a contraction of the ventricle and an increase of pressure. At the end of the S-phase, ejection of blood into the artery takes place. During the T-wave, the semilunar valves close, defining the diastole phase, which is the refractory period of the heart.

A8.12   Veins.

A8.13   The circulatory system is not a good example for classical hydrodynamics. The reasons are: (1) blood is a non-Newton fluid; (2) vessels are not rigid; (3) the blood flow is pulsatile and not continuous; (4) the flow can be turbulent and not laminar.

A8.14   Because of Kirchhoff's second law of parallel resistances.

A8.15   Suppression of turbulent flow.

A8.16   Information on cardiac performance.

A8.17   Hypotonic RBC may rupture; hypertonic RBC cannot carry oxygen.

A8.18   Hemoglobin consists of four side chains, each containing one heme molecule with Fe^{2+} at the center.

A8.19   The high–low-spin transition is due to Fe^{2+} sitting in a crystal electric field that changes with O_2 absorption. It is not important for oxygen transport, but it signifies high oxygen consumption in active parts of the brain during MRI.

A8.20 O_2 binding in myoglobin can be described by a simple Langmuir equation, whereas O_2 binding in hemoglobin has an S-shape. The S-shape signals a cooperative O_2 uptake with a positive feedback system.

A8.21   Ferritin stores iron as a buffer.

A8.22   Oxy-hem: 535 and 575 nm; de-oxy-hem: 560 nm.

A8.23   The color of blood is determined by the difference in absorption for oxy- and deoxyhemoglobin in the spectral range from 600 to 700 nm.

A8.24   Cardiopulmonary bypass takes over the tasks of heart and lung during open-heart or lung surgery.

Chapter 9

A9.1   Lung, thorax, and diaphragm.

A9.2   In the alveoli.

A9.3 O_2 versus CO_2.

A9.4   The pleural cavity maintains a pressure lower than the atmospheric pressure, which couples the thorax to the lung.

A9.5   The lung will collapse (pneumothorax).

A9.6   Chemical binding.

A9.7   By chemical reaction to bicarbonate.

A9.8   The vital capacity is composed of expiratory reserve volume, tidal volume, and residual volume.

A9.9   0.5 l.

A9.10   15/min.

A9.11   150 Pa.

A9.12   Because of the elastic properties of the lung and thorax acting like a combined elastic spring.

A9.13 The compliance increases in the case of emphysema; the compliance decreases in the case of fibrosis.

A9.14 Surfactants lower the surface tension and enable the alveoli to expand.

A9.15 Turbulent flow in the trachea.

A9.16 Four types are used: nasal cannulae, noninvasive ventilation, invasive ventilation, and extracorporeal membrane oxidation.

A9.17 Mainly three methods are available: Cryogenic distillation, pressure swing adsorption, and membrane diffusion.

Chapter 10

A10.1 Control of water level and electrolytes, removal of metabolites, in particular urine.

A10.2 Nephron.

A10.3 Glomerular capsule or Bowman's capsule.

A10.4 Plasma.

A10.5 The RPF rate is the total plasma flow through the kidneys; glomerulus filtration rate is that part of the RPF, which is filtered, amounting to 20% of RPF.

A10.6 The nephron consists of Bowman's capsule including glomerulus, proximal tubule, loop of Henle, distal tubule, and collecting duct.

A10.7 Sieve-like filter, allowing only small molecules to pass.

A10.8 Water, urea, creatinine, salts, and glucose.

A10.9 Their size is too big.

A10.10 Water, salts, and glucose.

A10.11 For balance of water and electrolytes controlling osmotic pressure.

A10.12 Water, urea, and creatinine.

A10.13 Fractional excretion: 0.001; total excretion: 1.5 l.

A10.14 Active secretion and reabsorption via ATP pump, diffusion, and active filtration via overpressure.

A10.15 Reduced production of insulin by the β-cells and enhanced insulin resistance of muscle cells.

A10.16 Hemodialysis is indicated as soon as glomerular filtration fails. Another possible answer: Hemodialysis is indicated as soon as the blood sugar level is too high and no longer controllable by insulin uptake.

A10.17 Parasympathetic efferent nerve controlling the detrusor muscle.

A10.18 Glucose should be absent in the urine.

Chapter 11

A11.1   Cornea, aqueous humor, lens, vitreous humor, and retina.

A11.2   A stiffening of the lens.

A11.3   With a tonometer pressing against the cornea.

A11.4   With a diverging lens.

A11.5   Cataract is a disease of the lens; glaucoma is a disease of the retina.

A11.6   Seven: Cones, rods, horizontal, bipolar, amacrine, Muller, and ganglion cells.

A11.7   Support of metabolism and recycling of neurotransmitters.

A11.8   Because it is highly absorbing and a supply layer for cones and rods, which would not function on the proximal side of the retina.

A11.9   Fovea is the center of the macula with the highest density of cones. The macula is the area on the retina with the highest sensitivity.

A11.10   Two systems: Photopic and scotopic. The photopic system is supported by cones, and the scotopic system by rods.

A11.11   The Kohlrausch kink occurs in a plot of light sensitivity versus time. It signifies the crossover from cone to rod sensitivity in the dark. It is an indication of two visual systems: scotopic and photopic.

A11.12   10 orders of magnitude in intensity.

A11.13   400–700 nm.

A11.14   In the fovea.

A11.15   Rhodopsin.

A11.16   Photoisomerization of the retinal from cis to trans within 100 fs.

A11.17   No. It leads to a graded potential, in fact to a hyperpolarization from −30 to −70 mV.

A11.18   In ganglion cells.

A11.19   The CCD chip takes a single response to intensity, and the retina has a dual system that measures both brightness and darkness.

A11.20   In the OFF state: off-bipolar and off-ganglion cells. In the ON state, the respective ON-bipolar and ON ganglion cells.

A11.21   Receptive field distinguishes between center and surround. The surround has an inhibitive effect on the center for contrast enhancement.

A11.22   Contrast enhancement is achieved by the horizontal connection of bipolar cells and by lateral inhibition.

A11.23   Rods and cones. Rods have a black-white or dark-bright sensitivity. Three types of cones mediate red, green, and blue color sensation.

A11.24   P-cells are in the center of the fovea. They have small receptive fields and they detect shape and color. M-cells are in the periphery; they have large receptive fields and they detect brightness and movement.

A11.25 Only by crossing some of the nerve fibers, the visual field does not get mixed up.
A11.26 Nerve fibers originating from the nasal part of the retina cross, and those from the temporal part do not cross.

Chapter 12

A12.1 Frequency, amplitude or loudness, and direction.
A12.2 Outer ear canal as transducer of sound waves in air to mechanical oscillations of the ear drum, middle ear for impedance matching and amplification; the inner ear for conversion of mechanical oscillations in action potentials.
A12.3 The acoustic impedance is defined as a product of density and sound velocity in which the sound wave travels.
A12.4 Water has an impedance higher by a factor of 3500 compared to air.
A12.5 Whenever the other medium has an acoustic impedance that is much higher or much lower than the medium from where the acoustic wave arrives.
A12.6 The mismatch is such that 99.9% of the sound intensity is reflected at the interface.
A12.7 Bel or decibel.
A12.8 The auditory level is a physical measurable quantity derived from the sound intensity. The loudness level takes the individual perception of sound intensity into account.
A12.9 Phone.
A12.10 Yes, they do. Only at 1000 Hz do the auditory level and the loudness level overlap.
A12.11 Hammer, anvil, and stapes.
A12.12 By hydraulics and levers.
A12.13 About 600.
A12.14 Acoustic impedance matching, transmittance of a wide frequency band, and protection against destructive noise levels.
A12.15 Transmission of mechanical vibration into fluid flow.
A12.16 The task of the round window is a pressure release that enables the fluid in the cochlea to move back and forth.
A12.17 The scala vestibule and the scala tympani.
A12.18 High frequencies are detected at the base and low frequencies at the apex.
A12.19 Tonotopic mapping of sound frequencies.
A12.20 Inner hair cells and outer hair cells.
A12.21 Potential gated.
A12.22 Mechanically gated.

A12.23 The outer hair cells amplify the fluid flow.

A12.24 The inner hair cells are responsible for the tonotopic detection of sound frequencies.

A12.25 4000 inner hair cells.

A12.26 The inner hair cells are connected to afferent nerve fibers, and the outer hair cells to efferent nerve fibers.

A12.27 The outer hair cells can stretch and contract with the frequency of the sound wave. This is achieved by prestin proteins that can change length.

A12.28 Depolarization by leaning to the longer neighbors, and hyperpolarization by bending toward the shorter neighbors.

A12.29 The receptor is a mechanoelectric transducer transforming mechanical stimulus into receptor potentials.

A12.30 The frequency is encoded tonotopically and the intensity is encoded by stimulating additional nerve fibers phase locked to the primary excitation.

A12.31 There are two systems that work at different frequencies. At low frequencies, localization is achieved by interaural time difference, and at high frequencies by intensity level differences.

A12.32 Tone is a pressure wave of a single frequency, sound may consist of several frequencies, and noise refers to a continuous distribution of frequencies.

A12.33 For high frequencies.

A12.34 Ear canal hearing aids, middle ear implants, and cochlea implants.

A12.35 From the glottis in the larynx being responsible for the frequency and the oral cavity for the color of the sound.

A12.36 Five areas

A12.37 Their purpose is to sense three rotational and three translational accelerations.

Chapter 13

A13.1 Orthopedics.

A13.2 The C-legs have to mimic the complete knee, leg, and foot movement. The motor control of all movements aims at reproducing a normal gait pattern.

A13.3 The main challenge is the training of remaining nerve fibers attached to new muscles to generate EMG signals, which can be translated into a smooth and accurate control of upper limb prostheses.

A13.4 The best combination with the lowest wear is a ceramic on ceramic combination. It is also the most expensive one.

A13.5 These are the tracts of the gall, of the tear duct, and the outflow duct of the aqueous fluid.

A13.6   A stent should not disturb the hemodynamics and should not initiate blood clotting. Stents should be plastically deformable and rigid in the expanded shape.

A13.7   The knee joint is a so-called modified hinge joint that allows a monoaxial extension and a limited rotation in the flexed position.

A13.8   After replacement, the knee is strictly monoaxial, allowing only flexion.

A13.9   Epiretinal microelectrode array and subretinal microphotodiode array.

A13.10   Blind patients, whose retina and, in particular, the ganglion cells together with the optic nerve bundle are still intact.

15 Solutions to exercises

Chapter 1

E1.1 **Answer: Cell types**
Eukaryotic cells have a nucleus that holds the genetic code. Prokaryotic cells do not have a nucleus. Their DNA is floating freely in the cytoplasm. Only bacteria and archaea are prokaryotes. All other living cells are eukaryotes, including all animals and plants.

E1.2 **Answer: Number of cells**
Assuming an average cell diameter of 20 µm, the cell volume is 4.2×10^{-15} m³. The body volume may be roughly 0.08 m³. So the ratio is the number of estimated cells: 1.9×10^{13}.

E1.3 **Answer: Density**
To solve the problem, one needs to know the body volume. This is rather difficult to get precisely because of the body's irregular shape. The question is answered in Chapter 2. The average density of the body is about 1.08 g/cm³. This is slightly larger than the density of water, and therefore, the body will sink in normal water.

E1.4 **Answer: Body measurements**
Periodically reoccurring properties: heartbeat, breathing frequency, eye blink, systolic and diastolic blood pressure, bladder filling, number of bladder voidings per day, sleeping period, and calories uptake. Constant properties: height and all other body measures of arm length, foot size, etc., weight (evening and morning), body temperature, and blood oxygen concentration (with the help of a pulse oximeter). With a bit more effort and the help of the Internet, you can determine your audible sensitivity level as a function of frequency, accommodation ability, visual acuity, time for dark adaptation, and color contrast sensitivity. More on these topics in later chapters.

E1.5 **Answer: Homeostasis**
Core temperature; blood glucose concentration; Ca, K, Na ion concentrations; Fe concentration; and a few more, to be discussed in later chapters.

E1.6 **Answer: Feedback systems.**
In a positive feedback system, a physical quantity is enhanced by an external field. For instance, the magnetization of a paramagnet is to enhance in an external magnetic field.

https://doi.org/10.1515/9783110756951-015

A negative feedback system tries to reduce a physical quantity in response to an external parameter. Examples are blood oxygen/CO_2 ratio, body temperature regulation, and the arterial vessel tension in kidneys, glucose level in the blood, and all topics in later chapters.

Chapter 2

E2.1 **Answer: Body density**

The difference $W_1 - W_2$ equals the weight of the displaced water, which is

$$W_{water} = \rho_{water} V_{body} g.$$

Now considering that
$V_{body} g = W_1 / \rho_{body}$.
Therefore,

$$W_1 - W_2 = \frac{\rho_{water} W_1}{\rho_{body}} \text{ or } \rho_{body} = \rho_{water} \frac{W_1}{W_1 - W_2}.$$

E2.2 **Answer: Body fat**

Measure the height H and the bodyweight W. Determine the $BMI_{meas} = W/H^2$. Compare with the ideal BMI for that particular height and calculate the ideal weight: $W_{ideal} = BMI_{ideal} H^2$. Now use Tab. 2.1 for calculating the weight of bones and muscles for this height. The remaining weight is the fatty part.

E2.3 **Answer: Force on the lumbar vertebra**

There are three torques counterbalanced by the lumbar vertebra:

$$\left(700 \text{ N} \times \frac{1}{2} + 100 \text{ N} \times \frac{2}{3} + 700 \text{ N} \times \frac{1}{4} \right) l \times \sin(60°) = F_m \frac{2}{3} l \times \sin(12°),$$

yielding F_m=3696.75 N. Without holding extra weight, the force would be 2184 N. The difference is 1512.75. Thus keeping the weight in the bent posture adds a force on the lumbar vertebra, which is 15 times higher than its weight. Lifting weight in this posture is therefore not a good idea.

E2.4 **Answer: Torque at forearm**

For the bent position, the torque of the weight is $T_W = F_W l_W \sin(120°)$, and for the biceps brachii $T_m = F_m l_m \sin(60°)$. In equilibrium, both torques are equal: $T_W = T_m$, and the sinus of the angle cancels. Therefore, the force on the muscle is independent of the angle.

E2.5   **Answer: Standing on the tiptoe**

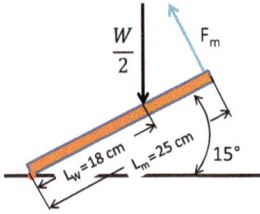

Standing on the tiptoe is a lever of the second kind. The fulcrum is at the toe, indicated by a red dot. In equilibrium, the torques due to the weight $\vec{T}_W$ and due to the muscle $\vec{T}_m$ must equilibrate:

$$\vec{T}_W - \vec{T}_m = 0 \quad \text{or} \quad \vec{T}_m = +\vec{T}_W.$$

$\vec{T}_W$ acts clockwise and therefore has a positive sign, and $\vec{T}_m$ acts counterclockwise and therefore has negative sign.

   Since we lift the body on both toes, each one lifts only half the body weight, and the corresponding torque is

$$T_W = W/2 \cdot L_W \sin(90° + 15°).$$

The weight arm is estimated to be about 18 cm. The muscle force makes an angle of 15° with respect to the normal, and the muscle arm is the total length from toe to heel, 25 cm. Taking this into account, the muscle torque is:

$$T_m = -F_m \cos(15°) \cdot L_m \sin(90° + 15°).$$

Considering only the magnitude, and inserting a body weight of 700 N, we find for the muscle force:

$$F_m = \frac{W/2 \times L_W}{L_m \cos(15°)} = \frac{700/2N \times 0.18\,m}{0.25\,m\cos(15°)} = 260N < W/2.$$

Since the muscle arm is longer than the weight arm, standing on the toe has a mechanical advantage. Responsible for lifting is the Achilles tendon together with the gastrocnemius muscle. This is by far not the limit of this muscle. Running and jumping add accelerating forces that can be much higher than the static forces.

E2.6   **Answer: Lifting speed**
With rapid muscle shortening, the myosin–actin filaments glide against each other very quickly. This requires that myosin–actin bonds are continuously and quickly broken in order to create new cross-bridges. Heavy weights can only be lifted when the tension increment is relatively small, which requires more time resulting in lower speed.

E2.7 **Answer: Myosin-actin pairs**

For holding a weight of 10 N in the hand, the force required on the biceps is $F_m = F_w l_w / l_m$. With l_w = 30 cm, l_m=3 cm, F_m = 100 N. One sarcomere unit produces by contraction a force of 10^{-9} N. Therefore, we need 10^{11} myosin-actin pairs to produce the required force.

Chapter 3

E3.1 **Answer: Moment of resistance**

$$M = \frac{1}{h}\int_{-h/2}^{+h/2} z^2 \, dA = \frac{1}{h}\int_{-h/2}^{+h/2} z^2 t \, dz = \frac{1}{3}\frac{t}{h}z^3\Big|_{-h/2}^{+h/2} = \frac{1}{3}\frac{t}{h}2\times\left(\frac{h}{2}\right)^3 = \frac{th^2}{12}.$$

Here t is the thickness of the beam and h is the height.

E3.2 **Answer: Approximation**

$$M = \frac{\pi}{2}\frac{(R^4 - r^4)}{R} = \frac{\pi}{2}\frac{(R^2 - r^2)(R^2 + r^2)}{R} \approx \frac{\pi}{2}\frac{(R-r)(R+r)(2R^2)}{R}$$
$$\approx \frac{\pi}{2}\frac{(\Delta R)(2R)(2R^2)}{R} \approx 2\pi R^2 \Delta R.$$

E3.3 **Answer: Contraction**

In Chapter 2, we learned that standing on one leg implies that it has to support 85% of the bodyweight W or in our case 595 N. The pressure on the leg (stick) is then 6.6×10^5 Pa. The Young's modulus for compressional tension of bones is $Y = 1.5 \times 10^{-5}$. Therefore, the relative contraction is $\varepsilon = \Delta l / l = \alpha / Y = 4.4 \times 10^{-5}$, and the absolute contraction turns out to be: $\Delta l = \varepsilon \times l = 4.4 \times 10^{-5} \times 0.7\text{m} = 0.03$ mm.

E3.4 **Answer: Beam bending**

Since $\delta = F/k$ and $\delta/L = F/Lk$, it follows that δ = 5 cm and δ/L = 1%.

E3.5 **Answer: Critical tensile stress**

According to eq. (3.8), there is a strain gradient from top to bottom. At the top, the stress is $\sigma_{max} = Yd/2R$. Therefore, $R = Yd/2\sigma_{max}$. Inserting numbers: $R = (15 \text{ GPa}/200 \text{ MPa}) \times 0.02$ m = 1.5 m.

E3.6 **Answer: Pressure in the body**

Blood pressure, pleural pressure, middle ear pressure, intraocular pressure, bladder and colon pressure, and filtration pressure in the kidneys.

E3.7   **Answer: Torque on the femur**

According to eq. (2.9), the torque is $T = 2 \times 0.35 BW \times l \times \sin(180° - CCD)$. Estimating $l = 0.05$ m, the torque at a young age is 1.8 Nm, at an advanced age: 2.6 Nm. The increase is about 40%.

E3.8   **Answer: Crystal components**

H^+, Ca^{2+}, and OH^- are abundant in the body. PO_4^{3-} can be identified as P_i, the inorganic phosphate group, which is split from ATP.

Chapter 4

E4.1   **Answer: Caloric oxygen equivalent**

a.   Efficiency: The total energy produced is composed of the energy for the production of ATP (129×48 kJ = 6.19 MJ) and heat (3.76 MJ), which sums up to 9.95 MJ. The efficiency is therefore 6.19 MJ/9.95 MJ = 0.62 or 62%.

b.   COE: The mole volume of O_2 is $22.4 \times 23 = 515.2$ l. Therefore, the COE is 9.98 MJ/515.2 l = 19.4 kJ/l.

c.   Energy density: The mol mass of $CH_3(CH_2)_{14}COOH = C_{16}H_{32}O_2$ is 256 g. Therefore, the energy density is 9.95 MJ/256 g = 39 kJ/g.

d.   RER: $16CO_2/23O_2 = 0.7$.

E4.2   **Answer: Burning fat**

The energy consumed results from the product of O_2 consumption per day (0.3 l/min $\times$ 1440 min = 432 l) and the caloric equivalent of fat (19.4 kJ/l O_2) and amounts to 8381 kJ/day. According to Exercise E4.1, the energy density of fat is 39 kJ/g so that 214 g of fat is burned in 1 day. You may also determine the number of mol O_2 (=19.28/day) and compare this to 23 mol O_2 needed for consuming 256 g of fat. The comparison also yields 214 g fat.

E4.3   **Answer: Heat conduction**

For the heat conduction, we use eq. (4.20) and a thermal conductivity of the tissue from Tab. 4.5. Inserting numbers $\dot{Q} = 0.1 \frac{W}{mK} \times 1.6\,m^2 \times 17\,K/0.05\,m = 54.4\,W$.

E4.4   **Answer: Radiative heat loss**

a.   The rate of heat loss $\dot{Q}_R = \varepsilon\,K\,A\,\Delta T$, with $K = 4\sigma T^3 = 6.4$ W/m² K. Therefore, $\dot{Q}_R = 6.4$ W/m² K $\times$ 2 m² $\times$ 12 K = 154 W.

b.   The heat production by metabolism is 80 W, and the heat loss is 154 W. The difference is 74 W. The heat loss can be determined using the specific heat of the body: $\dot{Q} = C\Delta T/\Delta t$, where T is the temperature and t is time. Therefore, $\Delta t = C\Delta T/\dot{Q} = $ (265 kJ/74 W) $\times$ 5 K = 1.8 $\times$ 10⁴ s or approx. 5 h.

E4.5 **Answer: Heat loss and gain**

A. Heat emission through radiation: $\varepsilon\sigma A T^4$ = 440 W. The radiation absorbed is 700 W/m^2 × 0.9 m^2 × 0.8 = 504 W. The difference between heat absorption and heat emission through radiation is: 504 W − 440 W = 64 W.

B. Heat loss due to the breeze is heat loss due to convection. Using eq. (4.26), K_C = 26.5 W/m^2 K; heat loss is then: $\dot{Q}_C = K_C \, A \, \Delta T$ = 26.5 W/m^2 K × 0.9 m^2 2 K = 47.7 W.

C. 90 W − 10 W + 64 W − 47.7 W = 96.3 W. Sweating should remove the remaining heat production of 96.3 W.

Chapter 5

E5.1 **Answers: The cell as a capacitor**

a. In a spherical capacitor, the field between the plates is not constant in contrast to a plate capacitor. Instead, the field has a $1/r^2$ dependence like for a charged sphere. Thus, the field between the inner and outer membranes at radii R_1 and R_2, respectively, drops off as follows:

$$E(Q,r) = \frac{1}{4\pi\varepsilon\varepsilon_0} \frac{Q}{r^2} = \frac{1}{\varepsilon\varepsilon_0} \frac{Q}{A},$$

where $R_1 < r < R_2$. Substituting numbers, we find

$$E(Q,r) = \frac{1}{5 \times 8.8 \times 10^{-12}} \frac{\text{Vm}}{\text{As}} \times 0.1 \frac{\text{C}}{\mu\text{m}^2} = 2.2 \times 10^7 \frac{V}{m}.$$

The field gradient between the outer and inner membranes is negligible for a membrane thickness of 5 nm at a spherical radius of 5 μm.

b. The electric field across the cell membrane is

$$E(Q,r) = -\frac{dU}{dx} = -\frac{70\text{mV}}{5\text{nm}} = 1.4 \times 10^7 \frac{V}{m}.$$

This result is quite comparable to the one obtained in a.

c. The stored energy in a capacitor E_C (including those with a spherical shape) is

$$E_C = \frac{1}{2}C(\Delta U)^2.$$

ΔU is the potential difference across the membrane. Using $C = \varepsilon\varepsilon_0 A/\Delta x$ and $\Delta U = E\Delta x$ (E = electric field), we find for the stored energy:

$$E_C = \frac{1}{2}\varepsilon\varepsilon_0 E^2 A \Delta x = \frac{1}{2}\varepsilon\varepsilon_0 E^2 V,$$

where $V = A\Delta x$ is the volume of the spherical shell, which is equivalent to the volume of the membrane. Therefore, the stored energy density is

$$\frac{E_C}{V} = \frac{1}{2}\varepsilon\varepsilon_0 E^2.$$

We already calculated the electric field in part (a). Substituting numbers, we obtain the energy density: $E_C/V = 1.1 \times 10^4 \, \text{J/m}^3$.

E5.2 **Answer: Time constant**
The time constant for charge and discharge is $\tau = RC$. Therefore, the resistance is $R = \tau/C = \tau/AC_A = \tau/\pi D^2 C_A$. Assuming that the time constant for discharging (depolarization) is about 0.5 ms, we obtain the ion channel resistance: $0.16 \times 10^9 = 0.16 \, \text{G}\Omega$. This gigaohm resistance is the one observed for single ion channels by the patch-clamp method.

E5.3 **Answer: Resting potential**
According to (5.33), the resting potential is linked to the ion channel potentials at rest and the respective conductances

$$\Delta U_{el} = \frac{g_K \Delta U_K + g_{Na} \Delta U_{Na} + g_L \Delta U_L}{g_K + g_{Na} + g_L}.$$

Substituting numbers, we find for $\Delta U_{el} = -62.3$ mV. If the Na^+ and Cl^- channels were completely closed, the cell's resting potential would correspond to the Nernst potential of the K^+ channel. However, the finite permeability of the other channels lowers the cell's resting potential.

E5.4 **Answer: ATP energy release**
We use the conversions: $1 \, \text{J} = 6.25 \times 10^{18}$ eV; $1 \, \text{mol} = 6 \times 10^{23}$ eV ions. Then substituting numbers, we find for the binding energy:

$$-30.5 \, \text{kJ/mol} = \frac{30.5 \times 10^3 \times 6.25 \times 10^{18} \text{eV}}{6 \times 10^{23}} = -0.31 \, \text{eV}.$$

The energy release per ATP$\rightarrow$ADP conversion is therefore $= -0.31$ eV.

Chapter 6

E6.1 Answer: Capacitance of axons

According to eq. (6.1), the capacitance can be expressed as follows:

$$\frac{C_m}{L} = \frac{2\pi\varepsilon\varepsilon_0 a}{d} = \frac{2\pi \times 6 \times 8.8 \times 10^{-12} \text{As/Vm} \times 10^{-6}\text{m}}{5 \times 10^{-9}\text{m}}$$

$$= 66.35 \times 10^{-9}\frac{\text{As}}{\text{Vm}} = 6.6 \times 10^{-8}\frac{\text{F}}{\text{m}}.$$

E6.2 Answer: Propagation velocity

The length of the axon from the brain to toe is estimated at 2 m full length when stretched. For unmyelinated and myelinated axons, the signal propagation velocity is 2 and 120 m/s, respectively.

The arrival time is $t = L/v$, which is 1 s and 17 ms for the unmyelinated and myelinated axons, respectively.

E6.3 Answer: Transmission lines

Technical transmission lines consist of various conductive elements, resistances, capacities, and also inductances. An example is shown below. The arrangement of these items varies depending on the actual situation. But real wires always exhibit inductive resistance at higher frequencies. The model transmission line describing neurons does not contain any inductive elements. The current amplitude and frequencies are not high enough to account for inductance.

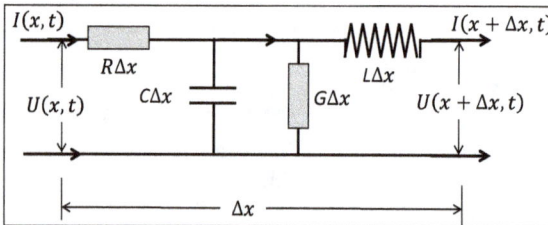

E6.4 Answer: Analog-to-digital conversion

Analog-to-digital converters (ADC) proceeds in three steps. 1. An analog signal such as temperature, pressure, voltage, and current is sampled with a constant frequency. 2. The scaling of the analog signal is broken down into incremental steps. If the measured value falls within the range of an increment, it is given that value. 3. The incremental value is represented as a binary code. This digital representation can then be processed, manipulated, computed, transmitted, or stored.

The signal transduction from receptor potential to action potentials has similarities to the ADC. The information carried by the graded receptor potentials is converted into the frequency and duration of action potentials in the axon. Higher intensity of an external parameter (loudness, pain, etc.) results in a higher frequency of action potentials. Thus, the signal amplitude is encoded in the output frequency. With some hand-waving, this conversion is similar to the ADC first and third step, but the second step is missing.

E6.5 **Answer: Propagation equation**

The harmonic wave equation in one dimension has the general form:

$$v^2 \frac{\partial^2 \xi}{\partial z^2} = \frac{\partial^2 \xi}{\partial t^2}$$

and the harmonic wave expressed by

$$\xi = \xi_0 \exp(kz - \omega t).$$

is a solution of the wave equation. Here $\xi(z,t)$ is any kind of amplitude and $v = z/t = \omega/k$ is the propagation velocity of the wave.

The harmonic wave equation cannot describe losses and gains of the wave amplitude due to dissipative processes. The harmonic wave has a constant amplitude. In contrast, the signal propagation of action potentials is characterized by a strong damping in space and time, which is described by the telegrapher's equation:

$$\lambda^2 \frac{\partial^2 \xi(z,t)}{\partial z^2} = \tau \frac{\partial \xi(z,t)}{\partial t} + \xi(z,t),$$

where λ is the decay length and τ is the decay time.

E6.6 **Answer: Piano player**

Assuming a speed of nerve signal on the order of 100 m/s and a distance from the brain to the fingers of 1 m, the signal arrives in 0.01 s = 10 ms. With 600 keystroke/min or 10 strokes/s, one stroke takes 100 ms. Therefore, there is a factor of 10 between signal speed and finger stroke, or the brain works 10 times faster than the fingers can follow.

Chapter 7

E7.1 **Answer: Dipole orientation**
In such a case, the dipole is directed from F to L. Such an orientation of the heart dipole does not occur during the cardiac cycle. It is, therefore, purely hypothetic.

E7.2 **Answer: Potential difference and dipole orientation**
According to eq. (7.8), $\Delta U_{III} = +0.3mV$. Therefore, $\Delta U_I = -\Delta U_{III}$. This is only possible for a dipole with angular orientation $\theta = 180°$. As ΔU_I is negative, the dipole must point from L to R.

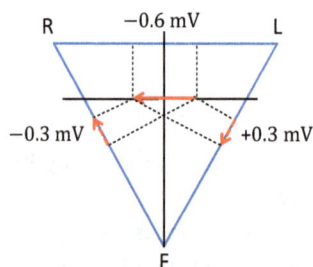

E7.3 **Answer: QRS complex**
A maximum in lead II is achieved if the vector lies parallel to lead II. aVL measures the potential perpendicular to lead II and is therefore at right angles to the projection recorded in lead II. Therefore, a maximum in lead II causes simultaneously a zero in lead aVL.

E7.4 **Answer: ST segment**
During the ST segment, the heart vector is oriented at an angle of about $-30°$, which is at right angle to lead II. Therefore, in lead II no potential difference is measured.

E7.5 **Answer: Goldberger leads**
The Goldberger leads are the perpendicular bisectors of all three sides of the equilateral Einthoven triangle. Therefore, the perpendicular bisector divides the 60° angle on each opposite corner in half. For instance, the Einthoven lead LF is rotated by 60° with respect to lead RL. The Goldberger lead from L to $RF/2$ is rotated by 30°. Same holds for all other leads.

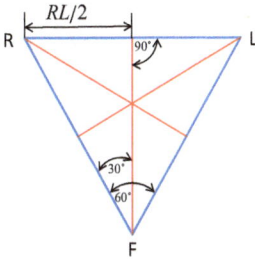

E7.6 Answer: Cartesian coordinates

Lead I: $\Delta U_I = \vec{H} \cdot \vec{l}_I = H_x$.

Lead II: $\Delta U_{II} = \vec{H} \cdot \vec{l}_{II} = \frac{1}{2}H_x + \frac{\sqrt{3}}{2}H_y$.

Lead III: $\Delta U_{III} = \vec{H} \cdot \vec{l}_{III} = -\frac{1}{2}H_x + \frac{\sqrt{3}}{2}H_y$.

E7.7 Answer: AV block

With an AV block of the first grade, the excitation of the AV node is delayed. This leads to a delayed contraction of the ventricles. It can be seen in an ECG recording by a lengthening of the PQ interval to more than 200 ms.

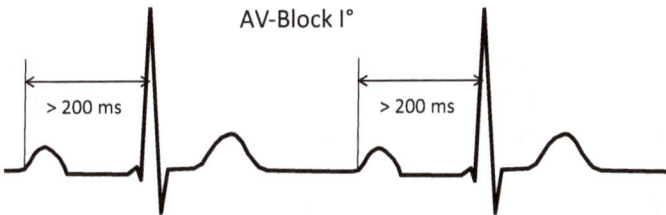

Chapter 8

E8.1 Answer: Number of blood cells

We have about 5×10^6 RBC in 1 µl of blood. Therefore, in 5 l blood, we have $(5 \times 10^6) \times (5 \times 10^6) = 2.5 \times 10^{13}$ RBC in the body. The total number of cells is then estimated to be: $5 \times 2.5 \times 10^{13} = 1.25 \times 10^{14}$.

E8.2 Answer: Compliance of the aorta

The compliance C is the ratio of the change in the blood volume V in the aorta and the change in the transmural pressure p, i.e., $C = \Delta V/\Delta p$.
Inserting it into the formula you get

$$C = \frac{\Delta V}{\Delta p} = \frac{180\ \text{ml} - 150\ \text{ml}}{15\ \text{kPa} - 10\ \text{kPa}} = 6\ \text{ml/kPa}.$$

E8.3 Answer: Flow rate

For a constant volume flow rate, the ratio of the cross sections and velocities must be

$$\frac{A_2}{A_1} = \frac{v_1}{v_2} = \frac{1}{2}.$$

With $A = \pi r^2$ follows:

$$\frac{A_2}{A_1} = \frac{\pi r_2^2}{\pi r_1^2} = \frac{r_2^2}{r_1^2} = \frac{1}{2}.$$

Therefore, the ratio of the radii is

$$\frac{r_2}{r_1} = \sqrt{0.5} \approx 0.71 = 71\%.$$

E8.4 Answer: Bypass

The volume flow rate: $I_V \sim r^4$. When the radius is doubled, the volume flow rate increases by a factor of 16. As both vessels run in parallel, the total volume flow rate is the sum flow rates of the individual vessels: $I_{V,\text{tot}} = I_{V,\text{disease}} + I_{V,\text{bypass}} = 17 I_{V,\text{disease}}$. Hence, the flow rate increases by a factor of 17.

E8.5 Answer: Giraffe

A pressure $\rho g h$ must be applied so that blood can rise from shoulder level, where the heart is located, to the giraffe's head. If we assume that the density of blood corresponds to that of water, then a pressure of 250 mbar = 250 hPa = 25 kPa is required for a rise of 2.5 m. This blood pressure is rather high and strenuous for the vessels. It is one of the reasons that the lifespan of giraffes is comparatively short at 25–30 years.

E8.6 Answer: Critical velocity

The Reynolds number is $\text{Re} = \rho v d / \eta$. Using Re = 1000 as critical value, and a radius of 1 cm for the aorta, then the critical velocity is

$$v_c = \text{Re}\, \eta / (\rho\, d) = 1000\, (4\,\text{mPa s}) / \left(10^3\,\text{kg/m}^3 \right) \left(10^{-2}\,\text{m} \right) = 0.4\,\text{m/s}.$$

E8.7 Answer: Frank's differential equation

Setting the right side to zero implies that $\dot{V}_{\text{in}} = 0$, and $\dot{V}_C = -\dot{V}_{\text{out}}$. The remaining differential equation cannot describe the inflow of the blood and the expansion of the Windkessel. However, it can describe the outflow from the Windkessel to the periphery. This is expressed in the exponential pressure drop with the time constant $\tau = RC$.

E8.8 Answer Number of RBCs

Knowing that the hematocrit value is about 0.5, we conclude that RBCs fill only 50% of the blood volume. Assuming a dense packing of the RBCs in the hematocrit volume, in 1 l of blood, there are 500 ml/100 fl $= 5 \times 10^{12}$ RBCs.

E8.9   **Answer: Foramen oval**

a.   Sketch:

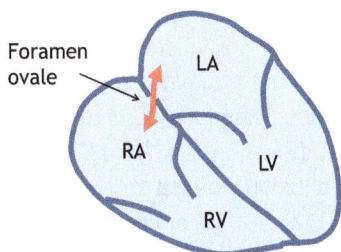

b.   The short-circuit connection is useful so that little or no blood flows through the pulmonary circulation, which has no function yet, since oxygen is only absorbed by the mother's blood.

E8.10   **Answer: Isotonic solution**

0.9% solution corresponds to 9 g of NaCl per 1000 ml of H_2O. The molar mass of NaCl is $m_{mol} = m_{mol}\left(^{23}Na\right) + m_{mol}\left(^{35,5}Cl\right) = 58.5$ g. Therefore, the mole fraction of 9 g of NaCl is: $n = 9$ g/58.5 g $= 0.15$. Because of the dissociation of NaCl in solution in Na^+ and Cl^-, the number of ions in solution is doubled, and therefore the mole fraction is $n = 0.3$. The osmotic pressure at a body temperature of $T = 310$ K is then

$$\Pi = \frac{nRT}{V} = \frac{0.3 \times 8.31\frac{J}{K} \times 310\,K}{10^{-3}\,m^3} = 772\,kPa = 7.7\,bar.$$

A pressure of 7.7 bar is still in the limits of an isotonic pressure.

E8.11   **Answer: Osmosis**

The mol mass of glucose is 180g. Five grams of glucose corresponds to a mol fraction of $n = 5/180 = 0.028$. Inserting into the equation for the osmotic pressure at 310 K yields

$$\Pi = \frac{nRT}{V} = \frac{0.028 \times 8.31 \times 310}{10^{-3}}\,Pa = 72\,kPa.$$

Therefore, by adding glucose, the osmotic pressure increases from 770 kPa to $770 + 72$ kPa $= 842$ kPa. Now we use the law of Boyle–Mariotte:

$p_1V_1 = p_2V_2$, or $V_2 = 842$ kPa $\times 1$ L/770 kPa $= 1.1$ l. Therefore, 100 ml of water must be added to achieve an isotonic solution again.

E8.12   **Answer: Reproduction rate**

The total number of erythrocytes in 6 l of blood is 6×10^6 μl $\times 5 \times 10^6/$μl $= 3 \times 10^{13}$ erythrocytes. One hundred days are 8.64×10^6 s. Therefore, the reproduction rate must be: $3 \times 10^{13}/8.64 \times 10^6 = 3.5 \times 10^6$ erythrocytes/s.

E8.13 **Answer: Total oxygen content of blood**

Each erythrocyte contains roughly 2.5×10^8 hemoglobin molecules with 4 oxygen binding sites each, resulting in 10^9 bonded O_2 molecules per erythrocyte. One erythrocyte has a volume of $v_{eryt} = \frac{\pi}{4} d^2 t$, where $d = 8$ μm is the diameter, and $t = 2$ μm is the thickness., yielding $v_{eryt} = 100$ fl (f = femto). In 1 l blood, only 50% is filled with hematocrit. Therefore, there are 0.5 l/100 fl = 5×10^{12} erythrocytes in 1 l of blood, binding $5 \times 10^{12} \times 10^9 = 5 \times 10^{21}$ O_2 molecules at full saturation. For the 2 + 3 l of blood we then obtain:

$$2 \times 5 \times 10^{21} + 3 \times 5 \times 0.75 \times 10^{21} = 2.12 \times 10^{22} \; O_2 \text{ molecules in 5 l of blood.}$$

One mole of oxygen contains 6×10^{23} molecules. Therefore, the fraction is about $2 \times 10^{22}/6 \times 10^{23} = 1/3 \times 10^{-1} = 0.03 = 3\%$. Thus, 5 l of blood carries about 3% of a mole of oxygen. The oxygen concentration (volume density) is then 6 mmol/l.

E8.14 **Answer: Venous blood flow**

In short, venous valves are valve-like structures in the veins that ensure that the blood flows exclusively in the direction toward the heart and does not flow backward. Valves are necessary because they act against the gravitational force.

E8.15 **Answer: Ultracentrifuge**

As the centrifugal force was 7000 times the gravitational force, we have the ratio:

$$\frac{F_c}{F_g} = \frac{\omega^2 r}{g} = 7000.$$

Therefore, the radius is given by

$$r = \frac{7000g}{(2\pi f)^2}.$$

About 12 000 rpm is equivalent to a frequency of $f = 200$ s^{-1}. Substituting numbers, we find:

$$r = \frac{7000 \times 10 \, \text{m/s}^2}{(2\pi \times 200 \; \text{s}^{-1})^2} = 0.0443 \, \text{m}.$$

The radius during centrifugation was 4.4 cm.

Chapter 9

E9.1 **Answer: Oxygen concentration**

A 1 µl blood contains about 5×10^6 RBC. Each erythrocyte carries about 10^9 O_2 molecules. Therefore, 1 l blood carries $5 \times 10^6 \times 10^6 \times 10^9 = 5 \times 10^{21}$ oxygen molecules. One mole of oxygen contains 6×10^{23} oxygen molecules. Therefore, 1 liter of blood contains about 1% mole of oxygen. If the physical solution of oxygen in the blood is the same as in water, then the solution is 2.7×10^{-4} mol/l, according to Tab. 9.3. Hence, the physical solution provides by a factor of 30 less oxygen than the chemical bonding.

E9.2 **Answer: Flow resistance and critical radius**

Using the continuity equation $I = Av = \pi r^2 v$, we find for the radius $r = \dfrac{2\rho_{air}I}{Re\,\eta\pi}$. Assuming Re = 2000 for the limit of laminar flow, the radius of the trachea should be less than 1.1 cm.

E9.3 **Answer: Smokers lung**

Since the inspiration time constant is $\tau = R_L C$, the time constant increases with increasing R_L. Assuming that the compliance is not affected, the smoker's time constant τ_{smoker} is four times bigger than for a healthy person $\tau_{healthy}$, i.e., 3 s instead of 0.75 s. A complete respiratory cycle requires roughly 6τ. Hence, a smoker needs 18 s for one respiration cycle compared to 4 s for a healthy person. Correspondingly, the breathing frequency drops from 15 to 3 per minute. If the tidal volume remains constant, there will be an undersupply of oxygen.

E9.4 **Answer: Alveoli volume and surface**

The vital volume is $7\,l = 7 \times 10^{-3}$ m³. Therefore, a rough estimate of the alveoli volume is 7×10^{-3} m³$/3 \times 10^8 = 2.33 \times 10^{-11}$ m³. This volume corresponds to a radius of 180 µm. The total surface, assuming a spherical shape, is $A = 3 \times 10^8 \times 4 \times \pi \times (180\,\mu m)^2 = 122$ m².

E9.5 **Answer: Diffusion in the alveoli**

a. Breathing air contains 21% oxygen, but in the alveolar air the proportion is only about 13% at a pressure of 13 kPa. The total amount of oxygen is thus $0.13 \times 0.3 \times 6 \times 10^{23} \times 0.5/22.4 = 5 \times 10^{20}$ oxygen molecules that cross the barrier during one breathing.

b. $R_D = \Delta t/V = 1\,s/0.5 \times 10^{-3}$ m³ $= 2 \times 10^3$ s/m³ (V is the tidal volume).

c. $D = (1/R_D)(\Delta x/A) = (2 \times 10^3\,s/m^3)^{-1} \times 1\,mm/80$ m² $= 6.2 \times 10^{-12}$ m²/s.

d. $P = D/\Delta x = 6.2 \times 10^{-12}$ m² s⁻¹$/10^{-6}$ m $= 6.2 \times 10^{-6}$ m/s or about 6 µm/s.

In fact, the diffusion constant is of the order of 10^{-11} m²/s, i.e., there is still a lot of leeway even with faster inhalation and oxygen exchange.

E9.6   **Answer: Time constant**

In order to get an exponential time dependence, R_L and C_L must be arranged in a series, like in an electric circuit. An example is discussed in Chapter 5 (eq. (5.17)).

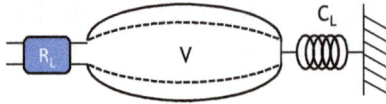

The transpulmonary pressure p_L to be applied consists of two components: the pressure loss due to the flow in the airways p_R, and the pressure due to the expansion and compression of the lungs p_C, where V_{res} is the residual lung volume. This yields the differential equations (9.7) and (9.8):

$$p_L = p_R + p_C = R_L \cdot \dot{V} + \frac{1}{C_L}(V - V_{res}).$$

In equilibrium,

$$\dot{V} + \frac{1}{R_L C_L}(V - V_{res}) = 0,$$

which has the solution: $V(t) = V_0 e^{-t/\tau} + V_{res}$.

E9.7   **Answer: Oxygen supply**

At constant temperature one can use Boyle–Mariotte's law:

$$p_b V_b = p_e V_e.$$

Here the subscript b stands for "bottle," and the e stands for "environment." The pressurized 50 l gas in the bottle would expand to an environmental volume of

$$V_e = \frac{p_b V_b}{p_e} = \frac{200\text{bar} \times 2\text{l}}{1\text{bar}} = 400\,\text{l}.$$

With a consumption of 1.5 l/min, the gas will last for 266.6 min or 4.4 h.

E9.8   **Answer: Cryogenic oxygen supply**

The density of liquid O_2 is 1.1 g/cm³. Therefore, 2 l of oxygen weighs 2.4 kg. The molar mass of O_2 is 32 g. Therefore, 2 l of liquid oxygen contains 75 mol, which corresponds to 1700 l oxygen gas. At a consumption of 1.5 l/min, the bottle lasts for 18.6 h. The expansion of the gas from 90 K to room temperature is taken into account by using a molar volume of 22.4 l at STP conditions.

Chapter 10

E10.1 **Answer: Water permeability**
On the thick part of the ascending Henle loop. Answer C is correct.

E10.2 **Answer: Clearance and filtration**
The situation in panel (g) is similar to those shown in panels (e) and (f). The filtration is partial and the reabsorption is partial. If the filtration is 20% of RPF, then GFR is 120 ml/min. Reabsorption is 20%, so 80% or 96 ml/min of the filtrate remains in the tubule. The secretion is 30% of RPF, which is 0.3×600 ml/min = 180 ml/min. Therefore, the clearance is C_{S6} = 96 ml/min + 180 ml/min = 276 ml/min. The fractional excretion FE = C_{S6}/C_{inulin} = 276 ml/min/120 ml/min = 2.3.

E10.3 **Answer: Renal processing**

Molecule S filtered at glomerulus	Renal processing	Example
$F_S > E_S$	Net reabsorption of S	Glucose, urea
$F_S < E_S$	Net secretion of S	PAH
$F_S = E_S$	No exchange of S	Inulin, creatinine
$C_S < C_{inulin}$	Net reabsorption of S	Glucose, urea
$C_S > C_{inulin}$	Net secretion of S	PAH
$C_S = C_{inulin}$	No exchange of S	Creatinine

E10.4 **Answer: Hemodialysis**
1. Inflammation of the kidney, bacterial or nonbacterial
2. Diabetes: Due to a lack of insulin, high blood sugar, which clogs the membranes.
3. If the blood pressure is permanently elevated, the fine capillaries in the nephrons break.

E10.5 **Answer: Voiding the bladder.**
$I_V = \Delta p/R$, $\Delta p = \rho g h$ for a cylinder with height of 25 cm, Δp=250 Pa. $R = (8\eta/\pi)(\Delta L/r^4) = 1.55 \times 10^7$ Pa · s/m^3. $I_V = 1.6 \times 10^{-5}$ m^3/s. Yielding a voiding time $t = V/I_V = 0.5$ l/I_V = 0.5×10^{-3} m^3/1.6 $\times 10^{-5}$ m^3/s = 30 s. The voiding time is about 30 s.

E10.6 **Answer: Bladder pressure**
At the detrusor point, the pressure is about 30 hPa. Assuming that the bladder has a height of 50 mm and the bladder is completely filled, the hydrostatic pressure is 5 hPa. Therefore, only 16% of the pressure is of hydrostatic nature, the rest is due to tension.

E10.7 **Answer: Bladder pressure at detrusor point**
The time it takes for the urine hitting the ground is $t = \sqrt{2y/g}$.
Here y is the height of the urine stream above the ground level, g is gravitational acceleration.
The range in x-direction is

$$x = v_x t = v_x \sqrt{2y/g}.$$

The x-component of the velocity is determined by the internal pressure:

$$p = \frac{1}{2}\rho v_x^2.$$

Eliminating the time t and solving for the pressure yields:

$$P = \frac{x^2 g\rho}{4y}.$$

A rough estimate using the following parameters: $y = 0.5\,\text{m}, x = 1\,\text{m}, g = 10\,\text{m/s}^2$, $\rho = 10^3\,\text{kg/m}^3$ yields a pressure of 50 hPa. This pressure is reasonable and close to the expected value of about 40–50 hPa at the detrusor point D in Fig. 10.18.

Chapter 11

E11.1 **Answer: Image size on the retina**
Applying the theorem of rays: $O/o = I/i$. (O, o are object size and distance; I, i are image size and distance. Therefore, $I = i \times O/o = 17\,\text{mm} \times 10\,\text{mm}/257\,\text{mm} = 0.66\,\text{mm}$. With this estimation, it is important that the node is taken into account for the ray theorem. This lies behind the cornea and the distance from the cornea to the nodal point must be added to the object distance. Thus, the image size is 0.66 mm large and inverted.

E11.2 **Answer: Corrective glasses with two refractive powers**
a. For the far sight in the upper part of the glasses, you need a diverging lens with the refractive power: $D_{\text{upper}} = -1/g_{\text{max}} = -1/2.5\,\text{m} = -0.4$ dpt.
b. For the near sight in the lower part of the lens, you need a converging lens to get sharp vision at a distance of 25 cm. This requires a refractive power of $D_{\text{lower}} = 1/g_0 - 1/g_{\text{min}} = (1/0.25\,\text{m}) - (1/0.75\,\text{m}) = 4\,\text{dpt} - 1.33\,\text{dpt} = +2.66$ dpt.

E11.3 **Answer: Accommodation width**
For the accommodation width, we have $\Delta A = 1/x - 1/1\,\text{m} = 2\,\text{m}^{-1}$. Therefore, $1/x = 2\,\text{m}^{-1} + 1\,\text{m}^{-1} = 3\,\text{m}^{-1}$. The person can see clear up to $x = 0.33\,\text{m}$. Without lens, the accommodation width ranges therefore from 0.33 m up to 1 m.

With the help of a concave lens, objects can be seen clear up to 0.2 m. The refractive power of the correction lens can be determined as follows:

$$D = \frac{1}{o - o_{min}} = \frac{1}{0.2\,m} - \frac{1}{0.33\,m} = 5\,dpt - 3\,dpt = +2\,dpt.$$

E11.4   **Answer: Complete hyperopia**

For the far point, we use the lens equation: $\frac{1}{\infty} + \frac{1}{i} = D_{eye} + D_{far, lens}$. Therefore, the refractive power of the eye is

$$D_{eye} = D_{far, lens} - \frac{1}{i} = \frac{1}{0.025\,m} - (-2)\,dpt = 40\,dpt + 2\,dpt = 42\,dpt.$$

For reading at the near point, we have to use the lens equation with correction lenses: $\frac{1}{o} + \frac{1}{i} = D_{eye} + D_{lens}$. Therefore:

$$\frac{1}{o} = D_{eye} + D_{lens} - \frac{1}{i} = 42\,dpt + 1\,dpt - 40\,dpt = +3\,dpt.$$

Thus the object distance is 0.33 m.

E11.5   **Answer: Chromatic aberration**

The focal distance f_2 inside the eye is according to eq. (11.4)

$$f_2 = \frac{n_2}{n_2 - 1} R.$$

a.   Here n_2 is the combined refractive index of cornea and lens, and R is the curvature of the cornea. Assuming for the curvature 7 mm, we obtain for the focal length:

$$f_{2, blue} = \frac{1.415}{1.415 - 1} 7\,mm = 23.86\ mm;\ f_{2, red} = 24.50\ mm.$$

The difference is 632 μm.

b.   The wavelength difference for green (534 nm) and red light (564 nm) according to Fig. 11.23 is only 30 nm instead of 300 nm assumed in part a. Therefore, we expect the focal length difference to be shorter by a factor of 10 or 63 μm.

The thickness of the retina is about 350 μm (11.19). The focal length difference is about 1/5 of the retinal thickness. Therefore, chromatic aberration in the fovea can safely be neglected.

E11.6   **Answer: Airy disk of a microscope**

a.   Using the lens equation for thin lenses: $\frac{1}{o} + \frac{1}{i} = \frac{1}{f}$, we find the image distance:

$\frac{1}{i} = \frac{1}{f} - \frac{1}{o}$. Therefore:

$$1/0.0033 - 1/0.0035 = 300\,dpt - 285\,dpt = 15\,dpt,$$

which corresponds to an image distance of 7 cm.

b. The diameter of the airy disk follows from:
$$\theta = \lambda/d = 500 \text{ nm}/0.002 \text{ m} = 2.5 \times 10^{-4} \text{ rad} = 0.25 \text{ mrad} = 0.014°.$$
At a distance of $L = 7$ cm where the image occurs, the diameter of the airy disk is $d = L\theta = 7 \times 0.25$ mrad $= 1.75 \times 10^{-3}$ cm $= 1.75 \times 10^{-5}$m $= 17.5$ μm.

E11.7   **Answer: Lateral enhancement**
A lateral inhibition via 100% inversion of the receptor potential generates a contrast enhancement by a factor of 4. After lateral inhibition, the receptor potentials are 5, −15, 5, 15. Without enhancement the lateral amplitude variation was 5; after lateral inhibition, the lateral amplitude is 20. See graph.

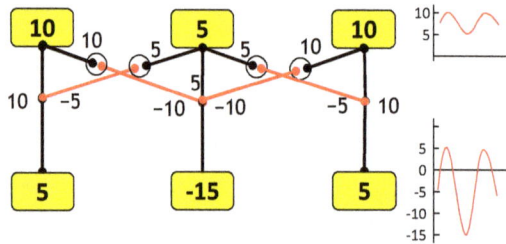

E11.8   **Answer: Optic nerve**
In MRI images, you may find that the optic nerve is bent slightly upward. This is necessary so that the nerve can pass through the optical canal in the skull.

E11.9   **Answer: Hyperpolarization**
The Na$^+$ channels in the receptors are open in the dark (dark current). When photons are absorbed, the Na$^+$ channels close in proportion to the light intensity. This leads to hyperpolarization. However, this does not trigger an action potential, rather the hyperpolarization is converted into a depolarization in the bipolar cells.

E11.10 **Answer: Visual field**
a. Failure of the optic chiasm corresponds to point "1" in the sketch. In this case, the nasal receptive area of the right eye is blocked corresponding to the right visual field. Similarly, the nasal receptive area of the left eye is blocked, corresponding to the left visual field.
b. Blocking of the right nerve in front of the chiasm causes the right eye to be blind. The left eye is not affected.
c. Blocking of the right optic nerve after the chiasm causes the left visual field of both eyes to be blind.

Respective drawing of nerve tracks and visual fields affected for cases (a) to (b) are shown:

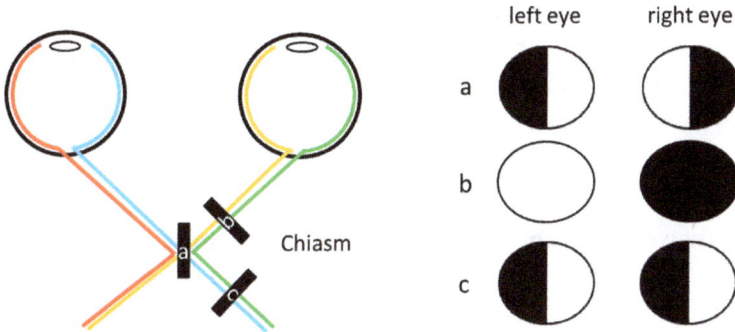

E11.11 **Answer: Color contrast**

Colored Mach bands are generated by incremental change of the RGB numbers. For instance, the eight red bands go from (255, 0, 0), (255, 25, 25), . . ., (255, 175, 175). Accordingly, for the other colors.

The perception of contrast is different for each person. However, green appears to produce the lowest contrast, blue shows the highest contrast, and red is between. Therefore, the lateral inhibition of green cones appears weaker than for red and blue.

Chapter 12

E12.1 **Answer: Sound intensity**

$\langle I \rangle = \frac{1}{2} B \omega k \xi_0^2$. Using the relation $B = v^2 \rho$, $k = \omega/v$, and $u_0 = \omega^2 \xi_0^2$, we find:

$\langle I \rangle = \frac{1}{2} (v\rho)(\omega \xi_0)^2 = \frac{1}{2} Z u_0^2$.

E12.2 **Answer: Hearing loss**

a. $\Delta L = 30$ dB $= 10 \lg(10^3)$, the sound intensity must be higher by the factor $I/I_{\min} = 1000$ to pass the threshold.

b. $\Delta p/p = (I/I_{\min})^{1/2} = (1000)1/2 = 31.6$.

E12.3 Answer: Thunderstorm
a. The distance between lightening and your observation point is velocity × time lap: 4 s × 330 m = 1320 m.
b. At the point of observation, the sensation level is 100 dB = $10 \log 10^{10}$. Therefore, $I/10^{-12} = 10^{10}$, and $I = 10^{-2}$ W/m^2.
c. The sound intensity is $I = 10^{-2}$ W/m^2 at a distance of 1320 m. Then the total intensity is the local intensity times the surface of a sphere at a distance of 1320 m. Therefore, the total intensity is $4\pi r^2 I = 4\pi\,(1320)^2$ m^2 × 10^{-2} W/m^2 = 219 kW.

E12.4 Answer: Doubling distance, doubling intensity
a. As the intensity decreases with the square of the distance: $I(d) \sim 1/d^2$, the ratio of the intensities $I_2/I_1 = 1/4$. The intensity level, therefore, drops by $\Delta I = \log(1/4) = 0.6$ and in decibel by 6 dB.
b. By doubling the intensity of sound sources at the same distance, the intensity increment is $\Delta I = \log(2) = 0.3$, or in decibel: 3 dB.

E12.5 Answer: Sensation threshold
By doubling the distance, the sound intensity level drops by 6 dB (see E12.4 (a)). At a distance of 2 m, the intensity level is therefore 34 dB, at 4 m distance it is 28 dB, etc. The threshold distance must be between 8 and 16 m.
Alternatively, the distance can be determined exactly; 40 dB implies that $I_1 = 10^{-8}$ W/m^2 at a distance of $r_1 = 1$ m. The auditory threshold is for $I_2 = 10^{-10}$ W/m^2 at a distance r_2. As $I_1/I_2 = (r_2/r_1)^2$, it follows that $r_2 = r_1 (I_1/I_2)^{1/2} = 10$ m.

E12.6 Answer: Localization
The difference in the path length for both ears is $\Delta s = d \sin 20° = 5.1$ cm, and the corresponding time difference is $\Delta t = \Delta s/v_{sound} = 0.15$ ms. This is by a factor of 10 shorter than the length of an action potential.

E12.7 Answer: Pressure amplitude
Displacement amplitude is $\xi = p\omega/z$. In air $\xi = 1.6$ μm. Due to the action of the middle ear, the pressure amplitude increases by a factor of 25, and the impedance z increases by a factor of $3500-1.5 \times 10^6$. As the frequency remains the same, the displacement $\xi = p\omega/z$ is reduced by the ratio 25/3500 = 0.007. Therefore, the displacement in water is 1.6 μm × 0.007 = 11 nm.

Chapter 13

E13.1 **Answer: Biocompatibility of alloys**

a. Biocompatibility is difficult to define, see 7.2.8 in Vol.3. It implies compatibility with body tissues in contact without releasing toxic substances or causing any irritations.

b. Alloys are mixtures of metals that are themselves metals and keep metal properties as concerns thermal and electrical conductance, and mechanical strength. However, alloys may display advantages not present in the parent metals, such as magnetic properties, melting temperature, or corrosion resistance.

c. Chromium-containing alloys are highly resistant to corrosion due to the spontaneous formation of a passivating oxide layer within the human body. The passivation is due to the formation of an ultrathin Cr-oxide layer at the surface that stops any further corrosion. This can be seen in stainless steel metal pieces that contain Cr. Furthermore, the Cr-oxide surface layer adheres to the metal substrate and, when scratched, forms a new protective layer.

E13.2 **Answer: Stents:**

With a diameter of 3 mm, the artery has an inside circumference of 9.4 mm. If the stent fills the inside wall, then the diagonal length of each square is 942 µm.

At the pulse maximum, the volume and cross section change are

$$\Delta p = Y(\Delta V/V) = Y(\Delta A/A) = Y(2\Delta r/r)$$

Therefore:

$$\Delta r/r = \Delta p/2Y = 10 \text{ hPa}/2 \times 20 \text{ kPa} = 0.25.$$

At the pressure maximum, the radius is 25% larger, i.e., the diameter is 3.75 mm. The inside circumference increases from 9.42 to 11.8 mm; the base length of the rhombi increases to $l = 1180$ µm. While the rhombi elongate, the angle y changes from 90° to 125°.

Each systolic pulse causes the coronary artery to widen by 25%, the stent follows by increasing its diameter by 238 µm, and the rhombohedral angle changes from 90° to 125°.

16 List of acronyms (used in all three volumes)

ACD	annihilation coincidence detection	DBT	digital breast tomosynthesis
ADC	analog digital converter	DCE	dynamic contrast enhancement
ADC	apparent diffusional constant	DES	drug elusion stent
ADP	adenosine diphosphate	DNA	deoxyribonucleic acid
ALARA	as low as reasonably achievable	DNP	dynamic nuclear polarization
amu	atomic mass units	DOV	depth of view
ANF	auditory nerve fiber	DSA	digital subtraction angiography
ATP	adenosine triphosphate	DSB	double strand break
AV	atrioventricular node	DTI	diffusion tensor imaging
AVV	atrioventricular valve	DTL	drift tube linac
BBB	blood-brain barrier	DWI	diffusion weighted imaging
BF	breath frequency	EBRT	external beam radiotherapy
BMD	bone mineral density	ECC	extra-corporal circulation
BMR	basal metabolic rate	ECG	electrocardiography
BMS	bare metal stents	EDP	end diastolic pressure
BNCT	Boron Neutron Capture Therapy	EDV	end diastolic volume
BOLD fMRI	Blood Oxygen Level Dependent fMRI	EEG	electroencephalography
BPS	biodegradable polymeric stents	EF	ejection fraction
BSA	beam shaping assembly	EM	electromagnetic
BSA	body surface area	EMG	electromyography
BW	body weight	EPI	echo planar imaging
CA	contrast agent	EPP	end plate potential
CAP	cardiac action potential	EPR	enhanced permeation and retention
CBF	cerebral blood flow	ERBT	external radiation beam therapy
CCC	continuous curvilinear capsulorhexis	ERV	expirational rest volume
CCD	caput-collum-diaphyseal angle	ESP	end systolic pressure
CCD	charge coupled device	ESV	end systolic volume
CD	coincidence detector	ETL	echo train length
CIRT	carbon ion radiation therapy	FCRM	fiber optic confocal reflectance microscope
CK	cyber knife	FDG	18F-fluoro-deoxy-glucose
CLE	confocal laser endoscopy	FE	fractional excretion
c.m.	center of mass	FEG	frequency encoding gradient
CNR	contrast to noise ratio	FET	18F-fluoro-ethyl-L-tyrosine
CO	cardiac output	FF	flattening filter
COE	caloric oxygen equivalent	FF	filtration fraction
CPA	charge particle activation	FFDM	full-field digital mammography
CPB	cardiopulmonary bypass	FFF	flattening filter free
CS	Compton scattering	FFT	fast Fourier transform
CSF	cerebrospinal fluid	FID	free induction decay
CT	computed tomography	FLACS	Femtosecond Laser Assisted Cataract Surgery
CTV	clinical target volume	FLAIR	fluid-attenuated inversion recovery
cw	continuous wave		
CZT	CdZnTe	fMRI	functional magnetic resonance imaging
dB	decibel		

https://doi.org/10.1515/9783110756951-016

FOV	field of view	MHR	metabolic heat production
FRC	fractional rest volume	MHT	magnetic hyperthermia
FSE	fast spin echo	MHz	megahertz
FT	Fourier transform	MI	mechanical index
GFR	glomerular filtration rate	MLC	multi-leaf collimator
GTV	gross tumor volume	MNP	magnetic nanoparticle
Hct	hematocrit value	mpMRI	multiparameter MRI
HD	hydrodynamic diameter	MRI	magnetic resonance imaging
HDR	high dose rate	MRSI	magnetic resonance
HEP	hemi endo prosthesis		spectroscopy imaging
hMRI	hyperpolarization magnetic	MRgRT	magnetic resonance-guided
	resonance imaging		radiation therapy
HSLS	high spin–low spin	MRT	magnetic resonance
IAEA	International Atomic Energy		tomography
	Agency	MSFP	mean systemic filling pressure
ICRP	International Commission for	MSO	medial superior olive
	Radiological Protection	MUAP	motor unit action potential
IGRT	image guided radiotherapy	MV	minute ventilation
IHC	inner hair cell	NBI	narrow band imaging
ILD	intensity level differences	NCT	neutron capture therapy
IMRT	intensity modulated radiation	NIR	near infrared
	therapy	NIRS	near infrared spectroscopy
IRT	internal radiation therapy	NMJ	neuromuscular junction
IRV	inspirational rest volume	NP	nanoparticle
ITD	interaural time difference	NRT	neutron radiation treatment
Kerma	kinetic energy release in matter	NTD	non-target dose
kHz	kilohertz	OAR	organ at risk
kV	kilovolt	OCT	optical coherence tomography
kVp	peak kilovoltage	OER	oxygen enhancement ratio
kW	kilowatt	OHC	outer hair cell
LASEK	laser epithelial keratomileusis	PAH	para-aminohippuric acid
LASER	light amplification by	PCI	percutaneous coronary
	stimulated emission of		intervention
	radiation	PCI	phase contrast imaging
LASIK	laser-assisted interstitial	PCV	packed cell volume
	keratomileusis	PD	proton density
LCI	low-coherence interferometry	PDD	percent depth dose
LDR	low-dose rate	PDR	proliferative diabetic
LED	light-emitting device		retinopathy
LET	linear energy transfer	PDR	pulse dose rate
Linac	linear accelerator	PDT	photodynamic therapy
LOR	line of response	PE	photoelectric effect
LQ	linear-quadratic	PEG	phase encoding gradient
LSO	lateral superior olive	PEG	polyethylene glycol
MC	Monte Carlo	PES	photoelectron emission
MEG	magnetoencephalography		spectroscopy
MEI	middle ear implants	PET	positron emission tomography
MET	mechanoelectric transduction	PHIP	para hydrogen induced
MeV	mega-electron volt		polarization

PI	pulsatility index	SPE	single photon emission
PMT	photomultiplier tube	SPECT	single-photon emission
PPD	percentage photon dose		computed tomography
PRF	pulse repeat frequency	SPIO	superparamagnetic iron oxide
PRK	photorefractive keratectomy	SPR	surface plasmon resonance
PRP	pan-retinal photocoagulation	SSB	single strand break
PRT	proton radiotherapy	SSD	source-to-surface distance
PRT	pulse repeat time	SSG	slice selection gradient
PSA	prostate-specific antigen	SUV	standard uptake value
PSMA	prostate-specific membrane	SV	stroke volume
	antigen	TCE	transient charged particle
PTV	planning target volume		equilibrium
PWV	pulse wave velocity	TE	time to (spin, acoustic) echo
PZT	$PbZrTiO_3$	TEP	total endo prosthesis
Q	quality factor	TERMA	total energy released per unit
QA	quality assurance		mass
RBC	red blood cell	TGC	time gain compensation
RBE	relative biological effectiveness	THz	terahertz
RBF	renal blood flow	TIPPB	transperineal interstitial
RC	respiratory coefficient		permanent prostate
Re	Reynolds number		brachytherapy
RES	reticuloendothelial system	TLC	total lung capacity
RF	radio frequency	TMR	targeted muscle reinnervation
RF	respiratory fraction	ToF	time of flight
RMR	resting metabolic rate	TPFR	total peripheral flow resistance
RPF	renal plasma flow	TPS	treatment planning system
RT	radiotherapy	TR	time of repetition
RTR	real time radiography	TRIGA	Training, Research, and
RV	residual volume		Isotopes, General Atomics
SA	sinoatrial node	TT	transfer time
SAD	source to axis distance	TV	tidal volume
SATP	standard ambient temperature	TV	target volume
	and pressure	UHMWPE	ultra-high molecular weight
SAXS	small angle x-ray scattering		polyethylene
SBRT	stereotactic body radiation	ULFMRI	ultra-low field magnetic
	therapy		resonance imaging
SCI	spinal cord injury	US	ultrasound
SE	spin echo	VC	vital capacity
SID	source to image distance	VEGF	vascular endothelial growth
SERS	surface-enhanced Raman		factor
	scattering	VOR	vestibulo ocular reflex
SGRT	surface-guided radiotherapy	XRR	x-ray radiography
SLAC	Stanford linear accelerator	XRT	x-ray radiotherapy
SNR	signal-to-noise ratio	YAG	yttrium-aluminum garnet
SOBP	spread out Bragg peak	ZFC	zero field cooling

17 Selection of fundamental physical constants, conversions, and relationships

Speed of light c	299 792 458 m/s ~3 × 10^8 m/s
Gravitational acceleration of the earth g	9.81 m/s^2
Planck constant h	6.623 × 10^{-34} Js
Compton wavelength λ_c	2.426 pm
Elementary charge e	1.602 × 10^{-19} As
Atomic mass unit u	1.66 × 10^{-27} kg
Electron mass m_e	9.109×10^{-31} kg
Avogadro number N_A	6.022×10^{23} mol^{-1}
Boltzmann constant k_B	1.38×10^{-23} JK^{-1}
Dielectric constant of the vacuum ε_0	8.854 ×10^{-12} As/Vm
Magnetic permeability μ_0	4π×10^{-7} Vs/Am = 1.256 × 10^{-6} Vs/Am
Faraday constant F	9.65 × 10^4 C/mol
General gas constant R	8.314 J/Kmol
Stefan–Boltzmann constant σ_{SB}	5.66×10^{-8} AV/m^{2}K^4
Bohr magneton μ_B	9.274×10^{-24} J/T
Proton gyromagnetic ratio γ_p	2.675×10^8 T^{-1}rad s^{-1}
Proton mass m_p	1.673×10^{-27} kg
Neutron mass m_n	1.675×10^{-27} kg
Nuclear magneton μ_N	5.05783×10^{-27}J/T.

Conversions

1 eV	= 1.602 × 10^{-19} J
1 Joule	= 6.242 × 10^{18} eV
1 Tesla	= 1 N/Am = 1 Vs/m^2 = 10^4 Gauss
1 Pascal	= 1 N/m^2
1000 hPa	= 1000 mbar
1 A/m	= 10^{-3} emu/cm^3
emu/cm^{3}ρ	= emu/g

Relationships

Avogadro number N_A =1 g/1 u	= 1 g/1.66 × 10^{-24} g
Faraday constant $F = N_A \times e$	= 6.022 × 10^{23} mol^{-1} × 1.602 × 10^{-19} As
General gas constant $R = N_A \times k_B$	= 6.022 × 10^{23} mol^{-1} × 1.38 × 10^{-23} JK^{-1}
Speed of light in vacuum $c = 1/\sqrt{\mu_0\varepsilon_0}$	$= \left(\sqrt{1.25 \times 10^{-6} \text{ Vs/Am} \times 8.85 \ 10^{-12} \text{ As/Vm}}\right)^{-1}$

https://doi.org/10.1515/9783110756951-017

18 List of scientists named in this volume

https://doi.org/10.1515/9783110756951-018

19 Glossary

Accommodation: The eye's ability to produce sharp images of objects located at far distance up to the near point.

Action potential: Depolarization of the cell's electric potential across the membrane, usually from −70 mV to +20 mV.

Aerobic energy consumption: Catabolic pathway using oxygen to generate energy through ATP (adenosine triphosphate).

Afferent neurons: Neurons that guide sensory signals from the periphery to the central nervous system (CNS).

Anabolism: A process that releases energy used in cells to synthesize molecules necessary for maintaining the cells' functions and performing cellular work.

Anaerobic energy: A catabolic pathway that uses inorganic compounds for producing energy.

ATP: The primary energy source of the body. The energy is released by splitting ATP into ADP (adenosine diphosphate) and a phosphate group.

AV node: The conductive connection between the atrium and the ventricles.

Basal metabolic rate (BMR): Energy consumption of the body at rest under well-defined conditions.

Basilar membrane: Membrane in the cochlea that allows a tonotopic frequency map of sound waves by position-sensitive bending.

Bipolar cells: Exhibit graded transmembrane potentials that continuously change between depolarization and hyperpolarization depending on the stimulus.

Caloric oxygen equivalent: The amount of energy produced by the consumption of 1 liter oxygen.

Cardiac action potential: The cycle of rapid depolarization, partial repolarization, prolonged plateau phase and repolarization of the myocardium.

Cardiac output: Product of stroke volume and heart frequency.

Catabolism: Enzymatic reaction degrading larger organic molecules into smaller molecular units.

Cell: Smallest biological building block that can multiply itself by dividing. The cell membrane separates the extracellular space from the cytoplasm.

Cell capacitance: Electric capacitance of the cell membrane, amounting to about $1\ \mu F/cm^2$.

Cellular respiration: The catabolic pathway of aerobic and anaerobic respiration which breaks down organic molecules to gain energy for the synthesis of ATP.

Compliance: The response (volume change) to an input parameter (pressure change).

Cytoplasm: Content of cells except the cell's nucleus, enveloped by the cell membrane.

Diffusion: Particle flow following a concentration gradient.

Efferent neurons: Neurons guiding action potentials from the CNS to organs and muscles for movement.

Electrocardiogram: Time-resolved recording of potential differences between the extremities reflecting the looping of the heart vector during a cardiac cycle.

Electromotility: Longitudinal length changes of cells induced by the receptor potential.

Eukaryotic cells: Cells with a nucleus containing the genes and a cell membrane enclosing further organelles.

Fick's law: Describes the steady-state flow of particles due to a concentration gradient.

Fovea: Small area in the center of the macula, containing only cones with very high density.

https://doi.org/10.1515/9783110756951-019

Glomerular filtration: Filtration of blood plasma and solutes with molecular weight below 20,000 atomic units.

Hematocrit: Fraction of erythrocytes to the total blood volume.

Homeostasis: The steady-state physiological condition of the body controlled by feedback systems.

Hyperpolarization: Change of a graded membrane potential from an intermediate level (−30 mV) to the polarization level (−70 mV).

Lateral inhibition: Causes enhanced contrast at the interface between areas of different shades via horizontal connection of bipolar cells in the retina.

Liposomes: Spherical shells consisting of phospholipid bilayers used as carriers in cells.

Macula: Area on the retina with the highest density of rods and cones resulting in the highest light sensitivity.

Metabolism: Combustion of organic compounds with oxygen in living cells for the production of energy. Metabolism consists of two parts: catabolism and anabolism.

Mitochondria: Organelle in eukaryotic cells that serves as the site of cellular respiration.

Motor unit: Consisting of a neuromuscular junction and all fibers that it can elicit.

Myelination: Piecemeal wrapping of nerve fibers with dielectric sheets to allow rapid saltatory signaling transmission along axons.

Nernst equation: Logarithm of the ratio of transmembrane potential difference to thermal energy.

Osmosis: Water flow through a membrane that separates areas of different ion concentrations.

Perfusion: Gas or liquid flow driven by a pressure difference.

Proteins: Organic molecule made up of amino acids and needed for the body to function.

Receptor potential: Bipolar graded potential without threshold and no refractory time.

Sacromere: Comprises thin actin filaments and thick myosin filaments, forming contractible molecular motors in skeletal muscle cells.

Refractory period: Resting time after repolarization during which new action potentials cannot be started.

Renal clearance: The volume of plasma cleared of a particular substance by the kidneys per unit of time.

Resilience: Energy density stored in materials under load up to the yield point.

Respiratory exchange rate: Number of CO_2 moles exhaled to the number of O_2 moles inhaled.

Resting potential: Transmembrane electrical potential due to a potassium ion concentration gradient between cytoplasm and extracellular space, usually −75 to −90 mV.

Retinal: Chromophore of light-sensitive protein opsin reacting by a cis–trans configurational change upon light absorption initiating vision.

Reynolds number: Dimensionless number used for estimating turbulent flow versus laminar flow.

Rhodopsin: Transmembrane protein composed of opsin and light-sensitive retinal.

SA node: Heart's primary pacemaker at a rate of 60–80 beats/min at rest.

Schwann cells: Sheet that wraps around axons to lower the axons' conductivity.

Stroke volume: Blood volume pumped by the heart during one contraction cycle.

Surfactants: Surface active agent coating surfaces to lower the surface tension, such as in alveoli.

Surround: Receptive field is divided into center and surround, distinguished by the response of ON and OFF receptor cells.

Synaptic gap (cleft): Gap in axons where the propagation of action potentials is converted into neurotransmitters.

Tidal volume: Inspirational-expirational air volume at rest.

Toughness: Energy storage stored in a material between the yield point and the point of fracture.

Transmembrane potential: Potential difference between cytoplasm and the membrane outer surface.

Transmural pressure difference: Pressure difference between cytoplasm and extracellular space.

Viscoelasticity: Combines properties of elastic and plastic materials, showing an immediate and a time-delayed response to external load.

Work-hardening: Entanglement of dislocations during plastic deformation that hinders further plastic flow.

Yield stress: Highest stress reached before the onset of plastic deformation.

20 Index of terms

https://doi.org/10.1515/9783110756951-020

www.ingramcontent.com/pod-product-compliance
Lightning Source LLC
Chambersburg PA
CBHW060955210326
41598CB00031B/4832